Laser-Induced Damage in Optical Materials

LASER-INDUCED DAMAGE IN OPTICAL MATERIALS

EDITED BY

DETLEV RISTAU

LEIBNIZ UNIVERSITY OF HANNOVER, INSTITUTE FOR QUANTUM OPTICS; AND
LASER ZENTRUM HANNOVER, DEPARTMENT OF LASER COMPONENTS
HANNOVER, GERMANY

CRC Press
Taylor & Francis Group
Boca Raton   London   New York

CRC Press is an imprint of the
Taylor & Francis Group, an **informa** business

CRC Press
Taylor & Francis Group
6000 Broken Sound Parkway NW, Suite 300
Boca Raton, FL 33487-2742

First issued in paperback 2016

© 2015 by Taylor & Francis Group, LLC
CRC Press is an imprint of Taylor & Francis Group, an Informa business

No claim to original U.S. Government works

Version Date: 20140929

ISBN 13: 978-1-138-19956-9 (pbk)
ISBN 13: 978-1-4398-7216-1 (hbk)

Visit the Taylor & Francis Web site at
http://www.taylorandfrancis.com

and the CRC Press Web site at
http://www.crcpress.com

To Dr. Arthur H. Guenther (1931–2007)
Founding Organizer of the Boulder Damage Symposium

Contents

SECTION I   Fundamentals

SECTION II   Measurements

Preface

Since the very beginning of laser technology, the limitations in the power-handling capability of optical components were always an obstacle for the development of laser systems and applications operating at high power levels and improved beam quality. In the early days of laser development, mainly the inclusions in laser rod materials were discussed as a major complication in the augmentation of the output power in solid-state laser systems. Nowadays, in the course of the development of optical materials with excellent quality and power-handling capability, the problem of laser-induced damage has shifted from the bulk to the surface of the optical component. During the last two decades, considerable progress has been achieved by an ever-growing scientific community working in various areas of high-power optical materials and components. For example, the development of stable coatings for ultrashort pulse systems was an essential milestone in the introduction of femtosecond (fs) lasers, which in turn initiated novel and exciting research fields and industrial and even medical applications. These advances were complemented by significant progresses in the theoretical modeling of fs laser–induced damage phenomena resulting in a deeper understanding of the underlying electronic excitation effects. Semiconductor lithography, with its trend toward shorter wavelengths, is another essential pacemaker in the development of high-power optics with long lifetimes. Nowadays, industrial stepper systems operating at the wavelength 193 nm of the ArF excimer laser depend on optics with extremely high quality and stability. Also, the first tools of the next generation of lithography based on imaging procedures in the extreme ultraviolet (EUV) spectral region have been implemented with the target of reaching even smaller half-pitch values well below 20 nm for integrated circuit designs. Some of the wide variety of prominent applications in fundamental research where high-power lasers play an important role are laser medicine, gravitational wave detection, and nuclear fusion. Looking back more than 50 years since the demonstration of the first ruby laser, it is apparent that the history of laser technology cannot be written without mentioning the advances in optical high-power materials and components.

The subject of this book is the emerging area of research in laser damage occurring in the bulk and on the surface or the coating of optical components. Far from covering all aspects of laser damage published in thousands of scientific papers during the last five decades, the present compilation is intended to offer a comprehensive overview on recent developments in this vivid field. The main idea is to provide a fundamental reference for readers interested in the various fields of laser technology, photonics, or fundamental research, as well as optical component, material, and coating production.

Focusing on the major topics of laser-induced damage in optical materials, this book is divided into four sections. The first section is dedicated to the fundamentals of laser damage and is introduced by a brief history including major definitions and developments in the field. In the four chapters that follow, the dominating laser damage mechanisms are discussed, including thermal degradation effects, defect-induced damage, and select nonlinear effects leading to catastrophic failures of optical media. The final chapter of the first section discusses the latest theoretical approaches in the description of damage phenomena in the ultrashort pulse regime.

The reliable measurement of laser-induced damage thresholds and related quality parameters of optical components is a major prerequisite for research in high-power optics and their optimization. In this context, the second section of the book gives an overview on the measurement and detection of laser damage as well as on the evaluation of statistical data recorded during the applied measurement protocols. The corresponding two chapters are of central importance for the understanding and ranking of laser damage data published in the literature and often also in the catalogs of optics suppliers. This section is rounded off by a comprehensive overview on the measurement of transfer properties and scatter losses of optics.

The third section of the book comprises detailed information on the involvement of materials, surfaces, and thin films in laser damage phenomena. In view of the enormous diversity of available scientific models and experimental results, a few current topics have been selected as examples representing the major branches of recent research as well as providing fundamental physics, relations, and data. The first three chapters are devoted to studies on materials for high-power lasers and concentrate on prominent applications mainly in the ultraviolet (UV), visible (VIS), and near-infrared (NIR) spectral regions. In this context, quartz and glasses are the most important material class and are considered within an extended chapter. The second chapter in the sequence discusses crystalline materials, especially for the UV spectral region and gives a comprehensive insight into related optical quality factors. Also of highest significance are optical materials for modern laser systems including active laser materials, frequency conversion crystals, and Q-switching materials, which are addressed in the third chapter of Section III. Following the production chain for optical components, the next step is the realization of an adequate surface quality, aspects of which are summarized in the fourth chapter of this section. After a general overview on optical coatings, which is the basis for all the following chapters of the book, two chapters on the optimization of coatings for NIR and fs lasers are included as examples illustrating the vast extent of research concentrated on high-power laser coatings. The final chapter of Section III compiles interesting aspects on the application of optical components for space-born laser systems and also addresses the contamination effects on optical surfaces under laser irradiation.

A book on laser-induced damage cannot be complete without a few insights into typical application areas of high-power coatings. In view of the huge impact of lithography on the development, a chapter on optics for deep ultraviolet (DUV) and EUV regions begins the last section, Section IV, which is on applications. Further applications considered exemplarily in the last two chapters are optics for free-electron lasers and for the large, high-power laser system PHELIX.

The authors of these chapters are associated with renowned research institutes and large-scale laser facilities active in the field for several decades. We hope that this complementary collection of scientific material, fundamental data, and practical information will be of general use and stimulate the interest of scientists, engineers, and technicians to gain deeper insights into the fascinating field of laser damage and its key role in laser technology.

Our thanks are due to the contributors, especially for their time and dedication in writing the chapters alongside their regular obligations and work schedules.

Editor

Detlev Ristau completed his study of physics at the University of Hannover in 1982 and was awarded a scholarship at Rice University in Houston, Texas, where he was working in the field of tunable color center lasers. He returned to Germany in 1983 and started his PhD research on the power-handling capability of optical coatings in the thin film group of the Institut für Quantenoptik at the University of Hannover. He earned a PhD in 1988 and was responsible for the thin film group at the institute until 1992. Professor Ristau is presently the director of the Department of Laser Components at the Laser Zentrum Hannover. After receiving a state doctorate in 2008, he was appointed professor for applied physics at the Leibniz University in Hannover in 2010, and he also assumed responsibility for a research group on ultrahigh-quality optical lasers.

Professor Ristau is the author or coauthor of more than 300 technical papers and has published several book chapters. His current research activities include the development and precise control of ion processes as well as the measurement of the power-handling capability and losses of optical components.

Contributors

Jonathan W. Arenberg
Northrop Grumman Space
    Technology
Vehicle Systems Engineering
    and Operations
Redondo Beach, California

Stefan Borneis
GSI Helmholtzzentrum für
    Schwerionenforschung GmbH
PHELIX/PP
Darmstadt, Germany

Philippe Cormont
Commissariat à L'Energie Atomique
Le Barp, France

Angela Duparré
Fraunhofer Institute for Applied Optics
    and Precision Engineering (IOF)
Jena, Germany

Luke A. Emmert
Department of Physics and Astronomy
University of New Mexico
Albuquerque, New Mexico

Vitaly Gruzdev
University of Missouri
Columbia, Missouri

Anne Hildenbrand-Dhollande
French-German Research Institute
    of Saint-Louis
Saint-Louis, France

Marco Jupé
Laser Zentrum Hannover
Department of Laser Components
Hannover, Germany

Laurent Lamaignère
Commissariat à L'Energie Atomique
Le Barp, France

H. Angus Macleod
Thin Film Center
Tucson, Arizona

Klaus R. Mann
Laser Laboratorium Göttingen
Göttingen, Germany

Christian Mühlig
Department of Microscopy—Laser
    Diagnostics
Leibniz-Institute of Photonic
    Technology (IPHT)
Jena, Germany

Jérôme Néauport
Commissariat à L'Energie Atomique
Le Barp, France

Semyon Papernov
Laboratory for Laser Energetics
University of Rochester
Rochester, New York

Wolfgang Riede
Deutsches Zentrum für Luft- und
    Raumfahrt e. V.
Institut für Technische Physik
Stuttgart, Germany

Detlev Ristau
Leibniz University of Hannover
Institute for Quantum Optics
and
Laser Zentrum Hannover
Department of Laser Components
Hannover, Germany

Wolfgang Rudolph
Department of Physics and Astronomy
University of New Mexico
Albuquerque, New Mexico

Sven Schröder
Fraunhofer Institute for Applied Optics
    and Precision Engineering (IOF)
Jena, Germany

Jianda Shao
Shanghai Institute of Optics and Fine
    Mechanics, CAS
Shanghai, People's Republic of China

Michelle Shinn
Accelerator Division Management
    Department
Thomas Jefferson National
    Accelerator Facility
Newport News, Virginia

M.J. Soileau
CREOL
The College of Optics and Photonics
University of Central Florida
Orlando, Florida

Christopher J. Stolz
Lawrence Livermore National
    Laboratory
Livermore, California

Wolfgang Triebel
Department of Microscopy—Laser
    Diagnostics
Leibniz-Institute of Photonic
    Technology (IPHT)
Jena, Germany

Frank R. Wagner
Institut Fresnel
Aix-Marseille Université
Marseille, France

Denny Wernham
European Space Research
    and Technology Center
Noordwijek, The Netherlands

Roger M. Wood (retired)
Cosolas Ltd.
Trimpley, Worcestershire,
    United Kingdom

Fundamentals

1. Laser-Induced Damage Phenomena in Optics

A Historical Overview

M.J. Soileau

1.1 Introduction: The First Laser-Induced Damage

On May 21, 1960, Ted Maiman demonstrated the first laser [1]. The elegantly simple device consisted of a helium flash lamp designed for photography and a metal cylindrical sleeve to help direct the light from the lamp to a ruby rod, the laser gain medium. The laser resonator was simply the ruby rod with silver coatings on both ends, with one coating thin enough to allow particle transmission to couple out the energy generated and stored in the resonator.

Laser-induced damage (LID) quickly followed. Though the archival record is sparse on this topic, conversations with pioneers of the laser indicate common problems with the output coupling films [2,3]. It was common for laser researchers to keep "*spare*" ruby rods for use when the output coatings started to fall off.

These early lasers had pulse width related to the time width of the pump lamp, about 0.4 ms, and output energies of up to 5 J from 3 mm radius ruby rods, or about 10,000 W of power. The output of these early devices had multispatial and temporal modes. Particle mode locking of temporal modes and interference among the multiple modes likely resulted in peak powers on the order of hundreds of kilowatts.

The invention of the Q-switch quickly followed the invention of the laser discovered in 1961 and published in 1962 [3]. Q-switching reduced laser pulse width to the order of tens of nanoseconds from values of hundreds of microseconds in ruby and other solid state lasers. This resulted in factors of 10,000 or more increases in peak power. The so-called

Laser-Induced Damage in Optical Materials. Edited by Detlev Ristau © 2015 CRC Press/Taylor & Francis Group, LLC. ISBN: 978-1-4398-7216-1.

Chapter 1

giant pulses easily damaged laser rods, cavity reflectors, Q-switch materials, and other optical components used to make the laser emitting the pulse. LID thus became a dominant feature in laser design and laser operations and remains half a century later. Partially transmitting metal film (sometimes called *soft coatings*) gave way to then called *hard* coatings made of transparent dielectrics and designs utilizing Brewster angle cut rods, external resonators, and, in some cases, reflectors and output couplers made of uncoated etalons. In addition, conventional material growth techniques had to be replaced to produce laser media with greatly reduced defects and impurities, all with the objective of reducing LID in the laser medium by the intense light produced within the medium.

Within 1 year after the invention of the ruby laser, McClung and Hellwarth reported difficulty of laser operation due to the fact that "output light burned holes in the silvered surfaces" on the ruby laser rod [3].

1.2   Development of High-Power Lasers

Precision optical components were rapidly developed in the World War II period, including multilayer, thin film coating to control reflection in complex optical systems. Quality control was essential for high resolution optics. Optical crystals and glasses needed to be free of strain to maintain resolution and avoid unwanted polarization effects. Smooth surfaces, without large amounts of scratches and digs, were produced to reduce scattering losses and increase image quality.

Thus, early laser builders had access to relatively high quality optics. However, while a surface defect may only marginally degrade an image, it can be the site of the initiation of catastrophic LID. Small amounts of absorption—say 1 part in 1000—do not diminish the functionality of an optical component in nonlaser applications but can easily result in damage to a surface, or delineation or ablation of a thin film. In fact, absorbing sites or defects sufficient to initiate catastrophic LID often escape detection using conventional techniques, for example, measurement of absorption by conventional spectrometers.

The demonstration of the first ruby laser served as a sort of *proof of existence* for lasers. An explosion of scientific and engineering efforts quickly followed as new laser sources were sought. This in turn drove the need for precision beam control and optical components needed for the many innovative ideas for laser applications that followed the laser's invention.

These new high-power optical devices became immediate tools to study nonlinear optics, and became victims of nonlinear optics. The first observation of second-harmonic generation by Peter Frankin [4] and coworkers sparked the beginning of the entire new field of nonlinear optics. Among the first reports of nonlinear refraction were LID tracks in ruby rods due to self-focusing in the medium [5].

One needs only to consider the electric field produced by a very modest Q-switched laser focused to a small spot; for example, a 1 mJ pulse of a 10 ns duration, focused into a 5 μm radius spot, inside a material of refractive index equal to 1.5 produces an electric field in excess of 5.5 mV/cm (root mean square field = RMS). Such an electric field exceeds the threshold needed for dielectric breakdown in many good dielectric materials, including many used to make components for lasers.

1.3 LID: An Electric Field–Dependent Phenomenon

James Clark Maxwell (1831–1879) established that light is electromagnetic radiation. Prior to Q-switched lasers, few phenomena were observed that were a direct manifestation of the electric fields in electric and magnetic (E&M) waves, other than as carriers of energy in the beam. In fact, LID is quite possibly among the first such field-dependent phenomena observed.

RMS electric fields in the megavolt range, bright sparks observed in LID in highly transparent materials, and the law linear absorption in such materials resulted in the conclusion that laser-induced breakdown (LIB) was the mechanism for failure in transparent materials (materials with band gaps in some cases an order-of-magnitude larger than the energy of photons from some lasers) [6]. However, a much simpler—even if perhaps more subtle—observation of the direct relation of the electric field associated with light with LID is the asymmetry observed in the laser damage fluence or irradiance for entrance and exit surfaces of materials.

Consider a typical dielectric window through which a laser beam is propagated. Fresnel reflection loss at air-window entrance surface clearly means that less energy reaches the exit surface of the window (here absorption and scattering losses are small and neglected). A reasonable conclusion is that it would take less laser input to the window to damage the entrance surface than the exit surface. However, the opposite is observed.

This observation was a subject of much speculation for a period of time until Crisp [7] produced the elegantly simple explanation: for a given irradiance into a window, the electric field associated with the beam is *higher* at the exit surface due to the basic fact that the Fresnel reflected field at the entrance surface is 180° out of phase with the incident field, whereas at the exit surface, the reflected field is in phase with the incident field. Thus, the net electric field is *higher* at the exit surface than at the entrance surface for a given input irradiance. The net result is the observed lower damage threshold for the exit surface, as measured in terms of the input irradiance.

This revelation led to much work on what is often called *field enhancement* due to materials defects, surfaces, and optical elements such as multilayer thin film systems where interference of the light field is key to the function of the device. This phenomenon is discussed in great detail in the chapters that follow.

There was a flood of ideas for laser applications following the invention of the ruby laser, including precision measurement, remote delivery of precise amounts of energy, and exotic applications such as laser weapons and laser-driven nuclear fusion for energy generation. Many promising applications, then as now, are limited by LID, either in the lasers themselves or in the various components needed to control and direct the light from the lasers.

1.4 Strategies for Minimizing LID

One simple way to manage problems associated with LID is to reduce the fluence or irradiance by simply expanding the laser beam. However, the cost—as well as weight and volume—increases as the surface area of optical components increases, often in a nonlinear fusion. In many applications, compact systems are essential to the intended use.

Chapter 1

One example is the laser source on the Mars exploration vehicle [8]. This remote space craft uses LIB spectroscopy to remotely analyze Martian rocks.

Therefore, practical considerations of cost and compactness often lead to laser systems and applications that are limited by LID to the various optical components associated with the laser systems. This, in turn, has led to international efforts to develop optical components more resistant to LID. Progress in this quest has demanded

1. Focused efforts to develop a fundamental understanding of physical processes that lead to LID
2. The development of measurement techniques to better identify materials characteristics that lead to LID and for the detection and diagnosis of LID events
3. The development of new materials and materials preparation techniques, especially the preparation of damage-resistant precision optical surfaces
4. Advancements in thin film materials and thin film deposition technology

These topics, along with examples of laser applications that are demanding improvement of optical performance, are the subject of the chapters that follow in this book. These topics have also been the focus in the annual symposium on LID to optical materials [8]. Over 40 years of scientific and technological progress are summarized in the proceedings of these meetings. The proceedings are available from the SPIE [9] and are available in searchable format online.

Solutions to the problem of LID are elusive in major part due to the broad scope of laser applications and the demands of various applications. Laser sources span the pulse width from continuous waves to pulses as short as 65 (18 orders of magnitude!) and terahertz region to x-rays (a factor of at least 1000 in photon energies) [10]. To this astronomical parameter space range, one needs to consider a variety of complex linear and nonlinear processes that can affect LID. Some applications demand novel materials for frequency conversation, polarization control and manipulation, beam combining, pulse width control, as well as precision beam pointing and focusing.

Taken together, all these factors make it impossible to answer an often posed question: *What is the LID threshold (LIDT) of this optical element?* Or, given that a certain optical component is known for a given set of conditions, can one predict or scale to another set of conditions? Experience has shown that precise scaling is speculative at best. (Scaling laws are discussed in Chapter 7.)

1.5 Rule of Thumb for LID Scaling

While detailed scaling of LID is not currently possible, there is a very simple *rule of thumb* for estimating limits imposed by LID [11].

$$E_d = (10 \text{ J/cm}^2)(tp/1 \text{ ns})^{1/2} \tag{1.1}$$

where
E_d is the damage threshold fluence in J/cm^2
tp is the laser pulse width in seconds

This equation gives a reasonable approximation of the LIDT to within plus or minus an order of magnitude for otherwise highly transparent materials over the wavelength range from the ultraviolet (UV) to the infrared spectral region. Laser systems expected to operate with fluences within this range or higher will likely be dominated by LID considerations, whereas fluences below this range will not likely result in LID problems.

There is limited theoretical basis for this rule of thumb except the assumption that in practical situations, LID is most often limited by defects at surfaces and in thin film coatings. Whatever the mechanism for coupling light into these defects, one can conclude that ultimately damage (melting, ablation, cracking, etc.) results from an impulse on energy deposited in a material. Resistance to damage is thus at least in part dependent on the diffusion of the deposited energy.

One should not use Equation 1.1 to predict exact values of LID thresholds, rather, it should only be used as an indicator of expected LID issues. If one needs to know the LIDT to greater precision than offered by Equation 1.1, then one must make careful measurements of the LID for the material system to be used and for the conditions of use (laser pulse width, wavelength, beam spot size, etc.).

References

1. Maiman, T.H., Stimulated optical radiation in http://spie.org/x1848.xmlRuby. *Nature*, 187(4736), 493–494.
2. Bass, M., Professor, CREOL—The College of Optics and Photonics, University of Central Florida, Orlando, FL, Private communication, 2012.
3. McClung, F.J. and Hellwarth, R.W., Giant optical pulsation from ruby. *Journal of Applied Physics*, 33(3), 828–829, 1962.
4. Frankin, P., Hill, A., Peters, C., and Weinreich, G., Generation of optical harmonics. *Physical Review Letters*, 7(4), 118, 1961.
5. Avizonis, P.V. and Farringt, T., Internal self-damage of ruby and ND-glass lasers. *Applied Physics Letters*, 7(8), 205–210, 1965.
6. Yasojima et al., Laser-induced breakdown in ionic crystals and a polymer. *Japanese Journal of Applied Physics*, 7, 552, 1968.
7. Crisp, M.D., in *Damage in Laser Materials*, Glass, A.J. and Guenther, A.H., eds., National Bureau of Standards (US) Special Publication 387, 1973.
8. Lanza, N.L., Wiens, R.C., Clegg, S.M., Ollila, A.M., Humphries, S.D., Newsom, H.E., and Barefield, J.E., Calibrating the ChemCam laser-induced breakdown spectroscopy instrument for carbonate minerals on Mars. *Applied Optics*, 49(13), C211–C217, 2010.
9. Bennett, H.E., Chase, L.L., Guenther, A.H., Newnam, B.E., and Soileau, M.J., *24th Annual Boulder Damage Symposium Proceedings—Laser-Induced Damage in Optical Materials: 1992*, Boulder, CO, September 28, 1992, http://spie.org/x1848.xml.
10. Zhao, K., Zhang, Q., Chini, M., Wu, Y., Wang, X., and Chang, Z., Tailoring a 67 attosecond pulse through advantageous phase-mismatch. *Optics Letters*, 37(18), 3891–3893, 2012.
11. Soileau, M.J., 40 Year retrospective of fundamental mechanisms, *Proceedings of SPIE*, Exarhos, G.J., Ristau, D., Soileau, M.J., and Stolz, C.J., eds., Vol. 7132, pp. 1–14, 2008.

Chapter 1

2. Laser-Induced Damage by Thermal Effects

Roger M. Wood

2.1 Introduction

Although most laser scientists instinctively link the term *laser-induced damage threshold* (LIDT) with dielectric breakdown and, thenceforth, quote its value in terms of the peak power density, W cm^{-2}, this is only half the story.

For all but the most highly transparent materials (e.g., fused silica, diamond, quartz, and sapphire), the damage threshold is more likely to be linked to thermal absorption in the laser pulse length region from 10^{-8} s to continuous wave (CW). In this scenario, the onset of damage, whether thought of as melting, catastrophic stress, drilling, or cutting, is related to the melting and/or vaporization of the material. These mechanisms vary between two pulse length–related regimes—the first regime is where the peak temperature is related to a steady-state process and is valid from CW to about 10^{-6} s. The second regime is where the peak temperature is governed by the relative size of the focused spot, the component diameter, and the thermal diffusivity of the sample under test. This second regime is valid for all pulse lengths below 10^{-6} s. However, for highly transmitting materials (e.g., fused silica, diamond, and sapphire), the thermal damage threshold is so relatively high that other damage mechanisms come into play before strict thermal effects take place. These are dielectric breakdown for pulse lengths of $\sim 10^{-8}$–10^{-10} s, avalanche ionization for pulse lengths of 10^{-10}–10^{-13} s, and multiphoton absorption for pulse lengths lower than $\sim 10^{-13}$ s. These mechanisms will be discussed in subsequent chapters.

Chapter 2

Laser-Induced Damage in Optical Materials. Edited by Detlev Ristau © 2015 CRC Press/Taylor & Francis Group, LLC. ISBN: 978-1-4398-7216-1.

The aim of this chapter is to allow the calculation of the theoretical maximum irradiation level below which no laser-induced damage (LID) occurs to a perfectly homogeneous material and to give some indication as to why this maximum LIDT is rarely reached in practice.

2.2 Theory

When a light beam is incident on the surface of a material or component, some of the energy is absorbed in the form of heat. The interaction depends on the laser parameters (wavelength, pulse length, and pulse repetition frequency [PRF]); the relative sizes of the beam and the material/component under test; the ambient conditions; the mounting conditions; and the optical, mechanical, and thermal properties of the material/component irradiated. The maximum temperature will normally be at the center of the irradiating beam on the surface of the irradiated material. Thermal damage will normally appear at this point unless the damage is stress related or resulting from localized absorbing imperfections within the material. A simplified schematic of the local peak temperature at the center of the beam on the surface of the material in terms of the pulse temporal shape and PRF is given in Figure 2.1a through e.

For short triangular pulses, see Figure 2.1a, the maximum temperature, at the center of the beam, occurs at or after the peak of the temporal pulse. For square and/or

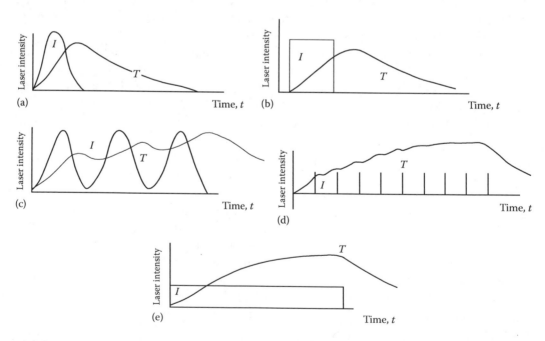

FIGURE 2.1 Laser beam intensity, I, and material temperature, T, versus time, t. (a) Short pulse, (b) long pulse, (c) low PRF pulse train, (d) high PRF pulse train, and (e) CW beam. PRF, pulse repetition frequency; CW, continuous wave. (From Wood, R.M., 2003, *Laser-Induced Damage of Optical Materials*, IOP Publishing/Taylor & Francis Group, Bristol, U.K., 2003.)

long pulses, see Figure 2.1b, the maximum temperature is close to the end of the pulse. For a low repetition series of pulses, see Figure 2.1c, the temperature oscillates in line with the repetition temperature of the sample and gradually rises. For a high PRF train of pulses, the temperature at the center of the sample gradually rises, see Figure 2.1d, and, apart from being slightly more spiky, shows very much the same pattern as the temperature measured for a CW beam, see Figure 2.1e. In all these figures, I stands for the laser beam intensity envelope and T stands for the temperature measured at the surface of the material at the center of the laser beam.

Absorption of energy will give rise to a rise in temperature, leading to thermal expansion, strain, birefringence, movement of internal defects, cracking, melting, and catastrophic shattering. High peak power densities may also give rise to the advent of nonlinear absorption and transmittance, electro-optic effects, second harmonic generation, optical parametric oscillation, and self focusing. These effects may add to the amount of energy absorbed and lower the observed LIDT. A combination of these mechanisms may add up to change the beam shape, induce birefringence, and shatter or melt the component. They can also be put to use to cut, drill, weld, and anneal.

There are four predominant scenarios where thermal damage may be initiated. The first is in transparent and semitransparent materials where the laser energy is deposited in a slightly truncated cylinder. This will apply to most laser window materials and will be highly wavelength sensitive. The second scenario is where the material in question is highly absorbing on the surface of the material and little energy reaches into the material (e.g., in the case of metal mirrors). The third scenario is where there are absorbing inclusions within a semitransparent material (e.g., metal particulates in glass or carbon particulates in diamond) or where there are small, highly mobile grain boundaries, which can move with a rise in temperature and can join together to form thermal barriers with subsequent cracking. There is a fourth scenario that is brought about by nonperfect input surface finish. This is the subject of Chapter 12. The third and fourth scenarios will drop the LIDT from that achieved for a perfect, homogeneous material.

2.3   Heat Transfer in Transparent Materials

When laser radiation is incident in a transmitting medium, it is absorbed in a cylinder, passing through the material on the axis of the laser beam. In the case of a semiabsorbing material, the cylinder changes to a truncated cone. Any absorption in the material at the laser wavelength causes both a temperature rise at the center of this cylinder and a radial strain between this center line and the edge of the component. In most cases, this can be seen as a two-dimensional temperature gradient between the center axis of the laser radiation and the edge of the component. Unless the laser beam is being focused inside the component, LID will occur at the surface of the component under test.

The thermally induced LIDT may therefore be defined as the point where the material surface at the center of the beam reaches the melting point of the material under test, T_m. The temperature rise, the spread of the heat, and the radial strain depend not only upon the material properties and the amount of heat absorbed but also on the beam diameter, the component diameter, and the laser pulse duration.

$$E_D = C\frac{dT}{\alpha},$$

(2.1)

where

α is the absorption coefficient of the material at the laser wavelength

C is the heat capacity of the component

$dT = T_m - T_a$ is the laser-induced temperature rise, where T_a is the ambient temperature and T_m is the melting point of the material

A threshold for catastrophic cracking can similarly be defined:

$$E_D = \frac{C\kappa S}{\beta\alpha},$$

(2.2)

where

S is the damaging stress

β is the volume expansion coefficient

It is useful to define a diffusion length, L, the distance the heat will travel out from the center of the beam in the duration of the laser pulse, τ:

$$L^2 = 4D\tau$$

(2.3)

where

D is the diffusivity (=$\kappa/\rho C$)

τ is the laser pulse duration

κ is the thermal conductivity

ρ is the material density

When the thermal conduction out of the irradiated volume becomes nonnegligible, that is, when

$$\frac{r^2}{\tau} \ll D \ll \frac{R^2}{\tau}$$

where

r is the laser spot radius

R is the component radius

This occurs for small radius laser spots, long laser pulse lengths (and high PRF trains of laser pulses), and large component diameters.

The illuminated region can now be treated as a continuous line source (Carslaw and Jaegar 1947).

The energy densities required for melting or catastrophic strain are then given by

$$E_D = \frac{4DCdT \ln\left(4D\tau/r^2\right)\tau}{\alpha r^2} \tag{2.4}$$

or

$$E_D = \frac{4\kappa CDS\tau}{\beta\alpha r^2}. \tag{2.5}$$

It will be seen from these equations that in this temporal regime, the damage threshold will be dependent on the power in the beam rather than on energy density or even on the more conventionally quoted power density.

In the extreme, in the case of long irradiation times (long pulse, multiple pulsing, and CW irradiation), the component temperature gradually rises to a maximum and the equation for the temperature rise, at the point of damage, reduces to

$$dT = T_m - T_a = \frac{P\alpha}{2\rho C\tau\pi D}, \tag{2.6}$$

which is the steady-state temperature profile in a semi-infinite solid (see Figure 2.1d and e).

LIDT will therefore, for long pulse, high-pulse repetition frequency pulse trains and CW irradiation, be governed by the equation

$$\frac{P}{\tau} = \frac{dT2\rho C\pi D}{\alpha}. \tag{2.7}$$

It will be noted, from perusal of this equation, that the LIDT will be proportional to the linear energy density, W cm^{-1}, if the laser pulse length, τ, is long enough for the heat to be transferred out of the beam center during the irradiation.

When the beam radius is much larger than the diffusion length, that is, $2r > L$, that is, for large laser spot sizes and for short duration pulses, there is negligible spread of the energy in the pulse duration. This means that for short irradiation times the damage thresholds will be pulse length independent. It will therefore be seen that in the short pulse regime the damage threshold is a constant value if expressed in terms of the peak power density, that is, in units of J cm^{-2}.

As should be expected from these equations, the LIDT for thermal damage can be calculated from the material parameters. Table 2.1 lists the expected LIDTs for a range of commonly used laser window materials. It may be noted, from this table, that many of the materials will suffer LID before they reach their true melting points as they exhibit other deleterious effects before they reach that point. For example, most glasses soften at temperatures below the melting point; chlorides, sulfides, and selenides dissociate; diamond graphitizes and all the semiconducting materials suffer from thermal runaway.

Chapter 2

Table 2.1   Summary of Thermal Properties of Commonly Used Laser Window Materials

Material	Specific Heat C_p (J g⁻¹ K⁻¹)	Conductivity K (W cm⁻¹ K⁻¹)	Density σ (g cm⁻³)	Temperature T_m (°C)	Absorption Coefficient α (cm⁻¹)	Thermal Diffusivity D (cm² s⁻¹)	LIDT $Cd T/\alpha$ (J cm⁻²)
1.064 µm							
Fused silica	0.81	0.014	2.21	1610 s	10^{-4}	0.0075	1.1×10^7
BK7	0.86	0.011	2.51	719 s	10^{-3}	0.0051	4×10^5
LiNbO₃	0.61	0.046	4.64	1170 m	10^{-2}	0.015	5×10^4
Al₂O₃	0.72	0.46	3.99	2015 m	2×10^{-4}	0.16	7×10^6
Diamond	6.1	22.0	3.51	1000 g	10^{-4}	1.03	6×10^7
				2000 m			1.5×10^8
10.6 µm							
KCl	0.69	0.07	1.99	776 d	10^{-4}	0.047	3×10^6
NaCl	0.88	0.065	2.17	801 d	10^{-4}	0.034	4×10^6
ZnS	0.49	1.72	4.09	700 tr	0.24	0.86	8×10^2
				1830 d			3×10^3

ZnSe	0.3	0.17	5.27	700 tr	0.2	0.11	7×10^2
				1520 d			2×10^3
BaF_2	0.44	0.11	4.89	1280 m	10^{-2}	0.055	4×10^4
CaF_2	0.86	0.095	3.18	1360 m	10^{-2}	0.035	10^5
GaAs	0.27	0.48	5.31	800 tr	2×10^3	0.33	7×10^{-2}
				800 tr	10^{-2}		10^4
				1238 m	10^{-2}		2×10^4
CdTe	0.18	0.075	5.85	1090 m	10^3	0.070	10^{-1}
					10^{-2}		10^4
Ge	0.31	0.67	5.32	370 tr	2×10^{-2}	0.41	5×10^2
				937 m			10^4
Si	0.71	1.68	2.33	1410 m	10^{-2}	1.02	8×10
HgCdTe	0.15	0.10	7.6	930 m	10^3	0.09	10^{-1}

Notes: m, melting point; s, softening point; d, dissociation point; g, graphitization point; tr, onset of thermal runaway.

Chapter 2

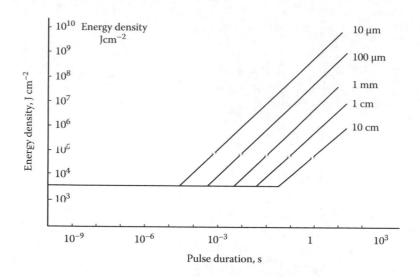

FIGURE 2.2   LIDT ($J\ cm^{-2}$) versus pulse length and spot size, ZnSe, 10.6 µm.

The damage thresholds for a ZnSe laser window material, at 10.6 µm, are plotted in Figures 2.2 and 2.3 using different abscissa units. In Figure 2.2, where the damage threshold is quoted in terms of the energy density ($J\ cm^{-2}$), it will be seen that the threshold, LIDT, is constant at short pulse lengths and dependent on both spot size and pulse length at longer pulse lengths. Figure 2.3, where the damage threshold is plotted in terms of the linear power density ($W\ cm^{-1}$), indicates the opposite. These two figures show that it is necessary, in quoting the damage thresholds of these materials, to be aware that the thermo-mechanical properties of the materials affect the relationship between the pulse length, the pulse energy, and the pulse shape. The break point between the pulse

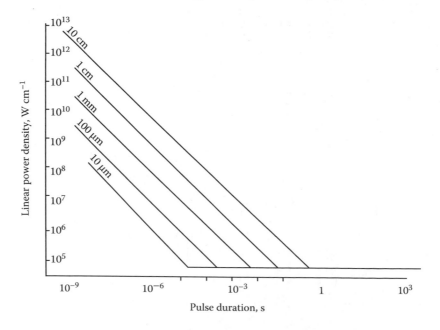

FIGURE 2.3   LIDT ($W\ cm^{-1}$) versus pulse length and spot size, ZnSe, 10.6 µm.

Table 2.2 Thermal Nonlinear Break Point

Thermal Diffusivity, D (cm² s⁻¹)	Uniform Beam Radius, r, or Gaussian Beam $1/e$ Diameter Pulse Length, τ (s)			
	10 µm	100 µm	1 mm	1 cm
10^{-3}	10^{-3}	10^{-1}	10^{1}	10^{3}
10^{-2}	10^{-4}	10^{-2}	10^{0}	10^{2}
10^{-1}	10^{-5}	10^{-3}	10^{-1}	10^{1}
10^{0}	10^{-6}	10^{-4}	10^{-2}	10^{0}

length-dependent equation (Equation 2.4) and the pulse length-independent equation (Equation 2.1) sections of the energy density versus pulse duration graphs for thermal damage is given by $\tau = r^2/D$. This transition point can be calculated for a range of spot radii and thermal diffusivity values, and these are listed in Table 2.2. The table lists the pulse lengths, τ, at which the transition occurs for a range of thermal diffusivities, D, and laser beam spot radii, r. It should be noted that although the equations have been derived, assuming a square-topped spatial energy density profile of radius, r, they are still correct if the value of the $1/e$ diameter of a Gaussian beam is substituted.

Figure 2.4 depicts the LIDT versus pulse length for a number of common optical and laser window materials, showing the break point. In practice, the transition is much more

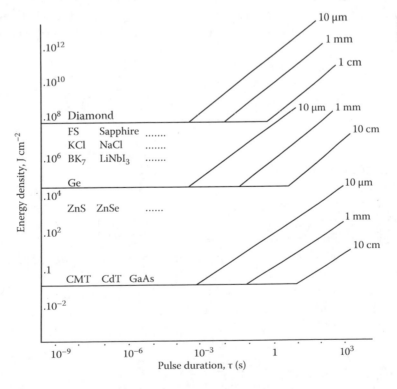

FIGURE 2.4 Thermally induced LIDT (J cm⁻²) versus pulse duration, τ (s).

curved. It must also be emphasized that these *relatively high LIDTs* are only observed in the highly transmitting region of the wavelength spectrum. Outside the transmission window, the LIDTs become progressively lower as the material absorption increases.

For all the semiconducting materials that form the main class of infrared-transmitting materials (e.g., germanium, zinc selenide, zinc sulfide, gallium arsenide, mercury cadmium telluride, etc.), it should be noted that thermal runaway occurs at fairly low temperatures, compared to the melting point (see Table 2.1). In practice, this means that the value for T_m in Equations 2.1 and 2.6 is a low value, the point at which thermal runaway starts, and not the true melting point of the material.

It should be noted that the only common infrared-transmitting materials that do not suffer from thermal runaway are the halides. As long as these materials are kept in a strictly dry environment, they exhibit relatively high LIDTs, which do not drop with laser PRF. However, they are all hygroscopic and, if left in the open, absorb water vapor, which then both dissolves the substrate and absorbs infrared radiation.

2.4 Absorbing Materials

When a laser beam is incident on a nontransmitting metallic surface, a small amount of radiation penetrates it to a distance called the skin depth, which is in turn a function of the electrical conductivity. Subsequent absorption of this radiation by free carriers within the material raises the temperature of the surface. As the temperature rises, stress and distortion of the surface will occur. In the limit, catastrophic damage may occur due to mechanical failure, to melting of the surface, or to a combination of both.

The temperature rise due to absorption of energy may be calculated using a one-dimensional heat flow calculation. At the center of the beam, the temperature rise, dT, is given by

$$dT(x,t) = 2\alpha I_o \frac{t}{\pi \kappa \rho C} \, \text{ierfc} \, \frac{x}{2\sqrt{\kappa t / \rho C}}, \tag{2.8}$$

where
    t is the irradiation time
    α is the absorption
    I_o is the peak intensity of the beam
    κ is the thermal conductivity
    ρ is the density of the material
    x is the depth at which the measurement takes place ($x = 0$ at the surface)

As long as the function ierfc(y) is small ($y > 2$), a sample thickness of $x > 4\sqrt{\kappa t / \rho C}$ is effectively infinitely thick as regards the effect on the maximum surface temperature. This leads to the expression

$$dT(t) = \frac{2\alpha I_o \sqrt{t}}{\sqrt{\pi \kappa \rho C}} \quad \text{at } x = 0. \tag{2.9}$$

If damage is due to surface melting, then

$$dT = T_m - T_a,$$

where

T_m is the temperature at which the surface melts
T_a is the ambient temperature

It should be noted that the absorption is the sum of the intrinsic absorption and the surface absorption. It has been proved that the roughness of the surface of a metal mirror affects the absorption. It should be noticed however that the roughness needs to be measured in terms of the wavelength of the incident laser radiation.

The expression $(T_m - T_a)\sqrt{\kappa \rho C}$ may be regarded as a figure of merit (FOM) for LID of metal surfaces. It may however be modified by anomalous absorption of the sample at the wavelength of interest. A table containing this FOM, with respect to copper, for a range of metal mirror candidates for use at 10.6 μm, is shown as Table 2.3. It should be recognized that all the constants involved may be temperature dependent. The thickness

Table 2.3 Figure of Merit for 10 μm Mirror Candidate Materials

Element	T_m (°C)	κ (W cm⁻¹ K⁻¹)	ρ (g cm⁻³)	C (J g⁻¹ K⁻¹)	FOM wrt Copper
Copper	1084	4.01	8.96	0.385	1.000
Aluminum	660	0.243	2.7	0.90	0.124
Beryllium	1287	2.00	1.85	1.825	0.832
Chromium	1907	0.937	7.15	0.449	0.827
Gold	1064	3.17	19.3	0.129	0.741
Hafnium	2233	0.230	13.3	0.144	0.371
Molybdenum	2623	1.38	10.2	0.251	1.236
Nickel	1455	0.907	8.9	0.444	0.400
Niobium	2477	0.537	8.57	0.265	0.234
Osmium	3033	0.876	22.5	0.130	1.219
Palladium	1555	0.718	12.0	0.246	0.565
Platinum	1768	0.716	21.5	0.133	0.632
Rhenium	3186	0.479	20.8	0.137	0.934
Rhodium	1964	1.50	12.4	0.243	1.044
Ruthenium	2334	1.17	12.1	0.238	1.356
Silver	962	4.29	10.5	0.235	0.774
Tantalum	3017	0.575	16.4	0.140	0.870
Titanium	1668	0.219	4.5	0.523	0.299
Zirconium	1855	0.227	6.52	0.278	0.324

Source: Lide, D.R., *CRC Handbook of Chemistry and Physics*, CRC Press, Boca Raton, FL, 1994.

Chapter 2

of any coating, dielectric or metallic, will also affect the LIDT. In the case of a coated substrate (e.g., a gold-coated copper mirror), if the coating thickness is much less than $\sqrt{\kappa\tau/\rho C}$, then the values for κ, ρ, and C are determined by the substrate material, while the value for α is that of the coating material and T_m is the lower of the melting points of the two materials.

Taking into account the correction that has to be made if the laser pulse is not square, the LIDT is given by

$$T_m - T_a = dT = \frac{I_o\alpha}{2\sqrt{\pi\rho\kappa C}} f(t), \tag{2.10}$$

where $f(t)$ is a time-dependent function, the precise form of which is determined by the input pulse shape. This is discussed in more detail in Wood (2003).

Note that the LIDT is inversely proportional to the temperature of the material under test and that it can therefore be raised by back cooling the substrate.

It is now necessary to consider when the peak temperature on the center axis of the laser beam reaches a maximum. The equation of the temperature curve can be approximated by

$$\Delta T = \frac{P\alpha}{2\rho Cr\pi D} \operatorname{erfc} \frac{r}{\sqrt{4Dt}},$$

where P is the power of the laser beam, W, which, for large values of t, reduces to

$$\Delta T = \frac{P\alpha}{2\rho Cr\pi D},$$

which is, again, the steady-state radial profile of the temperature in a semi-infinite solid.

It will be realized that the temperature of the material is inversely proportional to the radial distance, r, from the center of the beam.

For short, single pulses, the temperature rise is proportional to the incident energy density, that is, for short, single pulses, the LIDT is a constant when expressed in terms of J cm^{-2}.

When the pulse duration is longer than the material heat diffusion constant, that is, when the heat has time to diffuse out of the irradiated spot, the damage threshold becomes constant when measured in terms of W cm^{-1}.

The two scenarios discussed earlier can be seen easiest when the LIDT of an oxygen free high conductivity (OFHC) copper mirror sample is plotted versus the laser pulse duration (see Figures 2.5 and 2.6). It will be noticed that the shapes of these graphs are identical with those generated for transmitting materials.

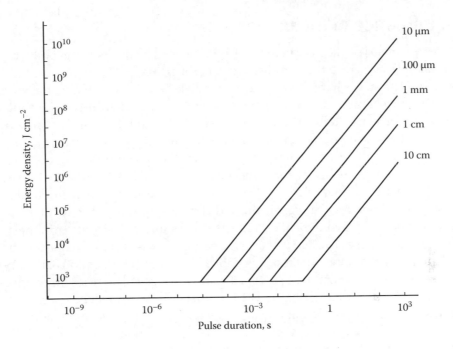

FIGURE 2.5   LIDT (J cm^{-2}) versus pulse duration, τ, and spot size, r. OFHC copper at 10.6 μm.

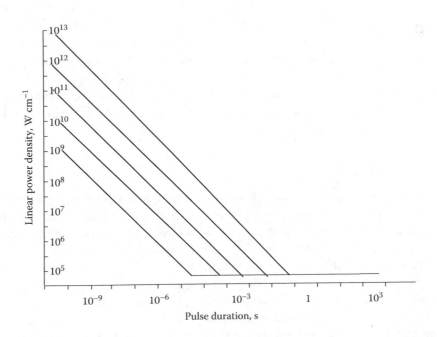

FIGURE 2.6   LIDT (W cm^{-1}) versus pulse duration, τ, and spot size, r. OFHC copper at 10.6 μm.

Chapter 2

2.5   Influence of Inclusions

There are two major scenarios where LIDT is influenced by the local material properties rather than by the average thermo-mechanical properties.

The first scenario is where there are localized absorbing specks of material within the lattice of a transparent or semitransparent material. The heat absorbed is a function of the speck cross section. When the speck is large, $r > 100$ µm, the temperature rise averages out over the surface of the speck and is easily transmitted to the transparent lattice. In this case, although the optical quality of the material is compromised, the absorption rarely causes damage. When the speck is of the order of 10–0.1 µm, the particle will absorb the radiation and damaging stress is likely to build up between the absorbing speck and the surrounding lattice due to the thermal expansion of the speck. A theoretical treatment has predicted the critical radius to be 0.4 µm for laser pulses in the ns pulse length range (Koldunov et al. 2002). When the particle size is <0.1 µm, any absorption will only add to the absorption and damage is unlikely to occur (Duthler and Sparks 1973).

This has been observed in the case of platinum or tungsten particles originating from the crucibles in which glass or Nd:YAG laser crystal has been melted. In this case, the speck was seen to have melted and exploded, leaving craters inside the bulk material. Another instance of absorbing speck has been observed inside substandard chemical vapor deposition (CVD) grown diamond window material. A third example is that of carbonization of thin glue layers used in cementing optical filters and/or dielectric polarizers. In all these cases, the absorption first causes melting/evaporation or carbonization and this leads to unacceptably high levels of stress and to catastrophic cracking.

The second scenario involves the movement of microscopic grain boundaries inside laser/optical crystals under steady and/or high PRF irradiation. This has been observed in germanium laser window crystal and in CVD-grown diamond. In both cases, the crystal showed little or no deterioration of transmittance during continuous irradiation until, quite suddenly, the crystal split in a circular ring round the laser entry point. In the case of the germanium window, there was evidence of melting at the center of the circular damage ring. In the case of the diamond substrate, the thermal conductivity suddenly underwent a sharp drop at this point (Sussman et al. 1993).

2.6   Discussion

Many laser engineers have been used to measure, and quote, the laser damage thresholds of their optical/laser materials in terms of W cm^{-2} because they are thinking in terms of dielectric breakdown under pulsed irradiation. It took many years, and the incidence of 10.6 µm, CO_2 lasers, for the realization that for all but highly transmitting materials, the damage mechanism might be melting and that in that case the units of measurement should be in terms of J cm^{-2}. It was only fairly recently realized that there was another commonly experienced regime—that of the long pulse, the high repetition rate train of pulses, and CW. In this temporal regime, it was noted early on that the laser damage level was constant for pulses longer than a few seconds, but it was not realized that the steady-state temperature could not be stated in terms of J cm^{-2} for many years. It was only realized that there was a serious problem with quoting the LIDT in terms of the peak energy density for long pulses and CW radiation when the use of beam scanners, fabricated from beryllium, was promoted because of the lightness of this material. The concern came about because of the extreme toxicity of Be vapor. An analysis of the LIDT showed that if the safe power loading was calculated in terms of J cm^{-2} for small spot sizes, then it could be catastrophic if the laser spot size was increased under the impression that the component could withstand higher laser powers at the same energy density, instead of making the calculations at W cm^{-1} (Wood 1997).

Although the use of W cm^{-1} measurement units had been cautiously highlighted some years previously (Wood 1986) (as very few people were really interested in CW testing at this stage), it came into focus when, in the process of an industry/university cooperative program, the results showed that the damage thresholds of zinc selenide window material under CW testing at 10.6 μm exhibited a constant value of 3000 W mm^{-1} (Greening 1994, 1997).

It has frequently been argued that LIDTs must be proportional to either the power or the energy density (i.e., either measured in terms of J cm^{-2} or in terms of W cm^{-2}) as the eye damage thresholds are measured in these units. This argument does not hold water once it is realized that all eye damage threshold measurements are quoted for a fixed iris diameter. If the diameter of the spot is left out of the calculation, then the parameter of interest is the power, W, and the Electrochemical Commission (IEC) standard does predict that the LIDT for CW beams is a constant power level. The same problem arises with safety goggles and with laser safety screens. If a preliminary damage measurement is made using a CW beam focused down to a small spot, then if the spot size is increased (i.e., allowance is made for defocusing the beam), the assumed safety level, scaled in terms of W cm^{-2}, is soon far above the true safety level, scaled using W cm^{-1}.

References

Carslaw, H.S. and Jaegar, J.C., 1947, *Conduction of Heat in Solids*, Oxford, U.K.: Clarendon Press.

Duthler, C.J. and Sparks, M., 1973, *Theory of Material Failure in Crystals containing Infrared Absorbing Inclusions*. NBS Special Publication, 387, pp. 208–216.

Greening, D., 1994, *Optics and Laser Engineering*, April, p. 25; 1977, *Optics and Laser Engineering*, June, p. 41.

Koldunov, M.F., Manenkov, A.A., and Pokotilo, I.L., 2002, Mechanical damage in transparent solids caused by laser pulses of different durations. *Quantum Electronics* 32, 335–340.

Lide, D.R., 1994, *CRC Handbook of Chemistry and Physics*, Boca Raton, FL: CRC Press.

Sussman, R.S., Scarsbrook, G.A., Wort, C.J.H., and Wood, R.M., 1993, Laser damage testing of CVD grown diamond windows. *Conference on Diamond, Diamond-Like and Related Coatings*, Albufeira, Portugal.

Wood, R.M., 1986, *Laser Damage in Optical Materials*, Bristol, U.K.: Adam Hilger.

Wood, R.M., 1997, Laser induced damage thresholds and laser safety levels. Do the units of measurement matter? *Journal of Optics and Laser Technology* 29(8), 517–522.

Wood, R.M., 2003, *Laser-Induced Damage of Optical Materials*, Bristol, U.K.: IOP Publishing/Taylor & Francis Group.

Chapter 2

3. Defect–Induced Damage

Semyon Papernov

Chapter 3

Laser-Induced Damage in Optical Materials. Edited by Detlev Ristau © 2015 CRC Press/Taylor & Francis Group, LLC. ISBN: 978-1-4398-7216-1.

3.1   Introduction

Ever since the appearance of a laser as a tool that permits extreme spatial and temporal concentration of light energy, it has been realized that localized absorbing defects in optical components are a major source of laser-induced damage. There is ongoing continuous effort to reduce the number and size of absorbers in optical materials used for laser applications. Still, with ever-increasing laser power densities (for instance, in laser fusion facilities), even nanoscale absorbing defects continue to be a major source of damage.

In this chapter, we consider physical processes that describe the interaction of pulsed-laser radiation with transparent dielectric materials containing isolated absorbing defects in the form of particles made of homogeneous material. Such an interaction has two main aspects in relation to laser damage: First, laser-energy absorption generates thermal and mechanical effects, like heating, melting, evaporation, and fracturing of the host material. The second aspect is that the light field of the laser pulse can be amplified in the vicinity of the absorbing or structural defect, which consequently affects modification of the host material.

The main part of this chapter is devoted to an analysis of thermomechanical processes, and effects of light intensity amplification are addressed toward the end in a special paragraph. First, laser-damage criterion based on reaching a critical temperature is introduced. The effects that the optical and thermal properties of both the defect and host material, along with defect size and shape, have on damage thresholds are considered within the thermal diffusion model. The damage-threshold pulse-length scaling is also evaluated within the thermal diffusion model's framework. These results are compared with experimental data and results of numerical simulations taking into account field enhancement, thermal stresses, and phase transitions for damage initiated by metallic and dielectric defects.

Special consideration is given to the concept of absorption delocalization during the laser pulse for absorbers much smaller than a laser wavelength and related thermal explosion theory. Mechanisms of the free-electron formation in the defect-surrounding material and experimental verification of absorption delocalization using model systems with well-characterized metallic particle absorbers are discussed.

The majority of the processes considered take place during the laser pulse (timescale $\sim 10^{-11}$ to 10^{-8} s) and consequently describe the state of matter during energy deposition. On the other hand, the damage process also has a second stage characterized by material melting, evaporation, and (in the case of brittle materials) cracking that may continue to evolve after the end of the laser pulse and defines the final damage morphology. In the bulk, the extent of damage is usually defined by the size of cracks formed during the cooling stage. We discuss conditions for crack formation and the effects of pulse length and absorber size.

When a defect located near the surface initiates damage, the damage morphology usually takes the form of a crater. The crater size may be linked to the defect depth using simple scaling relations. Effects of spallation caused by a shock wave originating at the defect location, as well as the difference in energy deposition for the two cases of front (free-surface side) and back irradiation will be considered here.

For most topics presented in this chapter, the assumption is made that the electric field of the light wave outside the absorber is the same as the electric field of the wave inside the homogeneous dielectric host material. Exceptions to this are made only for the few cases of numerical modeling. In fact, this field can be significantly larger in the vicinity of the structural defect, which could be either an absorber in the bulk or differently shaped and oriented surface cracks. We concentrate our analysis here on the effect of the light intensification and its relative role in extrinsic and intrinsic mechanisms of damage.

Localized absorbing defects exist in real optical materials as an ensemble characterized by distributions of sizes, geometry, optical properties, and other parameters. As a result, laser damage initiated by an ensemble of localized defects is probabilistic by its nature, and the probability of damage is linked to the probability of finding a defect within the laser beam area where the fluence exceeds the damage threshold. Consequently, it leads to the beam spot size dependence of the damage probability and related damage threshold. We discuss how damage probability studies can provide information on defect densities and, in some cases, on the defect's optical properties.

The chapter closes with considering laser conditioning as a method for improving the damage resistance of materials containing absorbing defects.

3.2   Thermal Diffusion Model of Absorbing Defect-Induced Damage

3.2.1   Damage Criterion

Modeling of laser-induced damage requires establishing a damage criterion, usually reaching some fixed value by an important damage process parameter. As an example, it could be thermal stress equal to or exceeding the strength of the material, or reaching some critical temperature. Historically, it has been the temperature of the defect-surrounding material reaching a melting point that was considered the decisive event establishing damage. Justification for such criterion in the case of transparent dielectric materials lies in fact that close to this point, structural changes are accompanied by a loss in transparency (band-gap collapse). One can argue that it might be insufficient to create irreversible changes in the optical properties of the material, but, in most cases, it allows one to predict damaging laser intensities and establish important rules governing the damage process.

3.2.2   Heat Equation and Boundary Conditions

The study of localized absorber-driven damage considering energy transfer from a laser-heated defect to the host material by heat conduction was first performed by Hopper and Uhlmann (1970) for the case of metallic absorbers and then explored by other authors (Walker et al., 1981; Lange et al., 1982, 1984; Fuka et al., 1990; Feit and Rubenchik, 2004a)

Chapter 3

for absorbers with different thermal and optical properties and various geometries. We consider here a spherical absorber inside an infinite transparent material. The energy of laser radiation is deposited linearly inside the absorber, and then, the heat diffuses from the absorber into the host material. The boundary between the absorber and the host is assumed to be perfect (no thermal resistance). The heat equation for spherically symmetric problem then takes the following form (Fuka et al., 1990):

$$\frac{1}{D_a}\frac{\partial T_a}{\partial t} = \frac{1}{r^2}\frac{\partial}{\partial r}\left(r^2\frac{\partial T_a}{\partial r}\right) + \frac{A}{K_a} \quad 0 \leq r \leq a \tag{3.1}$$

$$\frac{1}{D_h}\frac{\partial T_h}{\partial t} = \frac{1}{r^2}\frac{\partial}{\partial r}\left(r^2\frac{\partial T_h}{\partial r}\right) \quad r > a, \tag{3.2}$$

where
   a is the radius of the sphere
   T is the temperature
   t is the time
   r is the radial distance from the sphere center
   D is the thermal diffusivity
   K is the thermal conductivity
   A is the heat source term (heat generated in unit volume per unit time)
   The subscript a denotes absorber and h denotes host material

The interface between the absorber and the host is characterized by the boundary conditions

$$T_a(a) = T_h(a) \quad \text{and} \quad K_a\left(\frac{\partial T_a}{\partial r}\right)_{r=a} = K_h\left(\frac{\partial T_h}{\partial r}\right)_{r=a} \tag{3.3}$$

$$T_a(r, t = 0) = T_h(r, t = 0) = 0.$$

An exact solution for the temperature $T(r = a, t = \tau)$ at the end of the rectangular pulse τ with constant intensity I (W/cm²) using Equations 3.1 and 3.2 with these boundary conditions was derived by Goldenberg and Trantor (1952) and has a rather complex integral form:

$$T = \frac{3\sigma I}{4\pi a K_a}\left[\frac{1}{3}\frac{K_a}{K_h} - \frac{2b}{\pi}\int_0^\infty \exp\left(\frac{-y^2\tau}{\delta}\right)F(y)dy\right]$$

$$F(y) = \frac{(\sin y - y\cos y)\sin y}{y^2\left[(c\sin y - y\cos y)^2 + b^2 y^2 \sin^2 y\right]} \tag{3.4}$$

$$\delta = \frac{a^2}{D_a}; \quad b = \frac{K_a}{K_h}\sqrt{\frac{D_h}{D_a}}; \quad c = 1 - \frac{K_h}{K_a},$$

where σ is absorption cross section.

Solving the heat diffusion problem can be simplified (Chan, 1975; Feit and Rubenchik, 2004a), assuming that energy is homogeneously deposited inside the absorber, which is valid for very small, strong absorbers (radius is comparable to a skin depth; metals) and larger, weak absorbers (dielectrics). Consequently, as a reasonable approximation, the temperature inside the absorber can be considered homogeneous. This allows simple expression for the boundary condition that follows from energy conservation considerations: The energy deposited inside the absorber must be equal to increase in the absorber heat content plus heat leaving the surface of the absorber:

$$\gamma I(t) = \frac{4}{3}\rho_a C_a a \left(\frac{\partial T}{\partial t}\right)_{r=a} - 4K_h \left(\frac{\partial T}{\partial r}\right)_{r=a}, \tag{3.5}$$

where
   γ is the absorptivity ($\gamma = \sigma/\pi a^2$, σ is the absorption cross section)
   I is the light intensity
   ρ_a is the mass density
   C_a is the heat capacity of the absorber

Solving Equation 3.2 with boundary conditions (Equation 3.5) gives a simple expression (Chan, 1975) for the temperature of the absorber heated by a rectangular pulse of duration τ:

$$T = \frac{\gamma F_a}{4K_h \tau}\left[1 - \exp\left(-\frac{4D\tau}{a^2}\right)\right], \tag{3.6}$$

where
   F is the laser fluence (J/cm^2)
   $D = 3K_h/4\rho_a C_a$ is the effective diffusivity

In Figure 3.1a, the temperature is plotted as a function of absorber size normalized by the effective diffusion length $2\sqrt{D\tau}$.

One can see that there is an absorber size most susceptible to heating. A qualitative physical explanation for homogeneously heated absorbers is that for very small absorbers a high surface-to-volume ratio leads to efficient heat removal by heat conduction. For large absorbers, absorption cross-section scales (see Section 2.4.2) as a^2, but the volume as a^3, which leads to reduced absorbed energy per unit volume and lower temperatures. Consequently, the most efficient heating might be expected for some intermediate absorber size. At a fixed pulse length, it follows from Equation 3.6 that an absorber subjected to a maximum heating has a radius approximately equal to the effective thermal diffusion length, $a = 1.8\sqrt{D\tau}$. A corresponding temperature T_{max} can be calculated using Equation 3.6:

$$T_{max} = \frac{0.32\gamma F}{K_h}\sqrt{\frac{D}{\tau}}. \tag{3.7}$$

Chapter 3

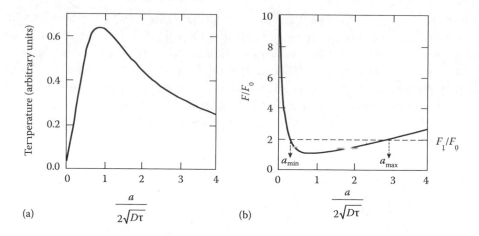

FIGURE 3.1 (a) Absorber temperature and (b) normalized threshold as a function of normalized absorber radius. (Reprinted figure with permission from Feit, M.D. and Rubenchik, A.M., Implications of nanoabsorber initiators for damage probability curves, pulse length scaling, and laser conditioning, in *Laser-Induced Damage in Optical Materials: 2003*, G.J. Exarhos, A.H. Guenther, N. Kaiser, K.L. Lewis, M.J. Soileau, and C.J. Stolz, eds., Vol. 5273, pp. 74–82, SPIE, Bellingham, WA. Copyright 2004 by SPIE.)

Assuming that T_{max} equals the critical temperature of the material, T_{cr}, the damage-threshold fluence F_0 takes the form

$$F_0 = \frac{3.1 T_{cr} K_h \sqrt{\tau}}{\gamma \sqrt{D}}. \tag{3.8}$$

Normalized to F_0, damage threshold versus normalized absorber size dependence (Feit and Rubenchik, 2004a) is plotted in Figure 3.1b. For each fluence $F > F_0$, there is a range of the defect sizes $a_{min} < a < a_{max}$ for which the host material is expected to fail. The resulting damage density may be found by integrating the absorber density between sizes a_{min} and a_{max}. The temperature behavior for the Gaussian pulse shape, $I = I_0 \exp[-(t/\tau)^2]$, with the fluence defined as $F = I_0 \tau$, is very similar to the case of the rectangular pulse, and Equations 3.7 and 3.8 hold (Feit and Rubenchik, 2004a).

3.2.3 Pulse–Length Scaling

The threshold fluence given in Equation 3.8 scales as $\sqrt{\tau}$, the result frequently used to quickly compare the threshold measurements performed with different pulse lengths. Strictly speaking, this result is valid only when absorptivity γ is not size dependent. An exact calculation (Bonneau et al., 2001) of absorptivity as a function of the absorber's size using Mie's theory (Mie, 1908; Bohren, 1983) for metallic absorbers embedded in SiO_2 is shown in Figure 3.2a. One finds strong changes in absorptivity for absorbers smaller than ~200 nm for a 1053 nm wavelength and smaller than ~60 nm for a 351 nm wavelength. For weak absorbers, like SiO, absorptivity is a weakly increasing function of the size (Feit and Rubenchik, 2004a). In the latter case,

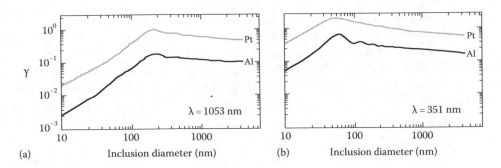

(a) Inclusion diameter (nm)   (b) Inclusion diameter (nm)

FIGURE 3.2 Metal particle absorptivity as a function of particle diameter for (a) $\lambda = 1053$ nm and (b) $\lambda = 351$ nm. (Reprinted figure with permission from Bonneau, F., Combis, P., Vierne, J., and Daval, G., Simulations of laser damage of SiO_2 induced by a spherical inclusion, in *Laser-Induced Damage in Optical Materials: 2000*, G.J. Exarhos, A.H. Guenther, M.R. Kozlowski, K.L. Lewis, and M.J. Soileau, eds., Vol. 4347, pp. 308–315, SPIE, Bellingham, WA. Copyright 2001 by SPIE.)

using a simple approximation $\gamma \sim a^q$, where $q < 1$, the threshold fluence scaling has a power coefficient that is always less than 0.5:

$$F_0 \sim \tau^{(1-q)/2}. \tag{3.9}$$

The pulse-length scaling given in Equations 3.8 and 3.9 assumes a broad distribution of absorber sizes, including those most susceptible to heating. In this case, the damage threshold is defined by the failure fluence for the most susceptible absorber. As an example, for absorbers in fused silica and effective diffusivity equal to that of fused silica, $D = 0.008$ cm²/s, and pulse length $\tau = 1$ ns, the most dangerous absorber has a radius $a \approx 50$ nm. Obviously, with increasing pulse length, the size of the absorber that causes failure also increases. Figure 3.3 provides an example of threshold fluence

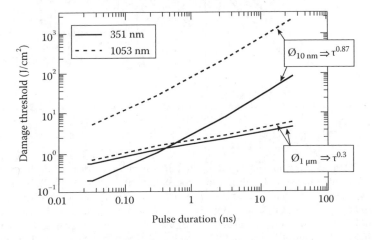

FIGURE 3.3 The 351 nm and 1053 nm damage thresholds as a function of pulse duration for 10 nm and 1000 nm Al-particle sizes. (Reprinted figure with permission from Bonneau, F., Combis, P., Vierne, J., and Daval, G., Simulations of laser damage of SiO_2 induced by a spherical inclusion, in *Laser-Induced Damage in Optical Materials: 2000*, G.J. Exarhos, A.H. Guenther, M.R. Kozlowski, K.L. Lewis, and M.J. Soileau, eds., Vol. 4347, pp. 308–315, SPIE, Bellingham, WA. Copyright 2001 by SPIE.)

Chapter 3

numerical calculations (Bonneau et al., 2001), using a thermal diffusion model for Al absorbers and taking into account the temperature dependencies of all important parameters (absorptivity, thermal conductivity, etc.). Here, a different pulse-length scaling exists for 10 and 1000 nm Al absorbers. Note that for metallic absorbers with diameters much larger than skin depth, an assumption of homogeneous absorption is not valid. For such absorbers, energy is deposited in only the layer adjacent to the surface, and heat diffusion initially takes place inside the absorber, leading to changes in the pulse-length scaling.

The scaling for absorbers much smaller or much larger than diffusion length can be predicted qualitatively by considering their temporal heating behavior (Feit et al., 2001b). Small absorbers very quickly reach thermal equilibrium when energy deposition is balanced by the conductive heat removal. This means that the peak temperature will be defined mainly by laser intensity. Consequently, damaging fluence $F = I \cdot \tau$ scales linearly with the pulse length. For very large absorbers, nearly all deposited energy goes into absorber heating (heat removal is small) and peak temperature depends roughly on the total absorbed energy. This results in a threshold fluence being independent of pulse length. Frequently observed $\sqrt{\tau}$ threshold scaling corresponds to an intermediate case of absorbers whose size is comparable to the diffusion length. The 1053 nm damage-threshold measurements for superpolished fused silica and calcium fluoride surfaces (Stuart et al., 1995) (see Figure 3.4a) provide an example of $\sqrt{\tau}$ scaling in the 20 ps to 1 ns pulse-length range. Figure 3.4b shows similar $\tau^{0.4:0.5}$ scaling obtained for fused-silica surfaces in earlier 1064 and 355 nm measurements (Campbell et al., 1991) covering a pulse range from 100 ps to 50 ns. Numerical calculations (Feit et al., 2001b) using

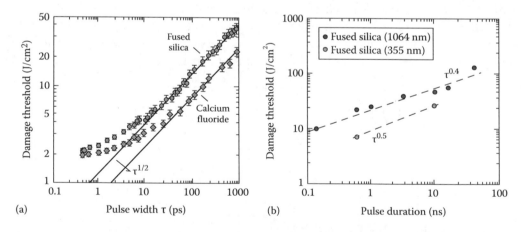

FIGURE 3.4 (a) Damage threshold as a function of pulse duration for fused silica and calcium fluoride, $\lambda = 1053$ nm. (Reprinted figure with permission from Stuart, B.C., Feit, M.D., Rubenchik, A.M., Shore, B.W., and Perry, M.D., Laser-induced damage in dielectrics with nanosecond to subpicosecond pulses, *Phys. Rev. Lett.*, 74, 2248–2251. Copyright 1995 by the American Physical Society.) (b) Damage threshold as a function of pulse duration for fused silica, $\lambda = 355$ nm, 1064 nm. (Reprinted figure with permission from Campbell, J.H., Rainer, F., Kozlowski, M.R., Wolfe, C.R., Thomas, I.M., and Milanovich, F.P., Damage resistant optics for a megajoule solid state lasers, in *Laser-Induced Damage in Optical Materials: 1990*, H.E. Bennett, L.L. Chase, A.H. Guenther, B.E. Newnam, and M.J. Soileau, eds., Vol. 1441, pp. 444–456, SPIE, Bellingham, WA. Copyright 1991 by SPIE.)

an exact solution of the thermal equations and hypothetical absorber size distribution dominated by small absorbers resulted in a scaling with a power of 0.35–0.45, which appears to be in very good agreement with the experimentally observed $\tau^{0.35}$ damage-threshold scaling for potassium dihydrogen phosphate (KDP) crystals (Runkel et al., 2000).

In the preceding analysis of pulse-length scaling, absorptivity was considered as a temperature-independent parameter. In fact, temperature dependence (known for the number of optical materials) may lead to thermal instability in a form of nonlinear temperature growth, the so-called thermal explosion (Danileiko et al., 1978; Koldunov et al., 1990, 1996). We will consider this damage mechanism later (see Section 3.3) but, in the context of pulse-length scaling, one might expect different kinetics of absorber heating and consequently different scaling. In this case, the absorber may be character-ized by some critical temperature T_{cr} above which the temperature growth cannot be balanced by heat diffusion. A solution of heat equations similar to Equations 3.1 and 3.2 with boundary conditions given by Equation 3.5 (homogeneous absorption) and temperature-dependent absorptivity $\gamma = \gamma(T)$ leads to the following threshold intensity scaling for a rectangular pulse (Koldunov et al., 1996):

$$I_{th}\left(\tau\right)=\frac{I_{cr}}{1-\exp\left(-\tau/2\tau_0\right)}, \tag{3.10}$$

where

$I_{cr} = I_{cr}\left(T_{cr}\right)$ is the threshold intensity at $\tau \gg \tau_0$
$\tau_0 = \rho_a C_a a^2/3K_h$ is the absorber temperature's relaxation time

It should be noted here that Equation 3.6, after substituting intensity I for F/τ, takes on a functional expression very similar to Equation 3.10. For short pulses when $\tau \ll \tau_0$, Equation 3.10 reduces to the form

$$I_{th}\left(\tau\right)=\frac{2\tau_0}{\tau}I_{cr}. \tag{3.11}$$

It follows from Equation 3.11 that in this case, damage threshold is characterized by flu-ence F_{th} (energy density):

$$F_{th}=\tau I_{th}\left(\tau\right)=2\tau_0 I_{cr}. \tag{3.12}$$

The threshold pulse-length curves calculated using Equation 3.10 for rectangular pulses and numerical calculations for a Gaussian temporal pulse shape (Koldunov et al., 1996) are shown in Figure 3.5. These curves, calculated using τ_0 as the only adjustable parameter, are in good agreement with experimental data for fused silica (Soileau et al., 1983) and alkali halide crystals (Garnov et al., 1993) in the pulse range of 4 ps to 31 ns.

Chapter 3

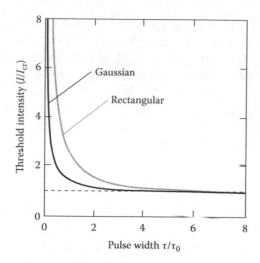

FIGURE 3.5 Normalized threshold intensity versus pulse duration normalized to absorber tempera-ture relaxation time. (Reprinted figure with permission from Koldunov, M.F., Manenkov, A.A., and Pokotilo, I.L., Pulse-width and pulse-shape dependencies of laser-induced damage threshold to transparent optical materials, in *Laser-Induced Damage in Optical Materials: 1995*, H.E. Bennett, A.H. Guenther, M.R. Kozlowski, B.E. Newnam, and M.J. Soileau, eds., Vol. 2714, pp. 718–730, SPIE, Bellingham, WA. Copyright 1996 by SPIE.)

3.2.4 Influence of the Thermal and Optical Properties of the Absorber and Host Material

3.2.4.1 Thermal Properties

The role of thermal conductivity in the context of damage-threshold calculation is dif-ferent for the host material and absorber (Walker et al., 1981; Lange et al., 1984). Reduced host thermal conductivity leads to slower heat removal from the heated absorber, which, in turn, leads to lower thresholds. In the case considered earlier of homogeneous heat deposition in the absorber, substituting $D = 3K_h/4\rho_a C_a$ into Equation 3.7 gives

$$F_0 = \frac{6.2\, T_c \sqrt{K_h}}{\gamma} \sqrt{3\rho_a C_a \tau}. \tag{3.13}$$

Here the threshold fluence varies as a square root of the host thermal conductivity. For obvious reasons, there is no dependence on absorber conductivity since the tem-perature is assumed homogeneous inside the absorber (no diffusion). Calculations (Fuka et al., 1990) using an exact solution of Equations 3.1 and 3.2 with the bound-ary conditions given by Equation 3.3, and accounting for inhomogeneous absorp-tion inside the absorber, showed a relatively weak threshold dependence on host and absorber thermal conductivities. The effect is different for different absorber sizes, but, on average, damage threshold changes roughly by a factor of 2 per order of magnitude change in the host thermal conductivity. In contrast to host conductiv-ity, thresholds increased on average for decreasing absorber conductivity. For a laser

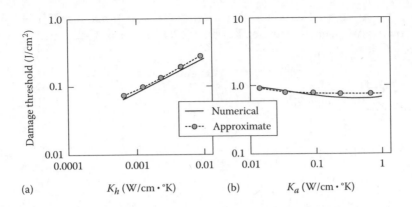

FIGURE 3.6   Damage threshold (a) for platinum impurity in glass as a function of host conductivity and (b) for ThO_2 impurity in a ThF_4 host as a function of absorber conductivity. (Reprinted figure with permission from Lange, M.R., McIver, J.K., Guenther, A.H., and Walker, T.W., Pulsed laser induced damage of an optical material with a spherical inclusion: Influence of the thermal properties of the materials, in *Laser Induced Damage in Optical Materials: 1982*, Nat. Bur. Stand. U.S. Special Publication, Vol. 669, pp. 380–386, U.S. Government Printing Office, Washington, DC. Copyright 1982 by SPIE.)

pulse length of $\tau \gg a^2/D_a$, an exact solution for the threshold F_0 given by Equation 3.4 can be simplified to (Lange et al., 1982, 1984)

$$F_0 = \frac{16T_c}{\pi}\sqrt{\rho_h C_h K_h \tau}\left[1 + \frac{\pi \rho_a C_a}{120\rho_h C_h}\left(\frac{15\rho_h C_h}{\rho_a C_a} - 10 - \frac{K_h}{K_a}\right)\right]^{-1}. \tag{3.14}$$

Figure 3.6 shows the damage threshold for platinum impurity in glass as a function of host conductivity and ThO_2 impurity in ThF_4 as a function of impurity conductivity (Lange et al., 1982) calculated using exact numerical (Equation 3.4) and approximate (Equation 3.14) solutions. One finds a weak threshold dependence on absorber conductivity. Finite element analysis calculations (Papernov and Schmid, 1997) of the temperature fields around Hf metal and HfO_2 cylindrical nanoabsorbers (radius 5–20 nm, height 11–27 nm) also showed small changes in peak temperatures with conductivity variation.

3.2.4.2   Optical Properties

Optical properties of the absorber and host materials in the context of a thermal diffusion model affect damage thresholds through the heat source term A in Equation 3.1:

$$A = \frac{I\sigma}{V},$$

where $V = 4\pi a^3/3$ is the absorber volume. The key parameter to be calculated here is an absorption cross section σ_{abs}, which has the following general expression:

$$\sigma_{abs} = \sigma_{ext} - \sigma_{sca}, \tag{3.15}$$

Chapter 3

where σ_{ext} and σ_{sca} are extinction and scattering cross sections, respectively. According to Mie's theory (we will use further Kerker's [1969] notation), σ_{ext} and σ_{sca} are defined as follows:

$$\sigma_{ext} = \frac{\lambda^2}{2\pi} \sum_{n=1}^{\infty} (2n+1)\left[\mathrm{Re}(a_n + b_n)\right],$$

$$\sigma_{sca} = \frac{\lambda^2}{2\pi} \sum_{n=1}^{\infty} (2n+1)\left(|a_n^2| + |b_n^2|\right),$$

(3.16)

where a_n and b_n are scattering coefficients that describe different electrical and magnetic multipole (dipole, quadrupole, etc.) contributions to light-wave scattering. These coefficients are expressed through combinations of the Riccati–Bessel functions and the Hankel functions of size parameters:

$$\alpha = \frac{2\pi a}{\lambda} \quad \text{and} \quad \beta = \frac{2\pi a m}{\lambda},$$

(3.17)

where
  λ is the wavelength in a medium
  m is the relative complex refractive index, $m = m_1/m_2$, and indices 1 and 2 denote absorber and host, respectively

The refractive index of homogeneous dielectric host m_2 is real, $m_2 = n_2$, and absorber index m_1, in a general case, is complex, $m_1 = n_1(1 - ik)$.

The absorption cross section, σ_{abs}, as defined earlier (Equations 3.15 and 3.16), requires a rather complex numerical calculation. The calculation can be significantly simplified if size parameters α and β are small:

$$\alpha \ll 1 \quad \text{and} \quad |\beta| \ll 1.$$

(3.18)

It should be noted here that laser-quality glass and crystalline materials (fused silica, laser amplifier glass, KDP, DKDP crystals, etc.) have practically no visible defects in the bulk as observed using dark-field microscopy with white-light illumination (~550 nm center wavelength or ~370 nm in the medium with $n = 1.5$). Such observation suggests that absorbers in these materials are smaller than ~100 nm ($a < 50$ nm). It has also been shown (Feit and Rubenchik, 2004a) that these absorbers are not metal-type, strong absorbers. This bolsters the validity of small-size approximation given by Equation 3.18. In this case, extinction is dominated by absorption and σ_{abs} can be expressed (Kerker, 1969) using only the leading term (dipole term) of expansions for σ_{ext} (Equation 3.16):

$$\sigma_{abs} = -4\pi a^2 \alpha \, \mathrm{Im}\left(\frac{m^2 - 1}{m^2 + 2}\right).$$

(3.19)

Introducing ε_1 as an absorber dielectric function and using substitutions,

$$m = \frac{m_1}{m_2}, \quad m_2 = n_2, \quad m_1^2 = \varepsilon_1, \quad \varepsilon_1 = \varepsilon_1' + i\varepsilon_1'', \quad \alpha = \frac{2\pi a}{\lambda}, \quad \lambda = \frac{\lambda_0}{n_2},$$

Equation 3.19 takes the following form:

$$\sigma_{abs} = -\frac{24\pi^2 a^3 n_2}{\lambda_0} \, \text{Im}\left(\frac{\varepsilon_1' + i\varepsilon_1'' - n_2^2}{\varepsilon_1' + i\varepsilon_1'' + 2n_2^2} \right). \tag{3.20}$$

Calculation of the imaginary part of the term in brackets gives

$$\sigma_{abs} = \frac{24\pi^2 a^3}{\lambda_0} \frac{\varepsilon_1'' n_2^3}{\left(\varepsilon_1' + 2n_2^2 \right)^2 + \varepsilon_1''^2}. \tag{3.21}$$

One can see that the absorption cross section in the first approximation is proportional to a^3, or the volume of the absorber, and can be calculated for known dielectric functions. It also shows that absorption is more effective for shorter wavelengths and for the host medium with higher refractive index n_2. Equation 3.21 can also be used for very small, strong metal-type absorbers (Kreibig and Vollmer, 1995) like gold particles with a size less than ~20 nm. For larger particles, the scattering contribution (~a^6) to σ_{ext} becomes large enough and $\sigma_{abs} \cong \sigma_{ext}$ is not valid.

For small dielectric absorbers, an alternative approximate expression for σ_{abs} can be used (Walker et al., 1981; Kerker, 1969) if indices $n = n_1/n_2$ and k are in the range of $1 < n \le 1.5$ and $0 < nk \le 0.25$:

$$\sigma_{abs} = \pi a^2 \left[1 + \frac{\exp\left(-8\pi n_1 k \dfrac{a}{\lambda}\right)}{8\pi n_1 k \dfrac{a}{\lambda}} + \frac{\exp\left(-8\pi n_1 k \dfrac{a}{\lambda}\right) - 1}{\left(8\pi n_1 k \dfrac{a}{\lambda}\right)^2} \right]. \tag{3.22}$$

Figure 3.7 shows an example of exact calculations of absorptivity for a wide range of size parameters and absorption indices. A strong dependence on size is apparent for high absorption indices and for a relatively small parameter $\alpha \le 1$. For large parameters $\alpha > 10$, absorptivity approaches unity, which means that the absorption cross section is approaching a value for a geometrical cross section. The earlier presented calculations (Bonneau et al., 2001) for metallic absorbers in Figure 3.2 show a behavior similar to that shown in Figure 3.7 for the highest absorption indices, with clear maximum and very slow changes for absorbers with a diameter $2a > 1000$ nm. In the same work (Bonneau et al., 2001), it was demonstrated that calculated damage threshold versus size dependences for Al absorbers at 351 and 1053 nm wavelength (see Figure 3.8) strongly

Chapter 3

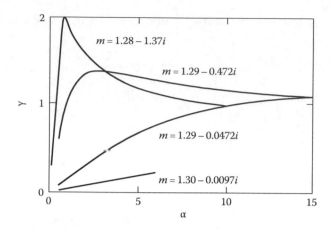

FIGURE 3.7   Absorptivity for various absorption indices as a function of parameter $\alpha = 2\pi a/\lambda$. (Reprinted from *The Scattering of Light and Other Electromagnetic Radiation*, Kerker, M. Copyright 1969, with permission from Elsevier.)

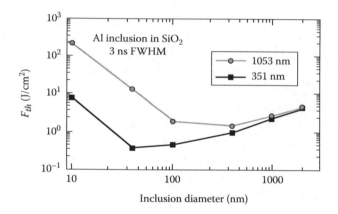

FIGURE 3.8   Damage threshold for Al absorber in SiO_2 and two laser wavelengths, 351 nm and 1053 nm, as a function of Al absorber size. (Reprinted figure with permission from Bonneau, F., Combis, P., Vierne, J., and Daval, G., Simulations of laser damage of SiO_2 induced by a spherical inclusion, *in Laser-Induced Damage in Optical Materials: 2000*, G.J. Exarhos, A.H. Guenther, M.R. Kozlowski, K.L. Lewis, and M.J. Soileau, eds., Vol. 4347, pp. 308–315, SPIE, Bellingham, WA. Copyright 2001 by SPIE.)

correlate with the behavior of absorptivity shown in Figure 3.2. Note the strong wavelength dependence (more than the order of magnitude difference) for small, ~10 nm absorbers, and vanishing wavelength dependence for large, >1000 nm absorbers.

3.2.5   Influence of Absorber Geometry and Orientation

It is of interest to compare the results from applying a thermal diffusion model to three standard absorber geometries of an absorber—plates, rods, and spheres. Such geometries correspond to one-, two-, and three-dimensional diffusion with a characteristic size parameter l equal to the half-thickness of the plate and the radius for rods and spheres.

Calculation of peak temperatures, assuming for simplicity the same material for absorber and host, and homogeneous energy deposition at a constant rate (constant intensity, flat-in-time laser pulse), yields the following results for plates (index 1), rods (index 2), and spheres (index 3) (Trenholme et al., 2006):

$$T_1 = \frac{\beta F}{C_V}\left[1 - \left(1 + \frac{2}{v}\right)erfc\left(\frac{1}{\sqrt{v}}\right) + \frac{2}{\sqrt{\pi v}}e^{-1/v}\right],$$

$$T_2 = \frac{\beta F}{C_V}\left[1 - e^{-1/v} + \frac{E_1(1/v)}{v}\right], \tag{3.23}$$

$$T_3 = \frac{\beta F}{C_V}\left[1 - \left(1 - \frac{2}{v}\right)erfc\left(\frac{1}{\sqrt{v}}\right) - \frac{2}{\sqrt{\pi v}}e^{-1/v}\right],$$

where

β is the absorption coefficient

F is the laser fluence

C_V is the heat capacity per unit volume

E_1 is the exponential integral

v is the normalized time, given by the actual time divided by the diffusion time for a characteristic distance l:

$$v = \frac{4Dt}{l^2}.$$

The quantities inside the square brackets are the diffusion factors, and the term in front gives temperature rise in the absence of diffusion ($t \ll l^2/4D$).

The peak temperature is always reached at the end of the pulse and, *for times much longer than the diffusion time $l^2/4D$*, scales as $t^{1/2}$ for plates and $\ln(t)$ for rods. For spheres, the peak temperature asymptotically approaches a constant value equal to the no-diffusion value and heating applied over time period twice the diffusion time. This result shows that the plate- and rod-shaped absorbers are more effectively heated than the balls and, consequently, should fail at lower laser fluences.

In the case of plates and rods, an additional parameter to be accounted for is the orientation relative to the beam-propagation direction. The results of calculations given by Equation 3.23 are for the laser beam propagating normal to either the plate or the rod axis. For different orientations, the energy deposition rate is proportional to the cosine of the angle between the laser beam and vector normal to the plate surface, and proportional to the sine of the angle with the rod axis. In the case of random absorber orientations, this will reduce the number of absorbers failing at particular laser fluence, exceeding the threshold value.

Chapter 3

3.3   Absorption Delocalization as a Major Damage Mechanism for Small Absorbers

The thermal diffusion–based model for damage initiated by absorbing inclusions considered the host material reaching the critical temperature to be a damage-initiation condition, and the corresponding laser fluence (J/cm²) to be a damage-threshold fluence. In the absolute majority of these studies, the melting point of the dielectric host material (typically 1500–3000 K) is taken as a critical temperature. On the other hand, melting of the thin layer adjacent to the absorber does not necessarily lead to structural changes on a scale relevant to changes in the performance of the optic (scattering, absorption, etc.), which can be considered as damage. For instance, several studies (Hopper and Uhlmann, 1970; Koldunov et al., 1997, 2002), where conditions for mechanical damage caused by metallic absorbers, particularly crack formation, are analyzed, suggest that required temperatures should approach ~10^4 K and minimum absorber radius should exceed 100 nm. Yet, the pin-point-type damage in glasses, crystals, and thin films is frequently formed at sites free from visible defects (as observed prior to irradiation), which points to absorbers with radius $a \ll 100$ nm.

In an attempt to explain the damage caused by such small absorbers, it has been suggested (Danileiko et al., 1973, 1978; Koldunov et al., 1990) that nonlinear behavior of thermal conductivity of both the absorber and host materials along with temperature dependence of the absorption coefficient $\alpha = \alpha(T)$ may lead to a thermal explosion manifested itself by exponential temperature growth inside the absorber. In this case, the onset of damage is attributed to the onset of exponential temperature growth. Calculations (Danileiko et al., 1973) performed for nickel particles with size $2a = 30$ nm embedded inside ruby crystal irradiated by 30 ns pulses of 0.63 µm light gave threshold intensities ranging from 2 to 9 GW/cm², depending on the chosen $\alpha = \alpha(T)$ dependence. In later work (Aleshin et al., 1976), it was pointed out that these results remain unable to explain experimentally observed damage morphology in the bulk of transparent dielectrics.

All this suggests that a mechanism exists by which absorption, once started in a small, nanoscale absorber, spreads out to the surrounding host matrix, thereby effectively increasing energy deposition in the material. Experimental evidence that such a process indeed takes place was obtained (Papernov and Schmid, 2002) using well-characterized artificial gold nanoparticle absorbers embedded in thin-film SiO_2 material (see Figure 3.9). After irradiation of such a sample by 351 nm, 0.5 ns single pulses with fluences slightly above damage threshold, submicrometer-size damage craters were discovered at exactly the nanoparticle locations (see Figure 3.10). Atomic force microscopy (AFM) mapping of a sample surface before and after irradiation made it possible to establish an exact correlation between each gold absorber and the crater formed. Using AFM analysis, crater volume and the corresponding amount of material removed during crater formation was determined and energy required for crater formation E_{cr} was calculated. For the absolute majority of craters investigated, this energy several times exceeded the energy absorbed by the gold absorber E_{abs}:

$$E_{cr} \gg E_{abs}. \tag{3.24}$$

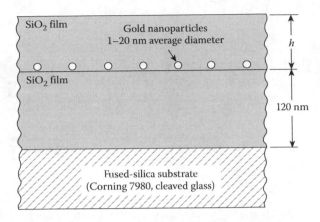

FIGURE 3.9   Schematic presentation of SiO_2 sample with embedded gold nanoparticles.

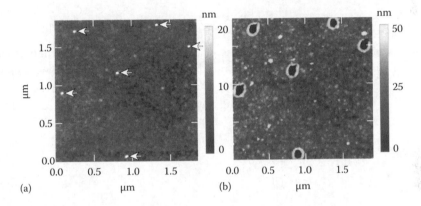

FIGURE 3.10   AFM images of a surface area containing six gold particles: (a) nanoparticle map and (b) same area after a 60 nm SiO_2 overcoat and 351 nm, ns pulse irradiation. (Reprinted with permission from Papernov, S. and Schmid, A.W., Correlations between embedded single gold nanoparticles in SiO_2 thin film and nanoscale crater formation induced by pulsed-laser radiation, *J. Appl. Phys.*, 92, 5720–5728. Copyright 2002, American Institute of Physics.)

The validity of Equation 3.24 is further supported by the fact that part of the absorbed energy is spent on heating of the host material beyond the crater boundary. The inequality given by Equation 3.24 clearly indicates that absorption must take place in a volume much larger than the absorber volume itself, thereby proving absorption delocalization.

In order to consider the possible physical mechanisms that might cause the delocalization effect, it is useful to first evaluate temperatures generated during the laser-damage event. It can be done by experimentally measuring the emission spectra from the sample site that may be well approximated by blackbody radiation. Such measurements (Carr et al., 2004) using nanosecond excitation pulses with wavelengths $\lambda = 355$, 532, and 1064 nm for six transparent dielectrics (band gap E_g range 6.4–13.7 eV) gave maximum temperatures in the range of 8,300–12,400 K. Taking into account that the experiment was conducted at laser fluences approximately twice the damage-threshold

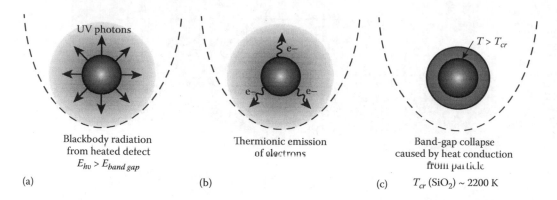

FIGURE 3.11 Schematic presentation of three mechanisms transforming absorber-surrounding matrix into absorbing medium: (a) photoionization by UV radiation from heated absorber, (b) thermionic emission of electrons, and (c) heat-transfer-induced band-gap collapse.

fluence, these temperatures should be considered as an upper limit for absorber temperatures necessary to initiate absorption in the surrounding matrix.

Several possible mechanisms of absorption delocalization are schematically presented in Figure 3.11. The first mechanism (Aleshin et al., 1976) (which was also the first historically) suggested that heating the medium in the absorber vicinity should induce additional absorption with corresponding absorption coefficient $\alpha(T)$ in the form

$$\alpha = \alpha_0 \exp\left(-\frac{E_g}{2kT}\right), \tag{3.25}$$

where
　E_g is the optical band gap of the host medium
　k is the Boltzmann constant

While such an approach would be able to explain absorption delocalization in moderate-band-gap semiconductors, it would require reaching extremely high temperatures $\gg 1 \times 10^4$ K (1 eV corresponds to $T \approx 1.15 \times 10^4$ K) by an absorber embedded inside wide band-gap (6–9 eV) dielectrics. This practically excludes any possible role for such a mechanism in wide band-gap dielectrics.

The second mechanism is called photoionization thermal explosion (Danileiko et al., 1973, 1978; Koldunov et al., 1988). It starts with a thermal explosion in the absorber (see the beginning of this section) characterized by exponential temperature growth. Upon reaching temperatures that approach 1×10^4 K, thermal radiation from the heated absorber can cause photoionization of the surrounding matrix. The calculation (Papernov and Schmid, 1997) shows that radiation from a blackbody with $T = 10,000$ K contains ~10% photons with energy $h\nu \geq E_g = 5.6$ eV (HfO$_2$ dielectric); for $T = 20,000$ K, such content is 57%. Once excited by this blackbody radiation into the dielectric conduction band, the free carriers are able to interact with the remaining laser pulse, resulting in electron heating, avalanche formation, and eventual damage. Temperatures close to 1×10^4 K, as mentioned earlier, were observed in the experiment (Carr et al., 2004), which suggests photoionization thermal explosion as a plausible damage mechanism.

The third mechanism is related to injection of electrons into the host matrix through thermionic emission resulting from irradiation of metal-type absorbers. Absorption of laser photons in such an absorber creates electron distribution with sufficiently hot electrons to be able to both overcome potential barrier and penetrate into the host dielectric conduction band. Calculations for 5.2 nm diameter gold nanoparticles embedded in a SiO_2 matrix (Grua et al., 2003) showed that more than 90% of all free electrons in the particle irradiated by 351 nm, 0.5 ns pulses with fluence of 0.3 J/cm² ($I = 0.6$ GW/cm²) can be transferred into the glass conduction band. Further heating of these electrons by the laser pulse leads to damage. The quasineutrality is preserved by transitions of glass valence band electrons into the particle conduction band. Such a mechanism might be important in metal-based oxide films where few-nanometer-sized metal clusters are suspected to be the source of absorption and damage (Papernov et al., 2011).

The last and probably the most versatile mechanism, similar to the first one, involves heating of the host matrix in the absorber vicinity and the temperature dependence of the matrix absorption coefficient. The difference here is that the temperature dependence of the absorption coefficient does not follow Equation 3.25, which contains a constant band gap E_g. The appropriate temperature dependencies for amorphous materials contain an optical energy gap that is temperature dependent. Measurements in fused silica confirmed (Saito and Ikushima, 2000) that an optical energy gap changes from 8.5 to 7.0 eV within a temperature range of 4–1873 K. In a related laser-damage experiment with fused silica (Bude et al., 2007), this material (which is normally transparent to 355 nm laser light), upon reaching ~2200 K, showed an absorption increase strong enough to generate exponential temperature growth leading to damage. Such behavior indicates thermally induced band-gap collapse.

Once absorption is initiated in the absorber-surrounding matrix, the next stage is the interaction of laser radiation with free electrons in the host matrix. Laser heating of these electrons is quickly (10^{-12} s) coupled to the lattice through electron–phonon interaction. This leads to a larger volume being affected by heating, followed by further absorption and ionization of this larger volume. Assuming that all energy absorbed by the electron plasma is spent to ionize the matrix material, growth of the matrix ionization wave can be described as follows (Feit et al., 1998):

$$nI_0 4\pi a^2 \frac{da}{dt} = \sigma I(t),\qquad(3.26)$$

where
   n is the atomic concentration
   I_0 is the ionization potential
   a is the radius of the plasma ball
   σ is the absorption cross section
   I is the laser intensity

Chapter 3

Absorption by the plasma can be described similarly to the particle absorption using Mie's theory (Mie, 1908; Bohren, 1983), where the dielectric coefficient is given according to Drude's model:

$$\varepsilon \approx -\frac{n_e}{n_c}\left(1-\frac{i\nu}{\omega}\right), \tag{3.27}$$

where

n_e is the electron density
n_c is the critical electron density
ν is the collision frequency
ω is the light angular frequency

Solving Equation 3.26 gives an exponential growth of the plasma ball with time:

$$a = a(t=0)\cdot\exp\gamma, \quad \gamma \sim F/nI_0.$$

Estimates (Feit et al., 1998) using realistic material and laser parameters ($I_0 = 10$ eV, $\lambda = 355$ nm, 3 ns pulses, $F = 10$ J/cm^2) give $\gamma \approx 10$, which ensures a very fast plasma ball growth to $a \sim 500$ nm. At this point, despite the fact that the plasma growth rate practically saturates as the plasma density exceeds critical density (Kruer, 1988), energy stored in a plasma ball exceeds by many times the energy required for material evaporation, which eventually leads to damage.

3.4 Mechanical Damage Resulting from Localized Absorption

High peak temperatures reached by the host material in the vicinity of the localized absorber are usually accompanied by high local stresses that lead to crack formation at the final stage of the damage process. It should be noted here that damage in transparent solids in the form of cracks is observed for a wide pulse-length range, from picoseconds to microseconds. For very long, millisecond pulses, it is mostly melting, and for very short, femtosecond-scale pulses, the dominating damage mechanism is ablation. The crack-formation process occurs on a characteristic timescale of $>10^{-8}$ s, longer than two other important time constants—electron–phonon relaxation $\sim 10^{-12}$ s and thermoelastic stress-formation time—since the crack forms after the stresses are established. As a first approximation, we will consider conditions for crack formation, neglecting dynamic effects and using results obtained within the framework of a quasistationary thermoelastic theory. The role of dynamic effects (shock-wave formation, phase transitions, material parameter modification with temperature, and pressure growth) in mechanical damage will be shown later using examples of numerical simulations.

3.4.1 Stress-Field and Energy Criteria of Crack Formation

The first study (Hopper and Uhlmann, 1970) of crack formation in the context of laser damage considered stress generated in a glass from thermal expansion of metallic (platinum) inclusions embedded in the glass matrix. The only criterion for crack initiation used in this work was exceeding the tensile strength of the material, usually estimated as ~0.1 E, where E is the Young's modulus. The average platinum inclusion temperature required to generate such stress was estimated at ~1×10^4 K. This result was obtained using a simplified model of a sphere of a size $R + \Delta R$ placed in a hole of size R in an infinite elastic solid.

A more rigorous treatment of crack formation caused by local laser energy deposition in the spherical region requires the solution of the thermoelastic equations (Boley and Weiner, 1960). In the case of an absorber embedded in the mechanically different host medium, the equations should be complemented by the boundary conditions. The temperature, heat flow, and radial component of the stress tensor should be continuous at the absorber/host boundary. For future reference, indices 1 and 2 denote absorber and host medium, respectively. The solution for tensile stresses takes a simple analytical form for two limiting cases (Koldunov et al., 1997): First, when a laser pulse τ_p is much shorter than the diffusion time $\tau_2 = R^2/D_2$, where R is the absorber radius and D is the thermal diffusivity. Then, heat transfer from the absorber to the host can be neglected and the expression for the tensile stress σ_ϕ in the host medium takes the form

$$\sigma_\phi(r) = \frac{\alpha_1 \theta}{3} \frac{E_1}{(1-v_1)} \frac{R^3}{r^3} \quad \text{for } r \geq R, \tag{3.28}$$

where
  r is the coordinate
  α is the thermal expansion coefficient
  E is the Young's modulus
  v is the Poisson ratio
  $\theta = T - T_0$ is the temperature rise above T_0 (the initial temperature) at the end of the laser pulse

One can see that the stress diminishes as $1/r^3$ over distance r, and the maximum tensile stress is located at $r = R$:

$$\sigma_\phi^{\text{max}} = \frac{\alpha_1 \theta}{3} \frac{E_1}{(1-v_1)}. \tag{3.29}$$

For a crack to form, the σ_ϕ^{max} value should exceed ultimate tensile strength of the material:

$$\sigma_\phi^{\text{max}} \geq \sigma_\phi^{cr}, \quad \text{where } \sigma_\phi^{cr} = 0.1 E_2. \tag{3.30}$$

Chapter 3

When Equation 3.29 is combined with the mechanical damage criterion, the mechanical damage condition becomes

$$\theta_{cr} \geq \frac{0.3\beta_{cr}}{\alpha_1}, \quad \text{where } \beta_{cr} = \frac{E_2}{E_1}(1-v_1). \tag{3.31}$$

The second limiting case concerns the laser pulse length (with constant intensity), which is much longer than the diffusion time $(\tau_p \gg \tau_r)$ when a steady-state temperature distribution is established. In this case (Koldunov et al., 1997), the mechanical damage condition is similar to that given by Equation 3.31, but with a modified expression for β_{cr}:

$$\beta_{cr} = \frac{E_2}{E_1}\frac{(1-v_1)}{(1-\delta)}, \quad \text{where } \delta = \frac{E_2}{E_1}\frac{(1-v_1)}{(1-v_2)}\frac{\alpha_2}{\alpha_1}. \tag{3.32}$$

The results of numerical calculations (Koldunov et al., 2002) for the temporal behavior of a maximum stress, assuming an absorber material has the same thermomechanical properties as the host, are presented in Figure 3.12. Here the maximum stress rises for a period $t \leq 1.15\,\tau$, where τ is the thermal diffusion time. At longer times, due to heat diffusion and reduced temperature gradient, stress falls monotonically.

The mechanical damage criterion given by Equation 3.30 is a necessary condition but not sufficient for crack formation. To produce a crack with a finite size, some energy should be spent to create an additional surface. In the case of heating a spherical region with a radius R, the amount of energy required is given by (Koldunov et al., 2002)

$$E_m \cong 39R^2\gamma, \tag{3.33}$$

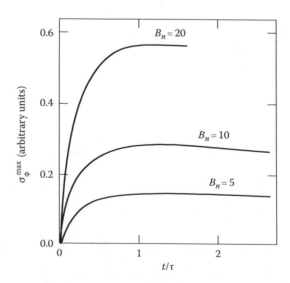

FIGURE 3.12   Time dependence of the maximum stress in the vicinity of a locally heated region. B_n is the normalized power density of heat sources. (Reprinted figure with permission from Koldunov, M.F., Manenkov, A.A., and Pokotilo, I.L., Mechanical damage in transparent solids caused by laser pulses of different durations, *Quant. Electron.*, 32, 335–340. Copyright 2002 by Institute of Physics.)

where γ is the surface energy density of the host material. It means that for damage to occur, the strain-field energy generated by absorption and heating should exceed E_m, which results in

$$\eta \sigma I \tau_p \geq E_m, \tag{3.34}$$

where

σ is the absorption cross section

I is the intensity of the laser pulse (flat in time)

τ_p is the pulse length

$\eta = (T_0/9C_V) \cdot \left[(1+\nu)/(1-\nu)\right]^2 \cdot \alpha_2^2 c_1^2$ is the connectivity coefficient

C_V is the specific heat at constant volume

c_1 is the longitudinal speed of sound

Equations 3.33 and 3.34 indicate that, for fixed radius R and pulse intensity I, pulses shorter than τ_p will not initiate crack formation. Also, in the case of a small interaction region ($R \ll \lambda$), since absorbed energy varies as R^3 (Equation 3.21) and crack formation energy as R^2 (Equation 3.33), reduction in absorber size quickly stops crack formation. Estimates of the critical radius using Equations 3.33 and 3.34, strong absorber approximation, and experimental damage threshold data (Stuart et al., 1995) for transparent dielectric surfaces irradiated by pulses ranging $20\text{ ps} < \tau_p < 1\text{ ns}$ gave an average critical radius $R \approx 0.4$ µm. Since mostly melting and no cracking represented damage morphology in the experiment referenced earlier, one can surmise that damage initiators were much smaller in size.

It should be noted here that the estimate given earlier carries a high level of uncertainty due to the absence of experimentally pertinent information on absorber material, geometry, size distribution, volume density, etc. Moreover, subsurface mechanical damage with trapped absorbing particles of abrasive material found near all polished dielectric surfaces may significantly affect mechanical strength of the near-surface layer. Finally, the theory presented earlier used a quasistatic approximation that neglected the impact of dynamic processes such as shock-wave formation and changes in material properties under high temperatures and pressures, characteristic for pulsed laser–induced damage.

3.4.2 Role of Dynamic Effects in Mechanical Damage

Both dynamic and high-temperature, high-pressure effects were addressed (Bonneau et al., 2004) in the numerical simulations of damage in a model system comprised of a 600 nm diameter gold particle embedded inside the silica matrix and irradiated by 3 ns pulses of 355 nm laser light. The one-dimensional (1-D) hydrodynamic Lagrangian code was used to describe energy deposition in the absorber, followed by the two-dimensional (2-D) hydrodynamic Lagrangian–Eulerian code for the study of fracture and fragmentation in the host material. The 1-D code accounted for the inhomogeneous, time-dependent absorption in the absorber and surrounding host material, electric-field amplification in the absorber vicinity, heat conduction and radiative heat transfer, phase transitions, and propagation of the shock waves. At each time step, the evolution of internal energy

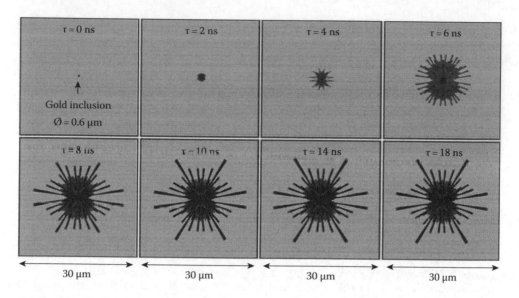

FIGURE 3.13   Evolution of stresses produced by 355 nm, 3 ns, 6 J/cm² irradiation of a 600 nm gold sphere in the bulk of fused silica. (Reprinted figure with permission from Bercegol, H., Bonneau, F., Bouchut, P. et al., Comparison of numerical simulations with experiment on generation of craters in silica by a laser, in *Laser-Induced Damage in Optical Materials: 2002*, G.J. Exarhos, A.H. Guenther, N. Kaiser et al., eds., Vol. 4932, pp. 297–308, SPIE, Bellingham, WA. Copyright 2003 by SPIE.)

calculated by the 1-D code was injected into the 2-D code, which then calculated the evolution of stresses and fracture/fragmentation for both absorber and host media. The results of the fracture kinetics simulations are presented in Figure 3.13. One can see that upon irradiation by 355 nm, 3 ns pulses with 6 J/cm² fluence, micron-scale fracture is observed already at a time $t = 2$ ns, and continues to grow up to $t = 8$ ns. From this moment on, damage grows much slower and achieves a final size of ~25 µm at time $t \approx 14$ ns, defined mostly by the attenuation and damping of the shock wave. This modeling confirms the importance of the dynamic factors and temperature/pressure dependences of the material parameters to realistically describe the evolution of laser-induced damage.

3.4.3   Fracture versus Melted and Deformed Material in Localized Damage Morphology

From previous discussions, it follows that the critical temperature of the host medium required to initiate damage exceeds the melting point for most optical materials. As a result, the central core of the localized damage morphology should always contain melted and plastically deformed material, while the periphery also may (or may not) show fractures. The relation between these two types of damage morphology may be understood by using dimensional analysis. Let one assume that locally absorbed energy E is large enough to generate plastic deformation and also material fracture. Then the size of the plastically deformed zone Rp can be expressed as (Feit and Rubenchik, 2004b)

$$R_p \sim \left(\frac{E}{P} \right)^{1/3},$$

(3.35)

where P is the compressive material strength. The corresponding fracture zone size R_f depends on fracture toughness K and deposited energy E:

$$R_f \sim \left(\frac{E}{K}\right)^{2/5}. \tag{3.36}$$

One can see that for a large enough deposited energy E (large damage), the fracture zone is always larger than the plastic deformation zone and, for close-to-threshold conditions, released energy may be too small to open the crack and the site will have only plastically deformed material. The energy boundary E_c between two regimes and a corresponding zone radius $R_c = R_f = R_p$ scales as

$$E_c \sim \frac{K^6}{P^5} \quad \text{and} \quad R_c \sim \left(\frac{K}{P}\right)^2. \tag{3.37}$$

Calculations (Feit and Rubenchik, 2004b) performed for a fused-silica host material using $P = 1.1$ GPa, $K = 0.75$ MPa-m$^{1/2}$, and appropriate proportionality coefficients in Equations 3.35 and 3.36 yielded $E_c \approx 0.1$ nJ and $R_c \approx 0.5$ μm. The corresponding dependences of the damage zone radius on deposited energy for the two regimes presented in Figure 3.14 clearly show that for energies $>E_c$, fracture zone and cracks extend beyond the plastic deformation zone.

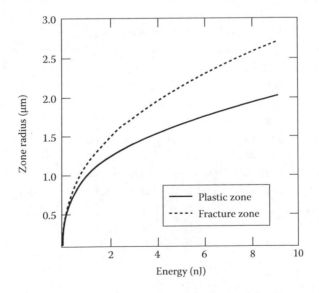

FIGURE 3.14 Plastic deformation zone and fracture zone radii in fused silica as a function of deposited energy. (Reprinted figure with permission from Feit, M.D. and Rubenchik, A.M., Influence of subsurface cracks on laser-induced surface damage, in *Laser-Induced Damage in Optical Materials: 2003*, G.J. Exarhos, A.H. Guenther, N. Kaiser, K.L. Lewis, M.J. Soileau, and C.J. Stolz, eds., Vol. 5273, pp. 264–272, SPIE, Bellingham, WA. Copyright 2004 by SPIE.)

Chapter 3

3.5 Role of the Free-Surface Proximity in Localized Absorber-Driven Damage

The well-known fact that surface damage resistance is lower than that of the bulk stems from the technological need for chemo-mechanical surface processing in meeting wavefront and roughness requirements. In that process, a near-surface layer undergoes structural modification due to chemical reactions, incorporation of micro- or nanoscale particulates, and partial cracking and deformation frequently referred to as subsurface damage. All this results in reduced mechanical strength of the material, enhanced absorption, and, consequently, lower damage thresholds as compared to the bulk. Another surface aspect is related to the possibility of light-intensity enhancement in the surface proximity caused by constructive interference between incident and reflected waves.

Finally, energy-dissipation processes are affected by the nearby free surface; this last aspect will be specifically addressed in this section. For this purpose, we will analyze processes following light absorption in a localized spherical absorber residing in a transparent, homogeneous, nonabsorbing dielectric at some distance beneath the surface. Since damage morphology in this case mostly takes the form of a crater, we will consider conditions for crater formation and crater characteristics (shape, depth, and lateral size) as a function of absorber location and amount of energy deposited.

3.5.1 Stresses Generated by Heated Absorber in the Proximity of a Surface

As a first approximation, we will neglect dynamic effects, such as plasma ball and shock wave formation, and consider a stress field formed around a heated absorber in the free-surface proximity as quasistatic effect (Duthler, 1974). This approximation holds for laser pulse lengths shorter than thermal diffusion time but still longer than the round-trip time for sound to travel from absorber to surface. As an example, for an absorber with a 0.5 μm radius embedded in fused silica (thermal diffusivity 8×10^{-3} cm²/s and sound velocity 5.9×10^5 cm/s) at 1 μm beneath the surface, the thermal diffusion time is 1.5×10^{-7} s and the round-trip travel time for sound is 1.7×10^{-10} s. For these experimental parameters, the nanosecond pulse range satisfies conditions for quasistatic stress formation and the heated absorber in the host can be treated as a spherical cavity within which quasistatic hydrostatic pressure is applied. Then heat conduction to the host material can be neglected, and stresses produced are exclusively caused by thermal expansion of the absorber. In this case, pressure P at the absorber/host boundary may be expressed as (Duthler, 1974)

$$P = 3\alpha_{ab} B_{eff} T,\tag{3.38}$$

where
   B_{eff} is the effective bulk modulus
   $B_{eff}^{-1} = B_{ab}^{-1} + B_h^{-1}$, ab and h denote absorber and host
   α_{ab} is the thermal expansion coefficient
   T is the temperature

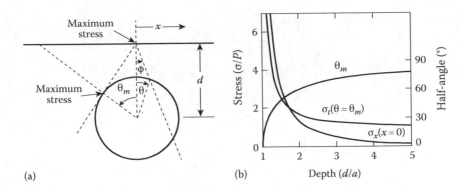

FIGURE 3.15 (a) Sketch of the inclusion of radius a at a distance d beneath the surface. (b) Tensile stresses and cone half-angle as a function of absorber depth. (Reprinted with permission from Duthler, C.J., Explanation of laser-damage cone-shaped surface pits, *Appl. Phys. Lett.*, 24, 5–7. Copyright 1974, American Institute of Physics.)

The absorber temperature reaches maximum at the end of the pulse and can be calculated using Equation 3.5 without the heat-conduction term. An exact solution for the equilibrium stresses is not available for sphere (3-D case), but 2-D approximation in the form of a long cylinder in a semi-infinite medium offers good insight into the problem whose geometry is shown in Figure 3.15a.

The tensile stress σ_t along the cylindrical surface is given by

$$\sigma_t = P\left(1 + 2\tan^2 \phi\right), \tag{3.39}$$

where $\tan \phi = a \sin \theta / (d - a \cos \theta)$. Maximum stress occurs at the angle $\theta_m = \cos^{-1}(a/d)$, where a is the radius of the cavity and d is the distance below the surface (see Figure 3.15a). Along the plane free surface, the tensile stress σ_x can be expressed as

$$\sigma_x = -4Pa^2 \left(\frac{x^2 - d^2 + a^2}{x^2 + d^2 - a^2}\right), \tag{3.40}$$

where x is measured along the plane from the point directly above the cavity. The maximum stress occurs at the point with $x = 0$, and $\sigma_x(0) > P$ for $d < \sqrt{5}a$.

The maximum tensile stresses $\sigma_t(\theta_m)$ and $\sigma_x(0)$ and cone half-angle θ_m are plotted as functions of d/a in Figure 3.15b. As d decreases, both $\sigma_t(\theta_m)$ and $\sigma_x(0)$ become larger than P, with $\sigma_x(x = 0) > \sigma_t(\theta_m)$ for $d < \sqrt{3}a$. For $d/a \approx 1$, the tensile stresses around the cavity are very asymmetrical (surface influence) and are much larger than applied pressure P. For large d, these stresses asymptotically approach values $\sigma_t(\theta_m) = P$ and $\sigma_x(0) = 0$, respectively.

Problem solutions for the spherical cavity are expected to be similar to those considered earlier. Maximum tensile stress along the spherical cavity surface is expected to also occur at the angle $\theta = \theta_m$, but the σ versus d/a curves will be quantitatively different.

Chapter 3

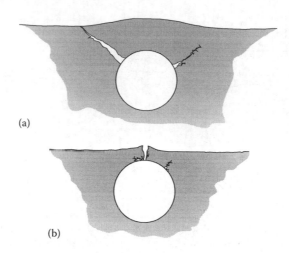

(a)

(b)

FIGURE 3.16   Two scenarios of crack initiation (a) at the inclusion surface and (b) at the plane surface.

For large d/a, stress outside the cavity will decrease as $1/r^3$, as compared to $1/r^2$ for the 2-D case, and $\sigma_t(\theta_m) = P/2$. In the case of d/a close to unity, $\sigma = \sigma(d/a)$ dependence is steeper than shown in Figure 3.15b, as expected.

As an illustration, two different scenarios of crack formation are presented in Figure 3.16. When $d/a > \sqrt{3}$, stress reaches a maximum on the boundary at a point corresponding to θ_m, and a crack will propagate from this point in the direction normal to the boundary, eventually forming a cone-shaped crater (Figure 3.16a). For absorbers close to the surface, $d/a < \sqrt{3}$, maximum tensile stress is found at the point $x = 0$, and a crack will propagate from this point to the cavity surface (Figure 3.16b).

3.5.2   Influence of Asymmetry in Plasma–Ball Growth on the Crater-Formation Process

Surface damage initiated by small ($a \ll \lambda$) near-surface absorbers involves the necessary ionization of the absorber-surrounding matrix and plasma-ball growth to a size sufficient for damage-crater formation. Since plasma-ball growth implies laser light absorption by the plasma, the plasma front partially screens downstream material in the case of one-sided irradiation. This should cause a departure from spherical symmetry and preferential plasma growth toward the laser-beam source. In this case, front irradiation generates plasma growing toward the matrix/air interface, thereby facilitating crater formation (see Figure 3.17). Back irradiation, on the other hand, causes plasma propagation away from the interface, which reduces the probability of crater formation. Consequently, different crater-formation thresholds are to be expected for two cases of front and back irradiation of the near-surface absorbers.

The effect of plasma-ball asymmetry was tested in an experiment (Papernov and Schmid, 2008) with 8 and 14 nm diameter gold absorbers buried in fused-silica film deposited on a fused-silica substrate (see Figure 3.18) and irradiated by 351 nm, 0.5 ns pulses at close to normal incidence. It was found (Papernov and Schmid, 2002) that gold absorbers of this size require plasma ball formation in order to produce craters in silica

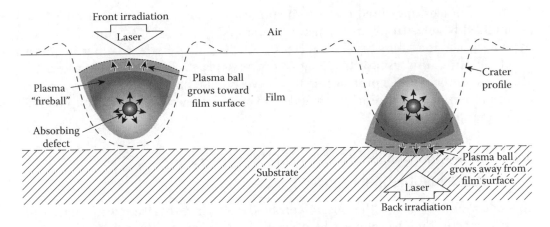

FIGURE 3.17 Illustration of the impact that plasma ball growth symmetry has on crater formation for front- and back-irradiation geometries. (From Papernov, S. and Schmid, A.W., *J. Appl. Phys.*, 104, 063101, 2008.)

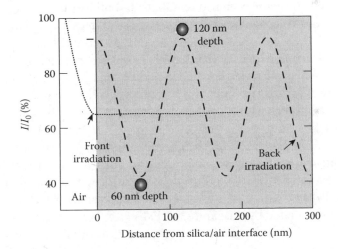

FIGURE 3.18 *E*-field intensity distribution inside fused silica containing gold nanoparticles at 60 nm depth and 120 nm depth, corresponding to minimum and maximum of the back-irradiation intensity, respectively. (From Papernov, S. and Schmid, A.W., *J. Appl. Phys.*, 104, 063101, 2008.)

film at these irradiation conditions. This makes the described model system very suitable for testing plasma-ball interaction with the laser beam.

To detect differences in such an interaction for two opposite beam-propagation geometries, it is important to account for a possible difference in light-field intensity inside the film. For the case of irradiation from the front, the intensity is constant inside the film, which corresponds to the traveling wave. For back irradiation, interference between the incident beam and the beam reflected from the film/air interface (~4%) leads to a standing-wave periodic pattern. Intensity patterns for both cases are shown in Figure 3.18. Two different types of samples were used, the first one with 8 nm absorbers located at the minimum of intensity pattern (60 nm depth) and the second with 14 nm absorbers located at the maximum of intensity pattern (120 nm depth).

Chapter 3

For both samples, measured crater-formation thresholds T_{front}^n and T_{back}^n, after being normalized to internal intensities, gave a ratio of $T_{back}^n / T_{front}^n \approx 1.2$, which supports a hypothesis about asymmetrical character of plasma-ball growth. Another important conclusion from this result is that it again points to plasma-ball formation as a necessary stage in the damage process initiated by very small nanoscale absorbers. No difference in normalized thresholds should be expected if energy deposition is confined to just the absorber.

3.5.3 Phenomenological Theory of Crater Formation Based on Thermal Explosion

To model crater formation, the most important parameters to be defined are total locally absorbed energy and the volume where absorption takes place. The following phenomenological approach (Feit et al., 2001a) allows one to establish useful scaling relations for damage craters initiated by near-surface absorbers. Assuming that nanoscale absorbers (radius $R \ll \lambda$) are the main type of absorbers in the material and the absorption process follows the thermal explosion scenario with plasma-ball formation (see Section 3.3), the absorbed energy and volume parameters can be estimated. The plasma ball formed in the matrix around the small absorber acquires energy from the laser pulse and quickly grows to a size comparable to laser wavelength λ. It is reasonable to assume that almost all incident energy E_i is absorbed in the formed plasma, $E_{abs} \approx E_i$, and E_{abs} is given by $E_{abs} \approx F\pi\lambda^2$ where F is the laser fluence. One can see that, using the effective absorber volume $4/3\ \pi\lambda^3$, the absorbed energy density is then of the order of $0.75\ F/\lambda$, which, for a typical glass surface damage fluence of ~5 J/cm², $\lambda = 0.35\ \mu m$, and few-ns pulses, has a value that exceeds by several times the glass evaporation energy density. This ensures not only melting within the plasma-ball volume but also fracturing of the material beyond the plasma ball by the strong shock wave resulting from the microexplosion. The crushed material may then be treated as an incompressible liquid, the hydrodynamic motion of which is initiated by the explosion. Estimates based on explosion experiments show that the fraction α of deposited energy going into hydrodynamic motion is ≈ 0.1. Analysis of the hydrodynamic potential allows one to find the normal component u of the material velocity at the surface:

$$u = \frac{2Ah}{\left(r^2 + h^2\right)^{3/2}}; \quad r^2 = x^2 + y^2, \tag{3.41}$$

where
    coefficient A is the following function of E_{abs}
    h is the lodging depth
    ρ is the material density
    a is the radius of the absorbing volume:

$$A = \sqrt{\frac{\alpha E_{abs} a}{2\pi\rho}}. \tag{3.42}$$

The crater radius $r = R$ is defined by the condition $u = u_c$, where u_c is the critical velocity below which material is not damaged (crashed), giving (see Equation 3.41) for the crater radius R:

$$R^2 = \left(\frac{2Ah}{u_c}\right)^{2/3} - h^2. \tag{3.43}$$

One can see that for a fixed explosion energy E, there is a maximal lodging depth h_d at which a crater is formed and a depth h_m at which a crater has maximum size R_m:

$$h_d = \sqrt{\frac{2A}{u_c}}; \quad h_m \approx 0.44 h_d; \quad R_m = \sqrt{2} h_m \approx 0.6 h_d. \tag{3.44}$$

As follows from Equation 3.44, the depth for maximum crater size h_m varies only as 1/4 power of the absorbed energy times the effective absorber radius. Combining Equations 3.44 and 3.43 gives

$$R = h^{1/3} \left(h_d^{4/3} - h^{4/3}\right)^{1/2}. \tag{3.45}$$

For high explosion energies or shallow absorbers, when $h_d \gg h$, Equation 3.45 takes the form

$$R \cong \left(h_d^2 h\right)^{1/3}, \tag{3.46}$$

which results in a very slow $E^{1/6}$ crater radius versus energy dependence.

The validity of Equation 3.45 in describing the crater's radius dependence on absorber lodging depth was tested in experiments (Papernov and Schmid, 2003, 2005) with gold nanoparticle absorbers (8.4 ± 0.9 nm and 18.5 ± 0.9 nm average diameter) embedded at different depths in a fused-silica film matrix and irradiated by 351 nm, 0.5 ns pulses. In these experiments, as a first step, the crater-formation threshold fluence F_{cr} for the absorber at the deepest location was determined, which effectively established the deepest location as $h = h_d$. Irradiation at fixed fluence of absorbers embedded at a depth $h < h_d$ generated craters (see Figure 3.19), the diameter of which was measured using AFM mapping. The measurement results and calculated (Equation 3.45) curve for normalized radius versus normalized depth dependence are shown in Figure 3.20. One can see that in the case of 8.4 nm particles with a lodging depth $h_d \leq 60$ nm (Figure 3.20a), experimental data are in qualitative agreement with model predictions. In contrast, results for 18.5 nm absorbers with a maximum lodging depth $h_d = 240$ nm (Figure 3.20b) show a strong deviation from the model predictions, in particular, a large increase in crater-radius size for deeper absorber locations. The AFM analysis also revealed a dramatic change in crater morphology (see Figure 3.19) for absorbers with a lodging depth >110 nm, as compared to craters initiated by shallow absorbers. The most notable

Chapter 3

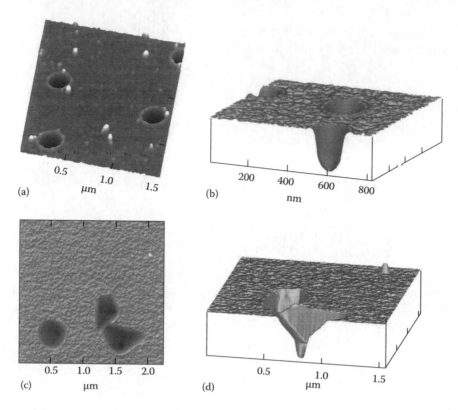

FIGURE 3.19 Characteristic crater morphologies produced by 351 nm, 0.5 ns irradiation of 18.5 nm gold particles lodged at different depths: (a) regular crater, surface plot, 60 nm absorber lodging depth (From Papernov, S. and Schmid, A.W., Damage behavior of SiO_2 thin films containing gold nanoparticles lodged at predetermined distances from the film surface, in *Laser-Induced Damage in Optical Materials: 2002 and 7th International Workshop on Laser Beam and Optics Characterization*, G.J. Exarhos, A.H. Guenther, N. Kaiser et al., eds., Vol. 4932, SPIE, Bellingham, WA, 2003, pp. 66–74.); (b) cross-sectional view of regular crater; (c) complex crater, surface plot, 190 nm absorber lodging depth; and (d) cross-sectional view of complex crater. (b–d: From Kudryashov, S.I. et al., *Appl. Phys. B*, 82, 523, 2006.)

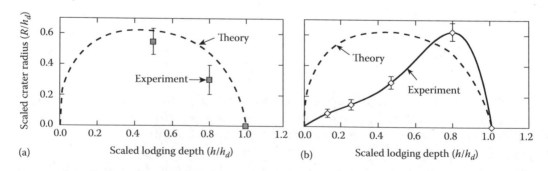

FIGURE 3.20 Scaled crater radius as a function of scaled particle-lodging depth for (a) 8.4 nm particles and maximum lodging depth $h_d = 60$ nm and (b) 18.5 nm particles and maximum lodging depth $h_d = 240$ nm. (From Papernov, S. and Schmid, A.W., Damage behavior of SiO2 thin films containing gold nanoparticles lodged at predetermined distances from the film surface, in *Laser-Induced Damage in Optical Materials: 2002 and 7th International Workshop on Laser Beam and Optics Characterization*, G.J. Exarhos, A.H. Guenther, N. Kaiser et al., eds., Vol. 4932, SPIE, Bellingham, WA, 2003, pp. 66–74; Papernov, S. and Schmid, A.W., *J. Appl. Phys.*, 97, 114906, 2005.)

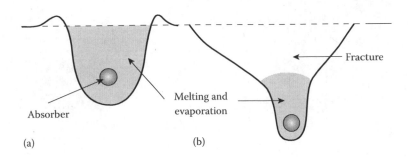

FIGURE 3.21 Schematic presentation of cross-sectional profiles and corresponding crater-formation mechanisms for (a) a regular crater and (b) a complex crater.

features here are random shape and sharp edges characteristic for fracture in the former case, as compared to a circular shape and elevated rim, characteristic for molten material flow and evaporation in the latter case. Cross-sectional profiles also reveal a double-cone, complex shape for deep craters and a bell-type shape for shallow craters.

Processes leading to such crater morphologies are schematically depicted in Figure 3.21. Craters initiated by shallow absorbers result from melting and evaporation within the plasma-ball volume. In the case of deep absorbers, the bottom part of the double-cone crater is also formed by melting and evaporation, but the wider upper cone may result from fracture initiated by tensile stresses around the plasma ball. The fracture may occur according to the mechanism discussed in Section 5.1 or from the spallation process (Kudryashov, 2006) (break-off of a near-surface layer) initiated by the plasma-ball-launched shock wave. Simultaneous contribution by both fracture mechanisms is also possible.

3.6 Shock-Wave-Driven Damage and Spallation

Formation of the shock waves is an integral part of the explosive energy release following the absorption of pulsed laser energy. When a shock wave is launched near a free surface, an initially compressive wave becomes a tensile wave after being reflected from the surface. The interference of these two waves inside the material defines the net stress generated at any particular point near the surface. For pulsed shocks (different from the quasistatic stress case where the net stress normal to the free surface is compressive in the near-surface layer [Duthler, 1974]), the net stress generated at the trailing edge of the pulse can be tensile and, if exceeding the tensile strength of the material, may cause spallation. The schematic in Figure 3.22a explains the time delay between the compressive wave and the reflected tensile wave (Feit and Rubenchik, 2007). The tensile wave propagates in the medium as a spherical wave originating at point B, which is the image of the source point A. At point C, located at distance z beneath the surface and distance s transversely, delay is defined by the time required for the wave to travel distance $r_i - r_s$.

The results of net stress calculation taking into account time delay and wave attenuation with distance r are presented in Figure 3.22b. The calculations used the following assumptions: The pressure pulse had a triangular form with a rise time equal to a laser pulse length τ and a decay time equal to $\beta \cdot \tau$, $\beta > 1$, a shock wave velocity

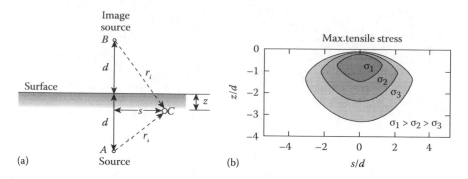

FIGURE 3.22 Tensile stresses induced by shock-wave propagation in the free-surface proximity. (a) Schematic explaining time delay between compressive wave propagating from the source A located at the depth d beneath the surface and tensile wave formed upon compressive wave reflection from the surface. (b) Contours of the maximum tensile stress due to shock wave reflection as a function of depth z beneath the surface and transverse location s. (Reprinted figure with permission from Feit, M.D. and Rubenchik, A.M., Importance of free surfaces for damage crater formation, in *Laser-Induced Damage in Optical Materials: 2006*, G.J. Exarhos, A.H. Guenther, K.L. Lewis, D. Ristau, M.J. Soileau, and C.J. Stolz, eds., Vol. 6403, 64030App., SPIE, Bellingham, WA. Copyright 2007 by SPIE.)

equal to sound velocity, and attenuation with distance was in the form $(1/r)^{\alpha}$, $\alpha = 1.85$. Calculation of the vertical component of the stress gradient by the same simulation (Feit et al., 2007) showed a sharp boundary between areas with tensile and compressive stress and predicted diameter to depth aspect ratios ~2:1 and larger for micron-scale absorber depths.

It should be noted here that plasma-ball formation during a laser pulse leads to melting and plastic flow of the material, which effectively reduces the material strength near the plasma ball and redistributes stresses. For this reason, the aforementioned stress modeling provides only a qualitative picture that supports a possible spallation mechanism. A more quantitative approach is provided by a numerical finite element analysis accounting for the temperature dependences for relevant parameters, phase transitions, and dynamic strength of the material.

To date, such modeling exists only for relatively large absorbers (>100 nm) and usually neglects plasma-ball formation. One example (Bonneau et al., 2004; Bercegol et al., 2003) is the modeling of damage kinetics in fused silica containing a 600 nm diameter gold particle, lodged either 2 or 5 μm below the free surface and irradiated by 355 nm, 3 ns pulses with 6 J/cm² fluence. Fracture evolution in fused silica in the case of a 2 μm particle lodging depth is shown in Figure 3.23. At time $t = 2$ ns, expansion of the heated particle initiates radial fractures that grow significantly at $t = 4$ ns, when a reflected shock wave starts contributing to the buildup of tensile stresses. At time $t = 6$ ns, the heavily damaged zone has a size of ~8 μm, and, as a vertical velocity field calculation shows (see Figure 3.24b and d), a few nanoseconds after the end of the pulse, most of the damaged material has a velocity high enough to be ejected. AFM cross-sectional analysis of craters (Bercegol et al., 2003) created in silica with 600 nm gold absorbers (Figure 3.24a and c) gives crater lateral size and depth values in good agreement with those predicted by the model. Note that the double-cone profile of these craters, similar to those initiated by small absorbers (see Section 5.2), becomes more pronounced with increasing lodging depth. This further supports fracture-driven mechanism of

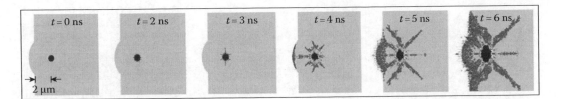

FIGURE 3.23 Simulation of the fracture kinetics in fused silica initiated by 355 nm, 3 ns, 6 J/cm² irradiation of 600 nm gold inclusion located 2 µm beneath the surface. (Reprinted figure with permission from Bercegol, H., Bonneau, F., Bouchut, P. et al., Comparison of numerical simulations with experiment on generation of craters in silica by a laser, in *Laser-Induced Damage in Optical Materials: 2002*, G.J. Exarhos, A.H. Guenther, N. Kaiser et al., eds., Vol. 4932, pp. 297–308, SPIE, Bellingham, WA. Copyright 2003 by SPIE.)

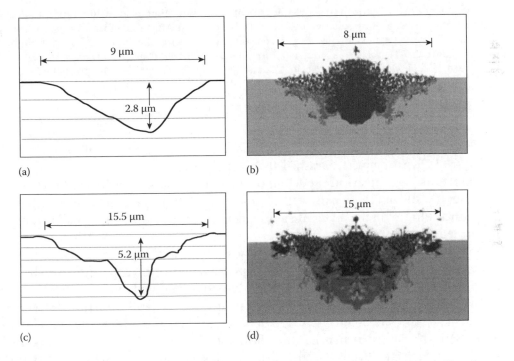

FIGURE 3.24 Damage craters produced by a 355 nm, 3 ns, 6 J/cm² irradiation of 600 nm gold inclusion embedded in silica: [(a),(b)] at 2 µm depth; [(c),(d)] at 5 µm depth; [(a),(c)] experimental AFM measurements; [(b),(d)] finite-element modeling. (Reprinted figure with permission from Bercegol, H., Bonneau, F., Bouchut, P. et al., Comparison of numerical simulations with experiment on generation of craters in silica by a laser, in *Laser-Induced Damage in Optical Materials: 2002*, G.J. Exarhos, A.H. Guenther, N. Kaiser et al., eds., Vol. 4932, pp. 297–308, SPIE, Bellingham, WA. Copyright 2003 by SPIE.)

the upper cone formation. Energy considerations favor fracture mechanism of crater formation, as compared to melting and evaporation. The former is associated with the breaking of molecular bonds only along the crater surface, and the latter in the entire volume of the crater.

Figure 3.25 shows another example of modeling (Feit et al., 1998) that takes into account shock-wave propagation and the dependence of the material's mechanical parameters on energy density and loading time. Here, a 100 nm cerium particle embedded 300 nm beneath the surface in fused-silica glass is irradiated by 355 nm, 3 ns pulses

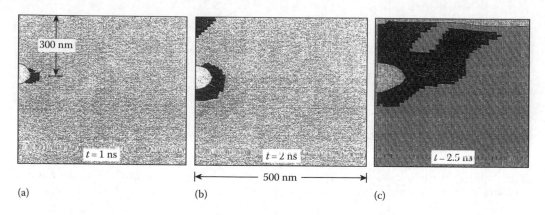

FIGURE 3.25 Fracture growth in fused silica initiated by absorption inside a 100 nm cerium particle irradiated by a 355 nm, 3 ns pulse with a fluence of 10 J/cm². (a) After 1 ns of irradiation, (b) 2 ns, and (c) 2.5 ns. (Reprinted figure with permission from Feit, M.D., Campbell, J.H., Faux, D.R. et al., Modeling of laser-induced surface cracks in silica at 355 nm, in *Laser-Induced Damage in Optical Materials: 1997,* G.J. Exarhos, A.H. Guenther, M.R. Kozlowski, and M.J. Soileau, eds., Vol. 3244, pp. 350–355pp, SPIE, Bellingham, WA. Copyright 1998 by SPIE.)

with 10 J/cm² fluence. Fractures are first initiated near the absorber's equator, then later ($t = 2$ ns) at the free surface, and eventually all material within the conical region shown in Figure 3.25c is completely crashed and ejected.

Unfortunately, numerical modeling of the damage process initiated by very small absorbers ($\ll$100 nm), which involves plasma-ball formation, remains a challenge due to large uncertainties in the material parameters within both the plasma-ball volume and its immediate vicinity.

To summarize, in the case of a localized absorber near the free surface, damage morphology takes the form of craters whose shape and aspect ratio depend on the absorber's lodging depth. For a shallow lodging depth, craters are formed mostly by the host material's melting and evaporation. With increasing lodging depth, fracture-driven crater formation leads to double-cone-shaped craters. Surface proximity substantially influences both the quasistatic tensile stress field and dynamic tensile stresses generated by shock-wave reflection from the free surface. Consequently, material with near-surface absorbers should have lower damage thresholds than the same material with similar absorbers in a bulk.

3.7 *E*-Field Intensity Enhancement by Localized Defects and Related Mechanisms of Damage

Early on in laser-induced damage studies, it was realized (Bloembergen, 1973) that structural defects like cracks, voids, and embedded absorbers can significantly modify the local electric field of a light wave and thereby lower the apparent damage threshold of a dielectric material. In Bloembergen's work, calculations of *E*-field enhancement for cracks, cylindrical grooves, and spherical voids (see Figure 3.26), with characteristic

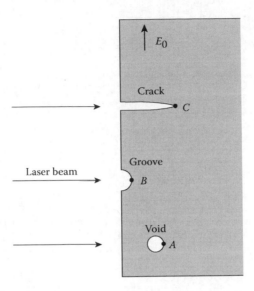

FIGURE 3.26 Typical void geometries considered in electric field–enhancement analysis.

sizes much smaller than the wavelength λ, gave the following enhancement factors for a linearly polarized and zero-degree incident wave:

$$\frac{E_{cr}}{E_0} = n^2; \quad \frac{E_{gr}}{E_0} = \frac{2n^2}{\left(n^2 + 1\right)}; \quad \frac{E_v}{E_0} = \frac{3n^2}{\left(2n^2 + 1\right)}, \tag{3.47}$$

where
   Indices cr, gr, and v correspond to crack, groove, and void, respectively
   n is the refractive index of the medium
   E_0 is the field inside the homogeneous dielectric

Enhancement factors are calculated at locations A, B, and C shown in Figure 3.26 for void, groove, and crack, respectively. Accounting for the fact that intensity $I \sim E^2$, one can see that for typical $n = 1.5$, intensity enhancement will be the largest (≈ 5) for cracks and the smallest (≈ 1.5) for voids.

   Even higher light intensification by cracks with widths w not limited by the condition $w \ll \lambda$ and oriented at various angles relative to the glass surface was obtained using numerical code solving the Maxwell equations (Génin et al., 2001). Figure 3.27 schematically shows planar crack and internal reflections that lead to intensity enhancement. The intensity enhancement factors (IEFs) were calculated for a fixed crack depth $d = 2$ μm, linear light polarization of the plane wave, $\lambda = 355$ nm, and for fused-silica glass as a material ($n = 1.47$). The varied parameters were crack angle θ, width w, and polarization mode (TE or TM mode). It should be noted that peak intensities were always higher for the output surface configuration, as compared to the input surface case. The selected results for the output surface are listed in Table 3.1 (all results can

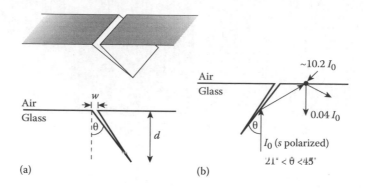

FIGURE 3.27　(a) Schematic of the planar crack and (b) a diagram illustrating intensity enhancement related to interference of the beams reflected from the crack and nearby surface. (Reprinted figure with permission from Génin, F.Y., Salleo, A., Pistor, T.V., and Chase, L.L., Role of light intensification by cracks in optical breakdown on surfaces, *J. Opt. Soc. Am. A*, 2001, 18, 2607–2616. With permission of Optical Society of America.)

Table 3.1　355 nm Light Intensity Enhancement Factors (IEFs) for 2 μm Deep Planar Cracks in Fused Silica and Light Incident on the Exit Surface

Crack Angle	Width		
θ (°)	w (nm)	Mode	IEF
0	50	TE	2.7
0	50	TM	4.9
0	120	TE	2.7
0	120	TM	3.9
20	50	TE	5.3
20	120	TE	6.5
30	50	TE	4.7
30	120	TE	9.1
30	120	TM	6.5
30	200	TE	10.7
40	50	TE	5.5
40	120	TE	3.9

Source:　Génin, F.Y. et al., *J. Opt. Soc. Am. A*, 18, 2607, 2001.

be found in Génin et al., 2001). IEFs as large as 10.7 can be produced by planar cracks in fused silica, and in the same work, for the less-probable conical cracks, IEFs can reach a value of ~100. These results indicate that the presence of scratches on dielectric surfaces can lead to reduction in damage thresholds, which indeed was observed in experiments (Salleo et al., 1998; Josse et al., 2006) with engineered scratches on fused-silica surfaces.

An important question then arises as to whether such intensity enhancements can initiate damage through intrinsic mechanisms (Glebov, 2002) relevant to the bulk damage of the defect-free material. Damage thresholds representative of intrinsic optical breakdown are usually obtained with tightly focused laser beams in the bulk damage experiments (Efimov, 2004; Smith and Do, 2008). Despite all efforts to exclude contributions by nonlinear effects by limiting the beam power to below critical power for both self-focusing and stimulated Brillouin scattering, damage thresholds measured in such experiments show large variations. For fused silica, $\lambda = 1.06$ μm, and nanosecond pulses, reported threshold power densities P are in the range of 200–1000 GW/cm². We will use a value of $P = 500$ GW/cm² for further analysis. Now consider typical 1.06 μm, 7 ns surface damage thresholds (Natoli et al., 2002) for fused-silica windows that are of the order of ~70 J/cm2, which gives $P = 10$ GW/cm². This implies 50-fold intensification to explain observed damage thresholds by intrinsic optical breakdown, which sounds unrealistic in light of more typical 5–10 IEF factors. Another important consideration is that in most cases surface damage investigated with small beam-spot sizes shows probabilistic behavior, which contradicts the very sharp, deterministic onset of damage driven by intrinsic mechanism. These considerations indicate that intensity enhancements in the vicinity of cracks and other structural defects lead to reduced surface damage thresholds not through intrinsic breakdown, but most probably through increased absorption by localized nanoscale absorbing particles trapped within scratches or subsurface fractures left by the surface-manufacturing process.

Another hypothesis regarding crack-promoted damage mechanism is related to the inherently different material electronic structure at the surface versus the bulk. Calculated electronic structure of the SiO_2 finite crystal (Bercegol et al., 2007) shows the presence of surface quantum states within the band gap with energies high enough to permit single-photon ionization by 351 nm (3.5 eV) photons. Further modeling (Bercegol et al., 2007) of the damage process is based on the assumption that initial energy acquisition from a 355 nm, 2.5 ns laser pulse can create a free-electron density equal to the density of states that initiate absorption. As a result, ~1 nm thick absorbing surface layer with strongly modified dielectric function is formed at the silica/vacuum interface, which can dramatically enhance the intensity of light experiencing multiple reflections inside a 0.5 μm wide crack with parallel walls. Corresponding enhancement factors strongly depend on electron plasma density and can exceed 100 for favorable light polarization state, crack width, and angle of incidence. Numerical modeling in the same work using a hydrodynamic code shows that irradiation of such cracks with absorbing walls by 2.5 ns pulses with 4 J/cm² fluence leads to temperatures exceeding 7000 K and, eventually, damage.

We will finalize the analysis of possible causes of intensity enhancement by considering effects of embedded absorbing (metallic) and nonabsorbing (dielectric) spherical particles. Table 3.2 presents the results of numerical calculations (Bonneau et al., 2001) for SiO_2 as a host material, Al and HfO_2 as embedded particles, and 1053 nm wavelength. One can see that enhancement factors E_{max}/E_0 (E_{max} is the maximum field in the particle vicinity) increase with diminishing metallic-particle size and with increasing size of dielectric particles. The latter is due to focusing of light by large dielectric spheres. Relatively large field enhancements for small absorbers may be a critical factor in facilitating plasma-ball growth (see Section 3.4).

Table 3.2 1053 nm E-Field–Enhancement Factors E_{max}/E_0 for Al (Metal) and HfO$_2$ (Dielectric) Particles Embedded in SiO$_2$ Host Material

Particle	Size (μm)	Enhancement Factor E_{max}/E_0
Al	0.1	3.6
	0.5	2.9
	4.0	1.8
HfO$_2$	0.1	1.4
	0.5	1.7
	4.0	7.1

Source: Bonneau, F., Combis, P., Vierne, J., and Daval, G., Simulations of laser damage of SiO$_2$ induced by a spherical inclusion, in *Laser-Induced Damage in Optical Materials: 2000*, G.J. Exarhos, A.H. Guenther, M.R. Kozlowski, K.L. Lewis, and M.J. Soileau, eds., Vol. 4347, SPIE, Bellingham, WA, 2001, pp. 308–315.

3.8 Probability of Damage Initiated by an Ensemble of Distributed Localized Defects

Laser-induced damage initiated by an ensemble of localized absorbing defects under typical damage test conditions utilizing mm-scale laser-beam spots is probabilistic in nature. The outcome (damage or no damage) is linked to the probability of finding an absorber within a portion of the beam spot where fluence exceeds the threshold for absorber failure (see Figure 3.28a). In the following paragraph, we will consider probability for damage initiated by surface defects, since the case for the bulk is conceptually very similar. There is a significant amount of work published on this subject (Fradin and Bua, 1974; Picard et al., 1977; Danileiko et al., 1981; Foltyn, 1984;

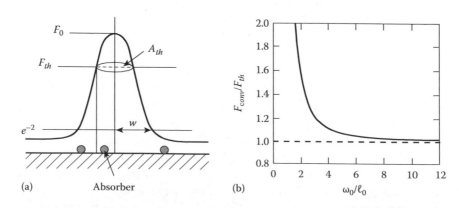

FIGURE 3.28 Damage probability modeling: (a) illustration of the probability of finding the defect within the region of Gaussian beam with energy density greater than threshold and (b) spot-size dependence of the conventional (50%) damage threshold. (Reprinted figure with permission from Foltyn, S.R., Spotsize effects in laser damage testing, in *Damage in Laser Materials: 1982*, H. E. Bennett, A. H. Guenther, D. Milam, and B. E. Newnam, eds., Nat. Bur. Stand. U.S., Special Publication, Vol. 669, pp. 368–379, U.S. Government Printing Office, Washington, DC. Copyright 1984 by SPIE.)

Porteus and Seitel, 1984; O'Connell, 1992; Krol et al., 2005; Gallais et al., 2004). We will start with the description given by Foltyn (1984) for the degenerate ensemble (all absorbers fail at the same fluence) and later consider the case of a power-law ensemble (Porteus and Seitel, 1984).

In the case of illumination by a Gaussian spatial-energy distribution, $F(r) = F_0 \exp[-2(r/w)^2]$, where F_0 is the maximum (axial) fluence, and w is the beam radius at the e^{-2} level, the area A_{th} of the beam with fluence larger than threshold fluence F_{th} is (Foltyn, 1984)

$$A_{th} = \frac{\pi w^2}{2} \ln\left(\frac{F_0}{F_{th}}\right).$$
(3.48)

According to the Poisson statistics, damage probability P for a surface with a defect density d can be expressed as follows:

$$P = 1 - \exp\left(-dA_{th}\right).$$
(3.49)

After combining Equations 3.48 and 3.49, probability P takes the following form:

$$P(F_0) = 1 - \left(\frac{F_{th}}{F_0}\right)^{\frac{\pi w^2 d}{2}}.$$
(3.50)

Similar formulas describe damage probability for volume defects, with the only difference being that in the exponent of Equation 3.49, the volume defect density must be used together with a beam volume where fluence exceeds F_{th}.

Equation 3.50 clearly shows that damage probability depends on the beam's spot size and, in particular, that this dependence vanishes for close-to-zero probabilities corresponding to damage onset. It is useful to compare two fluences—the laser fluence frequently used in damage evaluation that corresponds to 50% damage probability and the zero-probability fluence. Figure 3.28b shows the relation between these two types of thresholds as a function of beam radius w normalized to mean defect separation $l_0 = 0.5/\sqrt{d}$. It is apparent that the conventional threshold (related to 50% probability) exceeds the onset threshold by ~10% for a beam radius four times mean defect separation, and the discrepancy in thresholds quickly increases with diminishing beam radius. Figure 3.29 shows an example of damage probability measured for a sample containing two different types of absorbers (Krol et al., 2005) characterized by a different volume density d and threshold fluence F_{th}. The presence of the second type of absorber manifests itself by a change in the slope of the curve. The best fit of experimental probability curves provides data on threshold fluence and absorber density.

The aforementioned derivation assumes that all absorbers have the same failure fluence F_{th}. More realistic approaches use some kind of threshold distribution, for example,

Chapter 3

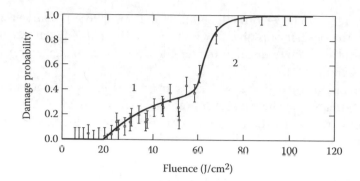

FIGURE 3.29 Experimental damage-probability data and theoretical best fit (solid line) for a Ta_2O_5 coating irradiated by 1064 nm, 5 ns pulses. Two kinds of defects are identified: $d_1 = 1.6 \times 10^4$ mm^{-3}, $F_{th_1} = 19$ J/cm^2, and $d_2 = 5 \times 10^5$ mm^{-3}, $F_{th_2} = 60$ J/cm^2. (Reprinted from *Optics Communications*, 256, Krol, H., Gallais, L. Grèzes-Besset, C., and Natoli, J.-Y., Investigation of nanoprecursors threshold distribution in laser-damage testing, pp. 184–189, Copyright 2005, with permission from Elsevier.)

power law (Porteus and Seitel, 1984; O'Connell, 1992) or Gaussian distribution (Krol et al., 2005). In the case of power law, the distribution is given by

$$f\left(F\right) = \frac{(p+1)N\left(F_n\right)}{\left(F_n - F_{th}\right)^{p+1}} \times K; \quad K = \begin{cases} 0; & F < F_{th} \\ \left(F - F_{th}\right)^p & F \ge F_{th} \end{cases}, \tag{3.51}$$

where

parameter p is in the range of $-1 < p \le 0$

F_n is the arbitrary normalization fluence

$$N\left(F_n\right) = \int_0^{F_n} f\left(F\right) dF \text{ is the cumulative density of defects that damage at } F_n > F_{th}. \text{ For}$$

a Gaussian spatial-beam profile with peak fluence F_0, damage probability has the form

$$P\left(F_0; F_{th}; N_w; p\right) = \begin{cases} 0; & F_0 < F_{th} \\ 1 - \exp\left[-N_w \int_1^{F_0/F_{th}} u^{-1}\left(u-1\right)^{p+1} du\right]; & F_0 \ge F_{th} \end{cases}, \tag{3.52}$$

where $N_w = \pi w^2 N(F_n)/2(F_n/F_{th}-1)^{p+1}$. If p is not an integer, Equation 3.52 requires numerical integration.

For the top-hat spatial-beam profile, damage probability has the analytical form

$$P\left(F_0; F_{th}; N_w; p\right) = \begin{cases} 0; & F_0 < F_{th} \\ 1 - \exp\left[-N_w \left(\frac{F_0}{F_{th}} - 1\right)^{p+1}\right]; & F_0 \ge F_{th} \end{cases}. \tag{3.53}$$

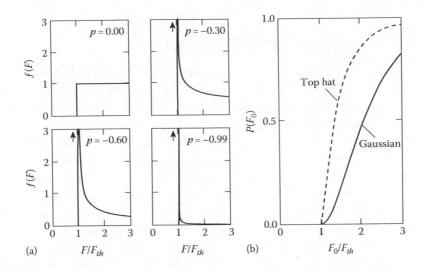

FIGURE 3.30 Damage probability in the case of power-law threshold distribution. (a) Distributed defect ensembles $f(F)$ for different parameters p; (b) comparison of damage probabilities for top-hat and Gaussian spatial-beam profiles, and parameter $p = 0$. (Reprinted figure with permission from Porteus, J.O. and Seitel, S.C., Absolute onset of optical surface damage using distributed defect ensembles, *Appl. Opt.*, 1984, 23, 3796–3805. With permission of Optical Society of America.)

Figure 3.30 shows an example of calculated damage probabilities for the exponent $p = 0$ (uniform distribution case) and both Gaussian and top-hat spatial-beam profiles. One can see that the top-hat profile gives a more precise determination of the damage onset. Formulas given by Equations 3.52 and 3.53 can be used for three-parameter (F_{th}, N_w, and p) fitting of the experimental damage probability data, which allows one to accurately determine the damage onset fluence F_{th}.

As discussed earlier, damage-probability studies offer information on both threshold fluence and absorber densities. It was shown recently (Gallais et al., 2004) that combining such data with integrated absorption measurements allows one to estimate the absorption by an individual absorber. Using these data in combination with thermal diffusion–based modeling of the damage process (see Section 3.2) allows one to detail information on absorber nature, size, and optical properties. This approach, implemented in SiO_2 film damage studies (Gallais et al., 2004) using 1064 nm, 5 ns pulses, produced ranges of possible absorber sizes (radius R) and imaginary refractive indices n_2: 10 nm $< R <$ 120 nm and $9 \times 10^{-3} < n_2 < 5$, respectively. There the smallest size, 10 nm, correlated with strong absorption (metal-type absorbers) and the largest, 120 nm, correlated with weakly absorbing dielectric. Large variations in these parameters are due to the lack of information on absorber thermal properties and real part of the refractive index. One can see that this methodology needs complementary information in order to reduce uncertainty in derived absorber parameters, and the development of new techniques for nanoscale absorber characterization remains an ongoing challenge.

Chapter 3

3.9 Laser Conditioning of the Materials with Localized Absorbing Defects

One of the material treatments aiming to improve damage resistance is irradiation by laser light at fluences below measured damage-threshold fluence, which is called laser conditioning. When damage is initiated by embedded nanoscale absorbers with a perfect absorber/host boundary (no voids), the conditioning process takes place if heating of the absorber results in temperatures exceeding the melting point of both materials (absorber and host). On the other hand, the temperature of the host should not exceed critical temperature for thermal explosion triggering plasma-ball formation and damage. Assuming that these conditions are satisfied, the conditioning process can be described as follows (Feit and Rubenchik, 2004a): With the appearance of the melting phase in both materials, dispersion of the inclusion takes place. This process is facilitated by the motion of molten material due to stresses caused by the difference in the density of molten and solid phases of the host material (Brener et al., 1999). Once the absorber material is dispersed within a larger volume with linear size that exceeds the diffusion length by several times, absorbed energy density drops dramatically. The result is a higher local damage threshold that manifests the conditioning effect.

When the host material is porous, which is typical for thin-film coatings, melted absorber material may be dispersed by diffusion into the void space without melting

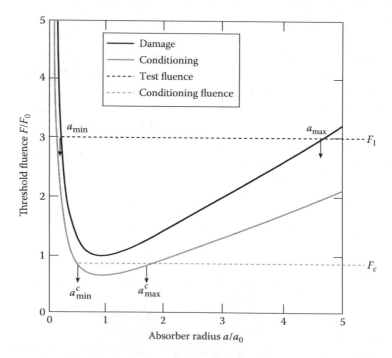

FIGURE 3.31 Damage threshold and conditioning threshold fluence curves illustrating how irradiation with conditioning fluence F_c effectively removes absorbers in the range $\{a_{min}^c; a_{max}^c\}$ from participation in the damage process. F_0 and a_0 are threshold fluence and radius for the most susceptible absorber, respectively. (Reprinted figure with permission from Feit, M.D., Rubenchik, A.M., and Trenholme, J. B., Simple model of laser damage initiation and conditioning in frequency conversion crystals, in *Laser-Induced Damage in Optical Materials: 2005*, G.J. Exarhos, A.H. Guenther, K.L. Lewis, D. Ristau, M.J. Soileau, and C.J. Stolz, eds., Vol. 5991, 59910W, SPIE, Bellingham, WA. Copyright 2006 by SPIE.)

the surrounding host material. For instance, irradiation of silica film containing few-nanometer-sized gold particles by subdamage-threshold laser fluence confirmed diffusion of molten gold in silica matrix to distances of the order of 130 nm (Bonneau et al., 2002). Consequently, an enhancement of the conditioning effect in select thin films may be expected as compared to the bulk materials.

Assuming that the conditioning effect, similar to the damage case, may be characterized by the critical temperature (Feit et al., 2006), conditioning threshold fluence should depend on an absorber size similar to the dependence shown in Figure 3.1b. Consequently, conditioning laser fluence $F_c < F_0$ (F_0 is the damage-threshold fluence) will make absorbers with radii between a_{min}^c and a_{max}^c benign (see Figure 3.31), which leads to reduced density of damage sites when the same volume is irradiated by fluence F_1 exceeding the original damage threshold.

3.10 Summary and Future Outlook

Various aspects of a pulsed-laser-induced damage process initiated by localized absorbing defects are considered, beginning with absorption inside the defect volume and ending with irreversible material modifications manifesting themselves as damage. The role of structural defects, such as scratches, is also analyzed in the context of light intensity enhancement and its link to damage. Mechanisms of damage initiated by embedded absorbers with different optical and thermomechanical properties and ranging in size from micrometers to a few nanometers can by now be considered as well understood. It is confirmed that even few-nanometer-sized absorbers can initiate the damage process through their interaction with the surrounding matrix, eventually leading to thermal explosion and damage of the latter. Similarly, light intensification effects in the proximity of structural defects, such as very small, few-tens-of-nanometer-wide surface scratches, can make even a weak local absorption very dangerous.

Continuous progress in optical technology leads to ever smaller sizes and number density of localized defects in optical materials. It is difficult, however, to expect their total elimination due to necessary surface processing in shaping bulk materials and because nanoscale defects are an inherent feature of thin-film coatings produced by physical vapor-deposition processes. For this reason, any further progress in improving damage resistance rests on the development of characterization tools that would be capable of detecting and analyzing individual nanoscale absorbers, their chemical composition and physical properties (optical, thermal, etc.), and their distribution in the material. This challenging task remains to be accomplished.

Acknowledgments

This work was supported by the U.S. Department of Energy Office of Inertial Confinement Fusion under Cooperative Agreement No. DE-FC52-08NA28302, the University of Rochester, and the New York State Energy Research and Development Authority. The support of DOE does not constitute an endorsement by DOE of the views expressed in this article. The author gratefully acknowledges his colleague Dr. A. Schmid, whose valuable comments immensely contributed to the manuscript quality.

Chapter 3

References

Aleshin, I. V., S. I. Anisimov, A. M. Bonch-Bruevich, Y. A. Imas, and V. L. Komolov. 1976. Optical breakdown of transparent media containing microinhomogeneities. *Soviet Physics-JETP* 43: 631–636.

Bercegol, H., F. Bonneau, P. Bouchut et al. 2003. Comparison of numerical simulations with experiment on generation of craters in silica by a laser. In *Laser-Induced Damage in Optical Materials: 2002*, eds. G. J. Exarhos, A. H. Guenther, N. Kaiser et al., Vol. 4932, pp. 297–308. Bellingham, WA: SPIE.

Bercegol, H., P. Grua, D. Hébert, and J.-P. Morreeuw. 2007. Progress in the understanding of fracture related laser damage of fused silica. In *Laser-Induced Damage in Optical Materials: 2007*, eds. G. J. Exarhos, A. H. Guenther, K. L. Lewis, D. Ristau, M. J. Soileau, and C. J. Stolz, Vol. 6720, 672003pp. Bellingham, WA: SPIE.

Bloembergen, N. 1973. Role of cracks, pores, and absorbing inclusions on laser induced damage threshold at surfaces of transparent dielectrics. *Applied Optics* 12: 661–664.

Boley, B. A. and J. H. Weiner. 1960. *Theory of Thermal Stresses*. New York: Wiley.

Bonneau, F., P. Combis, J. L. Rullier et al. 2002. Study of UV laser interaction with gold nanoparticles embedded in silica. *Applied Physics B* 75: 803–815.

Bonneau, F., P. Combis, J. L. Rullier et al. 2004. Numerical simulations for description of UV laser interaction with gold nanoparticles embedded in silica. *Applied Physics B* 78: 447–452.

Bonneau, F., P. Combis, J. Vierne, and G. Daval. 2001. Simulations of laser damage of SiO_2 induced by a spherical inclusion. In *Laser-Induced Damage in Optical Materials: 2000*, eds. G. J. Exarhos, A. H. Guenther, M. R. Kozlowski, K. L. Lewis, and M. J. Soileau, Vol. 4347, pp. 308–315. Bellingham, WA: SPIE.

Brener, E. A., S. V. Iordanskii, and V. I. Marchenko. 1999. Elastic effects on the kinetics of a phase transition. *Physical Review Letters* 82: 1506–1509.

Bude, J., G. M. Guss, M. Matthews, and M. L. Spaeth. 2007. The effect of lattice temperature on surface damage in fused silica optics. In *Laser-Induced Damage in Optical Materials: 2007*, eds. G. J. Exarhos, A. H. Guenther, K. L. Lewis, D. Ristau, M. J. Soileau, and C. J. Stolz, Vol. 6720, 672009pp. Bellingham, WA: SPIE.

Campbell, J. H., F. Rainer, M. R. Kozlowski, C. R. Wolfe, I. M. Thomas, and F. P. Milanovich. 1991. Damage resistant optics for a megajoule solid state lasers. In *Laser-Induced Damage in Optical Materials: 1990*, eds. H. E. Bennett, L. L. Chase, A. H. Guenther, B. E. Newnam, and M. J. Soileau, Vol. 1441, pp. 444–456. Bellingham, WA: SPIE.

Carr, C. W., H. B. Radousky, A. M. Rubenchik, M. D. Feit, and S. G. Demos. 2004. Localized dynamics during laser-induced damage in optical materials. *Physical Review Letters* 92: 087401.

Chan, C. H. 1975. Effective absorption for thermal blooming due to aerosols. *Applied Physics Letters* 26: 628–630.

Danileiko, Y. K., A. A. Manenkov, and V. S. Nechitailo. 1978. The mechanism of laser-induced damage in transparent materials, caused by thermal explosion of absorbing inhomogeneities. *Soviet Journal of Quantum Electronics* 8: 116–118.

Danileiko, Y. K., A. A. Manenkov, V. S. Nechitailo, A. M. Prokhorov, and V. Y. Khaimov-Mal'kov. 1973. The role of absorbing impurities in laser-induced damage of transparent dielectrics. *Soviet Physics-JETP* 36: 541–543.

Danileiko, Y. K., Y. P. Minaev, V. N. Nikolaev, and A. V. Sidorin. 1981. Determination of the characteristics of microdefects from statistical relationships governing laser damage to solid transparent materials. *Soviet Journal of Quantum Electronics* 11: 1445–1449.

Duthler, C. J. 1974. Explanation of laser-damage cone-shaped surface pits. *Applied Physics Letters* 24: 5–7.

Efimov, O. M. 2004. Self-optical breakdown and multipulse optical breakdown of transparent insulators in the femto-nanosecond region of laser pulse widths. *Journal of Optical Technology* 71: 338–347.

Feit, M. D., J. H. Campbell, D. R. Faux et al. 1998. Modeling of laser-induced surface cracks in silica at 355 nm. In *Laser-Induced Damage in Optical Materials: 1997*, eds. G. J. Exarhos, A. H. Guenther, M. R. Kozlowski, and M. J. Soileau, Vol. 3244, pp. 350–355. Bellingham, WA: SPIE.

Feit, M. D., L. W. Hrubesh, A. M. Rubenchik, and J. N. Wong. 2001a. Scaling relations for laser damage initiation craters. In *Laser-Induced Damage in Optical Materials: 2000*, eds. G. J. Exarhos, A. H. Guenther, M. R. Kozlowski, K. L. Lewis, and M. J. Soileau, Vol. 4347, pp. 316–323. Bellingham, WA: SPIE, 2001.

Feit, M. D. and A. M. Rubenchik. 2004a. Implications of nanoabsorber initiators for damage probability curves, pulselength scaling, and laser conditioning. In *Laser-Induced Damage in Optical Materials: 2003*, eds. G. J. Exarhos, A. H. Guenther, N. Kaiser, K. L. Lewis, M. J. Soileau, and C. J. Stolz, Vol. 5273, pp. 74–82. Bellingham, WA: SPIE.

Feit, M. D. and A. M. Rubenchik. 2004b. Influence of subsurface cracks on laser-induced surface damage. In *Laser-Induced Damage in Optical Materials: 2003*, eds. G. J. Exarhos, A. H. Guenther, N. Kaiser, K. L. Lewis, M. J. Soileau, and C. J. Stolz, Vol. 5273, pp. 264–272. Bellingham, WA: SPIE.

Feit, M. D. and A. M. Rubenchik. 2007. Importance of free surfaces for damage crater formation. In *Laser-Induced Damage in Optical Materials: 2006*, eds. G. J. Exarhos, A. H. Guenther, K. L. Lewis, D. Ristau, M. J. Soileau, and C. J. Stolz, Vol. 6403, 64030App. Bellingham, WA: SPIE.

Feit, M. D., A. M. Rubenchik, and M. J. Runkel. 2001b. Analysis of bulk DKDP damage distribution, obscuration, and pulse-length dependence. In *Laser-Induced Damage in Optical Materials: 2000*, eds. G. J. Exarhos, A. H. Guenther, M. R. Kozlowski, K. L. Lewis, and M. J. Soileau, Vol. 4347, pp. 383–388. Bellingham, WA: SPIE.

Feit, M. D., A. M. Rubenchik, and J. B. Trenholme. 2006. Simple model of laser damage initiation and conditioning in frequency conversion crystals. In *Laser-Induced Damage in Optical Materials: 2005*, eds. G. J. Exarhos, A. H. Guenther, K. L. Lewis, D. Ristau, M. J. Soileau, and C. J. Stolz, Vol. 5991, 59910Wpp. Bellingham, WA: SPIE.

Foltyn, S. R. 1984. Spotsize effects in laser damage testing. In *Damage in Laser Materials: 1982*, eds. H. E. Bennett, A. H. Guenther, D. Milam, and B. E. Newnam, Nat. Bur. Stand. U.S., Special Publication, Vol. 669, pp. 368–379. Washington, DC: U.S. Government Printing Office.

Fradin, D. W. and D. P. Bua. 1974. Laser-induced damage in ZnSe. *Applied Physics Letters* 24: 555–557.

Fuka, M. Z., J. K. McIver, and A. H. Guenther. 1990. Effects of thermal conductivity and index of refraction variation on the inclusion dominated model of laser-induced damage. In *Laser Induced Damage in Optical Materials: 1989*, eds. A. J. Bennett, L. L. Chase, A. H. Guenther, B. E. Newnam, and M. J. Soileau, NIST U.S. Special Publication 801, Vol. 1438, pp. 576–583. Bellingham, WA: SPIE.

Gallais, L., P. Voarino, and C. Amra. 2004. Optical measurement of size and complex index of laser-damage precursors: The inverse problem. *Journal of the Optical Society of America B* 21: 1073–1080.

Garnov, S. V., A. S. Epifanov, S. M. Klimentov, and A. A. Manenkov. 1993. Pulse-width dependence of laser damage in optical materials: Critical analysis of available data and recent results for nanopicosecond region. In *Laser-Induced Damage in Optical Materials: 1992*, eds. H. E. Bennett, L. L. Chase, A. H. Guenther, B. E. Newnam, M. J. Soileau, and M. J. Soileau, Vol. 1848, pp. 403–414. Bellingham, WA: SPIE.

Génin, F. Y., A. Salleo, T. V. Pistor, and L. L. Chase. 2001. Role of light intensification by cracks in optical breakdown on surfaces. *Journal of the Optical Society of America A* 18: 2607–2616.

Glebov, L. B. 2002. Intrinsic laser-induced breakdown of silicate glasses. In *Laser-Induced Damage in Optical Materials: 2001*, eds. G. J. Exarhos, A. H. Guenther, K. L. Lewis, M. J. Soileau, and C. J. Stolz, Vol. 4679, pp. 321–331. Bellingham, WA: SPIE.

Goldenberg, H. and C. J. Tranter. 1952. Heat flow in an infinite medium heated by a sphere. *British Journal of Applied Physics* 3: 296–298.

Grua, P., J. P. Morreeuw, H. Bercegol, G. Jonusauskas, and F. Vallée. 2003. Electron kinetics and emission for metal nanoparticles exposed to intense laser pulses. *Physical Review B* 68: 035424.

Hopper, R. W. and D. R. Uhlmann. 1970. Mechanism of inclusion damage in laser glass. *Journal of Applied Physics* 41: 4023–4037.

Josse, M., J. L. Rullier, R. Courchinoux, T. Donval, L. Lamaignère, and H. Bercegol. 2006. Effects of scratch speed on laser-induced damage. In *Laser-Induced Damage in Optical Materials: 2005*, eds. G. J. Exarhos, A. H. Guenther, K. L. Lewis, D. Ristau, M. J. Soileau, and C. J. Stolz, Vol. 5991, 599106pp. Bellingham, WA: SPIE.

Kerker, M. 1969. *The Scattering of Light and Other Electromagnetic Radiation*. New York: Academic Press, Copyright Elsevier (1969).

Koldunov, M. F., A. A. Manenkov, and I. L. Pokotilo. 1988. Theoretical analysis of the conditions for a thermal explosion and of a photoionization instability of transparent insulators containing absorbing inclusions. *Soviet Journal of Quantum Electronics* 18: 345–349.

Koldunov, M. F., A. A. Manenkov, and I. L. Pokotilo. 1990. Thermal explosion of absorbing inclusions as the mechanism of laser damage to insulator surfaces. *Soviet Journal of Quantum Electronics* 20: 456–460.

Chapter 3

Koldunov, M. F., A. A. Manenkov, and I. L. Pokotilo. 1996. Pulse-width and pulse-shape dependencies of laser-induced damage threshold to transparent optical materials. In *Laser-Induced Damage in Optical Materials: 1995*, eds. H. E. Bennett, A. H. Guenther, M. R. Kozlowski, B. E. Newnam, and M. J. Soileau, Vol. 2714, pp. 718–730. Bellingham, WA: SPIE.

Koldunov, M. F., A. A. Manenkov, and I. L. Pokotilo. 1997. Formulation of the criterion of thermoelastic laser damage of transparent dielectrics and the dependence of damage threshold on pulse duration. *Quantum Electronics* 27: 918–922.

Koldunov, M. F., A. A. Manenkov, and I. L. Pokotilo. 2002. Mechanical damage in transparent solids caused by laser pulses of different durations. *Quantum Electronics* 32: 335–340.

Kreibig, U. and M. Vollmer. 1995. *Optical Properties of Metal Clusters*. Springer series in *Materials Science*, Vol. 25. Berlin, Germany: Springer-Verlag.

Krol, H., L. Gallais, C. Grèzes-Besset, and J.-Y. Natoli. 2005. Investigation of nanoprecursors threshold distribution in laser-damage testing. *Optics Communications* 256: 184–189.

Krucr, W. L. 1988. *The Physics of Laser-Plasma Interactions*. Frontiers in Physics, Vol. 73, ed. D. Pines. Redwood City, CA: Addison-Wesley.

Kudryashov, S. I., S. D. Allen, S. Papernov, and A. W. Schmid. 2006. Nanoscale laser-induced spallation in SiO_2 films containing gold nanoparticles. *Applied Physics B* 82: 523–527.

Lange, M. R., J. K. McIver, and A. H. Guenther. 1984. The influence of the thermal and mechanical properties of optical materials in thin film form on their damage resistance to pulsed lasers. *Thin Solid Films* 118: 49–60.

Lange, M. R., J. K. McIver, A. H. Guenther, and T. W. Walker. 1982. Pulsed laser induced damage of an optical material with a spherical inclusion: Influence of the thermal properties of the materials. In *Laser Induced Damage in Optical Materials: 1982*, Nat. Bur. Stand. U.S., Special Publication Vol. 669, pp. 380–386. Washington, DC: U.S. Government Printing Office.

Mie, G. 1908. Beiträge zur optik trüber medien, speziell kolloidaler metallosungen. *Annalen der Physik (Germany)* 25: 377–445; Bohren, C. F. and D. R. Huffman. 1983. *Absorption and Scattering of Light by Small Particles*. New York: Wiley.

Natoli, J.-Y., L. Gallais, H. Akhouayri, and C. Amra. 2002. Laser-induced damage of materials in bulk, thin-film, and liquid forms. *Applied Optics* 41: 3156–3166.

O'Connell, R. M. 1992. Onset threshold analysis of defect-driven surface and bulk laser damage. *Applied Optics* 31: 4143–4153.

Papernov, S. and A. W. Schmid. 1997. Localized absorption effects during 351 nm, pulsed laser irradiation of dielectric multilayer thin films. *Journal of Applied Physics* 82: 5422–5432.

Papernov, S. and A. W. Schmid. 2002. Correlations between embedded single gold nanoparticles in SiO_2 thin film and nanoscale crater formation induced by pulsed-laser radiation. *Journal of Applied Physics* 92: 5720–5728.

Papernov, S. and A. W. Schmid. 2003. Damage behavior of SiO_2 thin films containing gold nanoparticles lodged at predetermined distances from the film surface. In *Laser-Induced Damage in Optical Materials: 2002 and 7th International Workshop on Laser Beam and Optics Characterization*, eds. G. J. Exarhos, A. H. Guenther, N. Kaiser et al., Vol. 4932, pp. 66–74. Bellingham, WA: SPIE.

Papernov, S. and A. W. Schmid. 2005. Two mechanisms of crater formation in ultraviolet-pulsed-laser irradiated SiO_2 thin films with artificial defects. *Journal of Applied Physics* 97: 114906.

Papernov, S. and A. W. Schmid. 2008. Testing asymmetry in plasma-ball growth seeded by a nanoscale absorbing defect embedded in a SiO_2 thin-film matrix subjected to UV pulsed-laser radiation. *Journal of Applied Physics* 104: 063101.

Papernov, S., A. Tait, W. Bittle, A. W. Schmid, J. B. Oliver, and P. Kupinski. 2011. Near-ultraviolet absorption and nanosecond-pulse-laser damage in HfO_2 monolayers studied by submicrometer-resolution photothermal heterodyne imaging and atomic force microscopy. *Journal of Applied Physics* 109: 113106.

Picard, R. H., D. Milam, and R. A. Bradbury. 1977. Statistical analysis of defect-caused laser damage in thin films. *Applied Optics* 16: 1563–1571.

Porteus, J. O. and S. C. Seitel. 1984. Absolute onset of optical surface damage using distributed defect ensembles. *Applied Optics* 23: 3796–3805.

Runkel, M., A. K. Burnham, D. Milam, W. Sell, M. Feit, and A. Rubenchik. 2000. The results of pulse-scaling experiments on rapid-growth DKDP triplers using the optical sciences laser at 351 nm. In *Laser-Induced Damage in Optical Materials: 2000*, eds. G. J. Exarhos, A. H. Guenther, M. R. Kozlowski, K. L. Lewis, and M. J. Soileau, Vol. 4347, pp. 359–372. Bellingham, WA: SPIE.

Saito, K. and A. J. Ikushima. 2000. Absorption edge in silica glass. *Physical Review B* 62: 8584–8587.

Salleo, A., F. Y. Genin, J. M. Yoshiyama, C. J. Stolz, and M. R. Kozlowski. 1998. Laser-induced damage of fused silica at 355 nm initiated at scratches. In *Laser-Induced Damage in Optical Materials: 1997*, eds. G. J. Exarhos, A. H. Guenther, M. R. Kozlowski, and M. J. Soileau, Vol. 3244, pp. 341–347. Bellingham, WA: SPIE.

Smith, A. V. and B. T. Do. 2008. Bulk and surface laser damage of silica by picosecond and nanosecond pulses at 1064 nm. *Applied Optics* 47: 4812–4832.

Soileau, M. J., W. E. Williams, E. W. Vanstryland, T. F. Boggess, and A. L. Smirl. 1983. Picosecond damage studies at 0.5-μm and 1-μm. *Optical Engineering* 22: 424–430.

Stuart, B. C., M. D. Feit, A. M. Rubenchik, B. W. Shore, and M. D. Perry. 1995. Laser-induced damage in dielectrics with nanosecond to subpicosecond pulses. *Physical Review Letters* 74: 2248–2251.

Trenholme, J. B., M. D. Feit, and A. M. Rubenchik. 2006. Size-selection initiation model extended to include shape and random factors. In *Laser-Induced Damage in Optical Materials: 2005*, eds. G. J. Exarhos, A. H. Guenther, K. L. Lewis, D. Ristau, and M. J. Soileau, Vol. 5991, 59910Xpp. Bellingham, WA: SPIE.

Walker, T. W., A. H. Guenther, and P. Nielsen. 1981. Pulsed laser-induced damage to thin-film optical coatings—Part II: Theory. *IEEE Journal of Quantum Electronics* QE 17: 2053–2065.

Chapter 3

4. Self-Focusing and Nonlinear Effects

Self-Focusing a

Vitaly Gruzdev

4.1   Introduction

Scaling of rate of laser-induced effects with laser irradiance* (i.e., surface density of power; W/cm²) is one of the fundamental features that control dynamics of laser–material interactions. At low irradiance, the scaling is linear for most interactions: (1) zero irradiance produces zero rate, and (2) increase in irradiance by factor of K leads to increase in rate by factor of K (Figure 4.1). Those properties of the linear scaling are mathematically expressed as follows:

$$R = SI, \tag{4.1}$$

where
    R is the rate of an effect
    I is the laser irradiance
    Coefficient S does not depend on irradiance (while it might depend on other laser parameters, e.g., wavelength)

---

* Many laser–material interactions, for example, laser damage, are characterized in terms of *fluence*, that is, surface density of energy measured in J/cm². Peak irradiance is related to peak fluence via a coefficient that depends on pulse duration only. Therefore, linear or nonlinear scaling with irradiance means correspondingly linear or nonlinear scaling with fluence. Due to this fact, we discuss the nonlinear effects in terms of scaling with irradiance. For effects in bulk materials, irradiance is also referred to as *intensity* following the recent trend in scientific publications.

Laser-Induced Damage in Optical Materials. Edited by Detlev Ristau © 2015 CRC Press/Taylor & Francis Group, LLC. ISBN: 978-1-4398-7216-1.

Chapter 4

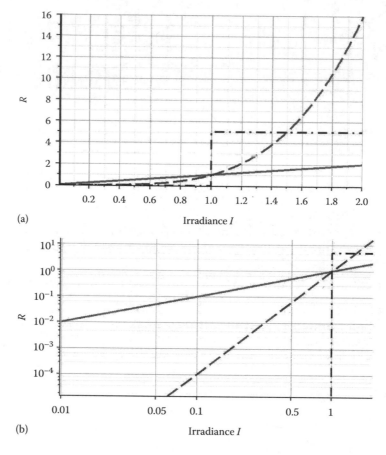

FIGURE 4.1 Linear (solid line), power I^4(dashed line), and threshold-type (dash-dotted line) scaling of rate R of a laser-induced process with laser irradiance I. (a) depicts the scaling in regular axes while (b) depicts the same scaling in log–log axes. Note that slope of the linear scaling is smaller than that of nonlinear power scaling in the log–log system shown in (b), while both are given by straight lines.

For example, the rate of light absorption scales according to Equation 4.1 at low irradiance and is therefore referred to as *linear absorption* (Ashcroft and Mermin 1976, Ridley 1993).

Strong linear absorption is believed to be a major driving mechanism of *laser-induced damage* (LID) of absorbing materials (e.g., metal surfaces of mirrors) (Anisimov et al. 1971, Ready 1971, Wood 2003). For those materials, LID occurs at surface at relatively low laser irradiance that is not favorable to induce significant nonlinear effects capable of initiating and driving LID. In contrary, laser irradiance required to produce LID of transparent materials (i.e., materials with negligible linear absorption at laser wavelength) is high enough to induce various effects that nonlinearly scale with irradiance. Among them, two major groups are processes of nonlinear absorption and effects of nonlinear propagation (that are of special importance for bulk damage). Their rates grow with irradiance faster than in linear case given by Equation 4.1 due to dependence of coefficient S on irradiance. For example, N-photon absorption scales as N-th power of irradiance I^N with $N \geq 2$ (Figure 4.1), and rate of absorption due to tunneling ionization exponentially scales with the square root of irradiance (Keldysh 1965).

The nonlinear effects of propagation and absorption make very significant contributions to LID since they are related to fundamental mechanisms of LID of transparent materials and determine its major parameter—LID threshold. Each of them includes fast and slow processes driven by fast electron excitations and slower atomic motion correspondingly. Accordingly, this chapter begins with an overview of elementary laser-induced microscopic processes in electronic and atomic subsystems of transparent solids (Section 4.2). Section 4.3 is devoted to fast mechanisms of nonlinear absorption featuring absorption due to multiphoton, tunneling, and avalanche ionization. Section 4.4 overviews nonlinear propagation effects including self-focusing and frequency conversion. We finish the chapter with summary (Section 4.5).

Throughout this chapter, transparent solids under consideration are assumed to be ideal dielectric crystals. Electronic defects (excitons and color centers) are not analyzed in detail, but their contribution to nonlinear absorption is discussed in Section 4.3. To be specific, we refer to direct-gap cubic crystals for the interpretation of the theoretical results with properties of alkali halides in mind as typical crystals of that group (Sirdeshmukh et al. 2001). Noncrystalline solids are not considered in this chapter because of lack of detailed and reliable microscopic models for understanding and description of their modification by laser radiation. Also, we consider LID made by a single laser pulse on a material spot not irradiated before to exclude the influence of laser-induced defects. The reader is assumed to be familiar with basic concepts of solid-state physics (Ashcroft and Mermin 1976, Kittel 1985, Ridley 1993).

4.2 Overview of Elementary Laser-Induced Processes

LID is associated with the permanent change of material structure (Ready 1971, Wood 2003) and implies significant displacement of atoms from their original location, for example, formation of voids in bulk materials (Schaffer et al. 2001, Juodkazis et al. 2008). On the one hand, laser radiation is considered as the only supply of energy for variations of material structure attributed to LID. On the other hand, the fundamental laws suggest that ultraviolet, visible, and infrared light is exclusively absorbed by electrons via excitation from lower to upper energy levels (Ashcroft and Mermin 1976, Kittel 1985, Ridley 1993). Therefore, there must be a chain of coupled processes of energy deposition and transfer in solids that originates from electron excitation and finishes up with the atomic motion to produce LID. Particular mechanisms of the energy transfer depend on structure and properties of specific solids, but major effects are universal for all high-power laser–material interactions including LID (Sundaram and Mazur 2002, Bulgakova et al. 2010, Gamaly 2011, Balling and Schou 2013). We analyze them later in this section from the view point of their duration.

Let us consider an ideal situation when all laser-induced processes of energy deposition and transfer are triggered simultaneously by a long flat-top pulse (see Figure 4.2). Due to the fundamental fact that electron mass is 10^3–10^4 times smaller than that of atoms, electrons respond to laser action much faster than atoms of a solid. This allows splitting all laser-induced processes into two groups: fast electronic effects that take place before atomic subsystem exhibits any significant reaction to laser action and slower effects that result from the motion of the atomic subsystem (Sundaram and Mazur 2002,

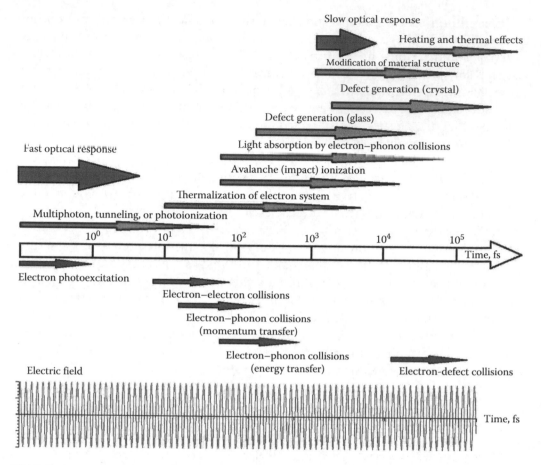

FIGURE 4.2 A scheme of time hierarchy of laser-induced microscopic processes (arrows in bottom part of figure), related particle dynamics (dark arrows in top part of figure), and optical response (two larger arrows).

Bulgakova et al. 2010, Gamaly 2011, Balling and Schou 2013). *Photoexcitation* (i.e., direct noncollision up-transitions of electrons via simultaneous absorption of several laser photons under action of electric field of laser light) is the fastest of the electronic processes with each individual transition taking less than 1 femtosecond (1 fs; 1 fs = 10^{-15} s) (Hentschel et al. 2001, Sundaram and Mazur 2002, Krausz and Ivanov 2009, Gamaly 2011). It is one of key contributors to nonlinear absorption. If the excited electrons are driven from bound states of valence band to unbound mobile states of conduction band, the photoexcitation is referred to as *photoionization* (Figure 4.3). Together with the photoionization, the photoexcitation includes various transitions of valence-band electrons to energy levels of bound states localized at native and laser-induced defects (Figure 4.3). In parallel to the photoexcitation, both bound and mobile electrons produce *fast electronic optical response* with sub-fs delay (Boyd 2008, Christodoulides et al. 2010). Irradiance scaling of the optical response is nonlinear at irradiance close to LID threshold in transparent solids.

The photoionization produces highly nonequilibrium energy distribution of conduction-band electrons (Kaiser et al. 2000). *Electron–particle collisions* promptly follow the photoexcitation and drive energy redistribution and transfer (Figure 4.2).

Each collision is an interaction of an electron with a particle (or quasiparticle, e.g., phonon) that leads to transfer of momentum and energy between them (Figure 4.3). If both colliding particles are electrons, *electron–electron collisions* are considered. They are delayed with respect to the photoexcitation because electrons perform non-collision laser-driven motion before each collision (Ashcroft and Mermin 1976). Therefore, to make first collisions and initiate the collision process after beginning of laser action, an average time of collision-free travel τ_{ec} (also referred to as *collision time*)

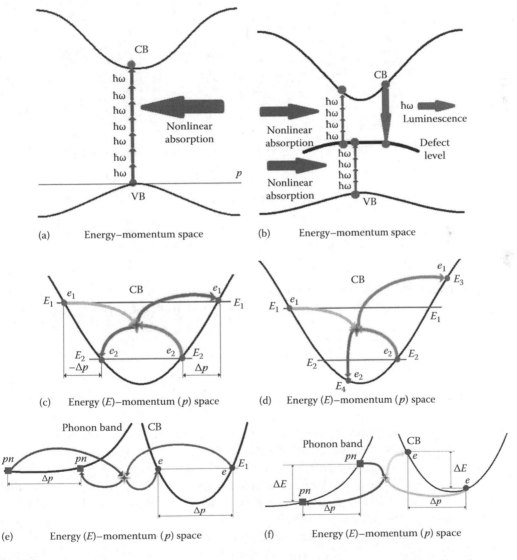

FIGURE 4.3 A sketch of major fast effects involved in laser–dielectric interactions: the photoionization (a), the photoexcitation to/from electron trapping states (defects) (b); elastic (c) and inelastic (d) electron–electron collisions; elastic (e) and inelastic (f) electron–phonon collisions. CB is for conduction band, VB is for valence band, e is for electron, *ph* is for photon, *pn* is for a phonon, and Δ shows the unmodified band gap.   *(Continued)*

Chapter 4

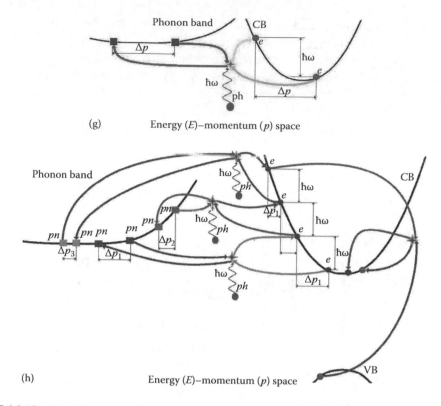

FIGURE 4.3 (Continued) A sketch of major fast effects involved in laser–dielectric interactions: one-photon absorption due to electron–phonon collisions (g); avalanche ionization (h).

is required. It is given by the inverse product of density of colliding electrons n_e, collision cross section σ, and average relative speed v_e of colliding electrons (Ashcroft and Mermin 1976):

$$\tau_{ec} = \frac{1}{v_{ec}} = \frac{1}{n_e \sigma v_e},\qquad(4.2)$$

where v_{ec} is the frequency of the collisions. Since electron–electron collisions are impossible if the conduction band is totally empty and the valence band is completely occupied (Ashcroft and Mermin 1976), the density of colliding electrons of Equation 4.2 equals the density of excited electrons departed from the valence band. The collision time τ_{ec} of Equation 4.2 reaches a lower limit at *critical density* of the conduction-band electrons that is believed to turn on LID (see Section 4.3). If critical density is close to free-electron density of metals (about 10^{22} 1/cm^3) (Ashcroft and Mermin 1976), the collision time reduces to 1–10 fs (Ashcroft and Mermin 1976, Gamaly 2013). The collisions result in momentum and energy transfer within the electron subsystem and drive it to quasiequilibrium state within 10^1–10^2 fs (this process is referred to as electron *thermalization* [Kaiser et al. 2000, Sundaram and Mazur 2002, Gamaly 2013]).

If the colliding particles are an electron and a phonon, *electron–phonon collisions* come in to play (Figure 4.2). Particular type of participating phonons (e.g., polar or nonpolar) and value of collision frequency significantly depend on specific material structure (Ashcroft and Mermin 1976, Jones et al. 1989). Two important contributions to the electron dynamics are made by the electron–phonon collisions (Ashcroft and Mermin 1976, Gamaly 2013): momentum transfer from electrons to phonons due to *elastic collisions* with average collision time of the order of 10 fs, and energy transfer due to less-frequent *inelastic collisions* with collision time about 10^2 fs (Figure 4.3). Also, the electron–phonon interactions support single-photon absorption by conduction-band electrons by providing momentum transfer for electronic intraband transitions via *electron–phonon–photon collisions* (Ashcroft and Mermin 1976, Gamaly 2013). At constant density of colliding electrons, this type of absorption scales linearly with laser irradiance, but if electron density is produced by laser light (i.e., the density depends on irradiance), the scaling of this absorption is nonlinear.

In parallel, interactions of conduction-band electrons with valence-band holes disturb atomic subsystem and result in formation of point defects, for example, color centers and self-trapped excitons (Jones et al. 1989, Saeta and Greene 1993, Audebert et al. 1994, Daguzan et al. 1995, Mao 2004). Duration of that process significantly depends on intrinsic properties and structure of transparent solids. For example, formation of self-trapped excitons takes 150–200 fs in glasses like fused silica (Audebert et al. 1994, Daguzan et al. 1995, Temnov et al. 2006) while it takes few picoseconds in wide-band-gap crystals like sapphire (Temnov et al. 2006). The defects form energy levels between the valence and conduction bands (Saeta and Greene 1993, Audebert et al. 1994, Daguzan et al. 1995 Menoni et al. 2012, Rudolph et al. 2013). They influence interband electron transitions, for example, by trapping valence electrons (Figure 4.3).

Together with establishing the equilibrium and energy transfer to phonons, inelastic electron–phonon collisions drive one more ionization mechanism—*avalanche* or *impact ionization* (Figure 4.3). It is associated with collision-induced interband up-transitions of valence electrons due to energy transfer from hot conduction electrons (Bloembergen 1974, Holway and Fradin 1975, Arnold et al. 1992, Stuart et al. 1996, Wood 2003, Starke et al. 2004, Jia 2006). Development of this process requires some mobile seed electrons in the conduction band and effective transport of electrons to high-energy levels by one-photon absorption due to electron–phonon–photon collisions (Holway and Fradin 1975, Arnold et al. 1992, Stuart et al. 1996, Mao 2004, Wood 2003, Starke et al. 2004, Jia 2006). The conduction-band electrons transfer their energy to valence-band electrons via inelastic collisions. If energy transferred to a valence-band electron during a collision exceeds *band gap* (i.e., minimum energy gap between the valence and conduction bands), the electron makes an interband transition to the bottom of the conduction band and can absorb laser energy to repeat that cycle. Multiple repetitions of those microscopic processes are believed to be quite similar to development of electron avalanche in gases during electric discharge (Seitz 1949, Sparks 1975, Wood 2003). Duration of development of the avalanche ionization is determined by the time of inelastic electron–particle collisions and duration of the electron promotion to high-energy levels (Kaiser et al. 2000, Rethfeld 2004), assuming enough density of seed conduction-band electrons. Overall time required to induce the avalanche ionization is estimated to vary from few tens to few hundred

Chapter 4

femtoseconds and strongly depends on laser irradiance, pulse width, and wavelength (Kaiser et al. 2000, Rethfeld 2004).

The electron–phonon interactions initiate energy transfer from electrons to atoms that feeds the second group of laser-induced processes with development time in picosecond (ps; 1 ps = 10^{-12} s) to nanosecond (ns; 1 ns = 10^{-9} s) domain (Figure 4.2). It includes all slower processes: significant change of material structure due to atomic displacement; relaxation of trapped electrons; formation of defects, for example, color centers; thermal processes; slow optical response due to atomic effects; and hydrodynamic effects, for example, motion of melted material and evaporation (Bulgakova et al. 2010, Balling and Schou 2013). The slowest of them are quasi-equilibrium effects while fast phase changes (e.g., ultrafast melting) occur under essentially nonequilibrium conditions (Sundaram and Mazur 2002, Rethfeld et al. 2004). It is believed that the fast processes trigger slower effects if the electronic excitation is strong enough (Bloembergen 1974, Holway and Fradin 1975, Arnold et al. 1992, Stuart et al. 1996, Sundaram and Mazur 2002, Wood 2003, Mao 2004, Rethfeld et al. 2004, Starke et al. 2004, Jia 2006, Bulgakova et al. 2010, Jasapara 2001, Gamaly 2013, Balling and Schou 2013). The strength of the excitation and the amount of laser radiation energy absorbed by the electrons serve as criteria for LID thresholds when LID is associated with intrinsic material properties (Holway and Fradin 1975, Arnold et al. 1992, Stuart et al. 1996, Wood 2003, Mao 2004, Starke et al. 2004, Jia 2006, Jasapara 2001, Balling and Schou 2013). Therefore, the electronic excitations result in modifications of material structure only if they supply enough energy to initiate the delayed processes.

Under real conditions, the laser-induced processes do not start all at once, but begin in a certain order determined by two factors. First, rates of those microscopic processes are characterized by strong nonlinear dependence on laser irradiance. Therefore, the fastest are the processes that combine the shortest time of development with the earliest start at low laser irradiance. If a process has short development time, but requires high irradiance for initiation (e.g., tunneling ionization), its development is delayed at least till peak irradiance is reached. Second, two basic properties of transparent solids significantly affect the dynamics of the laser-induced processes: (1) a wide band gap that is 2–10 times larger than laser-photon energy, and (2) almost zero density of mobile conduction-band electrons before laser action. The latter means that the valence band is almost completely occupied, the conduction band is almost completely empty, and density of colliding electrons is extremely low at the initial stage of laser action. This implies that the collision time is very large according to Equation 4.2 and can even exceed pulse duration. Under that condition, the only channel of energy deposition is the nonlinear absorption due to photoexcitation that might dominate over tens to hundreds of femtoseconds (Figure 4.4). The photoexcitation leads to generation of valence-band holes and conduction-band mobile electrons that are referred to as *electron–hole plasma* since they behave in the way similar to carriers of plasmas. Generation of defects (e.g., self-trapped excitons and color centers) and trapping electrons at nonmobile states localized at the defects starts in parallel with the photoionization. Since the photoionization tends to populate bottom of the conduction band (Kaiser et al. 2000), the absorbed laser

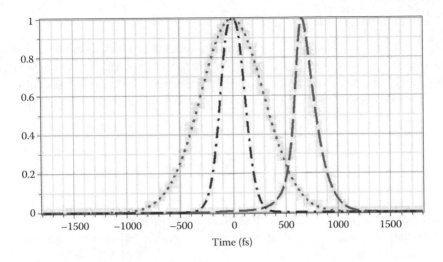

FIGURE 4.4 An example of time evolution of the photoionization rate (dash-dotted curve) and rate of avalanche ionization (dashed curve) with respect to time profile of laser irradiance (dotted curve). The photoionization rate follows $I^6(t)$ profile with zero delay with respect to the peak irradiance while the peak of the avalanche-ionization rate is delayed by almost 670 fs. Simulation is done according to equations from Section 4.3. Peak irradiance is 6.40 TW/cm^2, laser wavelength 800 nm, pulse width (full width at half maximum [FWHM] of irradiance) 706.45 fs, peak photoionization rate (cosine model) 2.0519×10^{16} fs/cm^3, and the peak rate of avalanche ionization is 4.2225×10^{18} fs/cm^3.

energy is mainly accumulated in the form of low-energy electron excitations (trapped localized electrons and electron–hole plasma) at the leading front of a laser pulse. Electron–particle collision time decreases fast with the increase in conduction-band electron density, but it can take hundreds of femtoseconds before it becomes short enough to provide electron thermalization and heating by electron–phonon–photon collisions. Therefore, the effects driven by electron–particle collisions can be negligible for significant part of leading front of a laser pulse and even for the most part of laser pulse if it is ultrashort (Figure 4.4). Time of their buildup is determined by the rate of conduction-band electron generation by the photoionization. Starting from the collision stage, laser energy is mainly deposited by one-photon absorption of the electron–hole plasma and is accumulated in the form of high kinetic energy of electrons. The delay between the fast electron excitation, the collision processes, and the slower atomic effects results in intensive heating of electron subsystem while atomic system can stay essentially cold (Kaiser et al. 2000, Sundaram and Mazur 2002, Gamaly 2011).

By the beginning of the stage of slow atomic effects, a significant density of hot electron–hole plasma is reached, energy of laser radiation is effectively absorbed by both interband and intraband transitions, the plasma gets to a thermal quasiequilibrium, and effective transfer of absorbed laser energy to phonons begins (Kaiser et al. 2000, Gamaly 2011, Balling and Schou 2013). If laser pulse is still not over and its peak intensity is above LID threshold, the slower effects start developing, for example, heating followed by melting, thermal stresses, and evaporation that change material structure (Woods 2003, Bulgakova et al. 2010, Balling and Schou 2013).

Chapter 4

4.3   Nonlinear Absorption

The overview of the microscopic laser-induced effects (Section 4.2) suggests that nonlinear absorption includes several very different contributions. Each of them is characterized by *absorption rate*, that is, the number of absorbed photons per unit time per unit volume. In general case, the rate can be represented as follows:

$$R = Bn(I)p(I),\qquad(4.3)$$

where

   B is a constant coefficient
   n is the density of absorbing particles (measured in cm^{-3})
   p is the average probability of absorption by a single particle per one unit of time (measured in inverse second)

The averaging is done over 1 period of electric field of laser radiation and over all absorbing particles of interaction volume. The contributions to nonlinear absorption can be classified according to Equation 4.3:

1. For the photoionization process, the absorbing particles are valence-band electrons whose density is about 10^{22}–10^{23} cm^{-1} (Ashcroft and Mermin 1976). It is believed that the density of conduction-band electrons required to produce LID (it is also referred to as *critical electron density*) is 100–1000 times less than the valence-band electron density (Stuart et al. 1996). Therefore, even if the photoionization is assumed to produce the critical electron density, it does not significantly reduce the density of available absorbing particles, and laser-induced variations of n can be neglected in Equation 4.3. This implies that the major contribution to nonlinear scaling of the photoionization rate is done by the absorption probability p as long as the photoionization results in excitation of a small part of initial valence-band electrons.

2. For the photoexcitations that involve electron transitions to or from bound defect states (e.g., self-trapped excitons or color centers), the density n of Equation 4.3 is density of the defects. For a high-quality transparent material, initial value of the density of the defects is so small that generation of laser-induced defects results in significant increase of n. Therefore, laser-induced variations of n are large enough to influence the nonlinear scaling of the rate in Equation 4.3 in case of the defect photoexcitation. On the other hand, absorption probability p can also be a nonlinear function of irradiance if energy gap between valence/conduction band and defect levels is larger than energy of single laser photon. That is the case for most wide-band-gap materials where two- or three-photon absorption is required to induce defect-to-band or band-to-defect electron transitions (Emmert et al. 2010, Menoni et al. 2012). Therefore, in case of the defect-related photoexcitation, both terms of Equation 4.3 provide significant contributions to nonlinear scaling.

3. For electron–hole plasma effects, the major absorption process is the one-photon absorption due to electron–phonon–photon collisions in the conduction band. Probability of the absorption scales linearly with laser irradiance, but density of the absorbing particles (that are conduction-band electrons) starts with almost zero and significantly grows during the ionization process. Therefore, the major contribution to nonlinear scaling of R is done by the term n of Equation 4.3.

In the following, we consider each of the contributions to nonlinear absorption bearing in mind that those three processes are driven by two competing modes of electronic dynamics. The photoionization is attributed to coherent electron oscillations driven by electric field of laser radiation (Keldysh 1965, Gruzdev 2004) that are frequently referred to as the *Bloch oscillations* (Nazareno and Gallardo 1989, Sudzius et al. 1998, Gruzdev 2004). The impact ionization is completely driven by laser-induced electron–particle collisions. Period of the electron oscillations is the same as period of laser light T_0 (e.g., 2.6 fs at laser wavelength 800 nm). If collision time significantly exceeds that period ($T_0 \ll \tau_{ec}$), the oscillatory motion dominates, and major contribution to the nonlinear absorption comes from the photoionization (Section 4.3.2) and the photoexcitation of defects (Section 4.3.3). The oscillations also result in modifications of the original energy bands and electronic properties of solids that cannot be neglected for proper description of the photoexcitation and nonlinear propagation effects at irradiance close to LID threshold (Keldysh 1965). If collision time approaches period of electron oscillations, electron–particle collisions significantly disturb the coherent oscillations and drive the electron subsystem toward chaotic collision-driven dynamics (Section 4.3.4). With the increase in electron–hole plasma density, collision time can become smaller than laser period, and the collisions can reduce the photoionization rate by dephasing the electron oscillations (Du et al. 1996). This process establishes domination of the avalanche ionization over the collision-suppressed photoionization.

Interrelation between those two types of the electron dynamics and domination of particular absorption mechanisms are still under discussion. In the following, we present a traditional view of the interplay between the absorption processes and their contributions to the nonlinear absorption.

4.3.1 Photo-Ionization and Multiphoton Absorption

Therefore, the major contribution to the nonlinear absorption of the leading front of a laser pulse is done by noncollision interband electron transitions induced by electric field of laser light. Since energy of laser photons is smaller than band gap of wide-bandgap materials for visible, UV, and IR light, the most natural concept is simultaneous absorption of $N > 1$ photons (Figure 4.5) whose total energy is enough to bridge the energy gap between initial states in the valence band and final states in the conduction band (Braunstein 1962, Bonch-Bruevich and Khodovoi 1965, Nathan et al. 1985). This type of interband electron transitions and the related absorption are referred to as *multiphoton*. An alternative concept considers tunneling of electrons from bound (valence band) to mobile state (conduction band) due to strong distortions of crystal potential by time-dependent electric field of laser radiation (Figure 4.5). This effect is referred to as *tunneling ionization*. It does not produce immediate light absorption since

Chapter 4

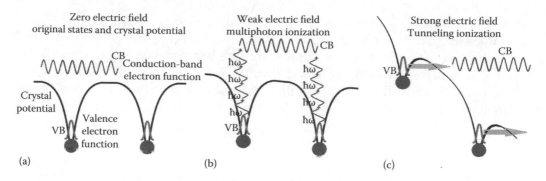

FIGURE 4.5 Illustration of the unperturbed electron states and crystal potential (a) as well as the traditional concept of the multiphoton (b) and tunneling (c) regimes of the photoionization that neglects laser-driven electron oscillations. CB is for conduction band, and VB is for valence band.

the energy of a tunneling electron is constant (Landau and Lifshitz 1958), but it generates conduction-band electrons that absorb light.

The multiphoton absorption is frequently characterized by coefficient of N-photon absorption α_N that can be evaluated from rate of electron transitions w (number of electron transitions per unit volume per second) and laser irradiance I (Nathan et al. 1985):

$$\alpha_N = \frac{2N\hbar\omega w}{I^N}, \tag{4.4}$$

where one-photon absorption is also included by assuming $N = 1$. Therefore, calculation of the absorption coefficient is reduced to evaluation of the transition rate w by methods of quantum mechanics (Landau and Lifshitz 1958). The calculations involve a *dispersion* (or *energy–momentum*) *relation* for electron–hole pairs generated by the transitions. It is the sum of energy of an electron $\varepsilon_C\left(\vec{p}_0\right)$ of the conduction band and energy $-\varepsilon_V\left(\vec{p}_0\right)$ of the corresponding hole produced by departure of the electron from the valence band considered as function of their quasimomentum p_0:

$$\varepsilon\left(\vec{p}_0\right) = \varepsilon_C\left(\vec{p}_0\right) - \varepsilon_V\left(\vec{p}_0\right). \tag{4.5}$$

For direct-gap cubic crystals, the minimum of the conduction band and the maximum of the valence band are in the center of the *Brillouin zone* (BZ) of quasimomentum space (see Ashcroft and Mermin 1976, Kittel 1985 for definition and properties of the zone), that is, at the point $p_0 = 0$. The most frequently utilized dispersion relations (Keldysh 1965, Jones and Reiss 1977, Nathan et al. 1985, Gruzdev 2007) are as follows (for simplicity, we assume that the bands are isotropic, and quasimomentum components are directed along the crystal principle axes [Ashcroft and Mermin 1976, Kittel 1985]):

- The parabolic relation

$$\varepsilon^P\left(\vec{p}_0\right) = \Delta + \frac{\vec{p}_0^2}{2m} = \Delta\left(1 + \frac{p_{0x}^2 + p_{0y}^2 + p_{0z}^2}{2m\Delta}\right), \tag{4.6}$$

where Δ is the band gap located at the point $p_0 = 0$ for the direct-gap crystals. This relation is a good approximation in a vicinity of the center of the BZ and is rigorous for a free electron (Ashcroft and Mermin 1976, Kittel 1985). Here, m is a *reduced effective electron–hole mass* that is evaluated from effective masses of an electron m_C at the bottom of the conduction band and a hole m_V at the top state of the valence band[*]:

$$\frac{1}{m} = \frac{1}{m_C} + \frac{1}{m_V}.$$
(4.7)

- The *Kane-type relation* (Kane 1957, Keldysh 1965)

$$\varepsilon^K\left(\vec{p_0}\right) = \Delta\sqrt{1 + \frac{\vec{p_0}^2}{m\Delta}} = \Delta\sqrt{1 + \frac{p_{0x}^2 + p_{0y}^2 + p_{0z}^2}{m\Delta}}.$$
(4.8)

It is a good approximation for real dispersion relations of narrow-gap semiconductors with strong spin—orbit interaction in vicinity of the center of the BZ.
- The *cosine relation* (Ashkroft and Mermin 1976)

$$\varepsilon_{CB}^{\cos 1}\left(\vec{p}\right) = \Delta \cdot \left(1 + \frac{\hbar^2}{m\Delta d^2} - \frac{\hbar^2}{m\Delta d^2}\cos\left(\frac{d}{\hbar}p_x\right)\cos\left(\frac{d}{\hbar}p_y\right)\cos\left(\frac{d}{\hbar}p_z\right)\right),$$
(4.9)

where d is the crystal-lattice constant (Ashcroft and Mermin 1976). It is valid for the entire BZ (including edges) of volume-centered cubic crystals under the approximations of tightly bound electrons and the dominating interaction of each atom with its nearest neighbors (that is the case for ionic crystals—Ashcroft and Mermin 1976).

In 1960s and early 1970s, analytical calculations of the multiphoton-absorption coefficient were done for $N = 2$, 3, and 4 based on the quantum perturbation theory for the parabolic relation (Equation 4.6) (Braunstein 1962, Yee 1971, 1972, Nathan et al. 1985). The multiphoton and tunneling effects were treated as completely independent, and analytical derivations for a larger number of simultaneously absorbed photons were not feasible due to significant technical problems. In 1964, Keldysh developed a unique approach that fixed the problems of the perturbation methods and derived the *Keldysh formula* for the relation (Equation 4.8) to evaluate the photoionization rate for arbitrary number of simultaneously absorbed laser photons in semiconductors (Keldysh 1965). For monochromatic laser radiation, he demonstrated that the multiphoton and tunneling effects were two limiting regimes of a general photoionization process. Currently, the Keldysh formula is the most comprehensive

---

[*] Effective mass determines local curvature of the band in energy-momentum space (Ashcroft 1976). The relation Equation 4.7 is for the center of the BZ. Moreover, one can assume $m^* = m_C$ with pretty good accuracy, since the highest valence band is almost flat for most transparent materials (e.g., see Sirdeshmukh 2001), that is, $m_V \gg m_C$.

Chapter 4

and popular tool for simulation of laser-induced electron excitations in various solids. It contains major exponential contribution:

$$w_K = 2 \cdot \frac{2\omega}{9\pi} \cdot \left(\frac{\sqrt{1+\gamma^2}}{\gamma} \cdot \frac{m\omega}{\hbar} \right)^{3/2} \cdot Q_K \left(\gamma, \frac{\tilde{\Delta}_{NP}}{\hbar\omega} \right) \cdot \exp\left[-\pi \left\langle \frac{\tilde{\Delta}_{NP}}{\hbar\omega} +1 \right\rangle \cdot \frac{K(\phi)-E(\phi)}{E(\theta)} \right], \quad (4.10)$$

and a slowly varying amplitude introduced as follows (Keldysh 1965)

$$Q_K(\gamma,x) = \sqrt{\frac{\pi}{2 \cdot K(\theta)}} \cdot \sum_{n=0}^{\infty} \exp\left\{ -\pi \cdot \frac{K(\phi)-E(\phi)}{E(\theta)} \cdot n \right\} \cdot \Phi \left\{ \sqrt{\frac{\pi^2 \left(\langle x+1 \rangle - x + n \right)}{2 \cdot K(\theta) \cdot E(\theta)}} \right\}. \quad (4.11)$$

Here, $K(x)$ and $E(x)$ denote the complete elliptic integrals (Abramowitz and Stegun 1972) whose arguments

$$\theta = \frac{1}{\sqrt{1+\gamma^2}}, \quad \phi = \frac{\gamma}{\sqrt{1+\gamma^2}}, \quad (4.12)$$

are expressed via the *Keldysh* (or *adiabatic*) *parameter* (Keldysh 1965):

$$\gamma = \frac{\omega\sqrt{m\Delta}}{eF}, \quad (4.13)$$

where
   Δ is the original band gap (Figure 4.5)
   $\omega = 2\pi/T_0$ is the angular frequency of laser light
   e is the elementary charge (1.6022×10^{-16} C)
   F is the amplitude of electric field of laser radiation
   The angle brackets $\langle x \rangle$ denote the integer part of x in Equations 4.4 and 4.5
   $\Phi(x)$ stays for the Dawson integral (Abramowitz and Stegun 1972):

$$\Phi(x) = \int_0^x \exp\left(\xi^2 - x^2 \right) d\xi. \quad (4.14)$$

The Keldysh formula includes laser-modified band gap (also referred to as *effective band gap*):

$$\tilde{\Delta}_{NP} = \frac{2}{\pi} \Delta \cdot \left[\frac{\sqrt{1+\gamma^2}}{\gamma} \cdot E\left(\frac{1}{\sqrt{1+\gamma^2}} \right) \right]. \quad (4.15)$$

Compared to original Keldysh's expression (Keldysh 1965), Equation 4.10 includes the factor of 2 to account for spin degeneracy of electrons in a solid, and Equation 4.12 includes corrected argument of the Dawson function to fix the misprint in the original Keldysh formula (Keldysh 1965).

Keldysh reduced the formula given by Equations 4.10 and 4.11 to simpler forms in two particular cases (Keldysh 1965). First, by assuming $\gamma \gg 1$, he arrived at the following equation:

$$
w_{MPH} = 2 \cdot \frac{2\omega}{9\pi} \cdot \left(\frac{m\omega}{\hbar}\right)^{3/2} \Phi\left\{\sqrt{2\left(\left\langle \frac{\tilde{\Delta}_{MP}}{\hbar\omega} + 1\right\rangle - \frac{\tilde{\Delta}_{MP}}{\hbar\omega}\right)}\right\}
$$

$$
\times \exp\left[2\left\langle \frac{\tilde{\Delta}_{MP}}{\hbar\omega} + 1\right\rangle \cdot \left(1 - \frac{e^2 F^2}{4m\omega^2\Delta}\right)\right] \left(\frac{e^2 F^2}{16m\omega^2\Delta}\right)^{\left\langle \frac{\tilde{\Delta}_{MP}}{\hbar\omega} + 1\right\rangle}. \tag{4.16}
$$

The effective band gap is represented by the sum of the original band gap and ponderomotive energy (i.e., average energy of oscillations driven by electric field F at frequency ω):

$$
\tilde{\Delta}_{MP} = \Delta \cdot \left[1 + \frac{1}{4\gamma^2}\right] = \Delta \cdot \left[1 + \frac{e^2 F^2}{4\omega^2 m\Delta}\right]. \tag{4.17}
$$

The last factor of Equation 4.16 includes the power dependence I^N on laser intensity, where

$$
N = \left\langle \frac{\tilde{\Delta}_{MP}}{\hbar\omega} + 1\right\rangle \tag{4.18}
$$

is the minimum number of laser photons to bridge the effective band gap (Equation 4.17). The same power scaling is obtained by calculating the rates of N-photon transitions based on the perturbation theory (Braunstein 1962, Yee 1971, 1972, Nathan et al. 1985). Therefore, this regime of the photoionization realized for $\gamma \gg 1$ should be attributed to interband transitions due to the multiphoton absorption. Large values of the Keldysh parameter are produced by large values of the ratio ω/F, that is, by high laser frequency and low laser intensity.

The second simple case of the general formula (Equation 4.10) is obtained by assuming $\gamma \ll 1$:

$$
w_{TUN} = 2 \cdot \frac{2}{9\pi^2} \frac{\Delta}{\hbar} \cdot \left(\frac{m\Delta}{\hbar^2}\right)^{3/2} \left(\frac{e\hbar F}{m^{1/2}\Delta^{3/2}}\right)^{5/2} \exp\left[-\frac{\pi}{2} \frac{m^{1/2}\Delta^{3/2}}{e\hbar F} \cdot \left(1 - \frac{m\omega^2\Delta}{8e^2 F^2}\right)\right]. \tag{4.19}
$$

Equation 4.19 includes the exponent term characteristic of tunneling under action of electric field F (Keldysh 1958, 1965). Therefore, the regime of the photoionization realized for small values of the Keldysh parameter should be associated with the tunneling mechanism. It takes place at high irradiance when ratio ω/F is small.

Chapter 4

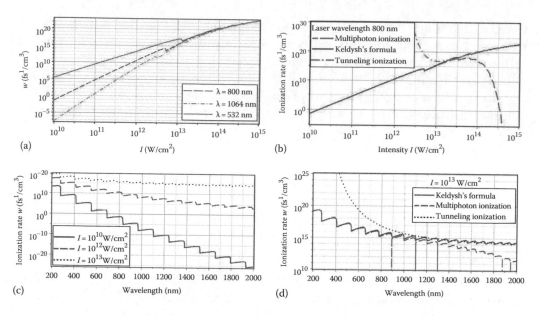

FIGURE 4.6 Photoionization rate evaluated by the Keldysh formula as a function of laser intensity I and laser wavelength λ. Calculations have been performed for material parameters of NaCl: effective mass is 0.6 of free-electron mass, and band gap is 8.97 eV. Depicted are (a) intensity scaling of the ionization rate for three values of laser wavelength (532, 800, and 1064 nm); (b) intensity scaling of the ionization rate for the Keldysh formula and its limiting cases—tunneling ionization and multiphoton ionization at fixed wavelength 800 nm; (c) wavelength scaling of the ionization rate for three values of laser intensity (0.01, 1.0, and 10.0 TW/cm²); (d) wavelength scaling of the ionization rate for the Keldysh formula, tunneling ionization, and multiphoton ionization at fixed intensity 10 TW/cm². (With kind permission from Springer Science+Business Media: *Alkali Halides: A Handbook of Physical* Properties, 2001, Sirdeshmukh, D.B., Sirdeshmukh, L., and Subhadra, K.G.)

The photoionization rate given by the Keldysh formula (Equation 4.10) exhibits special features as a function of laser irradiance and wavelength (Figure 4.6): specific singularities (referred to as *the Keldysh singularities*), gradual unlimited increase with irradiance, and step-wise decrease with wavelength. They can be interpreted in terms of laser-driven Bloch oscillations of electrons (Gruzdev 2004, 2007). That interpretation also explains the transition from multiphoton to tunneling ionization and predicts domain of laser and material parameters that provide good accuracy of evaluation of photoionization rate by the Keldysh formula (Equation 4.10). To develop the interpretation, we employ a frequently utilized *two-band approximation* in the following, that is, assume the electron transitions to happen from the highest valence band to the lowest conduction band.

Let us consider a direct-gap transparent crystal of cubic symmetry subjected to the action of electric field of monochromatic laser light at frequency ω (Keldysh 1965):

$$\vec{E}(t) = \vec{F}\cos(\omega t).$$

(4.20)

Its electrons and holes are driven by the superposition of intrinsic crystal potential and the time-dependent potential of laser electric field (Equation 4.20). Therefore, the absorption and optical response of the solid should be produced by laser-perturbed electron subsystem. At low intensity, the perturbation by the radiation field (Equation 4.20) is negligible compared to the unperturbed crystal potential, and only one-photon absorption is feasible in this case. It must be considered for electron transitions between unperturbed states. The states can be described, for example, by the Bloch functions frequently utilized for the analysis of electrons in crystals (Ashcroft and Mermin 1976, Kittel 1985):

$$\Phi\left(\vec{p}_0,\vec{r},t\right)=\varphi_b\left(\vec{p}_0,\vec{r}\right)\exp\left\{-i\frac{\varepsilon_b\left(\vec{p}_0\right)}{\hbar}t\right\},\quad \text{where } \varphi_b\left(\vec{p}_0,\vec{r}\right)=u_b\left(\vec{p}_0,\vec{r}\right)\exp\left\{\frac{i}{\hbar}\vec{r}\vec{p}_0\right\},$$

$$(4.21)$$

where
    u_b is the amplitude of the function that is periodic in space with period of the crystal
    p is the electron quasimomentum
    ε is the electron energy
    Subscript b denotes an energy band ($b = c$ for the conduction band and $b = v$ for the valence band)

At the intensity characteristic of nonlinear effects, the perturbation of electron states by time-dependent electric field of laser light must be taken into account. By neglecting electron–particle collisions (that is true as long as collision time is much longer than period of electric-field undulations), the following time-dependent function is obtained from the time-dependent Schrodinger equation for electron states of the crystal subjected to action of the monochromatic laser light (Equation 4.20):

$$\psi\left(\vec{r},t\right)=u_b\left(\vec{p}[t],\vec{r}\right)\exp\left(\frac{i}{\hbar}\vec{p}[t]\vec{r}-\frac{i}{\hbar}\int_0^t\varepsilon_b\left(\vec{p}[\tau]\right)d\tau\right).$$

$$(4.22)$$

This function is similar to the unperturbed Bloch one (Equation 4.21) but includes time-dependent quasimomentum modified by electric field of light (Keldysh 1965, Gruzdev 2004):

$$\vec{p}(t)=\vec{p}_0+\frac{e\vec{F}}{\omega}\sin\left(\omega t\right).$$

$$(4.23)$$

Each function (Equation 4.22) is not steady, but its instant space distribution is described by a Bloch function (Equation 4.21) at any fixed time. Therefore, the noncollision electron dynamics driven by laser radiation can be represented as a continuous transition from one Bloch state to another within an energy band (e.g., valence) as long as interband electron transitions are not too intensive. Periodic dependence of u_b on quasimomentum (Ashcroft and Mermin 1976, Kittel 1985) allows interpreting the dynamics of the laser-driven electrons described by functions (Equation 4.22) as specific Bloch oscillations in momentum space (Gruzdev 2004).

Since probability of the multiphoton absorption is the highest for the electron states near the minimum of the conduction band, it is reasonable to consider small values of p_0

Chapter 4

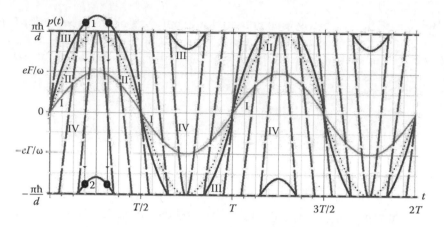

FIGURE 4.7 Trajectories of laser-driven Bloch oscillations of electrons in time-momentum space. The horizontal dashed lines depict edges of the BZ. Solid line curve I corresponds to low-amplitude harmonic oscillations, dotted curve II depicts harmonic oscillations with the amplitude given by Equation 4.26, solid curve III depicts the oscillations affected by the Bragg-type reflections at the zone edges, and dashed curve IV depicts almost saw-type oscillations with multiple perturbations by the Bragg reflections. The transition from state 1 to state 2 illustrates the Bragg-type reflections.

(Keldysh 1965). As long as oscillation amplitude $\delta_F = eF/\omega$ is smaller than the half-width of the BZ $\delta_B = \pi\hbar/d$, relations (Equations 4.22 and 4.23) suggest that quasimomentum harmonically varies at frequency of laser radiation ω within the BZ (Figure 4.7, curve I) from state

$$\psi_A\left(\vec{r},t\right) = u_b\left(\vec{p}_0 - \frac{e\vec{F}}{\omega},\vec{r}\right)\exp\left(\frac{i}{\hbar}\left[\vec{p}_0 - \frac{e\vec{F}}{\omega}\right]\vec{r} - \frac{i}{\hbar}\int_0^t \varepsilon_b\left(\vec{p}_0 + \frac{e\vec{F}}{\omega}\sin\tau\right)d\tau\right) \qquad (4.24)$$

to state

$$\psi_B\left(\vec{r},t\right) = u_b\left(\vec{p}_0 + \frac{e\vec{F}}{\omega},\vec{r}\right)\exp\left(\frac{i}{\hbar}\left[\vec{p}_0 + \frac{e\vec{F}}{\omega}\right]\vec{r} - \frac{i}{\hbar}\int_0^t \varepsilon_b\left(\vec{p}_0 + \frac{e\vec{F}}{\omega}\sin\tau\right)d\tau\right). \qquad (4.25)$$

For $\delta_F/\delta_B < 1$, the entire range of the states between ψ_A and ψ_B is inside the BZ and does not include all available states in the crystal. Maximum field amplitude of the harmonic oscillations is reached at $\delta_F/\delta_B = 1$ and corresponds to the motion of each oscillating electron across the entire BZ and passing through all physically different states (Figure 4.7). This happens at the following value of electric-field amplitude (Gruzdev 2004)

$$F_{TR} = \frac{\pi\hbar\omega}{ed}, \qquad (4.26)$$

at a given frequency and lattice constant. For NaCl ($d = 0.5628$ nm; refractive index $n_0 = 1.5356$ at wavelength 800 nm) (Sirdeshmukh et al. 2001), Equation 4.26 delivers intensity $I_{TR} = 5.087$ TW/cm^2. The value of the adiabatic parameter (Equation 4.13) corresponding to the field of Equation 4.26 reads as

$$\gamma_{TR} = \frac{d\sqrt{m\Delta}}{\pi\hbar}.$$

(4.27)

For the considered material parameters of NaCl (reduced effective electron–hole mass is 0.6 of free-electron mass; intrinsic bad gap is 8.97 eV (Sirdeshmukh et al. 2001)), it gives $\gamma = 1.5056$.

As soon as electric field exceeds F_{TR}, the range of states passed by the oscillating electrons exceeds the size of the BZ. It means that after having passed through all states inside the zone, each electron has to occupy a state outside the zone at certain instant (state 1 in Figure 4.7). That state is similar to the state inside the zone at its opposite edge (state 2 in Figure 4.7) (Ashcroft and Mermin 1976, Kittel 1985). Therefore, the transition to the state 1 is considered as a jump to a physically equivalent state 2 inside the zone near its opposite edge. This effect is referred to as the Bragg reflection of electrons at the edges of the BZ (Ashcroft and Mermin 1976, Nazareno and Gallardo 1989, Sudzius et al. 1998). It takes place each time when an oscillating electron crosses the zone edge. The reflection is 100% if the quasimomentum component normal to electric field is zero and the electric field is directed along one of the principal axes of the crystal. With the increase in electric field much above F_{TR}, the number of the reflections per period of electric field increases, and variations of quasimomentum approach saw-like oscillations (Gruzdev 2004) with period $T = 2\pi\hbar/(eFd)$ and slope given by the product of electric field and electron charge eF (Keldysh 1958, Nazareno and Gallardo 1989, Sudzius et al. 1998) (Figure 4.7).

Those oscillations of quasimomentum within the BZ do not always correspond to the motion of electron's mass center in real space, for example, they can be associated with a *breathing mode* of an electron wave packet with fixed position of its mass center (Sudzius et al. 1998). It is notable that a similar transition from the harmonic to the saw-type oscillations can be induced by variations of laser frequency at fixed (and rather high) electric field since amplitude of the oscillations δ_F is determined by the ratio F/ω. In a similar way to Equation 4.26, transition frequency ω_{TR} can be defined from the condition $\delta_F/\delta_B = 1$.

Therefore, perturbation of crystal potential by time-dependent electric field (Equation 4.20) results in electron oscillations in quasimomentum space. This is the fastest perturbation of the electronic subsystem of a solid introduced long before other perturbations (e.g., due to electron–hole plasma) come in to play. Laser radiation interacts with the perturbed crystal, and response to the laser action is produced by the perturbed crystal rather than by a crystal with original properties. In particular, the oscillations lead to modification of energy bands. Since instant energy of the nonsteady state (Equation 4.22) varies in time, initial amount of energy of each oscillating electron (that is given by the unperturbed band structure) must be modified by amount of average energy of the oscillations (referred to as *ponderomotive potential*). For monochromatic radiation (Equation 4.20) and any dispersion relation, the modified energy can be evaluated by averaging instant energy $\varepsilon_b(\vec{p}[t])$ of each electron over one period T_0 of the field oscillations (Keldysh 1965). The value of the average energy considered as the function of initial momentum p_0 forms *effective bands* instead of the original energy bands. Correspondingly, dispersion relation of an oscillating electron–hole pair should be considered for the effective bands and should be evaluated from the unperturbed energy–momentum relation as follows (Keldysh 1965):

$$\varepsilon_{eff}\left(\vec{p}_0\right)=\frac{1}{T}\int_0^T \varepsilon\left[\vec{p}_0-\frac{e\vec{F}}{\omega}\sin\left(\omega t\right)\right]dt. \tag{4.28}$$

Effective band gap is the minimum of $\varepsilon_{eff}\left(\vec{p}_0\right)$ within the BZ. Therefore, a minimum number of simultaneously absorbed photons should supply enough energy to bridge the original band gap and to provide the average oscillation energy for laser-driven motion of electron–hole pairs. This can be done by bridging the effective band gap that includes intensity-dependent ponderomotive potential. Table 4.1 summarizes effective dispersion relations and band gaps for the three relations given by Equations 4.6, 4.8, and 4.9.

Dependence of effective energy–momentum relation of electron–hole pairs on laser parameters is essentially dictated by unperturbed dispersion relation. For example, the effective relation stays parabolic with the increase in laser intensity for the initial parabolic relation (Equation 4.6) (Figure 4.8). Corresponding effective band gap monotonously grows with intensity according to Equation 4.17 (Figure 4.9). The simple type of

Table 4.1 Effective Dispersion Relations and Effective Band Gaps Induced by Monochromatic Laser Radiation for Several Dispersion Relations

Original Dispersion Relation	Effective Dispersion Relation	Effective Band Gap
Equation 4.6	$\varepsilon_{eff}^P\left(\vec{p}_0\right)=\Delta\left(1+\dfrac{1}{4\gamma^2}+\dfrac{p_{0x}^2+p_{0y}^2+p_{0z}^2}{2m\Delta}\right)$	$\tilde{\Delta}^P=\Delta\left(1+\dfrac{1}{4\gamma^2}\right)=\Delta+\dfrac{e^2F^2}{4\omega^2m\Delta}$
Equation 4.8	$\varepsilon_{eff}^K\left(\vec{p}_0\right)=\Delta\dfrac{2}{\pi}\left[\dfrac{\sqrt{1+\gamma^2}}{\gamma}E\left(\dfrac{1}{\sqrt{1+\gamma^2}}\right)+\dfrac{p_{0x}^2}{2m\Delta}\dfrac{\gamma}{\sqrt{1+\gamma^2}}E\left(\dfrac{1}{\sqrt{1+\gamma^2}}\right)+\dfrac{p_{0y}^2+p_{0z}^2}{2m\Delta}\dfrac{\gamma}{\sqrt{1+\gamma^2}}K\left(\dfrac{1}{\sqrt{1+\gamma^2}}\right)\right]$	$\tilde{\Delta}^K=\Delta\dfrac{2}{\pi}\dfrac{\sqrt{1+\gamma^2}}{\gamma}E\left(\dfrac{1}{\sqrt{1+\gamma^2}}\right)$
Equation 4.9	$\varepsilon_{eff}^{cos}\left(\vec{p}_0\right)=\Delta\left[1+\dfrac{\hbar^2}{md^2\Delta}\left(1-J_0\left(\chi\right)\cos\left(\dfrac{p_{0x}d}{\hbar}\right)\times\cos\left(\dfrac{p_{0y}d}{\hbar}\right)\cos\left(\dfrac{p_{0z}d}{\hbar}\right)\right)\right]$	$\tilde{\Delta}^{cos}\left(\vec{p}_0\right)=\Delta\left[1+\dfrac{\hbar^2}{md^2\Delta}\left(1-J_0\left(\chi\right)\right)\right]$

(see the following notation in this section)

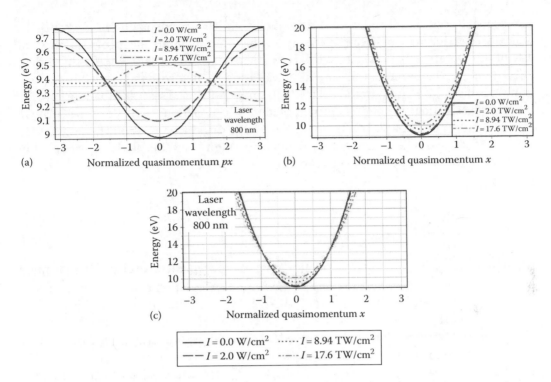

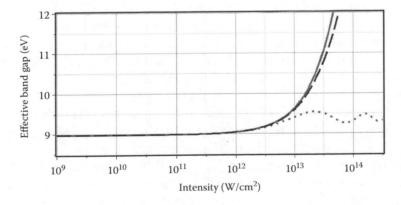

FIGURE 4.8 Variations of effective bands with the increase of laser intensity for the dispersion relations listed in Table 4.1: (a) cosine relation, (b) parabolic relation, (c) Kane-type relation. Laser wavelength is 800 nm, effective mass is 0.6 of free-electron mass, and initial band gap is 8.97 eV. On each part of this figure, a solid curve depicts the original bands (laser intensity is zero), dashed curve depicts effective bands at intensity 2×10^{12} W/cm^2, dotted curve depicts the effective bands at 8.94×10^{12} W/cm^2, and dash-dotted curve at intensity 17.6×10^{12} W/cm^2.

FIGURE 4.9 Variations of the effective band gap for the effective bands obtained from the original parabolic (solid curve), the Kane-type (dashed curve), and the cosine (dotted curve) energy–momentum relations. Laser and material parameters are the same as for Figure 4.8.

Chapter 4

the laser-induced modification of the parabolic effective bands is due to the independence of effective mass on electron energy for the original bands that provides constant rate of modification over the entire BZ. For dispersion relations with energy-dependent effective mass, the amount of laser-induced variation of effective bands depends on position within the BZ (i.e., on initial value of quasimomentum). This leads to more complicated expression for the ponderomotive potential and variations of shape of the effective energy bands. For example, the effective Kane-type dispersion relation (Equation 4.15) becomes flatter with the increase in laser intensity (Figure 4.8) due to the dependence of its effective mass on quasimomentum/energy.

$$m_K\left(\vec{p}_0\right)=m\left[1+\frac{\vec{p}_0^{\,2}}{m\Delta}\right]^{3/2}; \quad m_K\left(\varepsilon\right)=m\left[\frac{\varepsilon}{\Delta}\right]^3, \tag{4.29}$$

where $m=m_K\left(\vec{p}_0=0\right)=m_K\left(\varepsilon=\Delta\right)$ is the value of reduced effective electron–hole mass in the center of the BZ. The flattening of the Kane-type relation results from the faster upshift of the center of the original band (where the reduced effective mass is the smallest) and slower upshift of its peripheral part (where the reduced effective mass is larger) along energy axis. The flattening of the effective relation in energy-momentum space is quite mild because the effective mass (Equation 4.29) is positive over the entire BZ. For the cosine relation given by Equation 4.9, the effective mass changes sign from positive in the center to negative at the edges of the zone. This effect leads to very strong modification of the effective bands including flattening (Figure 4.8) because change of effective mass sign means switching from upshift to downshift, that is, different parts of the effective band shift in opposite directions. Corresponding effective band gap demonstrates damped oscillations with the increase in laser intensity (Figure 4.9).

In the connection with those modifications of the effective dispersion relation, one should note that unperturbed upper valence band of most wide-band-gap materials is quite flat (Sirdeshmukh et al. 2001) and stays flat (or becomes flatter) during laser-induced perturbation. Therefore, major contribution to the variations of shape of the effective dispersion relations is done by the lowest effective conduction band of the crystals. This implies that the plots of Figure 4.8 can be considered as illustrations of the variations of the effective conduction band. On the other hand, absolute values of up- and downshifts of the effective valence band can be quite large and can make significant contributions to the variations of the effective band gap.

Variations of the effective band gap with laser intensity explain the Keldysh singularities in Figure 4.6. With the increase in the effective energy gap between the laser-modified valence and conduction bands (Figure 4.10), the initial and final states involved in the N-photon interband transitions gradually shift toward the center of the BZ. At certain laser intensity, the shift arrives at the situation when the effective band gap rigorously equals the total energy of N laser photons, but this happens for the states with $p_0=0$, which are empty (since their density of states is zero—see Ashcroft and Mermin 1976, Kittel 1985). Since no electrons can make the transitions under those conditions, the N-photon ionization rate drops to zero and makes the Keldysh singularity, and absorption of $N+1$ photons starts to make the major contribution to the transitions, that is, the order of the multiphoton absorption increases by 1 right after the singularity.

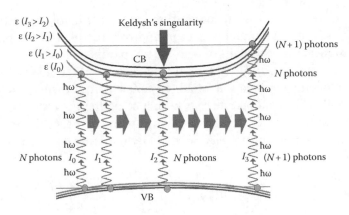

FIGURE 4.10 Illustration of the mechanism of the Keldysh-type singularities on dependence of the photoionization rate on laser parameters. N is number of laser photons absorbed before the singularity, and Δ shows the unmodified band gap.

Therefore, photoionization takes place in a laser-perturbed crystal. Perturbation comes through the laser-driven electron oscillations and results in the modification of energy bands of the crystal. For a proper evaluation of the photoionization rate, the modification of band structure should be taken into account. The Keldysh formula is derived by considering electron transitions between the perturbed electron states given by Equation 4.22 (Keldysh 1965). Therefore, the Keldysh approach automatically incorporates the laser-induced modifications of energy bands of a crystal that determines its multiple advantages over the classical perturbation methods.

The Bloch oscillations also provide an interpretation of the transition between the multiphoton and the tunneling ionization regimes (Gruzdev 2004). One can easily note that small-amplitude electron oscillations correspond to large value of the ratio ω/F that delivers large values of the Keldysh adiabatic parameter. Since large values of the adiabatic parameter are associated with the domination of the multiphoton ionization, small-amplitude electron oscillations should be attributed to the multiphoton regime of the photoionization. In turn, large values of the oscillation amplitude lead to small value of the ratio ω/F and, correspondingly, small value of the Keldysh adiabatic parameter. Since tunneling ionization dominates in this case, the saw-type electron oscillations should be associated with that ionization regime. The latter association is also supported by the fact that rigorous saw-type oscillations are produced by *dc* electric field (Gruzdev 2004) for which tunneling is the only noncollision ionization mechanism.

Together with the interpretation of the Keldysh formula, the concept of the Bloch oscillations allows determining the limits of validity of the formula (Gruzdev 2004, 2007). It is frequently believed that the Keldysh formula is a universal expression for the photoionization rate, and it is utilized for arbitrary materials including wide-band-gap crystal and amorphous glasses (Stuart et al. 1996, Kaiser et al. 2000, Wood 2003, Mao 2004, Rethfeld 2004, Starke et al. 2004, Jia 2006, Bulgakova et al. 2010, Jasapara 2001, Balling and Schou 2013). This is true only if oscillation amplitude eF/ω is small enough, that is, for $\delta_F/\delta_B \ll 1$ to keep entire range of states passed by oscillating electrons in the vicinity of the center of the BZ. For that domain of quasimomentum space, any dispersion relation is well approximated by the parabolic one (Equation 4.6) (Ashcroft and Mermin 1976, Kittel 1985). As just discussed, this situation corresponds to the multiphoton regime. Increase in the oscillation amplitude involves electron states that are beyond the area of the parabolic-relation approximation, and the photoionization rate must become sensitive to details of realistic nonparabolic band structures of particular materials. For the case $\delta_F/\delta_B = 1$ (i.e., at intensity about few TW/cm^2), proper averaging of electron energy by Equation I.4.28 requires rigorous dispersion relation that is valid over entire BZ. To demonstrate that influence of energy–momentum relation on the photoionization rate, we refer to the following analog of the Keldysh formula derived for the cosine energy–momentum relation given by Equation 4.9 (Gruzdev 2007):

$$w_{cos} = 2\frac{2\omega}{9\pi}\sqrt{\frac{\omega^3}{\hbar\Delta}}\frac{m}{dJ_0(\chi)}\exp\left\{-2\frac{u_c-1}{J_0(\chi)}\left(\left\langle\frac{\tilde{\Delta}_{cos}}{\hbar\omega}+1\right\rangle-\frac{\tilde{\Delta}_{cos}}{\hbar\omega}\right)f_2(\chi)\right\}$$

$$\times\exp\left\{-2\frac{\Delta}{\hbar\omega}\frac{u_c}{u_c-1}\left[\sinh^{-1}\left(\frac{u_0}{\chi}\right)-\frac{u_0}{u_c}f_1(\chi)\right]\right\}Q_{cos}\left(\chi,\frac{\tilde{\Delta}_{cos}}{\hbar\omega}\right), \tag{4.30}$$

where the slowly varying amplitude takes the following form:

$$Q_{cos}(\chi,s) = \sqrt{\sqrt{\frac{u_c-1}{u_c+1}\left(\chi^2+u_0^2\right)}}$$

$$\times\sum_{n=0}^{\infty}\exp\left\{-2n\frac{u_c-1}{J_0(\chi)}f_2(\chi)\right\}\Phi\left(\sqrt{\frac{2(u_c-1)}{J_0(\chi)}\left(\langle s+1\rangle-s+n\right)}\sqrt{\frac{u_c+1}{(u_c-1)\left(\chi^2+u_0^2\right)}}\right). \tag{4.31}$$

Here, the following notations are introduced (Gruzdev 2007):

$$f_0(\chi) = \int_0^1\frac{\sinh(u_0\xi)}{\sqrt{\chi^2+u_0^2\xi^2}}d\xi, \tag{4.32}$$

$$f_1(\chi) = \int_0^1\frac{\cosh(u_0\xi)}{\sqrt{\chi^2+u_0^2\xi^2}}d\xi, \tag{4.33}$$

$$f_2(\chi) = \frac{u_0}{u_c-1}f_1(\chi)-\sqrt{\frac{u_c+1}{(u_c-1)\left(\chi^2+u_0^2\right)}}, \tag{4.34}$$

$$u_0 = \cosh^{-1}u_c; \quad u_c = 1+\frac{md^2\Delta}{\hbar^2}, \tag{4.35}$$

where $J_0(\chi)$ is the Bessel function of the first kind of the zero order (Abramowitz and Stegun 1972)

Φ is the Dawson integral (Equation 4.14), and effective band gap reads as follows:

$$\tilde{\Delta}_{\cos} = \Delta\left[1 + \frac{\hbar^2}{md^2\Delta}\left(1 - J_0(\chi)\right)\right]. \tag{4.36}$$

Instead of the Keldysh adiabatic parameter of Equation 4.13, the following modified adiabatic parameter enters those equations:

$$\chi = \frac{eFd}{\hbar\omega}. \tag{4.37}$$

In contrary to the Keldysh parameter that decreases with the increase in oscillation amplitude F/ω, the modified parameter given by Equation 4.37 follows variations of the amplitude, that is, it increases with the transition from the multiphoton to the tunneling regime of the photoionization. For example, the multiphoton limit is reached at $\chi \ll 1$ that simplifies Equation 4.30:

$$w_{\cos}^{MP} = 2\frac{2\omega}{9\pi}\sqrt{\frac{\omega^3}{\hbar\Delta}}\frac{m}{d}\sqrt{u_0\sqrt{\frac{u_c-1}{u_c+1}}}\exp\left\{2\frac{\Delta}{u_0\hbar\omega}\sqrt{\frac{u_c+1}{u_c-1}} + \chi^2\frac{\sqrt{u_c^2-1}}{u_0^3}\left[\frac{\Delta}{\hbar\omega} - \alpha\right]\right\}$$

$$\times \exp\left\{\left[\frac{\Delta}{\hbar\omega}\frac{u_c}{u_c-1} - \alpha\right]\left(1 + \frac{\chi^2}{4}\right)\left(\frac{u_0^2}{2}\left(1 + \frac{u_0^2}{24}\right) - 2\frac{\sqrt{u_c^2-1}}{u_0} + \frac{u_0^6}{2160}\right)\right\}\left(\frac{\chi^2}{4u_0^2}\right)^\alpha \Phi(\beta), \tag{4.38}$$

where

$$\alpha = \left\langle\frac{\tilde{\Delta}_{MP}}{\hbar\omega} + 1\right\rangle, \tag{4.39}$$

$$\beta = \sqrt{2\frac{\sqrt{u_c^2-1}}{u_0}\left[\alpha - \frac{\tilde{\Delta}_{MP}}{\hbar\omega}\right]\left(1 + \chi^2\frac{u_0^2-2}{4u_0^2}\right)}, \tag{4.40}$$

and effective bad gap is given by Equation 4.17.

Figure 4.11 confirms the predictions based on the concept of the Bloch oscillations. Values of the multiphoton ionization rate evaluated from the Keldysh formula (Equation 4.10) coincide with those obtained from Equation 4.30 at low intensity, that is, at small amplitude of the laser-driven Bloch oscillations. Starting from intensity about 1 TW/cm², the rate given by formula (Equation 4.30) deviates from that given by the Keldysh formula, and the deviation increases with laser intensity. This fact demonstrates that high-intensity values of the photoionization rate and the corresponding coefficient of nonlinear absorption critically depend on nonparabolic

Chapter 4

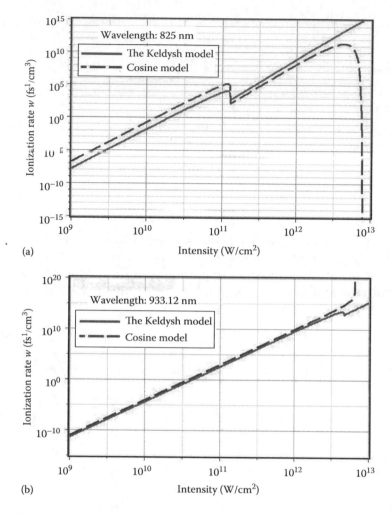

FIGURE 4.11 Photoionization rate evaluated by the Keldysh formula (solid curve) and by the formula for the cosine relation (dashed curve) as functions of laser intensity I for two values of laser wavelength. Calculations have been performed for material parameters of NaCl. The intensity scaling is depicted for laser wavelengths 825 nm (a) and 933.12 nm (b). Effective mass is 0.6 of free-electron mass, and band gap is 8.97 eV. (With kind permission from Springer Science+Business Media: *Alkali Halides: A Handbook of Physical Properties*, 2001, Sirdeshmukh, D.B., Sirdeshmukh, L., and Subhadra, K.G.)

details of band structure. The most striking result is that the cosine dispersion relation results in a strong singularity on dependence of the photoionization rate on laser parameters (Gruzdev 2007) at the following value of the adiabatic parameter (Equation 4.37):

$$\chi = \xi_1 = 2.4048, \tag{4.41}$$

where $\xi_1 = 2.4048$ is the first root of the Bessel function $J_0(x)$. Corresponding electric field,

$$F_{TH} = \frac{\hbar \cdot \omega}{e \cdot d} \cdot \xi_1, \tag{4.42}$$

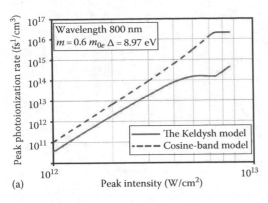
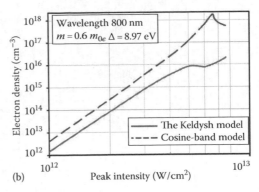

FIGURE 4.12 Dependence of the peak photoionization rate (a) and maximum conduction-band electron density (b) on peak laser irradiance for femtosecond laser pulses of variable peak irradiance. Calculations are performed for material parameters of NaCl. (With kind permission from Springer Science+Business Media: *Alkali Halides: A Handbook of Physical* Properties, 2001, Sirdeshmukh, D.B., Sirdeshmukh, L., and Subhadra, K.G.)

delivers irradiance $I_{TH} = 8.93 \times 10^{12}$ W/cm² at wavelength 800 nm. Approaching the singularity, the ionization rate either grows to infinity or decreases to zero depending on laser wavelength (Figure 4.11). The abrupt increase in the ionization rate with intensity approaching the singularity point can be associated with LID threshold (Gruzdev and Chen 2008), while the reduction of the rate means suppression of the photoionization (Gruzdev 2013) and nonlinear absorption (Figure 4.12). The mechanism of the singularity is quite different from that of the Keldysh singularities and is associated with flattening of the effective dispersion relation (see Figure 4.8) (Gruzdev 2007). Bearing in mind the fact that the upper valence band is quite flat in most wide-band-gap crystals (Sirdeshmukh et al. 2001), one can attribute the flattening of the dispersion relation to flattening of the effective conduction band. The flattening results from the simultaneous shift of the central part of the effective band (with positive initial effective mass) to higher energy and the shift of peripheral part of the band (where initial effective mass is negative) toward lower energy (Gruzdev 2007). In the case of monochromatic radiation and the cosine energy–momentum relation (Equation 4.9), the flattening is rigorous. For realistic nonmonochromatic radiation and dispersion relation with small corrections to the cosine term (Equation 4.9), the conduction band does not flatten but reduces to a minimum width. Even under the realistic conditions, the ionization suppression might take place for a certain range of laser wavelengths if the minimum depth of the effective conduction band is smaller than energy of laser photons (Figure 4.13).

The strong singularity at electric field (Equation 4.42) evidently demonstrates that the simple two-band model fails to describe the photoionization at that or larger electric field of laser radiation. The failure might result from two major reasons. The first one is the fast excitation of too many electrons (as suggested by the superexponential increase in the photoionization rate in Figure 4.11b) that significantly disturbs crystal potential and breaks interatomic bonds. In this case, the singularity must mean LID that is driven by the photoionization with negligible influence of the avalanche ionization, and the singularity threshold (Equation 4.42) is the LID threshold (Gruzdev and Chen 2008). The second reason is the need to include higher conduction bands when transitions to the lowest conduction band become impossible due to the flattening effect

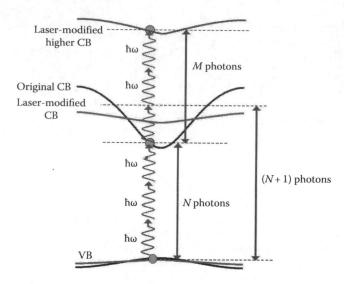

FIGURE 4.13 An illustration of possible ionization suppression due to partial flattening of effective CB in case where the energy of laser photon exceeds the width of the effective band. In that case, there is no state of the effective conduction band that could accept electrons from VB.

(Figure 4.11a). In this case, the probability of laser-driven electron transitions to higher conduction bands must be very small because they require simultaneous absorption of even more photons than for the lowest conduction band (Figure 4.13). It means that the photoionization suppression predicted from the singularity (Figure 4.12) still must be evident in experiments on time-resolved dynamics of laser-induced electron–hole plasma density. Therefore, the nonlinear effects associated with the singularity given by Equation 4.42 should be either the photoionization-driven LID or suppression of the photoionization observable in experiments.

4.3.2 Experimental Verifications of the Photo–Ionization Models

Experimental verification of the photoionization mechanism underlying nonlinear absorption is a challenge due to the fact that a direct measurement of the photoioniza-tion rate is not feasible. Therefore, the photoionization mechanisms can be judged only by detecting results of the photoionization, for example, by measuring LID threshold and density of electron–hole plasma.

If the photoionization is the major mechanism of electron–hole plasma generation in transparent solids, LID threshold and density of the plasma plotted as functions of laser wavelength and intensity must include the Keldysh singularities and abrupt steps due to increase in the number of absorbed laser photons (Figure 4.6). Abrupt steps on wavelength scaling of LID threshold have been confirmed by measurements of wave-length dependence of LID threshold for nanosecond (Carr et al. 2003) and femtosecond (Jupe et al. 2009) pulses (Figure 4.14). The steps are in good agreement with predictions of the Keldysh formula and clearly demonstrate that the photoionization and related nonlinear absorption are the major mechanisms of LID under reported conditions (Carr et al. 2003, Jupe et al. 2009). On the other hand, similar abrupt steps of density

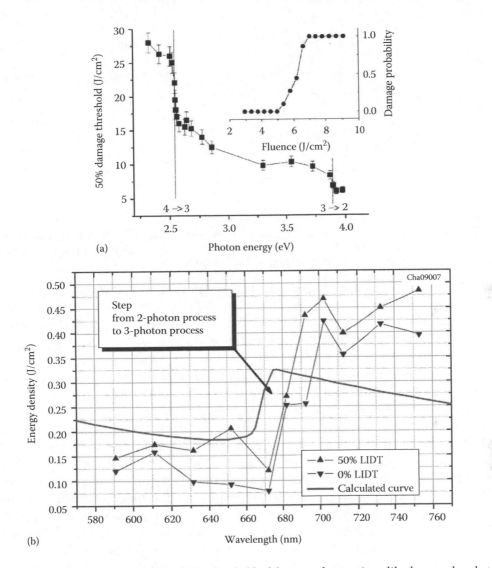

FIGURE 4.14 Measured dependence of LID threshold of deuterated potassium dihydrogen phosphate (DKDP) crystal (a) and a thin titanium-oxide film (b). (a: Reprinted with permission from Carr, C.W., Radousky, H.B., and Demos, S.G., Wavelength dependence of laser-induced damage: Determining the damage initiation mechanism, *Phys. Rev. Lett.*, 91, 127402, 2003. Copyright 2003 by the American Physical Society; b: From Jupe, M., Jensen, L., Melninkaitis, A., Sirutkaitis, V., and Ristau, D., Calculations and experimental demonstration of multi-photon absorption governing fs laser-induced damage in titania, *Opt. Express*, 17, 12269–12278, 2009. With permission of Optical Society of America.)

of electron–hole plasma plotted as function of irradiance of ultrashort laser pulses are not observed (Quere et al. 2001, Temnov et al. 2006). The latter fact might be attributed to several effects that kill the abrupt variations of the plasma density. They include averaging of electron response over interaction volume and over entire pulse duration during measurements, and saturation of plasma response due to increased absorption (Quere et al. 2001, Temnov et al. 2006). The averaging can also lead to saturation of peak density of laser-induced electron–hole plasma with laser irradiance approaching LID threshold that has been observed in experiments (Quere et al. 2001, Temnov et al. 2006).

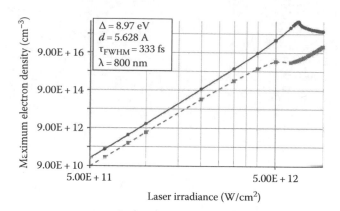

FIGURE 4.15　Simulated dependence of electron-plasma density induced by femtosecond laser pulses in NaCl. (From Gruzdev, V., How the laser-induced ionization of transparent solids can be suppressed, *Proc. SPIE*, 8885, 88851S, 2013. Published with permission from SPIE.)

On the other hand, modeling predicts that the saturation of plasma density can result from suppression of the photoionization observed in simulations for the cosine dispersion relation at intensity approaching the level of the strong singularity (Equation 4.42) (Gruzdev 2013) (Figure 4.15).

The latter point deserves special discussion in connection with recent experiments on LID and electron-plasma generation by ultrashort pulses (Daguzan et al. 1995, Stuart et al. 1996, Jasapara et al. 2001, Quere et al. 2001, Mao 2004, Temnov et al. 2006, Jupe et al. 2009, Rudolph et al. 2013). The monochromatic light generates conduction-band electrons by simultaneous absorption of several similar laser photons (Figure 4.16). This implies that electrons mainly make the interband transitions between two participating energy levels—one in the conduction and the other in the valence band—that satisfy the energy conservation law with a minimum number of simultaneously absorbed photons. Therefore, there exists only one channel of the interband transitions determined by the fixed (for given photon energy $\hbar\omega_0$) total energy of the absorbed photons. However, a single-frequency line spectrum of a monochromatic radiation (Figure 4.16) is a poor approximation for spectra of ultrashort laser pulses that have significant bandwidth $\Delta\omega$. For spectral-limited pulses, the bandwidth can be evaluated from the pulse duration τ (Diels and Rudolph 2006): $\Delta\omega = K/\tau$, where K is close to $\pi/2$ and is determined by temporal profile of a laser pulse. For a 40 fs pulse, this relation delivers $\Delta\omega \approx 10^{14}$ Hz that is about 10% of central frequency ω_0 for near-infrared radiation (e.g., at central wavelength 800 nm). Therefore, there are many wavelengths around a central wavelength that carry enough intensity to participate in the photoionization. It means that photons of slightly different energies are absorbed simultaneously (Figure 4.16) in contrast to the monochromatic approximation. There are many combinations of the simultaneously absorbed photons that bridge band gap of a transparent material and satisfy the energy conservation law with a minimum number of the absorbed photons for each combination (Figure 4.16 shows just few combinations as an example). Each of the combinations corresponds to a certain value of the total absorbed energy for a particular transition, and those values continuously vary within a limited range whose lower limit is the band gap. Therefore, the photoionization by nonmonochromatic laser pulses is a multichannel process and involves a continuous range of initial levels of the valence

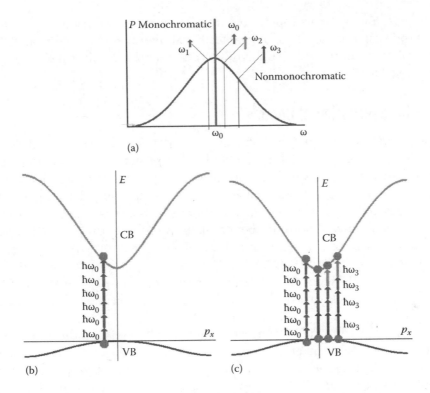

FIGURE 4.16 A sketch of nonmonochromatic laser spectrum versus monochromatic spectrum (a) and the multiphoton transitions induced by the monochromatic light (b) versus the multiphoton transitions induced by nonmonochromatic light (c). Frequencies shown in part (c) are the same as shown in part (a).

band and corresponding range of final levels of the conduction band. Since different combinations of simultaneously absorbed laser photons provide different values of interband transition probability, all channels of the photoionization transitions should be nonlinearly coupled and compete with each other. This qualitative consideration leads to important conclusions: (1) the photoionization rate by significantly nonmonochromatic pulses should be much higher than the rate for the monochromatic approximation, and (2) the photoionization must result in specific modifications of laser spectrum. So far, the photoionization of transparent solids by nonmonochromatic laser pulses has not been considered in spite of recent attempts to step beyond the monochromatic approximation for propagation simulations (Gulley 2012).

Meanwhile, many experimental data are in very good agreement with the photoionization models. In particular, wavelength dependence of LID threshold of wide-band-gap crystals by mid-infrared pulses (Simanovskii 2003) demonstrates monotonous decrease in the threshold with the increase in wavelength as it is predicted by Equation 4.42 (Gruzdev and Chen 2008). Moreover, values of LID threshold evaluated from the singularity of the ionization rate for the cosine dispersion relation (Gruzdev and Chen 2008) are very close to estimations obtained in experiments on LID of wide-band-gap solids by short and ultrashort pulses (Efimov 2004, Juodkazis et al. 2008). Predicted reduction of LID threshold with the increase in crystal-lattice constant (Equation 4.42) also correlates very well with measured dependence of LID threshold on average interatomic distance in several wide-band-gap solids (Juodkazis 2004, Juodkazis et al. 2008).

Chapter 4

Together with nonmonochromatic nature of laser pulses and realistic noncosine features of band structure, there are other effects that interfere with nonlinear absorption due to the photoionization: absorption by defects and collision-driven electron dynamics with impact ionization.

4.3.3 Photoexcitation of Defects

Any transparent material contains submicrometer defects that are either native or laser induced (Braunlich et al. 1981, Emmert et al. 2010, Menoni et al. 2012, Rudolph et al. 2013). The native defects result from violations of structure or composition of a solid and are specific of each particular material. For example, many defects of crystals are formed by lack of atoms at their regular positions in the crystal lattice (*vacancies*) or by extra atoms occupying positions between atoms of the crystal lattice (*interstitials*). Combinations of vacancies and interstitials can form the Schottky-type or Frenkel-type defects (Ashcroft and Mermin 1976). The Schottky-type defects are very common for alkali halides while the Frenkel-type defects are characteristic of silver halides (Ashcroft and Mermin 1976). The native defects exist before laser action and influence the nonlinear effects at the earliest stages of laser–solid interactions.

In contrary, the laser-induced defects result from the laser-induced photoionization followed by displacement of atoms from their equilibrium positions. Common examples of the laser-induced defects are excitons, polarons, self-trapped excitons, and color centers (Ashcroft and Mermin 1976, Mao 2004). Those defects result from strong electron–hole interaction that is the characteristic of wide-band-gap solids. Formation of the defects includes distortions of the atomic subsystem and can take from 150–200 fs in glasses (e.g., fused silica—see Saeta and Greene 1993, Daguzan et al. 1995, Mao 2004, Temnov et al. 2006) to few picosecond in crystals (Mao 2004, Temnov et al. 2006).

Both native and laser-induced defects have several common features that play important role in the nonlinear absorption and laser-induced electron dynamics. First, the defects form energy levels (and corresponding states) within the energy gap between the valence and the conduction bands (Figure 4.17) (Ashcroft and Mermin 1976, Ridley 1993). The defects with energy levels close to the bottom of the conduction band or ceiling of the valence band are referred to as *shallow defects* or *shallow traps* and can be ionized via one-photon absorption. Other defects (referred to as *deep traps*) form energy levels deeper in the interband energy gap and can be ionized only via multiphoton absorption. The defect energy levels significantly affect the photoionization: instead of a single direct interband transition, electrons can jump from the valence band to a defect level and then from the defect level to the conduction band (Figure 4.17). Those band-to-defect and defect-to-band transitions require simultaneous absorption of fewer photons compared to the band-to-band transition. Therefore, the defect levels reduce the multiphoton absorption of high order to a chain of at least two low-order multiphoton processes, that is, change the average probability of the photoexcitation, but still keep nonlinear scaling of the probability with intensity in Equation 4.3. Under similar conditions, probability of a chain of the low-order multiphoton transitions is higher than probability of a single high-order multiphoton process (Keldysh 1965). Moreover, total probability of interband transition with participation of intermediate defect levels increases by several orders of magnitude if either band-to-defect or defect-to-defect energy gap exactly fit $N = 1, 2, \ldots$ photons (Keldsyh 1965).

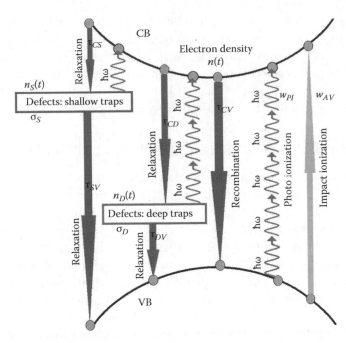

FIGURE 4.17   Illustration of shallow-trap (ST) and deep-trap (DT) defect levels in the energy gap between valence (VB) and conduction (CB) bands. Vertical arrows depict possible one-photon and multiphoton transitions between the bands and the defect levels.

Second, the electron states of the defects are localized, and electrons trapped by the defects are not mobile (Ashcroft and Mermin 1976). Absorption of some extra energy (referred to as *mobility activation energy*) is required to promote the trapped electrons to mobile conduction-band states. Shallow-trap exciton levels located just beneath the bottom of the conduction band of most wide-band-gap materials exhibit the lowest activation energy close to the energy of single photon of visible and infrared light (Mao 2004). Nonetheless, the defects states should be considered as traps for conduction-band electrons that reduce electron–hole plasma density and, therefore, delay the development of the fast electronic collision processes.

Third, density of the native defects is usually very limited in high-quality optical materials. Deep native traps provide only limited supply of traps for capturing the valence-band electrons while shallow native defects provide very limited supply of electrons for the electron–hole plasma. For example, with density of the deep defects at the level of 1 ppm, there are about 10^{15}–10^{16} traps per cubic cm. Since rate of the photoionization reaches values of 10^{16} fs/cm^3 at relatively low irradiance (see Figure 4.6), those traps get populated within few femtoseconds of the photoionization at the leading edge of a laser pulse. After that, they do not affect the photoionization by a single laser pulse if its duration is smaller than the time of laser-driven defect modification (that is typically around 10^2–10^3 fs—see Mao 2004). In contrary to the case of ultrashort pulses, the native traps significantly influence laser–solid interactions in the case of nanosecond pulses as well as continuous-wave and multi-pulse irradiation (Emmert et al. 2010, Menoni et al. 2012, Rudolph et al. 2013).

Fourth, density of the laser-induced defects varies in time and depends on laser irradiance in nonlinear way. Since their formation is delayed, they make almost no contribution to the action of a single ultrashort laser pulse (Emmert et al. 2010, Rudolph et al. 2013). For pulses longer than the time of defect formation and for multipulse action, the laser-induced defects provide a very significant contribution to the nonlinear effects and to initiation of LID (Braunlich et al. 1981, Woods 2003, Mao 2004, Emmert et al. 2010, Menoni 2012, Rudolph et al. 2013). In particular, they affect dynamics of the electron–hole plasma as discussed in the following section.

4.3.4 Nonlinear Effects due to Collision–Driven Electron Dynamics

More nonlinear effects are attributed to the laser-induced electron–hole plasma. Various electron–particle collisions are believed to be a major driving process of plasma dynamics. The collisions might suppress the photoionization by dephasing the coherent laser-driven electron oscillations (Du et al. 1996). On the other hand, frequent electron–particle collisions must support the alternative mechanism of interband electron transitions—impact ionization. A simple (and very traditional) concept of the impact ionization considers beginning of the process with certain amount of seed electrons in the conduction band that are provided by the photoionization (Stuart et al. 1996, Kaiser et al. 2000, Wood 2003, Mao 2004, Rethfeld 2004, Starke et al. 2004, Jia 2006). The seed electrons are promoted to higher energy levels (i.e., they are heated) by intraband transitions via absorption of single laser photons (Figure 4.3). The absorption requires participation of an electron, a photon, and a third heavy particle to observe simultaneous conservation of energy and momentum. The third particle is a phonon, since it can carry little energy, but a large amount of momentum. Thus, the heating of the conduction-band electrons takes place via electron–phonon–photon collisions that result in single-photon absorption. That process is referred to as the *inverse Bremsstrahlung* and can promote electrons to energy levels with kinetic energy as high as band gap. The high-energy conduction electrons transfer their kinetic energy to the valence-band electrons via inelastic collisions and promote them over the band gap to the bottom of the conduction band. This process increases density of the conduction electrons and creates holes in the valence band. The following cycle of the impact ionization involves the newly arrived conduction-band electrons. The entire impact-ionization process is believed to be similar to electron avalanche in gas. This mechanism of the ionization is believed to dominate at laser irradiance approaching LID threshold. It is considered as one of the major processes of the laser-induced electron dynamics.

With the increase in conduction-band electron density, a specific mechanism of nonlinear absorption comes into play. Probability of a single elementary absorption act by conduction-band electrons linearly scales with intensity since it is single-photon absorption. The major effect that provides nonlinear scaling of the overall absorption by electron–hole plasma is attributed to nonlinear scaling of the conduction-band electron density n_e with laser irradiance I. Therefore, dynamics of the conduction-band electron density provides a cornerstone contribution to all possible plasma-induced nonlinear effects.

The most intuitive approach suggests that time variations of conduction-band electron density n_e result from two contributions. On the one hand, the density is increased by supply of electrons from the valence band due to the photoionization and the avalanche ionization. On the other hand, the density is reduced by trapping conduction electrons at defect levels, and by various recombination and relaxation effects. Those effects are reflected by corresponding terms of a *rate equation* of the following type (Stuart et al. 1996, Mao 2004, Starke et al. 2004, Jia 2006, Jing et al. 2009):

$$\frac{\partial n_e(t)}{\partial t} = w_{PI}\left(I[t]\right) + w_{AV}\left(I[t]\right)n_e(t) - w_R\left(n_e[t],t\right), \tag{4.43}$$

where
  w_{PI} is the rate of photoionization
  w_{AV} is the rate of avalanche ionization
  w_R is the rate of electron relaxation to trapping defect levels

The photoionization rate is given by the Keldysh formula (Keldysh 1965) in most simulations of laser-induced electron dynamics in transparent solids (see, e.g., Stuart et al. 1996, Tien et al. 1999, Jasapara et al. 2001, Mao 2004, Starke et al. 2004, Jia 2006, Jupe et al. 2009, Jing et al. 2009). The relaxation term is usually described within the relaxation-time approximation (Ashcroft and Mermin 1976):

$$w_R = \frac{n_e(t)}{\tau_R}, \tag{4.44}$$

where relaxation time τ_R includes contributions from all processes that support decay of the conduction-band electron density. It is usually extracted from experimental data on laser-induced electron dynamics in solids (Audebert et al. 1994, Daguzan et al. 1995, Quere et al. 2001, Temnov et al. 2006). The evaluation of the second term of Equation 4.43, that is, the impact-ionization rate is one of the most challenging problems in simulation and understanding of the laser-driven electron-plasma dynamics. The most common approaches are as follows:

1. The constant-rate model proposed first by Stuart et al (1996) considers linear scaling of the impact-ionization rate with laser intensity I:

$$w_{AV\_St} = \alpha_{AV} I(t), \tag{4.45}$$

   where $I(t) = 0.5\, c\, n_0\, \varepsilon_0\, |F(t)|^2$ in the simplest case of plain wave (ε_0 is the electric constant, c is the speed of light in vacuum, and n_0 is the constant part of refractive index), and $F(t)$ is the slowly varying amplitude of electric field of laser pulse (Born and Wolf 2003, Boyd 2008). Coefficient α_{AV} is of the order of 10^{-2} cm²/ps/GW and is evaluated either from fitting experimental data on LID or from numerical solving a Fokker–Plank equation for electrons of the conduction band. The linear scaling of Equation 4.45 is valid only at relatively low laser intensity where the multiphoton regime of the photoionization takes place. At higher intensity approaching LID, a better nonlinear approximation is required.

2. The model of Sparks (1975, Sparks et al. 1981) considers microscopic mechanisms of the electron-avalanche development and takes into account electron

Chapter 4

collisions with longitudinal optical phonons. A simplified equation for the avalanche rate (Sparks 1975),

$$w_{AV\_Sp} = \frac{Ae^2\tau_K^2 F^2(t)}{m_C\Delta\left(1+\omega^2\tau_K^2\right)} - \frac{A\hbar\omega_{ph}}{\Delta\tau_L}, \tag{4.46}$$

suggests linear scaling of the rate with laser irradiance I and includes a constant coefficient A, average phonon energy $\hbar\omega_{ph}$, total scattering rate $1/\tau_L$ that includes all types of electron-particle scattering, and time of electron-momentum relaxation τ_K for large-angle scattering on phonons. Evaluation of the relaxation rate and time of Equation 4.46 is quite challenging and is usually done either by fitting experimental data on LID or by numerical solving a Fokker–Plank equation for conduction-band electrons.

3. Some publications (e.g., Tien et al. 1999) suggested utilizing of the Thornber model:

$$w_{AV\_Th} = \frac{v_S e F(t)}{\Delta} \exp\left\{-\frac{F_I}{F(t)\left(1+F(t)/F_{PHON}\right)+F_{kT}}\right\}, \tag{4.47}$$

where

v_S is the saturation velocity of electron drift (estimated to be about 2×10^7 cm/s)

F_I, F_{PHON}, and F_{kT} are the electric fields required for electrons to overcome the decelerating effects of ionization scattering, optical-phonon scattering, and thermal scattering within a single mean free path correspondingly (Thornber 1981)

In contrary to the models of Stuart and Sparks, the Thornber model predicts non-linear scaling of the impact-ionization rate with laser irradiance (Jing et al. 2009).

4. The Drude model of the impact ionization (Starke et al. 2004, Jupe et al. 2009, Jing et al. 2009) assumes that heating of the conduction-band electrons via one-photon absorption is quite similar to heating of free electrons in metals (Ashcroft and Mermin 1976). The corresponding absorption coefficient for intraband transitions is expressed via single-photon absorption cross section σ that involves average total time τ_C of momentum relaxation due to collisions:

$$w_{AV\_Dr} = \frac{\sigma}{\Delta} I(t), \tag{4.48}$$

where the cross section σ reads as

$$\sigma = \frac{e^2}{\alpha_0 n_0 m} \frac{\tau_C}{1+\omega^2\tau_C^2}, \tag{4.49}$$

and the relaxation time depends on the density of conduction-band electrons $n_e(t)$:

$$\tau_C = \frac{16\pi\varepsilon_0^2\sqrt{m(0.1\Delta)^3}}{\sqrt{2}e^4 n_e(t)}. \tag{4.50}$$

Equation 4.50 for the relaxation time is derived by estimating average collision cross section (Starke et al. 2004). Dependence of the relaxation time on electron density introduces strong nonlinear scaling of the impact-ionization rate with laser irradiance for the entire Drude model. Also, the Drude model contains a minimum number of uncertain material parameters compared to the other models of the avalanche-ionization rate.

Simulations of the electron–hole dynamics by the rate equation (Equation 4.43) and the Drude model for the avalanche ionization (Figure 4.18) demonstrate that several scenarios are possible for the nonlinear effects. First, if laser parameters are not favorable for development of the avalanche ionization (e.g., pulse width is small enough and/or peak irradiance is low enough), a major contribution to the nonlinear effects is done by the photoionization (Rethfeld 2004, 2006). In this case, the temporal profile of the ionization rate is peaked at maximum of laser irradiance and is symmetrically distributed with respect to the peak. Correspondingly, electron-plasma density increases monotonously throughout the laser pulse and reaches maximum level at its tail. Both front and tail parts of laser pulse are almost equally absorbed via the photoionization mechanism. Peak of the nonlinear absorption is reached at the peak of laser pulse, and high-intensity part of the pulse experiences the strongest absorption. In this regime, scaling of the nonlinear absorption and laser-induced electron-plasma density with irradiance is determined by the photoionization. For example, electron-plasma scaling

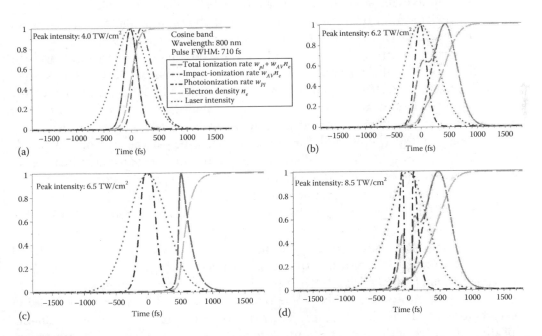

FIGURE 4.18 Examples of simulations of time variations of the photoionization rate (darker dashed dotted), avalanche ionization rate (dashed), total ionization rate (lighter dashed dotted), electron-plasma density (long dashed), and slow amplitude of laser pulse (dotted) for several values of peak laser irradiance. Laser parameters are shown in the figure. Material parameters correspond to NaCl. Time evolution of those values is shown for four values of peak irradiance: (a) 4.0 TW/cm^2—photoionization dominated regime; (b) 6.2 TW/cm^2—impact ionization comparable to the photoionization; (c) 6.5 TW/cm^2—impact-ionization dominated regime; (d) 8.5 TW/cm^2—strong impact ionization combined with one of the Keldysh singularities. (With kind permission from Springer Science+Business Media: *Alkali Halides: A Handbook of Physical Properties*, 2001, Sirdeshmukh, D.B., Sirdeshmukh, L., and Subhadra, K.G.)

Chapter 4

reproduces the power scaling with laser irradiance characteristic of the multiphoton regime (Quere et al. 1999, Temnov et al. 2006).

With the increase in pulse width and/or peak irradiance, contributions of the impact ionization to the overall ionization process and electron-plasma density grow and become comparable to those of the photoionization. In this case, the photoionization can dominate at the leading front of laser pulse while the avalanche ionization dominates at the tail part of the pulse. A general trend is that the avalanche ionization breaks the symmetry of time profile of the total ionization rate, increases total electron-plasma density, and shifts maximum of the total rate toward the tail of laser pulse. That trend is significantly affected by the Keldysh-type singularities of the photoionization rate that start developing at peak of laser irradiance, shift toward front and rare parts of the pulse, and suppress the ionization effects at the high-intensity part of laser pulse (Figure 4.18). Interplay between the singularity-dominated photoionization and the impact ionization leads to multipeak pattern of the total ionization rate and total nonlinear absorption. Also, those effects form peaks of electron-plasma density at the front and tail parts of laser pulse. The peaks are separated by a gap of almost constant plasma density within the high-intensity part of laser pulse. Scaling of total nonlinear absorption and peak plasma density with laser irradiance can significantly vary under this regime due to the complicated interplay of the mentioned effects and the transition from dominating photoionization to domination of the impact ionization. It is reasonable to expect that the scaling deviates from the scaling dictated by the photoionization but also does not reproduce the scaling expected from the electron-avalanche mechanism.

At high peak irradiance, the avalanche ionization dominates (Figure 4.18), and the major electron-plasma effects develop at the tail part of laser pulse if pulse width is larger than the duration of electron-avalanche development (Rethfeld 2004). In particular, the peak of the avalanche ionization and all related nonlinear absorption effects can be delayed with respect to maximum of laser irradiance by as much as a few hundreds of femtoseconds. For longer pulses (pulse width close to or larger than 1 ps), the peak of time profile of the avalanche ionization is reached around maximum of laser intensity (Stuart et al. 1996). In that case, the impact ionization makes a dominating contribution to overall ionization and to the nonlinear absorption throughout entire laser pulse. The Keldysh-type singularities do not produce any remarkable effect on electron plasma and ionization-rate variations in time. The photoionization makes negligible contribution to the nonlinear absorption just by providing seed electrons for the impact ionization at the front edge of laser pulse. Scaling of nonlinear absorption and peak electron-plasma density with laser irradiance is completely dictated by the impact ionization under this regime.

The strongest advantages of the single-rate equation are its simplicity and ease of interpretation of the results it delivers. For example, Equation 4.43 can be solved analytically in the case of three-photon ionization and the Stuart model of the impact ionization (Jasapara et al. 2001). Those advantages have stimulated multiple efforts focused on improving of the single-rate-equation approach. Among them, the most significant are introduction of multiple-rate equations (Rethfeld 2004, 2006, 2010) and extension of the model of Equation 4.43 by coupling it to rate equations for electron populations of shallow and deep defects (Emmert et al. 2010).

The need to introduce the multiple-rate equations instead of the single-rate equation is motivated by the fact that equation (Equation 4.43) delivers variations of the total

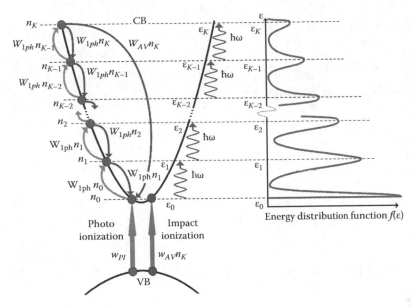

FIGURE 4.19   Illustration of the system of energy levels within the conduction band chosen to derive the multiple-rate equations. All notations are shown in the text.

amount of conduction-band electrons under the assumption of steady energy distribution of the electrons. One of the major reasons for variations of the distribution function is the intraband one-photon absorption that creates peaks of electron density at equidistant energy levels (Kaiser et al. 2000). The peaks start forming as soon as significant electron density appears in the conduction band and can exist as long as few tens of femtosecond (Kaiser et al. 2000). The peak positions in energy space are separated by gaps that are equal to the energy of one laser photon (Figure 4.19). This result can be interpreted, so that the absorbing electrons move to higher energy levels of the conduction band by climbing from one of the equidistant levels to another. Moreover, only electrons populating the highest energy level contribute to the impact ionization (Rethfeld 2004) while the model of Equation 4.43 assumes that all conduction-band electrons equally contribute to the avalanche development. Average time required for a conduction electron to climb up the ladder of energy levels determines the delay of development of the impact ionization that can be very significant depending on laser parameters (Kaiser et al. 2000, Rethfeld 2004). To mimic that effect, a simple model was proposed by Rethfeld (2004, 2006) that splits the total conduction-band electron density n_e into $K + 1$ contributions n_i made by each energy level ε_i (Figure 4.19). The lowest level ε_0 accepts electrons supplied by the photoionization and by the impact ionization. Each level ε_i accepts electrons promoted from lower level $\varepsilon_{i-1} = \varepsilon_i + \hbar\omega$ and supplies electrons to upper level $\varepsilon_{i+1} = \varepsilon_i + \hbar\omega$, so that the total dynamics of electron heating in the conduction band is described by the following system of coupled-rate equations (Rethfeld 2004, 2006):

$$\frac{\partial n_0(t)}{\partial t} = w_{PI}\left(I[t]\right) + 2w_{AV}\left(I[t]\right)n_K(t) - W_{1ph}\,n_0(t), \tag{4.51}$$

Chapter 4

$$\frac{\partial n_1(t)}{\partial t} = W_{1ph} \, n_0(t) - W_{1ph} \, n_1(t), \tag{4.52}$$

$$\frac{\partial n_2(t)}{\partial t} = W_{1ph} \, n_1(t) - W_{1ph} \, n_2(t), \tag{4.53}$$

$$\frac{\partial n_{K-1}(t)}{\partial t} = W_{1ph} \, n_{K-2}(t) - W_{1ph} \, n_{K-1}(t), \tag{4.54}$$

$$\frac{\partial n_K(t)}{\partial t} = W_{1ph} \, n_{K-1}(t) - w_{AV}\left(I[t]\right) n_K(t), \tag{4.55}$$

where W_{1ph} is the rate of one-photon intraband absorption, and

$$K = \left\langle \frac{\tilde{\Delta}_{CR}}{\hbar \omega} + 1 \right\rangle \tag{4.56}$$

is the number of photons required to cover the critical-energy gap between valence and conduction bands rather than a band gap (Rethfeld 2004). The critical energy of Equation 4.56 is determined from the effective band gap of Equation 4.17 as follows (Kaiser et al. 2000):

$$\tilde{\Delta}_{CR} = \frac{1+2\mu}{1+\mu} \tilde{\Delta}_{MP}, \tag{4.57}$$

where $\mu = m_C/m_V$. Equation 4.57 is obtained by taking into account both momentum and energy conservation during electron–particle collisions (Kaiser et al. 2000). It is notable that the multiple-rate Equations 4.51 through 4.55 have demonstrated excellent agreement with results of elaborated simulations of electron dynamics based on kinetic equation (Rethfeld 2004).

The other direction of improvement of the rate-equation approach is to include contributions of defect energy levels into electron dynamics (Emmert et al. 2010). For example, if one group of deep traps and one group of shallow traps are taken into account (Figure 4.17), the rate equation (Equation 4.43) should be replaced with the following system of three rate equations (Emmert et al. 2010):

$$\frac{\partial n(t)}{\partial t} = w_{PI}\left(I[t]\right) + w_{AV}\left(I[t]\right)n(t) - \frac{n(t)}{\tau_{CV}} - \frac{n(t)}{\tau_{CS}}\left\{1 - \frac{n_S(t)}{n_{SMAX}}\right\} + \sigma_S n_S(t)\left\{\frac{I[t]}{\hbar \omega}\right\}$$

$$- \frac{n(t)}{\tau_{CD}}\left\{1 - \frac{n_D(t)}{n_{DMAX}}\right\} + \sigma_D n_D(t)\left\{\frac{I[t]}{\hbar \omega}\right\}^{ND}, \tag{4.58}$$

$$\frac{\partial n_S(t)}{\partial t} = \frac{n(t)}{\tau_{CS}}\left\{1 - \frac{n_S(t)}{n_{SMAX}}\right\} - \sigma_S n_S(t)\left\{\frac{I[t]}{\hbar\omega}\right\} - \frac{n_S(t)}{\tau_{SV}}, \tag{4.59}$$

$$\frac{\partial n_D(t)}{\partial t} = \frac{n(t)}{\tau_{CD}}\left\{1 - \frac{n_D(t)}{n_{DMAX}}\right\} - \sigma_D n_D(t)\left\{\frac{I[t]}{\hbar\omega}\right\}^{ND} - \frac{n_D(t)}{\tau_{DV}}, \tag{4.60}$$

where

  n_S and n_D are the density of electrons captured by the shallow traps and by the deep traps, respectively

  σ_S and σ_D are the absorption cross sections of electrons in shallow and deep traps, respectively

  τ_{CV}, τ_{CD}, and τ_{CS} are the relaxation time constants for the respective transitions: conduction band to valence band (CV), conduction band to deep trap (CD), and conduction band to shallow trap (CS)

  τ_{SV} and τ_{DV} are the shallow-trap and deep-trap life-time constants, respectively

  n_{SMAX} and n_{DMAX} are the maximum concentrations of the shallow and deep traps, respectively (Emmert et al. 2010)

Equations 4.58 through 4.60 include many material parameters that characterize the two considered groups of defects. Most of them are obtained by fitting experimental data on LID (e.g., the photoionization and impact-ionization rates) and from independent measurements by pump-probe methods (Emmert et al. 2010). Being applied to thin films, the model of Equations 4.58 through 4.60 has demonstrated excellent fitting of experimental data on LID (Emmert et al. 2010). One of its major advantages is that it can be easily modified for the case of more than two groups of defects by adding corresponding transition rates to the right-hand parts of Equations 4.58 through 4.60 and by coupling them to rate equations for population of each added type of defects.

In spite of the improvements, the rate-equation approach to description of electron–hole plasma suffers from several fundamental issues associated with the approximations underlying it. In particular, the rate equation (4.43) assumes that energy distribution function of electrons does not change during the laser-induced excitation while only absolute amount of conduction-band electrons varies in time. This is not true at early stages of electron–hole plasma generation that is accompanied by very significant variations of the distribution function (Kaiser et al. 2000). Therefore, the rate-equation models can hardly provide correct description of the nonlinear absorption effects at the beginning of the ionization process. Reliable description of the collision electron dynamics and related nonlinear effects requires more fundamental approaches that consider time-dependent energy distribution function. Among them, various modifications of the Fokker–Plank equation (Holway and Fradin 1975, Epifanov 1981, Sparks et al. 1981, Stuart et al. 1996, Apostolova and Hahn 2000), quantum kinetic equations (Melnikov 1969, Epstein 1970), and the Boltzmann integral–differential equation (Kaiser et al. 2000) all formulated for energy distribution function of conduction-band electrons can be mentioned. Also, the Monte-Carlo stochastic simulations (Arnold et al. 1992, Nikiforov et al. 2012) can be a very effective tool to simulate electron avalanche.

Chapter 4

The major drawback of those methods is that they include electron–particle collision integrals that are evaluated under the assumption of constant effective electron mass for any energy of conduction-band electrons. The effective mass is constant only for the parabolic energy–momentum relation (Equation 4.6) that is a good approximation for very bottom part of the conduction band (Ashcroft and Mermin 1976). Therefore, those approaches neglect realistic nonparabolic features of energy bands and related energy dependence of effective mass and collision time. Proper incorporation of those features into simulations of the laser-driven electron dynamics is one of the major challenges for the field of fundamental mechanisms of high-power laser–solid interactions.

4.4 Nonlinear Propagation Effects

Together with the major contribution to deposition and transfer of energy of laser radiation, the nonlinear absorption makes a very significant contribution to the nonlinear propagation effects. The propagation processes are the second major group of the nonlinear effects that control the earliest stage of LID. They significantly affect bulk LID while their contribution might be less important in the case of LID at surface and in thin films due to zero or negligible propagation path.

The nonlinear propagation includes a variety of processes that have been extensively described and analyzed in many textbooks (see, e.g., Akhmanov et al. 1992, Agraval 1995, Sutherland 2003, Diels and Rudolph 2006, Boyd 2008). In this section, we focus only on two major effects (self-focusing and modifications of spectrum) since they are the most relevant to LID and substantially affect measured/observed LID thresholds. To avoid overlapping with the books on nonlinear optics, we pay most attention to several special features of those propagation effects (e.g., polarization dependence of self-focusing) that have impact on initiation of LID.

4.4.1 Nonlinear Refraction and Self-Focusing

Self-focusing is a nonlinear propagation effect that results in reduction of cross section of high-power laser beams that propagate through transparent material (Figure 4.20). Mechanism of that effect is attributed to laser-induced nonlinear modification of refractive index. In the simplest case of isotropic solid and negligible dispersion of the optical response (Sutherland 2003, Boyd 2008), the modification is proportional to irradiance I:

$$n_{ref} = n_0 + n_2 I(t), \tag{4.61}$$

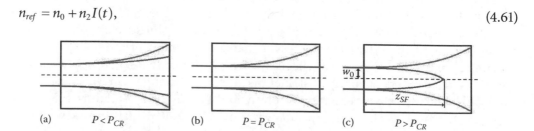

FIGURE 4.20 An illustration for the three cases of the competition between diffraction-driven divergence and self-focusing-driven convergence: (a) diffraction dominates, (b) diffraction is rigorously compensated by self-focusing, (c) and self-focusing dominates and drives the laser beam to collapse.

where *coefficient of nonlinear refraction* n_2 is expressed via components of a tensor of nonlinear susceptibility (Sutherland 2003, Boyd 2008), and its value is about 10^{-20} to 10^{-19} m²/W. The change of refractive index that scales linearly with laser irradiance is frequently referred to as the *Kerr effect* (Sutherland 2003, Boyd 2008). There are several contributions to that effect (Christodoulides et al. 2010). The fastest of them is nonlinear electronic response (Figure 4.2) that is called the *electronic Kerr effect*. It makes the smallest contribution to nonlinear coefficient of refraction and consists of polarization and current components produced by oscillating valence electrons and current of conduction-band electrons, respectively. On the other hand, the Kerr-type variations of refractive index can result from electrostriction effect associated with modifications of local density of solids induced by electric field of laser radiation (Boyd 2008, Christodoulides et al. 2010). That mechanism is much slower compared to the electronic response, but value of its contribution to the nonlinear coefficient is much larger than that of electrons.

Nonlinear coefficient of most optical materials is positive. Therefore, a propagating laser beam modulates refractive index of a transparent material according to transverse distribution of irradiance within the beam. That modulation is favorable for focusing effect that reduces beam cross section and concentrates power of laser beam at beam's axis. That process competes with diffraction that increases laser spot size due to diffraction divergence. Depending on the amount of power carried by laser radiation, the following options are possible (Figure 4.20):

- If focusing effect is weaker than diffraction, laser beam diverges due to diffraction, but rate of the diverging is smaller compared to that of linear propagation (when $n_{REF} = n_0$).
- If focusing rigorously compensates diffraction-induced divergence, the laser beam neither diverges nor collapses into a focal point, but propagates within a waveguide-type channel formed by self-induced modification of refractive index. This effect is referred to as *self-trapping*. To produce it, power of laser beam must equal the following value referred to as *critical power of self-focusing*:

$$P_{cr} = \frac{\pi(0.61)^2 \lambda_0^2}{8 n_0 n_2},$$
(4.62)

where λ_0 is the laser wavelength (Boyd 2008). Depending on nonlinear refraction coefficient, value of the critical power varies from 0.2 to 3 MW for visible and near-infrared light.

- If power of laser beam exceeds the critical power of Equation 4.62, self-induced focusing dominates and collapses the laser beam into a single focal spot referred to as *self-focus* (Boyd 2008, Woods 2003). This effect of beam collapsing is referred to as *self-focusing*. It takes place if beam has traveled certain distance within the transparent material that is called *self-focusing distance* z_{SF}:

$$z_{SF} = \frac{2 n_0 w_0^2}{\lambda_0} \frac{1}{\sqrt{P/P_{CR} - 1}},$$
(4.63)

Chapter 4

where w_0 is the beam radius at the entrance surface of transparent material. Depending on radiation power P, the self-focusing distance can vary from tens of centimeters down to few tens of micrometers. It is important to note that position of self-focus is not steady and fixed in space. As soon as the first self-focus is formed by a collapsing self-focused beam at the distance z_{SF} from front surface of a solid, another self-focus is formed on axis of the beam. The second self-focus is followed by the third, and that process is repeated to make a track of the self-focus points. The overall dynamics of self-focusing looks like the self-focus appears at the largest distance given by Equation 4.63 and moves toward the front surface of transparent solid (Akhmanov et al. 1992, Agraval 1995, Boyd 2008). This process is frequently referred to as the *effect of moving self-focus*.

- If power of laser beam exceeds the critical power of self-focusing very much ($P \gg P_{CR}$), the laser beam experiences *beam breakup*, that is, smooth longitudinal profile of laser irradiance splits into many (approximately $M = P/P_{CR} \gg 1$) self-focusing channels (Boyd 2008). For the breakup to happen, the beam must travel the self-focusing distance given by Equation 4.63 in a transparent material.

Reduction of laser spot size by self-focusing and beam breakup leads to local increase in laser irradiance to the levels that can easily exceed LID threshold and initiate LID. Those effects do not allow reliable measurements of threshold of bulk LID (Soileau et al. 1989, Glebov 2002, Woods 2003), because LID driven by self-focusing takes place as soon as irradiance reaches threshold of self-focusing that is lower than threshold of LID in most cases. Due to this reason, significant efforts have been made to develop approaches to control and suppress self-focusing in experiments on LID. The simplest of them utilized specific features of self-focusing-induced damage tracks in solids. Self-focusing produces specific damage morphology patterns stretched along direction of laser beam propagation, that is, they have evident orientation in space. In many cases, they look like filaments or tiny damage channels. With reduction of peak power of laser pulse toward critical power and slightly below it, the filaments shorten and turn into a short damage track that connects two enlarged star-like damage sites (Glebov 2002, Woods 2003). The star-like damage sites are attributed to locations of the moving self-focuses. That special structure of the LID patterns produced by self-focusing is evidently distinct from that produced under negligible self-focusing. In the latter case, the damage morphology is completely isotropic and does not demonstrate any preferable orientation in the direction of laser beam propagation (Glebov 2002).

However, experiments demonstrate that the absence of the filament-like damage patterns does not guarantee suppression of self-focusing during bulk LID. To provide a more reliable control on contribution of self-focusing, Soileau et al. (1989) noted that self-focusing would depend on polarization of laser light according to Equation 4.62. That dependence results from the fact that the components of nonlinear susceptibility tensor that build the nonlinear refraction coefficient n_2 depend on light polarization (Sutherland 2003, Boyd 2008), that is, they are different for linear and circular polarizations. This immediately implies that the ratio of critical self-focusing power P_{CR}^C for circular polarization to that for linear polarization P_{CR}^L is not 1.0 and depends on symmetry and structure of a solid. For example, the ratio is about 1.5 in the case of fused silica (Soileau et al. 1989). Therefore, if LID thresholds are measured for circularly and linearly polarized light in the same sample of fused silica at the same values of other laser parameters, and their ratio is 1.5, the LID is driven by self-focusing. Measurements for linear and circular polarizations of light clearly demonstrate that the ratio of LID thresholds for the two polarizations is 1.5 for spots with radius larger than approximately 20–30 μm. This can be interpreted as domination of self-focusing in initiation of LID at large focal spots. For smaller laser spots, the ratio of LID thresholds deviates from 1.5 and approaches 1.0 at spot radius slightly below 5 μm (Soileau et al. 1989).

Nonetheless, the absence of self-focusing of laser beams preliminary focused into the small spots is not guaranteed by the fact that the ratio of LID thresholds for linear and circular polarizations is 1. The reason for that is the assumption that the polarization of laser light is constant (e.g., linear polarization stays linear) during self-focusing (Akhmanov et al. 1992, Agraval 1995, Boyd 2008). Under that approximation, self-focusing is analyzed using a scalar wave equation (Akhmanov et al. 1992, Agraval 1995, Boyd 2008) that is acceptable within *paraxial approximation*, that is, if laser beam converges or diverges at small angles to its axis. Large angles of convergence should be expected when laser beam approaches a self-focus. Theoretical analysis of full vector wave equation (Gruzdev and Libenson 1994, 2000) shows that the paraxial approximation is significantly violated in the case of high-power laser light focused into a spot whose radius is smaller than 10 laser wavelengths. In particular, energy of initial vector components of electric field of laser light is transferred to all the other components by nonlinear coupling via the intensity-dependent term of refractive index. The nonlinear coupling of different electric-field components results in transformation of any initial polarization (either linear or circular) into elliptical polarization with specific distribution of polarization parameters at self-focus point (Figure 4.21). This effect becomes stronger with the increase in peak laser

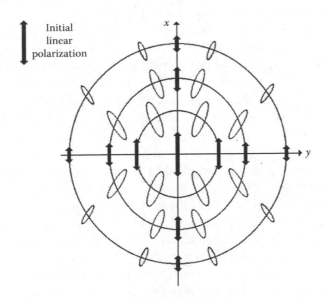

FIGURE 4.21   Space distribution of polarization at focal plane of the Gaussian TEM_{00} beam with initial linear polarization along x-axis. Focal-spot radius $\varpi_0 = 7.5\lambda$ makes 6.0 μm at laser wavelength $\lambda = 0.8$ μm. Peak laser intensity is $I = 2.5$ TW/cm². Calculated parameters of polarization are shown by arrows and ellipses. Coordinate axes and circles that correspond to $s^2 = (x^2 + y^2)/\varpi_0^2 = 0.3, 0.7, 1.0$ are also shown. (From Gruzdev, V.E. and Libenson, M.N., Polarization state of sharply focused light beam, *Proc. SPIE*, 2428, 444–457, 2000. Published with permission from SPIE.)

irradiance and reduction of laser spot size toward single laser wavelength (Gruzdev and Libenson 1994, 2000). That depolarization might be responsible for the specific polarization dependence of LID threshold reported for small laser spots (Soileau et al. 1989) since it tends to produce similar polarization patterns for both initial linear and initial circular polarizations at self-focus area. Therefore, reduction of the ratio of LID thresholds for the two different polarizations with decrease of laser spot radius down to few micrometers does not mean suppression of self-focusing but might result from the depolarization effect.

This problem is a part of a broader group of issues related to lack of reliable information on structure of irradiance distribution around self-focus points. Simulation of space distribution of laser irradiance and electric field requires utilizing of full vector wave equation coupled with equations for laser-driven electron dynamics at a self-focus point. In spite of technical challenges of such simulation, it is vitally important to understand the processes that initiate and drive LID by self-focused laser beams. For example, generation of electron–hole plasma makes negative contribution to refractive index that depends on plasma density n (Boyd 2008):

$$n_{REF} = n_0 + n_2 I(t) - \frac{2\pi e^2}{n_0 \, m \, \omega^2} n\big(I[t]\big).$$ (4.64)

Therefore, the laser-induced ionization supports the arrest of collapse of a self-focused laser beam and blocks the increase in laser irradiance at the self-focus point. The ionization and propagation effects are nonlinearly coupled and can exhibit very nontrivial nonlinear dynamics at a self-focus (e.g., see Gulley 2012).

Since no accurate simulations have been done for the self-focus points so far, most researchers rely on empirical approaches and experimental results to suppress self-focusing. For example, specific focusing systems with numerical aperture of 1.5 and larger are employed to focus laser beams into ultrasmall spots with subwavelength radius (Schaffer et al. 2001, Glebov 2002, Juodkazis et al. 2008). Full angle of beam convergence produced by those focusing systems is close to 90°, so that diffraction divergence of such beams approaches its fundamental limit. This method utilizes the fact that LID threshold is determined by irradiance (i.e., power divided by spot area) or fluence (energy divided by spot are), while self-focusing is controlled by the power of laser beam. Since irradiance and fluence scale as inverse cross-section area of laser beam, reduction of cross section by focusing of a beam into ultra-small spot allows reaching high levels of irradiance (or fluence) keeping peak power of laser pulse below the threshold of self-focusing. For example (Schaffer et al. 2001, Glebov 2002, Juodkazis et al. 2008), ultra-small spots produced by focusing systems with numerical aperture above 1.6 allow reaching irradiance above 10 TW/cm² at peak power 3–10 times smaller than the critical power of self-focusing.

An alternative approach to suppress self-focusing is to employ thin and ultrathin samples of transparent materials (Jasapara et al. 2001, Temnov et al. 2006, Jupe et al. 2009, Emmert et al. 2010, Rudolph et al. 2013). If sample thickness is much smaller than self-focusing distance given by Equation 4.63, self-focusing has no enough distance to develop into any remarkable collapse even if peak laser power is close or slightly above the critical power of Equation 4.62.

4.4.2   Self-Induced Modifications of Laser Spectrum

Another group of nonlinear propagation effects associated with self-induced modifications of spectrum also significantly influences initiation of LID and its threshold. This group includes two major processes—generation of higher harmonics and generation of super-continuum—that are briefly overviewed as follows:

1. Generation of higher harmonics (Sutherland 2003, Boyd 2008) considers enrichment of laser spectrum initially centered at frequency ω by components at higher frequencies, for example, 2ω (second harmonic) or 3ω (third harmonic). Fundamental regulations of higher-harmonic generation and experiments demonstrate that efficiency of energy conversion from the fundamental frequency to higher harmonics drops down very much with the increase in harmonic order (Agraval 1995, Boyd 2008). Therefore, generation of second and third harmonics is of major significance for LID.

    Generation of those harmonics can be intentional or accidental. The former case is related to nonlinear crystals utilized with the purpose of conversion of laser light into second or third harmonic, for example, potassium dihydrogen phosphate (KDP) (Sutherland 2003, Boyd 2008). Conversion efficiency can reach 20%–40%, and, therefore, very significant intensity of a higher harmonic is combined with high intensity of the fundamental harmonic in bulk of the crystals. Generation of second harmonic takes place at interfaces of most transparent solids (Boyd 2008), but also it can happen in bulk. It is believed that isotropic solids do not support second-harmonic generation because of specific symmetry properties of their tensor of nonlinear susceptibility (Sutherland 2003, Boyd 2008). That is not true for laser irradiance approaching LID threshold, because strong electric field of laser light significantly distorts the original symmetry of a solid and can induce local time-varying anisotropy. Due to this reason, unintended second-harmonic generation is observed round the focal area of high-power laser beams both in isotropic and anisotropic materials. Generation of the third harmonic can be rather efficient (at least—comparable to generation of the second harmonic) in isotropic solids and can supply small amounts of high-frequency light that is coherent with high-power radiation of fundamental harmonic.

    Those effects significantly affect the photoionization rate and can generate laser-induced defects in solids (e.g., color centers). For example, if six photons are required to bridge band gap of a transparent solid at fundamental frequency, three photons and two photons can bridge it at second-harmonic and third-harmonic frequencies correspondingly. Rate of three-photon absorption exceeds the rate of six-photon absorption by a factor of 10^4–10^5 while the rate of two-photon absorption is 10–100 times larger than the rate of three-photon absorption (Boyd 2008). Therefore, the presence of higher harmonics can significantly increase the photoionization rate. Another important effect can arise from the fact that the radiation of higher harmonics is coherent with the radiation of fundamental harmonic. Therefore, one should expect that photons of higher harmonics are effectively coupled with photons of fundamental harmonic to induce interband transitions via simultaneous absorption of several different photons (see Figure 4.16 as an example). Also, one

Chapter 4

should note that the third harmonic of visible and near-infrared light is always ultraviolet light that can easily produce point defects like self-strapped excitons and color centers along laser-beam path (Glebov 2001, Siiman et al. 2009). Therefore, another contribution of higher-harmonic generation to initiation of LID can potentially come from enhanced defect generation by the third and the second harmonics. It is notable that the ionization processes by multifrequency light (that contains, e.g., the fundamental and the second harmonics) have not been theoretically analyzed in proper detail.

2. Another important effect of nonlinear propagation is self-induced spectrum broadening that results in *super-continuum generation* or *white-light generation* (Agraval 1995, Brodeur and Chin 1999, Boyd 2008, Siiman et al. 2009). This effect is associated with a very essential broadening of initial spectrum of laser light during propagation in transparent media (Brodeur and Chin 1999). The broadened spectrum exhibits specific spectral patterns (e.g., asymmetric structure of spectrum wings and characteristic peaks) and extends on both sides of the original laser spectral line. In most cases (Agraval 1995, Brodeur and Chin 1999, Boyd 2008), short-wavelength wing of the broadened spectrum is smaller than the long-wavelength wing. A remarkable point is that super-continuum generation is produced exclusively by high-power ultrashort (picosecond and femtosecond) laser pulses. The super-continuum generation can take place in bulk materials as well as in fibers. Evidently, white-light generation in bulk materials is of interest for analysis of LID. For that case, available experimental data indicate that the super-continuum generation is coupled with self-focusing (Brodeur and Chin 1999, Siiman et al. 2009). Due to this special feature, the super-continuum generation is utilized as a signal effect to detect development of self-trapping and self-focusing of ultrashort laser pulses. Coupling of the two effects naturally leads to a conclusion that the development of super-continuum generation requires certain minimum power and certain distance to be traveled by laser beam within a transparent medium. Those expectations have been confirmed by experiments (Brodeur and Chin 1999, Siiman et al. 2009), but the simple estimations of the critical power (Equation 4.62) and development distance (Equation 4.63) do not provide acceptable accuracy for the case of super-continuum generation. The major reason for that might be attributed to significant role of dispersion that is one of the major effects for ultrashort pulses, but it is neglected in estimations (Equations 4.62 and 4.63).

Mechanisms of the super-continuum generation are associated with the formation of electron–hole plasma, dispersion of linear and nonlinear parts of refractive index, dispersion of absorption, and self-phase modulation (Brodeur and Chin 1999). The latter effect is attributed to intensity-dependent modulation of pulse phase φ by the Kerr and plasma effects in laser-induced variations of refractive index given by Equation 4.64. To illustrate self-phase modulation, one can consider the simplest case of plane-wave propagation with coordinate x measuring position along the direction of propagation (Born and Wolf 2003). The wave phase reads as follows:

$$\varphi = \omega_0 t - \frac{2\pi n_{REF}}{\lambda} x = \omega_0 t - \frac{2\pi x}{\lambda} \left\{ n_0 + n_2 I(t) - \frac{2\pi e^2}{n_0 m \omega^2} n\big(I[t]\big) \right\}, \tag{4.65}$$

where ω_0 is the initial laser frequency. Enrichment of instant spectrum with extra frequencies due to the plasma and Kerr effects is clearly demonstrated by the following equation:

$$\omega(t) = \frac{\partial \phi}{\partial t} = \omega_0 - \frac{2\pi x}{\lambda} \left\{ n_2 \frac{\partial I(t)}{\partial t} - \frac{2\pi e^2}{n_0 m \omega^2} \frac{\partial n(I[t])}{\partial t} \right\}. \tag{4.66}$$

Equation 4.66 shows that the extra frequencies are controlled by the propagation distance x, laser intensity, and electron-plasma density. Therefore, the effect of super-continuum generation must accompany all high-power interactions of ultrashort pulses with transparent materials that result in electron-plasma generation. The effect can be weaker or stronger depending on balance between the Kerr effect and plasma generation.

From the viewpoint of LID, the super-continuum generation supports reduction of LID threshold, because it can produce very intensive ionization by short-wavelength wing of the broadened spectrum (Siiman et al. 2009). Due to increased energy of the photons of the short-wavelength wing, a minimum number of simultaneously absorbed photons can be reduced compared to that of original frequency. It means that, nonlinear absorption rate for the short-wavelength wing of broadened spectrum might be significantly higher than the rate at initial fundamental wavelength. Also, the short-wavelength wing of the broadened spectrum can produce defects (e.g., color centers) and further increase the rate of nonlinear absorption as it is discussed earlier. The contribution of those defects to LID is significant in two cases: (1) if laser pulse width exceeds the duration of defect formation (i.e., 150–200 fs for fused silica [Daguzan et al. 1995, Temnov et al. 2006]) and (2) if LID is produced by multiple pulses so that each pulse interacts with the defects produced by all preceding pulses (Siiman et al. 2009). Therefore, influence of that spectrum-broadening effect might be weak in case of single ultrashort pulse. For LID by multiple pulses and/or by pulses longer than time of defect generation, the short-wavelength tail of pulse spectrum can produce very significant influence via defect generation.

4.5 Summary

This chapter contains overview of two major groups of nonlinear processes associated with LID of transparent solids: nonlinear absorption and nonlinear propagation effects. Fundamental mechanisms of those effects are attributed to fast laser-driven electronic processes and to much slower laser-induced atomic motion. In this chapter, the fast electronic processes are considered in detail due to their essential contribution to the nonlinear absorption and initiation of LID by both short (nanosecond and picosecond) and ultrashort laser pulses.

The nonlinear absorption includes contributions from the photoionization and impact ionization. The photoionization is a zero-inertia effect that begins at the front edge of a laser pulse. At high irradiance, the photoionization takes place in the form of high-frequency tunneling ionization. At low irradiance, the photoionization becomes the multiphoton ionization. The rate of photoionization and the rate of associated nonlinear absorption exhibit strong nonlinear dependence on laser parameters with characteristic singularities.

The impact ionization starts after the photoionization has provided enough seed electrons in the conduction band. Its development is delayed due to the need to generate the

Chapter 4

seed electrons and due to the specific inertia of the avalanche process. Compared to the photoionization, its rate is very high and scales exponentially with laser irradiance.

Both native and laser-induced defects contribute to the nonlinear absorption and initiation of LID. They can increase the photoionization rate by providing intermediate energy levels for interband electron transitions, but also can reduce it by capturing valence-band electrons at trap states. More significant influence is produced by the defects on the impact ionization, in particular, shallow traps provide seed conduction-band electrons for impact development more effectively than interband photoionization transitions. Influence of specific defects should be accurately analyzed bearing in mind defect and material parameters for each particular case.

Among various nonlinear propagation effects, self-focusing and self-induced modification of light spectrum including generation of higher harmonics and super-continuum make the most significant contribution to the initiation of LID. Basic mechanisms, special features, and contribution of those processes to initiation of LID of transparent solids are discussed.

References

Abramowitz M., Stegun I. A. 1972. *Handbook of Mathematical Functions*. 10th ed. Washington, DC: National Bureau of Standards.

Agraval G. P. 1995. *Nonlinear Fiber Optics*. New York: Academic Press.

Akhmanov S. A., Vysloukh V. A., Chirkin A. S. 1992. *Optics of Femtosecond Laser Pulses*. New York: American Institute of Physics.

Anisimov S. I., Imas, Ya. A., Romanov G. S., Khodyko Yu. V. 1971. *Action of High-Power Laser Radiation on Metals*. Springfield, VA: National Technical Information Service.

Apostolova T., Hahn Y. 2000. Modeling of laser-induced breakdown in dielectrics with subpicosecond pulses. *J. Appl. Phys.* 88: 1024–1034.

Arnold D., Cartier E., DiMaria D. J. 1992. Acoustic-phonon runaway and impact ionization by hot electrons in silicon dioxide. *Phys. Rev. B* 45: 1477–1480.

Ashkroft N. W., Mermin, N. D. 1976. *Solid State Physics*. New York: Saunders College Publishing.

Audebert P., Daguzan Ph., Dos Santos A. et al. 1994. Space-time observation of an electron gas in SiO_2. *Phys. Rev. B* 73: 1990–1993.

Balling P., Schou J. 2013. Femtosecond-laser ablation dynamics of dielectrics: Basics and applications for thin films. *Rep. Prog. Phys.* 76: 036502.

Bloembergen N. 1974. Laser-induced electric breakdown in solids. *IEEE J. Quant. Electron.* QE10: 375–386.

Bonch-Bruevich A. M., Khodovoi V. A. 1965. Multiphoton processes. *Sov. Phys. Usp.* 8: 1–38.

Born M., Wolf E. 2003. *Principles of Optics*. Cambridge, U.K.: Cambridge University Press.

Boyd R. W. 2008. *Nonlinear Optics*. 3rd ed. New York: Elsevier-Academic Press.

Braunlich P. F., Brost G., Schmid A., Kelley P. J. 1981. The role of laser-induced primary defect formation in optical breakdown of NaCl. *IEEE J. Quant. Electron.* QE-17: 2034–2041.

Braunstein R. 1962. Nonlinear optical effects. *Phys. Rev.* 125: 475–477.

Brodeur A., Chin S. L. 1999. Ultrafast white-light continuum generation and self-focusing in transparent condensed media. *J. Opt. Soc. Am. B* 16: 637–650.

Bulgakova N. M., Stoian R., Rosenfeld A. 2010. Laser-induced modification of transparent crystals and glasses. *Quant. Electron.* 40: 966–985.

Carr C. W., Radousky H. B., Demos S. G. 2003. Wavelength dependence of laser-induced damage: Determining the damage initiation mechanism. *Phys. Rev. Lett.* 91:127402.

Christodoulides D. N., Khoo I. C., Salamo G. J., Stegeman G. I., Van Stryland E. W. 2010. Nonlinear refraction and absorption: mechanisms and magnitudes. *Adv. Opt. Photon.* 2: 60–200.

Daguzan Ph., Martin P., Guizard S., Petite G. 1995. Electron relaxation in the conduction band of wide-band-gap oxides. *Phys. Rev. B* 52:17099–17105.

Diels J.-C., Rudolph W. 2006. *Ultrashort Laser Pulse Phenomena: Fundamentals, Techniques, and Applications on a Femtosecond Time Scale.* 2nd ed. New York: Academic Press.

Du D., Liu X., Mourou G. 1996. Reduction of multi-photon ionization in dielectrics due to collisions. *Appl. Phys. B* 63: 617–621.

Efimov O., Juodkazis S., Misawa H. 2004. Intrinsic single- and multiple-pulse laser-induced damage in silicate glasses in the femtosecond-to-nanosecond region. *Phys. Rev. A* 69: 042903.

Emmert L. A., Mero M., Rudolph W. 2010. Modeling the effect of native and laser-induced states on the dielectric breakdown of wide band gap optical materials by multiple subpicosecond laser pulses. *J. Appl. Phys.* 108: 043523.

Epifanov A. S. 1981. Theory of electron-avalanche ionization induced in solids by electromagnetic waves. *IEEE J. Quant. Electron.* QE-17: 2018–2022.

Epstein E. M. 1970 Scattering of electrons by phonons in a strong radiation field. *Sov. Phys. Solid State* 11: 2213–2217.

Gamaly E. G. 2011. The physics of ultra-short laser interaction with solids at non-relativistic intensities. *Phys. Reports* 508: 91–243.

Gamaly E. G., Rode A. V. 2013. Physics of ultra-short laser interaction with matter: From phonon excitation to ultimate transformations. *Prog Quant Electron* 37: 215–323.

Glebov L. B. 2001. Optical absorption and ionization of silicate glasses. *Proc. SPIE* 4347: 321–330.

Glebov L. B. 2002. Intrinsic laser-induced breakdown of silicate glasses. *Proc. SPIE* 4679: 321–330.

Gruzdev V. 2007. Photoionization rate in wide band-gap crystals. *Phys. Rev. B* 75: 205106.

Gruzdev V. 2013. How the laser-induced ionization of transparent solids can be suppressed. *Proc. SPIE* 8885: 88851S.

Gruzdev V. E. 2004. Analysis of the transparent-crystal ionization model developed by L. V. Keldysh. *J. Opt. Technol.* 71: 504–508.

Gruzdev V. E., Chen J. K. 2008. Laser-induced ionization and intrinsic breakdown of wide band-gap solids. *Appl. Phys. B* 90: 255–261.

Gruzdev, V. E., Libenson M. N. 1994. Polarization state of sharply focused light beam. *Proc. SPIE* 2428: 444–457.

Gruzdev V. E., Libenson M. N. 2000. Propagation of high-power tightly focused laser beams and self-depolarization effect. *Proc. of SPIE* 4065: 884–895.

Gulley J. R. 2012. Ultrafast laser-induced damage and the influence of spectral effects. *Opt. Eng.* 51: 121805.

Hentschel M., Kienberger R., Spielman Ch., Reider G. A., Milosevic N., Brabec T., Corkum P., Heinzmann U., Drescher M., Krausz F. 2001. Attosecond metrology. *Nature* 414: 509–513.

Holway L. H., Fradin D. W. 1975. Electron avalanche breakdown by laser radiation in insulating crystals. *J. Appl. Phys.* 46: 279–291.

Jasapara J., Nampoothiri A. V. V., Rudolph W., Ristau D., Starke K. 2001. Femtosecond laser pulse induced breakdown in dielectric thin films. *Phys. Rev. B* 63: 045117.

Jia T. Q., Chen H. X., Huang M., Zhao F. L., Li X. X., Xu S. Z., Sun H. Y., Feng D. H., Li C. B., Wang X. F., Li R. X., Xu Z. Z., He X. K., Kuroda H. 2006. Ultraviolet-infrared femtosecond laser-induced damage in fused silica and CaF2 crystals. *Phys. Rev. B* 73: 054105.

Jing X., Shao J., Zhang J., Jin Y., He H., Fan Zh. 2009. Calculation of femtosecond pulse laser induced damage threshold for broadband antireflective microstructure arrays. *Opt. Express* 17: 24137–24152.

Jones H. D., Reiss H. R. 1977. Intense-field effects in solids. *Phys. Rev. B* 16: 2466–2473.

Jones S. C., Braunlich P., Casper R. T., Shen X.-A., Kelley P. 1989. Recent progress on laser-induced modifications and intrinsic bulk damage of wide-gap optical materials. *Opt. Eng.* 28: 1039–1068.

Juodkazis S., Kondo T., Rode A., Gamaly E., Matsuo S., Misawa H. 2004. Three-dimensional recording and structuring of chalcogenide glasses by femtosecond pulses. *Proc. SPIE* 5662: 179–186.

Juodkazis S., Mizeikis V., Matsuo S., Ueno K., Misawa H. 2008. Three-dimensional micro- and nano-structuring of materials by tightly focused laser radiation. *Bull. Chem. Soc. Jpn.* 81: 411–448.

Jupe M., Jensen L., Melninkaitis A., Sirutkaitis V., Ristau D. 2009. Calculations and experimental demonstration of multi-photon absorption governing fs laser-induced damage in titania. *Opt. Express* 17: 12269–12278.

Kaiser A., Rethfeld B., Vicanek M., Simon G. 2000. Microscopic processes in dielectrics under irradiation by subpicosecond laser pulses. *Phys. Rev. B* 61: 11437–11450.

Kane E. O. 1957. Band structure of indium antimonide. *J. Phys. Chem. Solids* 1: 249–261.

Keldysh L. V. 1958. Behavior of non-metallic crystals in strong electric fields. *Sov. Phys. JETP* 6: 763–770.

Keldysh L. V. 1965. Ionization in the field of a strong electromagnetic wave. *Sov. Phys. JETP* 20: 1307–1314.

Kittel Ch. 1985. *Quantum Theory of Solids.* New York: John Wiley & Sons Inc.

Krausz F., Ivanov M. 2009. Attosecond physics. *Rev. Mod. Phys.* 81: 163–234.

Landau L. D., Lifshitz E. M. 1958. *Quantum Mechanics.* New York: Pergamon Press.

Chapter 4

Mao S. S., Quere F., Guizard S., Mao X., Russo R. E., Petite G., Martin P. 2004. Dynamics of femtosecond laser interactions with dielectrics. *Appl. Phys.* A 79: 1695–1709.

Melnikov V. I. 1969. Quantum kinetic equation for electrons in a high-frequency field. *Sov. Phys. JETP Lett.* 9: 120–122.

Menoni C. S., Langston P. F., Krous E., Patel D., Emmert L. A., Makosyan A., Reagan B. et al. 2012. What role do defects play in the laser damage behavior of metal oxides? *Proc. SPIE* 8530: 85300J.

Nathan V., Guenther A. H., Mitra S. S. 1985. Review of multiphoton absorption in crystalline solids. *J. Opt. Soc. Am.* B 2: 294–316.

Nazareno H. N., Gallardo J. C. 1989. Bloch oscillations of an electron in a crystal under action of an electric field. *Phys. Status Solidi (b)* 153: 179–184.

Nikiforov A. M., Epifanov A. S., Garnov S. V. 2012. Peculiarities of the electron avalanche for the case of relatively large photon energy. *Opt. Eng.* 51: 121813.

Quere F., Guizard S., Martin Ph. 2001. Time-resolved study of laser-induced breakdown in dielectrics. *Europhys. Lett.* 56: 138–144.

Ready J. F. 1971. *Effects of High-Power Laser Radiation*. New York: Academic Press.

Rethfeld B. 2004. Unified model for the free-electron avalanche in laser-irradiated dielectrics. *Phys. Rev. Lett.* 92: 187401.

Rethfeld B. 2006. Free-electron generation in laser-irradiated dielectrics. *Phys. Rev.* B 73: 035101.

Rethfeld B., Sokolowski-Tinten K., von der Linder D., Anisimov S. I. 2004. Timescales in the response of materials to femtosecond laser excitation. *Appl. Phys.* A 79: 767–769.

Rethfeld B., Brenk O., Medvedev N., Krutsch H., Hoffmann D. H. H. 2010. Interaction of dielectrics with femtosecond laser pulses: application of kinetic approach and multiple rate equation. *Appl. Phys.* A 101: 19–25.

Ridley B. K. 1993. *Quantum Processes in Semiconductors*. 3rd ed. New York: Oxford University Press, Inc.

Rudolph W., Emmert L., Sun Z., Patel D., Menoni C. 2013. Laser damage of thin films: What we know and what we don't. *Proc. SPIE* 8885: 888516.

Saeta P. N., Greene B. I. 1993. Primary relaxation processes at the band edge of SiO_2. *Phys. Rev. Lett.* 70: 3588–3591.

Schaffer C., Brodeur A., Mazur E. 2001. Laser-induced breakdown and damage in bulk transparent materials induced by tightly focused femtosecond laser pulses. *Meas. Sci. Technol.* 12: 1784–1794.

Seitz F. 1949. On the theory of electron multiplication in crystals. *Phys. Rev.* 76: 1376–1393.

Siiman L. I., Lumeau J., Glebov L. B. 2009. Nonlinear photoionization and laser-induced damage in silicate glasses by infrared ultrashort laser pulses. *Appl. Phys.* B 96: 127–134.

Simanovaskii D. M., Schwettman H. A., Lee H., Welch A. J. 2003. Midinfrared Optical Breakdown in Transparent Dielectrics. *Phys. Rev. Lett.* 91: 107601.

Sirdeshmukh D. B., Sirdeshmukh L., Subhadra K. G. 2001. *Alkali Halides: A Handbook of Physical Properties*. Berlin, Germany: Springer-Verlag.

Soileau, M. J., Williams, W. E., Mansour, N., VanStryland, E. W. 1989. Laser-induced damage and the role of self-focusing. *Opt. Eng.* 28:1133–1145.

Sparks M. 1975. Theory of electron-avalanche breakdown in solids. *NBS Spec. Publ.* 453: 331.

Sparks M., Mills D. L., Warren R., Holstein T., Maradudin A. A., Sham L. J., Loh E., King D. F. 1981. Theory of electron-avalanche breakdown in solids. *Phys. Rev.* B 24: 3519–3536.

Starke K., Ristau D., Welling H., Amotchkina T. V., Trubetskov M., Tikhonravov A. A., Chirkin A. S. 2004. Investigation of the nonlinear behavior of dielectrics by using ultrashort pulses. *Proc. SPIE* 5273: 501–514.

Stuart B. C., Feit M. D., Herman S., Rubenchik A. M., Shore B. W., Perry M. D. 1996. Nanosecond-to-femtosecond laser-induced breakdown in dielectrics. *Phys. Rev.* B 53: 1749–1761.

Sudzius M., Lyssenko V. G., Löser F. et. al. 1998. Optical control of Bloch-oscillation amplitudes: from harmonic spatial motion to breathing modes. *Phys. Rev.* B 57: R12693–R12696.

Sundaram S. K., Mazur E. 2002. Inducing and probing non-thermal transitions in semiconductors using femtosecond laser pulses. *Nat. Mater.* 1: 217–224.

Sutherland R. S. 2003. *Handbook of Nonlinear Optics*. 2nd ed. New York: Marsel Dekker, Inc.

Temnov V. V., Sokolowski-Tinten K., Zhou P., El-Khamhawy A., von der Linde D. 2006. Multiphoton ionisation in dielectrics: Comparison of circular and linear polarization. *Phys. Rev. Lett.* 97: 237403.

Thornber K. K. 1981. Applications of scaling to problems in high-field electronic transport. *J. Appl. Phys.* 52: 279–290.

Tien A.-C., Backus S., Kapteyn H., Murnane M., Mourou G. 1999. Short-pulse laser damage in transparent materials as a function of pulse duration. *Phys. Rev. Lett.* 82: 3883–3886.

Wood R. M. 2003. *Laser-Induced Damage in Optical Materials*. Bristol, U.K.: Institute of Physics Publishing.

Yee J. H. 1971. Four-photon transition in semiconductors. *Phys. Rev.* B 3: 355–360.

Yee J. H. 1972. Three-photon absorption in semiconductors. *Phys. Rev.* B 5: 449–458.

5. Femtosecond Laser–Induced Damage in Dielectric Materials

Luke A. Emmert and Wolfgang Rudolph

5.1   Introduction

In this chapter, we focus on femtosecond (fs) laser damage in dielectric materials. Fs laser damage studies of wide bandgap dielectric materials have gained importance for many practical applications such as laser micromachining and for the development of optical components for ever more powerful continuous wave (cw) and pulsed lasers. Fs laser pulses have also been used to mitigate damage pits produced by another laser, which is particularly important for extending the life of expensive large-scale optical components (Wolfe et al. 2011). Although this chapter focuses on thin films, many conclusions apply also to bulk materials.

Laser-Induced Damage in Optical Materials. Edited by Detlev Ristau © 2015 CRC Press/Taylor & Francis Group, LLC. ISBN: 978-1-4398-7216-1.

Chapter 5

The onset of optical damage is preceded by energy transfer from the laser pulse to the material. This process as well as the response of the material that leads to visible damage, for example, damage craters, strongly depends on the class of material, that is, whether the material is a metal, a semiconductor, or a dielectric. Metals can be excited by linear absorption within a thin skin layer near the material surface. In semiconductors, linear absorption is observed when the photon energy exceeds the bandgap energy. The optical energy is initially deposited into the electronic system. This is followed by electron–lattice (phonon) interactions, leading to an increase in the material temperature. If the density of energy deposited (J cm^{-3}) is larger than some critical value, depending on the material's phase diagram, damage proceeds by melting and vaporization or ejection of clusters from the surface (spallation) (Rethfeld et al. 2004). The threshold fluence, F_{th} (J cm^{-2}), or *laser-induced damage threshold* (LIDT), is related to the product of the critical energy density and absorption skin depth.

In high-quality dielectric materials with no or negligibly small linear absorption due to impurities and thermally exited carriers, the initial energy deposition must proceed through a multiphoton ionization (MPI) of the valence band (VB) in order to create quasi-free electron population in the conduction band (CB). Once in the CB, the electrons have the freedom to increase their energy by absorption of the latter part of the pulse. These hot electrons can impact ionize electrons from the valence band leading to a rapid increase in the CB electron density (avalanche ionization). When the electron density reaches the critical density N_{crit}, at which the plasma frequency equals the laser frequency, the material becomes strongly absorbing (N_{crit} ~10^{21} cm^{-3} for 800 nm excitation). The remainder of the pulse can thus deposit its energy very efficiently into a skin layer resulting in material damage. In dielectric materials, the actual processes resulting in visible damage are the Coulomb explosion, ablation, melting, and fracture, depending on pulse duration and repetition rate (Stuart et al. 1996, Stoian et al. 2000, Bulgakova et al. 2010). In the presence of a DC electric field of sufficient magnitude, the increase in conductivity due to CB electrons leads to the classical dielectric breakdown. In analogy, the initiation of optical damage is often called *laser-induced dielectric breakdown*, and the required AC field can be related to the DC critical field (Bloembergen 1974). Being independent of the specific material-dependent damage mechanism, the critical electron density is a useful quantity to explain common features of breakdown in dielectric materials.

Compared with picosecond (ps) and nanosecond (ns) pulses, fs laser damage appears to be much more deterministic (Joglekar et al. 2003). The reason is that long pulse damage is often initiated by the absorption at randomly distributed material defects even if their concentration is small (Jones et al. 1989). This defect absorption can lead to damage spots by direct heating and by providing seed electrons for impact ionization with or without subsequent heating (Sparks et al. 1981). In contrast, because of their higher intensities, fs pulses can excite band-to-band transitions in dielectric materials efficiently via multiphoton ionization, which dominates defect absorption in high-quality bulk materials and coatings. As a result, single fs-pulse damage thresholds of uniform high-quality dielectric films exhibit uncertainties of only a few percent, and damage studies on dielectric materials have been extended to pulse durations as short as 5 fs (Lenzner et al. 1998).

While the fundamental mechanisms leading to a critical electron density in the CB and to subsequent laser damage were identified early on (Keldysh 1965, Bloembergen 1974), it remained a challenge to predict LIDTs of dielectric materials reliably and to identify scaling laws for ns and ps damage. The main reason was the role that material defects played in the damage initiation and the subsequent degree of randomness (Jones et al. 1989). Another reason was that early experimental studies did not always clearly distinguish experimentally between multiple-pulse (S-on-1) and single-pulse (1-on-1) damage. This is partly true also for early fs damage studies. These two testing methodologies probe different material properties and give different results. In addition, damage in bulk materials is also affected by a variety of nonlinear and propagation effects depending on the pulse duration and bulk geometry such as the Brillouin scattering, self-focusing, and self-phase modulation (Smith and Do 2008). Fs pulses are more likely to probe intrinsic material properties, and thin films can minimize the impact of undesired effects associated with pulse propagation. Figure 5.1 summarizes the principal processes controlling single- and multiple-pulse damage.

In this chapter, we first discuss the excitation mechanisms leading to a critical electron density in the CB of dielectric materials. General scaling laws of the single-pulse fluence-producing damage, F_{th}, are derived and compared with experimental data. This is followed by a discussion of multiple-pulse (S-on-1) damage in terms of the accumulation of occupied defect and trap states during the pulse train. Finally, we will describe how the environment (atmosphere composition and pressure) can affect the damage behavior. Whenever possible we will keep the discussion applicable to dielectric materials in general. Most of the experimental results refer to dielectric metal oxides and thin films.

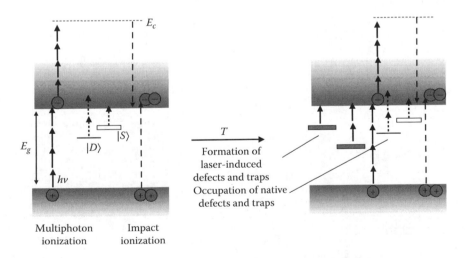

FIGURE 5.1 A single excitation event (pulse) excites electrons from the VB to the CB via a combination of multiphoton ionization and impact ionization. In a real material, initially populated deep ($|D\rangle$) and shallow ($|S\rangle$) defects are present at low densities that can also be ionized. Between pulses (separation T), the material does not relax completely, leading to occupied midgap levels (native traps and laser-induced defect states). Excitation of these states during subsequent pulses increases the CB electron density seeding impact ionization and thus lowers the damage threshold.

5.2 Modeling Femtosecond Laser Damage

5.2.1 General Considerations

Let us consider first what happens when an fs laser pulse of duration τ_p interacts with a dielectric (transparent) material and how this interaction changes if the fluence is increased (see Figure 5.2). At low fluences, the interaction is governed by the dielectric constant (refractive index) and pulse propagation is only affected by dispersion and diffraction. The laser field excites harmonic dipole oscillations (almost) instantaneously, and the material recovers to its initial state after the pulse. With increasing fluence (intensity), the dipole oscillations become anharmonic and nonlinear optical processes like self-phase modulation and harmonic generation may occur (Boyd 2003, Diels and Rudolph 2006). Depending on the material, nonlinear interactions, for example, multiphoton absorption (MPA), can also lead to permanent material modifications such as refractive index changes. Such processes have been explored and utilized for waveguide structures and optical modulators in certain glasses and polymers; see, for example, Hirao et al. (2001). All of these processes proceed over a certain range of pulse fluences. As the fluence is increased further, a sharply defined threshold is reached at which visible surface damage (crater) or bulk damage (void or crack) is observed. This is the optical damage we will be concerned with in this chapter. Short–laser pulse material removal (ablation) at fluences larger than this damage threshold has become attractive for micro- and nanostructuring (Gattass and Mazur 2008). The range of threshold fluences for most dielectric materials is relatively narrow and resides between regions of

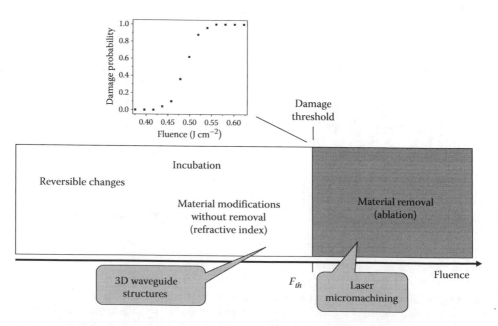

FIGURE 5.2 Possible outcomes of laser—dielectric material interactions as a function of fluence of a fs pulse. For a given material, the fs LIDT fluence, F_{th}, is sharply defined. Shown in the upper part is the measured damage probability for a hafnia film and 50 fs pulses. The threshold fluences for typical oxides are in a relatively narrow range about an average value F_{th} for a given pulse duration.

typical nonlinear optical interactions and where the incident laser field approaches the Coulomb field between electron and nucleus in an atom, enabling tunneling ionization.

Besides the pulse intensity (fluence), the time scale of the light matter interaction defines the material processes that affect the initial energy deposition. Two principal time scales are of importance—(1) the pulse duration ($\tau_p \leq 1$ ps) and (2) the pulse repetition period (typically $T \geq 1$ ns). Electron–electron and electron–lattice scattering typically proceed on time scales smaller than a few hundred fs; energy transfer to the lattice occurs within a few ps; relaxation into traps, formation of self-trapped excitons (STEs), and band-to-band relaxation have characteristic times ranging from fs to hundreds of ps depending on the material (Martin et al. 1997, Diels and Rudolph 2006). On much longer time scales, one can expect the formation and decay of color centers and heat diffusion, for example.

Using dielectric breakdown (i.e., a critical CB electron density) as damage criterion, models of laser damage predict the damage threshold by calculating the CB population as a function of pulse fluence and duration. The most comprehensive models, based on the Boltzmann and Fokker–Planck equations, attempt to calculate the electron distribution function (occupation vs. energy) of the CB (Stuart et al. 1996, Apostolova and Hahn 2000, Kaiser et al. 2000, Mero et al. 2005d, Christensen and Balling 2009). In addition to the interband transitions, these models include electron–electron as well as electron–phonon interactions. They are able to reproduce experimental trends, but absolute predictions are difficult, if not impossible, because of the many material parameters that are not well known. While lacking the rigor of the computer models, rate equations that treat the CB as a single state and use phenomenological cross sections and relaxation times have been used to describe LIDTs in terms of a few material constants (Du et al. 1994, Stuart et al. 1996, Tien et al. 1999, Mero et al. 2005b). Often, the so-obtained results are able to reproduce experimentally observed LIDT scaling laws and as such are a useful tool for practical applications. The following section develops the basic rate equations for band-to-band excitations and presents useful analytical solutions by assuming a square pulse shape.

5.2.2 Rate Equations for the Electron Density

Independent of the actual damage mechanism, energy must be transferred from the laser to the dielectric material, which involves electronic transitions from the valence band (VB) to the CB. If the CB is described as a single state with population density N, the generation rate of CB electron density can be written as

$$\frac{dN}{dt} = K(I) + A(I, N) - L(N). \tag{5.1}$$

The first term is the multiphoton ionization (MPI) rate, the second term describes the impact ionization, and the third term takes into account relaxation out of the CB. The depletion of the VB can often be ignored because the maximum density of electrons removed, N_{crit}, is several orders of magnitude smaller than a typical VB electron density.

To reach the CB, electrons must acquire an *effective bandgap* energy $\tilde{E}_g$, which is the sum of the bandgap energy E_g and the ponderomotive energy U_p. The latter is the time-averaged quiver energy of an electron in an external oscillating field, $U_p = e^2 E/(4m\omega_0^2)$,

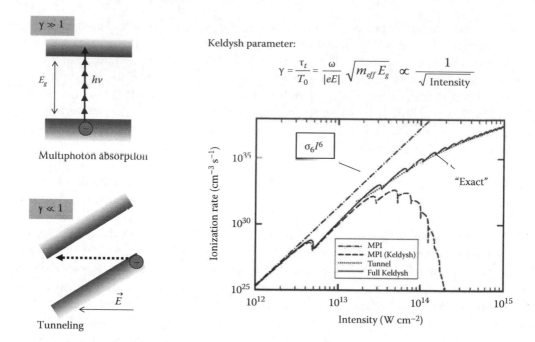

FIGURE 5.3   Band-to-band multiphoton ionization according to Keldysh for a bandgap energy $6\hbar\omega_0 < E_g < 7\hbar\omega_0$. The photoionization rate is produced by an interplay of two processes—MPA and tunneling, the relative weight of which is controlled by the Keldysh parameter γ. In the low-intensity limit, the excitation reduces to an m-photon absorption, at high intensities tunneling ionization dominates. (Adapted with kind permission from Springer Science+Business Media: *Strong Field Laser Physics*, Springer Series, vol. 134, 2009, p. 249, Lenzner, M. and Rudolph, W., Figure 5.5. Copyright Springer Science+Business Media, LLC.)

where E is the (linearly polarized) electric field amplitude; e and m are the electron charge and effective mass, respectively; and $\omega_0/2\pi$ is the field's carrier frequency.

The direct photo excitation rate, $K(I)$, from the VB to CB is often described by the theory of Keldysh (1965) developed for a monochromatic laser field. Figure 5.3 illustrates the ionization rate according to Keldysh and several limiting cases.

In the low- and high-intensity limits, the Keldysh formula reduces to MPA of order m and tunneling, respectively. In the MPA limit, the ionization rate is proportional to I^m. The value of the Keldysh parameter γ (cf. Figure 5.3) determines the relative weight of each of these processes in the MPI process. γ is the ratio of the tunneling time, τ_t, and the period of the laser field, $T_0 = 2\pi/\omega_0$. The full Keldysh theory (solid line) shows a series of steps where the ionization rate suddenly drops with increasing intensity. These steps occur when the quiver energy (and effective bandgap) increases by one quanta of energy, $\hbar\omega_0$.

Once in the CB, electrons can absorb energy from the still present excitation pulse. This process is sometimes called free-electron absorption and is also known as inverse bremsstrahlung. Note that a totally free electron is unable to absorb photons because energy and momentum conservation cannot be satisfied simultaneously. However, electrons in the CB interact with the lattice during the interaction with the light pulse (collide with phonons), which enables momentum conservation and subsequent electron heating. The average rate at which an individual electron absorbs energy is given by $\sigma_e I$, where σ_e is the absorption cross section and I is the intensity of the field. The absorption cross section is related to the imaginary part of the refractive index n_i by $\sigma_e = 2\omega_0 n_i / cN$, where c is the speed of light in vacuum. Using the Drude model for the refractive index of a quasi-free electron gas (see, e.g., Klein and Furtak, 1986) and assuming an optically thin plasma, which is valid at electron densities below dielectric breakdown, one can derive

$$\sigma_e = \frac{e^2}{m\epsilon_0 c \gamma_e} \frac{1}{1 + (\omega_0/\gamma_e)^2}, \tag{5.2}$$

where
  γ_e is an effective electron collision rate
  ϵ_0 is the vacuum permittivity

Note that the effective electron mass m and collision rate γ_e can depend on the electron energy relative to the CB band edge, although this is neglected in many applications of the Drude model.

When an electron has reached a critical energy E_c, collision with a VB electron resulting in two electrons near the bottom of the CB becomes possible. This process is called impact ionization and can lead to an avalanche-like increase in the CB electron density. The critical energy E_c is higher than the effective bandgap $\tilde{E}_g$ due to energy and momentum conservation. For the case of parabolic bands, E_c is 50% higher than $\tilde{E}_g$. The impact ionization rate, $A(I, N)$, of Equation 5.1 can be estimated analytically using the "flux doubling model" (Stuart et al. 1996). According to this approximative approach, when an electron reaches energy E_c, it immediately collides with a VB electron resulting in two electrons at the bottom of the CB, and the process repeats. The impact ionization rate is therefore proportional to the rate at which electrons reach the critical energy, $\sigma_e I/E_c$, and the density of CB electrons, and it often appears in literature in the form

$$A(I, N) = aNI, \tag{5.3}$$

where $a = \sigma_e/E_c$ is the impact ionization coefficient. Note that the so-derived coefficient a in fact depends on the light intensity I through the quiver energy.

Within the frame of these approximations and neglecting relaxation processes during the interaction, the overall rate of increase in the CB electron density N is the sum

$$\frac{d}{dt} N = K(I) + a(I)NI, \tag{5.4}$$

where $K(I)$ is the Keldysh ionization rate. For a square pulse of duration τ_p and intensity I, the electron density immediately after excitation becomes

$$N(I, \tau_p) = \frac{K(I)}{a(I)I} \left[e^{a(I)I\tau_p} - 1 \right] + N_0 e^{a(I)I\tau_p}, \tag{5.5}$$

where N_0 is the initial electron density. This background electron density N_0 can arise from thermally excited CB states. Excitation of electrons from states near the band edge by a one-photon absorption will have a similar effect. Such states are associated with native traps and defects, and N_0 is expected to be small for high-quality wide-gap materials.

If the MPA limit $[K(I) = \beta_m I^m]$ and the approximation $a(I) \approx a_0$ are applicable, Equation 5.5 yields for the electron density in terms of the pulse fluence $F = I\tau_p$

$$N(F) = \frac{\beta_m}{a_0} \left(\frac{F}{\tau_p} \right)^{m-1} \left(e^{a_0 F} - 1 \right) + N_0 e^{a_0 F}. \tag{5.6}$$

The damage fluence F_{th} can be obtained numerically from Equation 5.6 using the condition $N(F = F_{th}) = N_{crit}$.

The second term in Equation 5.6 describes avalanche ionization seeded by the background electrons. The first term accounts for the electrons produced by a combination of avalanche ionization and MPA. As is obvious from Equation 5.6, for short-pulse durations, the MPA dominates and background carriers are less relevant for the initial production of seed electrons. For longer pulses, the background carriers have to provide the seed electrons. This is one of the reasons why long-pulse LIDT is more statistical in nature.

Another consequence from the structure of Equation 5.6 is the fact that the damage fluence is roughly only logarithmically dependent on the actual value of N_{crit}. Figure 5.4a exemplifies this for two different pulse durations. An order of magnitude change of N_{crit} changes F_{th} only by 10%.

In the absence of background carriers, larger fluences are required for longer pulses to produce seed electrons for avalanche ionization because of their comparatively lower

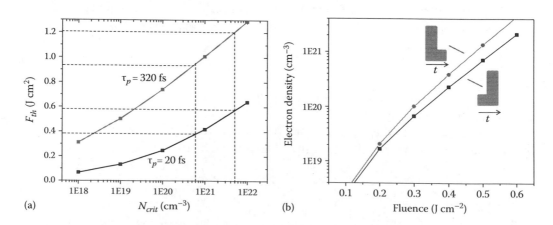

(a) (b)

FIGURE 5.4 (a) Damage fluence as a function of the value of the critical electron density for two different pulse durations. A four-photon MPA process was assumed; a_0 and β_4 values of HfO_2 films (Mero et al. 2005b) were used in Equation 5.6 to obtain the electron density. (b) CB electron density as a function of pulse fluence for the two pulse shapes shown exciting hafnia. The pulse duration of the main and satellite pulses are 50 fs. For a given total fluence, pulses with larger intensities at earlier times produce greater electron densities.

intensity (F/τ_p). The result is a relatively larger contribution from the impact ionization to the overall electron density (exponential term). Impact ionization is seeded by CB electrons that are produced by MPA. This makes the electron density sensitive to pulse shape (asymmetry) (see Figure 5.4b). Corresponding experimental trends were observed (Englert et al. 2007).

5.2.3 Additional Considerations

The previously introduced model explains fs damage threshold observations and trends in dielectric materials remarkably well despite the many simplifications and approximations that were necessary. Great progress has been made in the last decade to develop more sophisticated models that describe various aspects of energy deposition more accurately from a fundamental materials science point of view. However, due to the lack of knowledge of many material parameters that are needed in these models, broad application to predicting LIDTs remains elusive. We will briefly list some of the issues addressed by these (mostly) numerical simulations.

5.2.3.1 Local Fields

When discussing damage fluences or intensities, one has to distinguish between incident fields and fields that act at the location of damage. It was recognized early on that damage in dielectric slabs occurs preferentially at the output face (Crisp 1973). The superposition of incident and (partially) reflected field leads to a standing wave with maxima that exceed the field of the propagating wave. Dielectric films and stacks thereof are also examples where the incident laser produces characteristic standing wave patterns. When these local field enhancement factors are taken into account, the damage threshold can be predicted when the LIDT of the individual materials is known (Jasapara et al. 2001). For a given reflection behavior $R(\lambda)$, one can tailor the coating sequence to minimize the local field enhancements in the materials with the lowest LIDT (Starke et al. 2004, Chen et al. 2011) to increase the damage threshold of the stack. Similar field calculations are done for diffraction grating designs (Neauport et al. 2007).

During the excitation, the dielectric constant of the material changes from its initial value $\epsilon_i = n_0^2$. This leads to a time-dependent reflection coefficient R. To first order, this change is controlled by the time-dependent CB electron density and the Kerr nonlinearity. The relative permittivity of the medium

$$\epsilon(N) = 1 + \chi = n_0^2 - \frac{N/N_{crit}}{1 + i\gamma_e/\omega_0} + 2n_0 n_2 I, \tag{5.7}$$

where

  χ is the susceptibility
  $N_{crit} = \omega_0^2 \epsilon_0 m_e / e^2$ is the (critical) electron density at which the plasma frequency equals the laser frequency
  n_2 is the coefficient of the intensity-dependent refractive index $n(I) = n_0 + n_2 I$

Numerical methods can be used to study the time-dependent reflection in the presence of a standing wave (Jasapara et al. 2002).

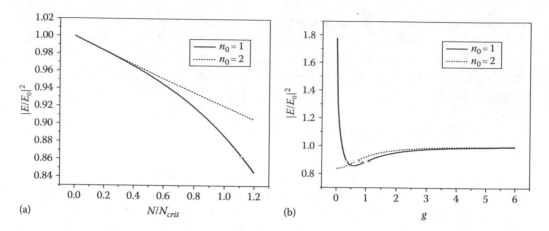

FIGURE 5.5 (a) Square of field enhancement as a function of the electron density for two different refractive indices n_0 of the nascent sample and $g = \gamma_e/\omega_0 = 1$. (b) Square of the field enhancement as a function of the normalized electron scattering rate g for two different refractive indices of the material and $N = N_{crit}$. The Lorentz field correction and Fresnel reflection were considered.

To illustrate the principal effects, let us assume a situation where we can neglect these Fabry–Perot effects and where the field entering the sample, E_s, is controlled by the Fresnel reflection at the air–sample interface

$$\frac{E_s}{E_0} = \frac{2}{1+\sqrt{\epsilon}}. \tag{5.8}$$

If we consider the plasma just as a perturbation of the optical properties of the sample, additional effects affecting the local field must be considered. The local field E acting on the material can be obtained from E_s taking into account the Lorentz field correction known from electrodynamics (Boyd 2003)

$$\frac{E}{E_s} = \frac{\epsilon+2}{3}. \tag{5.9}$$

Since ϵ depends on the electron density and the field intensity, cf. Equation 5.7, the field correction factor $|E/E_0|$ changes during excitation. Figure 5.5 shows examples of this effect.

It should be noted that the field correction becomes more pronounced at larger electron densities, that is, late in the pulse at the damage threshold. If the plasma forms as a thin skin layer, reflection at the plasma–sample material interface may have to be taken into account in addition. The total field in the plasma is then the sum of two counter-propagating waves $E' = E + E_{ref}$.

5.2.3.2 Keldysh Ionization

The Keldysh formalism (Keldysh 1965) of high-field ionization was originally developed for monochromatic electric fields with constant amplitude and the band model of Kane (1957). Effects like a finite pulse bandwidth, complex (realistic) band structures, and the dependence of the electron's effective mass on its energy above the band edge need to be taken into account. Reviews and description of recent developments can be found in Popov (2004), Gruzdev (2007), Jupé et al. (2009), and Gulley (2010).

5.2.3.3 Electron Heating and Impact Ionization

During the interaction with the fs pulse, CB electrons can scatter off other electrons, holes, and phonons, as well as relax into trap and defect states. In addition, the phonon distribution also changes. Numerical simulations based on the Boltzmann equation produce the temporal evolution of the phonon and electron (hole) distributions by solving collision integrals (Apostolova and Hahn 2000, Kaiser et al. 2000, Mero et al. 2005d, Petrov and Davis 2008, Christensen and Balling 2009). The difficulty in describing real materials and obtaining absolute damage fluences lies in the uncertainty of material parameters under high excitation conditions. In addition, in particular for multiple-pulse damage, the interaction of nascent and laser-induced defect states with the laser is very difficult to treat from first principles. Theoretical predictions about the presence and nature of material defects exist only for some materials.

In single-pulse damage, to keep the rate equation approach, one may assume that the impact ionization term becomes intensity dependent, for example, $a(I) = a_0 + a_1 I + a_2 I^2$, and add linear and bimolecular relaxation processes with effective rates k_1 and k_2, respectively, in Equation 5.4. The so-modified rate for the CB electron density becomes

$$\frac{d}{dt} N = K(I) + \left[a_0 + a_1 I + a_2 I^2 \right] IN - k_1 N - k_2 N^2. \tag{5.10}$$

In our simplified approach, the intensity dependence of the impact ionization parameter stems from the intensity dependence of the critical energy $E_c(I)$ and indirectly from the dependence of the collision rate γ_e and the effective mass on the electron energy. The former two causes can be analyzed with the help of the numerical approaches mentioned earlier. The latter would require band structure calculations combined with high-field light–matter interactions.

There have been attempts to measure $a(I)$ with fs pulses in fused silica (Rajeev et al. 2009). Data interpretation, however, is difficult because of the presence of relaxation processes and their affect on N and a during the pulse (see Equation 5.10).

The approximations made for the flux doubling model become increasingly unreliable for pulse durations below 100 fs (Stuart et al. 1996). One reason is that the electron excitation is considered to be instantaneous, neglecting transient processes in the interaction with other carriers and phonons while climbing the energy ladder in the CB. An overestimation of the avalanche ionization can thus be expected at early times, that is, for short pulses. This problem has been overcome by splitting the CB band into several sublevels spaced $\hbar\omega_0$. The electron heating up to the critical energy E_c then follows a step-like process (Rethfeld et al. 2010). This so-called multiple rate equation (MRE) model mimics the numerical simulations of the Boltzmann equation.

For very short pulses (<10 fs), observations of avalanche were explained using the "forest fire" model (Gaier et al. 2004). Enhanced impact ionization is predicted from the DC field surrounding the hole immediately after the creation of an electron–hole pair until the material can screen the charge.

Chapter 5

5.2.3.4 Material Relaxation and Defect States

Relaxation of electrons out of the CB during the pulse lowers N and leads to higher single-pulse damage thresholds (Mero et. al. 2005b). In S-on-1 experiments, the relaxation between pulses largely controls the multiple-pulse thresholds as will be explained later. For some common optical dielectric bulk materials, transient processes after excitation are well known. For example, in fused silica, CB electrons can form STEs very efficiently on a time scale of 150–250 fs, which decay into an E' color center within a few ten ps (Li et al. 1999, Petite et al. 1999, Grojo et al. 2010). Pump-probe measurements on dielectric oxide films also suggest initial relaxation steps on the order of a few hundred fs followed by a longer-lived state in the bandgap (Mero et al. 2005c).

The source of midgap trap states are defects in the material, and we will distinguish between the native (intrinsic property of the material) and laser-induced defects. First principle calculations of native point defects in dielectric wide-gap bulk materials, such as vacancies and interstitials, predict electronic states throughout the bandgap; see, for example, Foster et al. (2002) and Zheng et al. (2007) for hafnia films. Laser-induced defects based on the aforementioned STEs in ("perfectly") crystalline wide-gap materials, such as NaCl, SiO_2, and TiO_2, are well researched experimentally and theoretically (Song and Williams 1996, Ueta et al. 2001). They can have binding energies on the order of one eV and can form on fs time scales (Guizard et al. 1995, Martin et al. 1997). STEs are caused by lattice distortions around an excited electron. In fused silica, for example, an STE comprises a CB electron and a VB hole localized at a Si–O–Si (distorted) bridge (Saeta and Greene 1993, Guizard et al. 1996). During the relaxation of the STE, there is a small probability that it will displace an atom to form an interstitial–vacancy pair (color center) (Guizard et al. 1996). Since an atom has been displaced, the color center is long lived, lasting up to months.

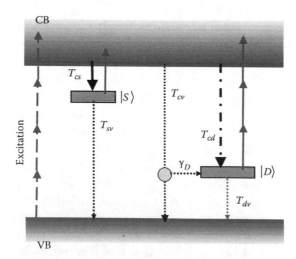

FIGURE 5.6 Excitation and relaxation processes involved in the interaction of dielectric materials with laser pulses. CB electrons can relax to shallow ($|S\rangle$) and deep traps ($|D\rangle$) and directly back to the VB. There is a certain possibility γ_D that during the relaxation new defect states form (laser-induced defects). Corresponding time constants T_i are shown. If the relaxation is bimolecular in nature, the rate $k = (T_i N_{max})^{-1}$, where N_{max} is the maximum defect density.

Important properties of these defects/states are their density, trapping rate, lifetime, and photoionization cross sections. Except for a few cases in bulk materials (e.g., fused silica and single crystals), little is known about the optical properties of defects in optical materials, particularly in films.

Figure 5.6 summarizes the basic generalized material processes involved in the interaction with a laser pulse and in the subsequent relaxation. While the physical origin and energy of the midgap states vary with the actual material, we distinguish for a general discussion only between native shallow traps ($|S\rangle$) that are within $\hbar\omega_0$ of the CB and deep traps ($|D\rangle$) with ionization energies $>\hbar\omega_0$. The latter can be either native or laser induced (i.e., created during a train of pulses).

5.3   Single-Pulse and Multiple-Pulse Damages

While mostly of academic interest, single-pulse (1-on-1) fs damage probes essentially the nascent material state, that is, the material without laser-induced modifications that accumulate during the exposure to a train of pulses. Historically, one of the most studied dependencies is the damage fluence as a function of pulse duration for a given material (Guenther and McIver 1993, Du et al. 1994, Stuart et al. 1996, Tien et al. 1999, Mero et al. 2005b). For pulses of duration of several ps and longer, a $F_{th} \propto \sqrt{\tau_p}$ relationship was often concluded from experimental studies. Such a square root dependence can be derived if one-dimensional transport of the deposited energy occurs during the pulse, and if damage is initiated when the material reaches the melting temperature T_m. One-dimensional transport can be expected if the absorption occurs in a layer whose thickness d is much smaller than the excitation beam diameter D. In this case, one-dimensional heat diffusion perpendicular to the layer dominates. To first order, the absorbed energy $W \approx \alpha F D^2$ is distributed during the pulse to a volume $V \approx D^2 \sqrt{\kappa_h \tau_p}$ with κ_h being the heat diffusion coefficient. The temperature increase thus scales as $\Delta T \propto W/V \propto F/\sqrt{\tau_p}$, and the damage fluence for which $\Delta T = T_m - T_0$ becomes $F \propto \sqrt{\tau_p}$.

For most practical applications, multiple-pulse damage, where the same sample site is illuminated multiple times, is important. In an S-on-1 damage test, a train of S equally spaced, identical pulses illuminates a single spot of the sample. The measured damage threshold, $F_{th}(S)$, is found to decrease with increasing number of pulses. This result implies laser-induced modifications to the material prior to damage, a process referred to as *incubation* (Rosenfeld et al. 1999). Later, we will provide, in detail, incubation results from the native and laser-induced defects and traps that become occupied and created during the pulse train.

5.3.1   Single-Pulse Damage (1-on-1)

Single-pulse damage experiments require the sample to be moved after each exposure to avoid incubation effects. Damage is thus determined by the excitation of the nascent sample.

Figure 5.7a shows the LIDT for a series of dielectric oxide films as a function of pulse duration (Mero et al. 2004, 2005b). All films were ion beam sputtered under similar conditions, and they differ in their refractive index and bandgap energy. Figure 5.7a

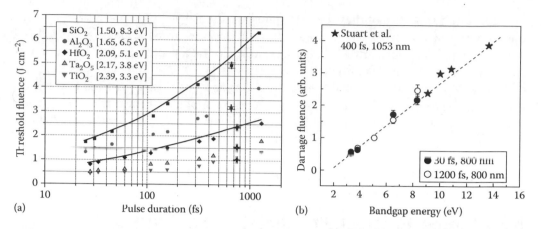

FIGURE 5.7 (a) Single-pulse damage fluence as a function of pulse duration for oxide thin films measured with pulses at 800 nm. Refractive index and bandgap are shown in brackets. The solid lines represent predictions from the scaling law Equation 5.11. (b) Breakdown fluence as a function of bandgap energy obtained with 30 fs (solid circles) and 1.2 ps (open circles) laser pulses. The data points are normalized to the damage fluence at $E_g = 5.1$ eV. The data shown by asterisks (bulk BaF_2, CaF_2, MgF_2, and LiF) were taken from Stuart et al. (1996). (Adapted Figures 5.1 and 5.3 with permission from Mero, M. et al., *Phys. Rev. B*, 71, 115109, 2005. Copyright 2005 by the American Physical Society.)

shows the LIDT as a function of pulse duration for these films. For a given material, the damage fluence increases with increasing pulse duration. For a given pulse duration, the damage threshold increases with the bandgap energy (see Figure 5.7b). The experimental data suggest a scaling law for the single-pulse damage threshold according to

$$F_{th}(E_g, \tau_p) \approx (c_1 + c_2 E_g)\tau_p^{\kappa}, \tag{5.11}$$

where
$\kappa = 0.30 \pm 0.03$
$c_1 = -0.16 \pm 0.02$ J cm^{-2} fs$^{-\kappa}$
$c_2 = 0.074 \pm 0.004$ J cm^{-2} fs$^{-\kappa}$ eV^{-1}

These scaling laws have been confirmed with binary oxide films, $Ti_xSi_{1-x}O_2$ (Nguyen et al. 2008) and $Hf_xSi_{1-x}O_2$ (Jensen et al. 2010), where the composition and bandgap could be tuned continuously by changing the composition parameter x. The coefficients c_1 and c_2 depend weakly on the film deposition and postdeposition treatment, for example, annealing. It is likely that these coefficients are different for other types of materials, for example, fluorides. It is interesting to note that a linear scaling of the LIDT with E_g was also found in bulk (crystalline) fluorides (Juodkazis et al. 2004).

Equation 5.11 is an empirical scaling law. It can be shown (Mero et al. 2005b) that the interplay of multiphoton and impact ionization as described by Equation 5.4 to reach a critical electron density at which damage occurs leads to the trends described by the aforementioned scaling law. In particular, the linear dependence of F_{th} on E_g results from the MPI bandgap dependence of the Keldysh multiphoton ionization rate. The exponent κ is weakly dependent on the impact ionization parameter a_0 and the MPI rate.

5.3.2   Multiple-Pulse Damage (S-on-1)

5.3.2.1   General

In the S-on-1 damage test, the measured damage threshold, $F_{th}(S)$, will typically be lower than that observed for a single-pulse $F_{th}(1)$ (from here on written as F_1). For large numbers of pulses, the threshold approaches a constant value, F_∞, the *multiple-pulse damage threshold*. Safe operation of optical components is possible for fluences $F < F_\infty$. Logically, damage must occur during the last pulse and all the preceding pulses change the material. Plots of the $F_{th}(S)$ versus S are sometimes called the *characteristic damage curve*, and they reflect the nature of the incubation. Figure 5.8a shows $F_{th}(S)$ for crystalline quartz as an example.

The fact that $F_{th}(S) \le F_1$ illustrates that the material does not fully recover between pulses for reasons suggested in Figure 5.1. The incubated material introduces additional excitation pathways increasing dN/dt (cf. Figure 5.6). Two principal mechanisms contribute to the incubation—(1) the occupation of existing (native) defects and traps, and (2) the creation of laser-induced defects. These occupied midgap states can be excited by subsequent pulses in addition to VB–CB transitions, increasing dN/dt and thus lowering the damage fluence.

The predamage modifications during the pulse train can be observed as changes to the optical properties of the material. As an example, Figure 5.8b shows the change in the film reflectance as a function of the pulse number in a train exciting the sample. The drop in reflectance indicates that the sample is modified although visible damage is not observable for pulse numbers $S < 14,000$.

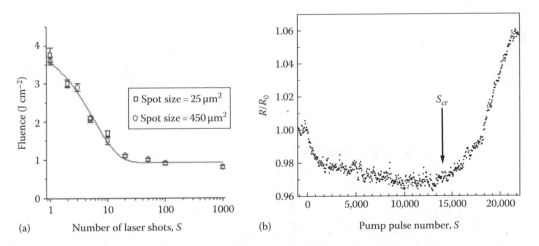

(a)   Number of laser shots, S

(b)   Pump pulse number, S

FIGURE 5.8   (a) Damage fluence F_{th} as a function of the number of pulses illuminating one and the same sample spot of a crystalline quartz sample with 100 fs pulses at 800 nm. (With kind permission from Springer Science+Business Media: *Appl. Phys. A*, Ultrashort-laser-pulse damage threshold of transparent materials and the role of incubation, 69, 1999, S373, Rosenfeld, A., Lorenz, M., Stoian, R., and Ashkenasi, D., Figure 5.2a. Copyright Springer-Verlag 1999.) (b) Reflectance of a tantala film on fused silica as a function of pulse number in a 1 kHz, 40 fs pulse train at 800 nm. The pulse fluence was chosen to be between the single- and multiple-pulse damage thresholds. The reflectance was probed by a weak, time-delayed femtosecond pulse (800 nm). The delay was negative, so that the sample response was monitored about 1 ms after each pump pulse. Damage occurred at pulse number $S = S_{cr}$. (Reprinted Figure 5.8b with permission from Mero, M. et al., *Opt. Eng.*, 44, 051107, 2005. Copyright 2005 by SPIE.)

Chapter 5

To understand how damage can occur in S-on-1 experiments, let us consider a sample that is illuminated by a train of equally spaced (period T) identical pulses of fluence $F < F_1$ and duration τ_p. The first pulse of the train produces a CB electron density according to Equation 5.6 of

$$N_1 = \frac{\beta_m}{a_0}\left(\frac{F}{\tau_p}\right)^{m-1}\left(e^{a_0 F} - 1\right) + N_0 e^{a_0 F} = Q + N_0 e^{a_0 F}, \tag{5.12}$$

where Q is the electron density produced by MPA and impact ionization. Assuming a lifetime of these electrons of T_{cv} (of the various processes in Figure 5.6 only band-to-band relaxation is considered), the population decays to $N_{02} = N_1 e^{-T/T_{cv}}$ before the second pulse arrives and N_{02} acts now as background electron density. The CB electron density N_2 after the second pulse is thus

$$N_2 = Q + N_{02} e^{a_0 F} = Q + Qp + N_0 p e^{a_0 F}, \tag{5.13}$$

where the factor $p = \exp(a_0 F - T/T_{cv})$ describes the balance between amplification and losses due to avalanche ionization and relaxation, respectively. Repeating this procedure one obtains that after S pulses the CB electron density takes on the form of a geometric progression

$$N_S = Q\sum_{n=0}^{S-1} p^n + N_0 e^{a_0 F} p^{S-1} = Q\frac{1-p^S}{1-p} + N_0 e^{a_0 F} p^{S-1}. \tag{5.14}$$

Figure 5.9b illustrates the CB population as a function of time during the pulse train according to Equation 5.14.

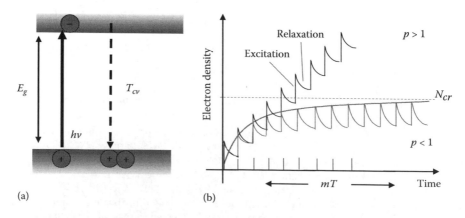

(a)   (b)

FIGURE 5.9   (a) Energy-level diagram showing the generation of CB electrons and their relaxation with a characteristic decay time T_{cv}. (b) CB electron growth versus time when the sample is excited by a train of pulses of period T. Depending on the parameter p, the CB electron density saturates or continues to grow.

With each subsequent pulse, the peak population gets larger. If the control parameter $p < 1$, the series converges for $S \to \infty$ to $N_\infty = Q/(1 - p)$. If $N_\infty < N_{crit}$, the material does not fail independent of the pulse number. If $N_\infty > N_{crit}$, there is a number of pulses S_{cr} for which $N_{Scr} \approx N_{crit}$ resulting in damage. On the other hand, if $p > 1$, the material will always damage at a certain pulse number because the electron density does not saturate. It should be mentioned that the value of p for given fluence is controlled by the material (a_0) and the pulse period T. If this period is too short, interpulse relaxation becomes less important and $p > 1$.

This simple discussion explains the principal feature of S-on-1 damage, that is, $F_{th}(S > 1) < F_1$, by using the simplest form of material incubation: the incomplete relaxation of CB electrons between pulses. Before we introduce a more comprehensive and accurate material model based on the relaxation processes and defect states sketched in Figure 5.6, we will describe damage studies with a pulse pair of adjustable delay.

5.3.2.2 Two-Pulse Excitation with Adjustable Pulse Spacing

The effect of material relaxation between pulses becomes obvious in the damage fluence $F_{th}(2)$ of a pulse pair (2-on-1) as a function of the pulse separation. $F_{th}(2)$ for two silica glasses is shown in Figure 5.10a (Li et al. 1999). For pulse separations <200 fs, $F_{th}(2)$ is up to 30% lower than the near constant value at later times.

This time scale corresponds to the STE formation time in silica glasses (see Section 2.3). Another example that spans a longer time scale and shows (almost) complete relaxation is shown in Figure 5.10b (Nguyen et al. 2010). The data are normalized to $F_{th}(1)$. For pulse separations $T > 1$ s, the material does not remember the first pulse when the second pulse arrives. The plateau in the ns and µs delay region indicates that the material recovers in two steps: a fast process in about 100 ps and a slow step in 100 ms.

The solid line in Figure 5.10b is a fit using a rate equation model that considers a single additional state between CB and VB (Nguyen et al. 2010). The first transient

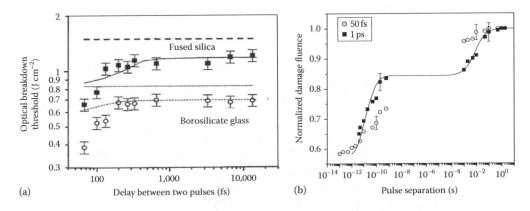

(a)   (b)

FIGURE 5.10 Damage fluence of a pair of pulses as a function of the time delay between pulses: (a) for fused silica and borosilicate glasses. (Reprinted Figure 5.3 with permission from Li, M., Menon, S., Nibarger, J.P., Gibson, G.N. et al., *Phys. Rev. Lett.*, 82, 2394, 1999. Copyright 1999 by the American Physical Society.) (b) For hafnia film (normalized to the single-pulse damage fluence) (Reprinted with permission from Nguyen, D.N., Emmert, L.A., Patel, D., Menoni, C.S., and Rudolph, W., $Ti_xSi_{1-x}O_2$ optical coatings with tunable index and their response to intense subpicosecond laser pulse irradiation, *Appl. Phys. Lett.*, 97, 191909, 2010. Copyright 2010, American Institute of Physics.)

in Figure 5.10b arises from the relaxation of electrons out of the CB while the second increase of $F_{th}(2)$ occurs when electrons relax out of the midgap state back to the VB. The height of the plateau region for the 1 ps excitation is determined by the absorption cross section of the midgap state and the branching ratio of relaxation into VB and the midgap state out of the CB. The difference in the 50 fs data indicates that additional states are photo excited by the second pulse. A closer inspection shows that the material does not recover completely even after minutes, which, however, affects damage thresholds only for larger pulse numbers (Emmert et al. 2010a). As will be shown in the next section, this is caused by the creation of laser-induced defects.

5.3.3　Multiple–Pulse Damage and Material Incubation

The damage fluence measured as a function of pulse number $F_{th}(S)$ in an S-on-1 experiment contains information about the properties of the native midgap states and the generation rate of laser-induced defects (cf. Figure 5.6). Combined with appropriate models based on rate equations that consider the formation, occupation and re-excitation of midgap states and predict the CB electron density during the pulse train, measured $F_{th}(S)$ curves can provide insight into complex material properties following short pulse excitation. This is discussed in this section.

The principal features of a typical $F_{th}(S)$ curve, cf. Figure 5.8a, have been characterized with fit functions. According to an ISO standard (ISO 11254-2:2011),

$$F_{th}(S) = F_\infty + \frac{F_1 - F_\infty}{1 - \Delta_D^{-1} \log_{10}(S)}. \tag{5.15}$$

This function describes $F_{th}(S)$ purely mathematically in terms of the single- and multiple-pulse threshold fluences and a fit parameter Δ_D. A functional form of

$$F_{th}(S) = F_\infty + (F_1 - F_\infty) \exp[-k(S-1)] \tag{5.16}$$

was used to model the quartz data from Figure 5.8a. Here, k is an empirical constant describing the rate of trap generation with pulse number S (Rosenfeld et al. 1999). It was assumed that the change in the threshold after the $(S + 1)^{th}$ pulse $[F_{th}(S + 1) - F_{th}(S)]$ is proportional to $[F_{th}(S + 1) - F_\infty]$.

A simplified material model can be used to derive $F_{th}(S)$ and relate fit parameters to basic material properties. Of the processes sketched in Figure 5.6, we consider only the excitation, cf. Equation 5.6, the formation and occupation of a laser-induced defect with time constant T_{LD}, relaxation from the CB to the VB with time constant T_{cv}, and shallow traps of number density N_{ST} that are completely ionized by each pulse and reoccupied between pulses. With these simplifications, one can derive the threshold fluence for a train of S pulses (Mero et al. 2005a):

$$[F_{th}(S)]^m \approx F_\infty^m + (F_1^m - F_\infty^m)\left(1 - \frac{T_{cv}}{T_{LD}}\frac{N_1}{N_{ST}}\right)^{S-1}, \tag{5.17}$$

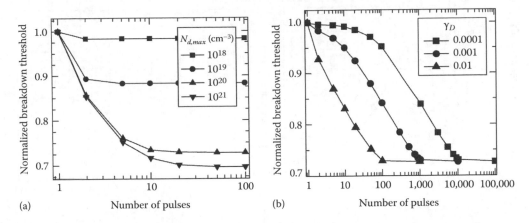

FIGURE 5.11   Numerical simulations of characteristic damage curves using rate equations that model the material processes sketched in Figure 5.6. (a) Increasing the maximum density of native traps ($N_{d,max}$) lowers the ratio (F_∞/F_1). (b) Changing the creation rate of laser-induced traps (γ_D) affects the pulse number at which saturation occurs. (Reprinted with permission from Emmert, L.A., Mero, M., and Rudolph, W., Modeling the effect of native and laser-induced states on the dielectric breakdown of wide band gap optical materials by multiple subpicosecond laser pulses, *J. Appl. Phys.*, 108, 043523, 2010. Copyright 2010, American Institute of Physics.)

where m is the order of the MPA process for band-to-band excitation. The last term in brackets controls the asymptotic approach $F_{th}(S) \rightarrow F_\infty$. It is determined by the relaxation branching ratio T_{cv}/T_{LD} and the ratio of CB electron density generated by a single pulse in the nascent material and the number density of shallow traps N_1/N_{ST}.

In reality, the material response is more complex than what has been considered so far, and $F_{th}(S)$ curves show features whose explanation requires the full set of processes sketched in Figure 5.6.

Because of the multitude of processes, it is instructive to look first at the effect of certain material parameters one at a time. Figure 5.11 shows results of a numerical study (Emmert et al. 2010b) on the dependence of $F_{th}(S)$ on the maximum density of native deep traps $N_{d,max}$ and the efficiency of defect creation γ_D during relaxation. The former shows that the ratio F_∞/F_1 decreases with a higher density of native defects. The latter shows that the multiple-pulse threshold is reached at larger pulse numbers for smaller values of γ_D.

The rate equation model can be applied to explain multiple-pulse damage behavior of dielectric materials for a broad range of excitation conditions (Emmert et al. 2010b). Figure 5.12a shows as an example the measured $F_{th}(S)$ curves for a hafnia coating at two different pulse durations and corresponding theoretical results.

The shapes of these curves for trains of 50 fs and 1 ps excitation pulses are distinctly different and reflect the effect of deep and shallow traps, respectively. The solid lines are simulations of damage threshold with a large maximum density of laser-induced deep traps and a smaller density of native shallow traps. In the case of 50 fs pulses, the intensity at the damage fluence is higher and the photoionization of the deep traps is more efficient than at long pulse durations. Therefore, the shape of $F_{th}(S)$ is dominated by the slow accumulation of these deep traps, because they have a higher density than the shallow states. The shape of the curve for 1 ps

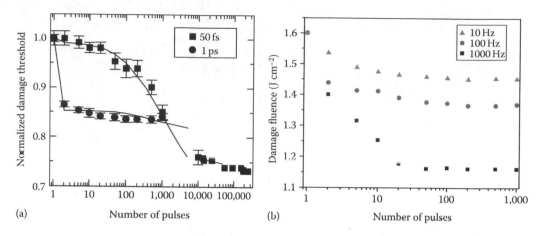

FIGURE 5.12   (a) Experimental values for the characteristic damage curve of hafnia at two different pulse durations. The curves are normalized to F_1. The solid lines are simulated curves using rate equations for the material processes sketched in Figure 5.6: a low-density (~10^{18} cm^{-3}) native shallow state and a high-density (~10^{20} cm^{-3}) laser-induced deep state were assumed. (Reprinted with permission from Emmert, L.A., Mero, M., and Rudolph, W., Modeling the effect of native and laser-induced states on the dielectric breakdown of wide band gap optical materials by multiple subpicosecond laser pulses, *J. Appl. Phys.*, 108, 043523, 2010. Copyright 2010, American Institute of Physics.) (b) Characteristic damage curve $F(S)$ of hafnia for different pulse repetition rates (τ_p = 1 ps).

pulse illumination is dominated by the occupation of shallow traps because of the lower excitation intensity and the subsequently lower efficiency of deep trap ionization through a multiphoton process. Because of the longer pulse duration, the relative contribution of avalanche ionization is larger (see Section 2.2). Even a small number of occupied shallow traps can provide a larger fraction of the required number of seed electrons compared to shorter pulses.

In conclusion, an ideal optical material would show no incubation effects. As illustration, Figure 5.12b shows the characteristic damage curve of hafnia films for different pulse repetition rates. At 10 Hz, the multiple-pulse damage threshold is much higher than for 1 kHz. This result is consistent with the two-pulse damage data, cf. Figure 5.10 that shows that shallow traps of hafnia eventually depopulate on a time scale somewhat below 1 s. As another example, doping with nitrogen ions, which is known to change the properties of shallow traps (Xiong et al. 2006), has been shown to eliminate incubation in hafnia films for long (800 fs) pulses (Nguyen et al. 2009).

5.4   Femtosecond Laser Damage and the Ambient Atmosphere

Most optical materials are used under atmospheric conditions, and damage thresholds are typically quoted and compared for this situation. High-intensity ps and fs laser systems, however, require that certain optical components be operated at reduced pressure (vacuum) to avoid undesired nonlinear optical processes in gases (air). Experiments with ns pulses motivated by space-based optical systems suggest that the ambient atmosphere has an effect on LIDTs (Riede et al. 2005). This observation was explained by the accumulation of absorbing graphitic layers driven by photolysis

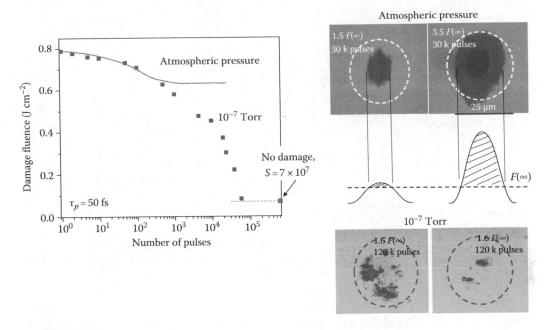

FIGURE 5.13 Characteristic damage curve of ion-beam-sputtered hafnia films in air and under vacuum when tested with 50 fs, 800 nm pulses. Damage morphology (Nomarski microscope) at atmospheric and vacuum conditions for different fluences and number of excitation pulses. The laser beam profile is also shown for comparison.

of background organics, mostly aromatic compounds. Normally, these deposits are removed by oxygen in the air when excited by the laser pulse.

It turns out that the fs LIDTs of oxide films also change with ambient gas composition and pressure (Nguyen et al. 2011), although for reasons different than those discussed for ns pulse exposure. Figure 5.13 shows that the multiple-pulse damage fluence in vacuum drops to about 10% of F_1 at atmospheric pressure. The single-pulse thresholds are unaffected by pressure, and there is little difference in the $F_{th}(S)$ curves for $S < 1000$. Together with the LIDT, the damage morphology changes. While at atmospheric pressure damage is very deterministic starting in the laser beam center where the fluence is maximum as discussed previously, damage is initiated at randomly distributed sites within the beam spot in vacuum.

Figure 5.14a shows the damage threshold for trains of 300,000 pulses at different pressures of pure atmospheres of nitrogen, oxygen, and water vapor (the three most prevalent components of air). Obviously, starting from vacuum condition, only the addition of water vapor can restore the damage threshold to the value in air at atmospheric pressure, oxygen has a small effect, and nitrogen does not change the LIDT. It should also be mentioned that pressure does not change the multiple-pulse damage threshold of bulk fused silica surfaces, but silica, alumina, and tantala films are affected (see Figure 5.14b) and that the shapes of these curves are material dependent.

The multiple-pulse damage model introduced in the previous section predicts that the damage threshold decreases when the density of defects becomes greater. It has been hypothesized (Nguyen et al. 2011) that the creation of oxygen vacancies by the laser pulse is responsible for incubation in oxides. The maximum density of vacancies in the film can

Chapter 5

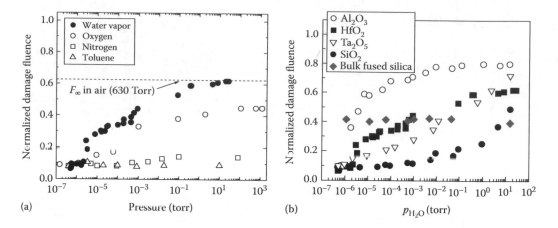

FIGURE 5.14   (a) Measured multiple-pulse ($S = 300{,}000$) damage thresholds as a function of pressure of various ambient gases (Nguyen et al. 2011). (b) LIDT ($S = 300{,}000$) of a variety of oxide films and bulk fused silica as a function of water vapor pressure (p_{H_2O}).

be related to the pressure of oxygen and water in the ambient atmosphere. Below a certain pressure (6×10^{-5} Torr), water vapor cannot form a monolayer and the defect density (and LIDT) saturates. These processes occur preferentially at randomly distributed sample sites. The latter can be grain boundaries between crystallites or different material phases.

5.5   Summary

Because of the deterministic behavior, fs laser damage studies in dielectric materials are a powerful tool to unravel the nature of nascent and laser-induced material properties important in high-field laser–matter interactions. There is a good empirical understanding of single- and multiple-pulse damage and the scaling of the LIDT with pulse duration and certain material properties. Experiment and theoretical results based on simple models are in good qualitative agreement. There are still open questions concerning the exact multiphoton ionization process in solids taking into account the band structure and its change under high-field excitations. There is clear evidence that multiple-pulse LIDTs are controlled by the accumulation and occupation of nascent and laser-induced defects, where relevant processes proceed on time scales from hundreds of fs to seconds and longer. A dramatic reduction in the multiple-pulse LIDT under vacuum conditions appears to be a combination of oxygen diffusion and subsequent defect formation at predisposed sample sites. Work is in progress to reevaluate the relative role of impact and multiphoton ionization and to study in more detail under which conditions the critical electron density should not be used as the damage criterion (Karras et al. 2011, Mouskeftaras et al. 2013, Rudolph et al. 2013).

Acknowledgments

We thank Prof. D. Ristau (Laser Zentrum Hannover e.V., Hannover, Germany) and Prof. C. S. Menoni (Colorado State University, Fort Collins, CO, USA) and their respective groups for the high-quality samples that made the experiments possible, as well as for many useful discussion. We are also grateful to former and current group members M. Mero, A. Sabbah, J. Liu, J. Zeller, D. Nguyen, R. Weber, Z. Sun, C. Rodriguez, and C. Karras for carrying out most of the experiments and simulations. The work was possible through funding from NSF (PHY-0722622), ONR/JTO (N00014-06-1-0664 and N00014-07-1-1068), and ARO/JTO (W911NF-11-1-007).

References

Apostolova, T. and Hahn, Y. 2000. Modeling of laser-induced breakdown in dielectrics with subpicosecond pulses. *J. Appl. Phys.* 88: 1024–1034.

Bloembergen, N. 1974. Laser-induced electric breakdown in solids. *IEEE J. Quan. Electron.* QE-10: 375–386.

Boyd, R. W. 2003. *Nonlinear Optics*, 2nd edn. San Diego, CA: Academic Press.

Bulgakova, N. M., Stoian, R., and Rosenfeld, A. 2010. Laser-induced modification of transparent crystals and glasses. *Quant. Electron.* 40: 966–985.

Chen, S., Zhao, Y., He, H., and Shao, J. 2011. Effect of standing-wave field distribution on femtosecond laser-induced damage of HfO_2/SiO_2 mirror coating. *Chin. Opt. Lett.* 9: 083101.

Christensen, B. H. and Balling, P. 2009. Modeling ultrashort-pulse laser ablation of dielectric materials. *Phys. Rev. B* 79: 155424.

Crisp, M. D. 1973. Some aspects of surface damage that can be explained with linear optics. *Damage in Laser Materials: 1973, Nat. Bur. Stand. (U.S.) Spec. Publ.* 387: 1–4.

Diels, J.-C. and Rudolph, W. 2006. *Ultrashort Laser Pulse Phenomena*, 2nd edn. New York: Academic Press.

Du, D., Liu, X., Korn, G., Squier, J., and Mourou, G. 1994. Laser-induced breakdown by impact ionization in SiO_2 with pulse widths from 7 ns to 150 fs. *Appl. Phys. Lett.* 64: 3071–3073.

Emmert, L. A., Mero, M., Nguyen, D. N., Rudolph, W., Patel, D., Krous, E., and Menoni, C. S. 2010a. Femtosecond pulse S on 1 LIDT in dielectric materials: Comparison of experiment and theory. *Proc. SPIE* 7842: 784211.

Emmert, L. A., Mero, M., and Rudolph, W. 2010b. Modeling the effect of native and laser-induced states on the dielectric breakdown of wide band gap optical materials by multiple subpicosecond laser pulses. *J. Appl. Phys.* 108: 043523.

Englert, L., Rethfeld, B., Haag, L., Wollenhaupt, M., Sarpe-Tudoran, C., and Baumert, T. 2007. Control of ionization processes in high band gap materials via tailored femtosecond pulses. *Opt. Expr.* 15: 17855–17862.

Foster, A. S., Gejo, F. L., Shluger, A. L., and Niemenen, R. M. 2002. Vacancy and interstitial defects in hafnia. *Phys. Rev. B* 65: 174117.

Gaier, L. N., Lein, M., Stockman, M. I., Knight, P. L., Corkum, P. B., Ivanov, M. Y., and Yudin, G. L. 2004. Ultrafast multiphoton forest fires and fractals in clusters and dielectrics. *J. Phys. B: At. Mol. Opt. Phys.* 37: L57–L67.

Gattass, R. R. and Mazur, E. 2008. Femtosecond laser micromachining in transparent materials. *Nat. Photon.* 2: 219–225.

Grojo, D., Gertsvolf, M., Lei, S., Barillot, T., Rayner, D. M., and Corkum, P. B. 2010. Exciton-seeded multiphoton ionization in bulk SiO_2. *Phys. Rev. B* 81: 212301.

Gruzdev, V. E. 2007. Photoionization rate in wide band-gap crystals. *Phys. Rev. B* 75: 205106.

Guenther, A. H. and McIver, J. K. 1993. To scale or not to scale. *Proc. SPIE* 2114: 488–499.

Guizard, S., Martin, P., Daguzan, P., Petite, G., Audebert, P., Geindre, J. P., Dos Santos, A., and Antonnetti, A. 1995. Contrasted behaviour of an electron gas in MgO, Al_2O_3 and SiO_2. *Europhys. Lett.* 29: 401–406.

Guizard, S., Martin, P., Petite, G., D'Oliveira, P., and Meynadier, P. 1996. Time-resolved study of laser-induced colour centers in SiO_2. *J. Phys.: Condens. Matter* 8: 1281–1290.

Gulley, J. R. 2010. Frequency dependence in the initiation of ultrafast laser-induced damage. *Proc. SPIE* 7842: 78420U.

Hirao, K., Mitsuyu, T., Si, J., and Qiu, J., eds. 2001. *Active Glass for Photonic Devices: Photoinduced Structures and Their Application (Springer Series in Photonics)*. New York: Springer.

ISO 11254-2:2011 Test methods for laser induced damage threshold of optical surfaces. Part 2: S on 1 test. *International Standard, 2001.* International Organization for Standardization.

Jasapara, J., Mero, M., and Rudolph, W. 2002. Retrieval of the dielectric function of thin films from femtosecond pump-probe experiments. *Appl. Phys. Lett.* 80: 2637–2639.

Chapter 5

Jasapara, J., Nampoothiri, A. V. V., Rudolph, W., Ristau, D., and Starke, K. 2001. Femtosecond laser pulse induced breakdown in dielectric thin films. *Phys. Rev. B* 63: 045117.

Jensen, L. O., Mende, M., Blaschke, H., Ristau, D., Nguyen, D. N., Emmert, L. A., and Rudolph, W. 2010. Investigations on SiO_2/HfO_2 mixtures for nanosecond and femtosecond pulses. In *Laser-Induced Damage in Optical Materials: 2010*, eds. G. J. Exarhos, V. E. Gruzdev, J. A. Menapace, D. Ristau, and M. J. Soileau, SPIE, Bellingham, WA, Vol. 7842, 784207.

Joglekar, A. P., Liu, H., Spooner, G. J., Meyhofer, E., Mourou, G., and Hunt, A. J. 2003. A study of the deterministic character of optical damage by femtosecond laser pulses and applications to nanomachining. *Appl. Phys. B* 77: 25–30.

Jones, S. C., Braunlich, P., Casper, R. T., Shen, X.-A., and Kelly, P. 1989. Recent progress on laser-induced modifications and intrinsic bulk damage of wide-gap optical materials. *Opt. Eng.* 28. 1039 1068.

Juodkazis, A., Kondo, T., Rode, A., Matsuo, S., and Misawa, H. 2004. Three-dimensional recording and structuring of chalcogenide glasses by femtosecond pulses. In *Laser-Precision Microfabrication*, eds. I. Miyamoto, H. Helcajan, K. Itoh, K. F. Kobayashi, A. Ostendorf, and K. Sugioka, SPIE, Bellingham, WA, Vol. 5662, pp. 179–184.

Jupé, M., Jensen, L., Starke, K., Ristau, D., Melninkaitis, A., and Sirutkaitis, V. 2009. Analysis in wavelength dependence of electronic damage. *Proc. SPIE* 7504: 75040N.

Kaiser, A., Rethfeld, B., Vicanek, M., and Simon, G. 2000. Microscopic processes in dielectrics under irradiation by subpicosecond laser pulses. *Phys. Rev. B* 61: 11437–11450.

Kane, E. O. 1957. Band structure of indium antimonide. *J. Phys. Chem. Solids* 1: 249–261.

Karras, C., Sun, Z., Nguyen, D. N., Emmert, L. A., and Rudolph, W. 2011. The impact ionization coefficient in dielectric materials revisited. In *Laser-Induced Damage in Optical Materials: 2011*, eds. G. J. Exarhos, V. E. Gruzdev, J. A. Menapace, D. Ristau, and M. J. Soileau, SPIE, Bellingham, WA, Vol. 8190, pp. 819028.

Keldysh, L. V. 1965. Ionization in the field of a strong electromagnetic wave. *Sov. Phys. JETP* 20: 1307–1314.

Klein, M. V. and Furtak, T. E. 1986. *Optics*, 2nd Edn. New York: Wiley.

Lenzner, M., Kruger, J., Sartania, S., Cheng, Z., Spielmann, C., Mourou, G., Kautek, W., and Krausz, F. 1998. Femtosecond optical breakdown in dielectrics. *Phys. Rev. Lett.* 80: 4076–4079.

Lenzner, M. and Rudolph, W. 2009. Laser-induced optical breakdown in solids. In *Strong Field Laser Physics*, ed. T. Brabec Springer Series, Vol. 134, pp. 243–257.

Li, M., Menon, S., Nibarger, J. P., and Gibson, G. N. 1999. Ultrafast electron dynamics in femtosecond optical breakdown of dielectrics. *Phys. Rev. Lett.* 82: 2394–2397.

Martin, P., Guizard, S., Daguzan, P., Petite, G., D'Oliveira, P., Meynadier, P., and Perdix, M. 1997. Subpicosecond study of carrier trapping dynamics in wide-band-gap crystals. *Phys. Rev. B* 55: 5799–5810.

Mero, M., Clapp, B., Jasapara, J. C., Rudolph, W., Ristau, D., Starke, K., Krüger, J., Martin, S., and Kautek, W. 2005a. On the damage behavior of dielectric films when illuminated with multiple femtosecond laser pulses. *Opt. Eng.* 44: 051107.

Mero, M., Liu, J., Rudolph, W., Ristau, D., and Starke, K. 2005b. Scaling laws of femtosecond laser pulse induced breakdown in oxide films. *Phys. Rev. B* 71: 115109.

Mero, M., Liu, J., Zeller, J., Rudolph, W., Starke, K., and Ristau, D. 2004. Femtosecond pulse damage behavior of oxide dielectric films. In *Laser Induced Damage in Optical Materials*, eds. G. J. Exarhos, A. H. Guenther, N. Kaiser, K. L. Lewis, M. J. Soileau, and C. J. Stolz, SPIE, Bellingham, WA, Vol. 5273, pp. 8–16.

Mero, M., Sabbah, A. J., Zeller, J., and Rudolph, W. 2005c. Femtosecond dynamics of dielectric films in the pre-ablation regime. *Appl. Phys. A: Mater. Sci. Process.* 81: 317–324.

Mero, M., Zeller, J., and Rudolph, W. 2005d. Ultrafast processes in highly excited dielectric thin films. In *Femtosecond Laser Spectroscopy*, ed. P. Hannaford, Springer Science + Business Media, Inc., Boston, MA, 305.

Mouskeftaras, A., Guizard, S., Federov, N., and Klimentov, S. 2013. Mechanism of femtosecond laser ablation of dielectrics revealed by double pump-probe experiment. *Appl. Phys. A* 110: 709–715.

Neauport, J., Lavastre, E., Razé, G., Dupuy, G., Bonod, N., Balas, M., de Villele, G., Flamand, J., Kaladgew, S., and Desserouer, F. 2007. Effect of electric field on laser induced damage threshold of multilayer dielectric gratings. *Opt. Expr.* 15: 12508–12522.

Nguyen, D., Emmert, L. A., Cravetchi, I. V., Mero, M., Rudolph, W., Jupe, M., Lappschies, M., Starke, K., and Ristau, D. 2008. $Ti_xSi_{1-x}O_2$ optical coatings with tunable index and their response to intense subpicosecond laser pulse irradiation. *Appl. Phys. Lett.* 93: 261903.

Nguyen, D. N., Emmert, L. A., Patel, D., Menoni, C. S., and Rudolph, W. 2010. Transient phenomena in the dielectric breakdown of HfO_2 optical films probed by ultrafast laser pulse pairs. *Appl. Phys. Lett.* 97: 191909.

Nguyen, D. N., Emmert, L. A., Rudolph, W., Patel, D., and Menoni, C. S. 2009. The effect of nitrogen doping on the multiple-pulse subpicosecond dielectric breakdown of hafnia films. *Proc. SPIE* 7504, Art. No. 750402.

Nguyen, D. N., Emmert, L. A., Schwoebel, P., Patel, D., Menoni, C. S., Shinn, M., and Rudolph, W. 2011. Femtosecond pulse damage thresholds of dielectric coatings in vacuum. *Opt. Exp.* 19: 5690–5697.

Petite, G., Guizard, S., Martin, P., and Quere, F. 1999. Comment on "ultrafast electron dynamics in femtosecond optical breakdown of dielectrics". *Phys. Rev. Lett.* 83: 5182.

Petrov, G. M. and Davis, J. 2008. Interaction of intense ultra-short laser pulses with dielectrics. *J. Phys.B: At. Mol. Opt. Phys.* 41: 025601.

Popov, V. S. 2004. Tunnel and multiphoton ionization of atoms and ions in a strong laser field (Keldysh theory). *Phys.-Usp.* 47: 855–885.

Rajeev, P. P., Gertsvolf, M., Corkum, P. B., and Rayner, D. M. 2009. Field dependent avalanche ionization rates in dielectrics. *Phys. Rev. Lett.* 102: 083001.

Rethfeld, B., Brenk, O., Medvedev, N., Krutsch, H., and Hoffmann, D. H. H. 2010. Interaction of dielectrics with femtosecond laser pulses: Application of kinetic approach and multiple rate equation. *Appl. Phys. A.* 101: 19–25.

Rethfeld, B., Sokolowski-Tinten, K., von der Linde, D., and Anisimov, A. S. 2004. Timescales in the response of materials to femtosecond laser excitation. *Appl. Phys. A: Mater. Sci. Process.* 79: 767–769.

Riede, W., Allenspacher, P., Schröder, H., and Wernham, D. 2005. Laser-induced hydrocarbon contamination in vacuum. *Proc. SPIE* 5991: 59910H.

Rosenfeld, A., Lorenz, M., Stoian, R., and Ashkenasi, D. 1999. Ultrashort-laser-pulse damage threshold of transparent materials and the role of incubation. *Appl. Phys. A* 69: S373–S376.

Rudolph, W., Emmert, L. A., Sun, Z., Patel, D., and Menoni, C. S. 2013. Laser damage in thin films—What we know and what we don't. *Proc. SPIE* 8885: 888516.

Saeta, P. N. and Greene, B. 1993. Primary relaxation processes at the band edge of SiO_2. *Phys. Rev. Lett.* 70: 3588–3591.

Smith, A. V. and Do, B. T. 2008. Bulk and surface laser damage of silica by picosecond and nanosecond pulses at 1064 nm. *Appl. Opt.* 47: 4812–4832.

Song, K. S. and Williams, R. T. 1996. *Self-Trapped Excitons.* 2nd edn. New York: Springer-Verlag.

Sparks, M., Mills, D. L., Warren, R., Holstein, T., Maradudin, A. A., Sham, L. J., E. Loh, J., and King, D. F. 1981. Theory of electron-avalanche breakdown in solids. *Phys. Rev. B* 24: 3519–3536.

Starke, K., Ristau, D., Welling, H., Amotchkina, T. V., Trubetskov, M., Tikhonravov, A. A., and Chirkin, A. S. 2004. Investigations in the nonlinear behavior of dielectrics by using ultrashort pulses. In *Laser-Induced Damage in Optical Materials: 2003*, eds. G. J. Exarhos, A. H. Guenther, N. Kaiser, K. L. Lewis, M. J. Soileau, and C. J. Stolz, SPIE, Bellingham, WA, Vol. 5273, pp. 501–513. SPIE.

Stoian, R., Ashkenasi, D., Rosenfeld, A., and Campbell, E. E. B. 2000. Coulomb explosion in ultrashort pulsed laser ablation of Al_2O_3. *Phys. Rev. B* 62: 13167–13173.

Stuart, B. C., Feit, M. D., Herman, S., Rubenchik, A. M., Shore, B. W., and Perry, M. D. 1996. Nanosecond-to-femtosecond laser-induced breakdown in dielectrics. *Phys. Rev. B* 53: 1749–1761.

Tien, A.-C., Backus, S., Kapteyn, H., Murnane, M., and Mourou, G. 1999. Short-pulse laser damage in transparent materials as a function of pulse duration. *Phys. Rev. Lett.* 82: 3883–3886.

Ueta, M., Kanzaki, H., Kobayashi, K., Toyozawa, Y., and Hanamura, E. 2001. *Excitonic Processes in Solids.* New York: Springer.

Wolfe, J. E., Qiu, S. R., and Stolz, C. J. 2011. Fabrication of mitigation pits for improving laser damage resistance in dielectric mirrors by femtosecond laser machining. *Appl. Optics* 50: C457–C462.

Xiong, K., Robertson, J., and Clark, S. J. 2006. Passivation of oxygen vacancy states in HfO_2 by nitrogen. *J. Appl. Phys.* 99: 044105.

Zheng, J. X., Ceder, G., Maxisch, T., Chim, W. K., and Choi, W. K. 2007. First-principles study of native point defects in hafnia and zirconia. *Phys. Rev. B* 75: 104112.

Chapter 5

Measurements

6. Measurement and Detection of Laser-Induced Damage

Jianda Shao

6.1 Introduction

In principle, the concept of laser-induced damage refers to the occurrence of permanent changes in the optical materials. In practice, this change results in the degradation of the optical performance of the system. Therefore, it is highly imperative to measure the power- and energy-handling capabilities of the optical components comprising the optical/laser system in order to ensure the lifetime and reliable operation of such systems.

The phenomena of laser-induced damage in optics have been observed in a variety of materials and optical systems. Laser scientists are actively looking for strategies to

Laser-Induced Damage in Optical Materials. Edited by Detlev Ristau © 2015 CRC Press/Taylor & Francis Group, LLC. ISBN: 978-1-4398-7216-1.

Chapter 6

accurately measure and evaluate the laser damage resistance of different optical components so as to define repeatable and reliable methods to determine the upper threshold limit of the optical materials to optical radiation. Such reliable measurements are highly important in order to understand the damage mechanisms and to improve the laser power-/energy-handling capabilities of these systems. Furthermore, they play an important role in setting forth the criteria for the selection of the optical materials for use in the optical systems. In general, the laser damage resistance of optical components is quantitatively estimated by laser-induced damage threshold (LIDT) and by the damage probability for specific laser fluences.[1,2] Ever since 1960, several studies showed that the measured LIDTs from different laboratories were not in agreement with each other, primarily due to the fact that the LIDT measurement is influenced by a variety of factors. These include the testing laser parameters, the damage detection techniques, and the damage evaluation methods.[3-9] These factors were actively discussed at the annual Boulder Damage Symposiums (1969 onward). To this end, Round-Robin laser damage test sequence was proposed in 1984 to sort out the reasons for the discrepancies among measurements.[10,11] Following that, the international standard for LIDT measurement, ISO 11254, was published in the 1990s,[1,2] which was later on updated to ISO 21254 in 2011.[12-15] The contents of ISO 21254 include general principles,[12] threshold determination,[13] laser power-/energy-handling capabilities,[14] and detection and measurement methods.[15] The implementation of these principles is of fundamental importance, especially for the control of beam quality and measurement of laser parameters.

Laser damage threshold of the optical components can be determined by small spot sampling methods. First, fast and precise detection techniques are required to determine where damage occurs. Typically, techniques such as microscopy,[16-18] scattered light diagnostics,[19,20] plasma spark monitor,[21-23] transmission/reflection alteration,[24,25] photoacoustic detection,[26,27] and photothermal deflection technique[28-30] are being used to identify the damage occurrence. However, laser damage from different kinds of optics may vary markedly in their "signature," making it necessary to choose the proper detection technique or a combination of techniques, depending on the requirements of the given test. Second, the sampling ratio and the laser radiation modes may influence the test results. So far, several distinct protocols have been developed to conduct the test procedure. The most commonly used procedures, as defined in ISO 21254, are single-shot (1-on-1) and multishot (S-on-1) tests. The LIDT values obtained using the ISO 21254 standard procedure are extrapolated by the damage probability curve, which in turn indicates the concentration of the laser damage precursors. Other protocols like the ramp test[31,32] and the raster scan test[33,34] have also been developed to evaluate the laser conditioning effects on the laser damage resistance of optics. Furthermore, the analysis of the characteristics of laser-induced damage, for example, the topography, can also serve as a powerful tool to deduce information about the damage precursors. They are particularly useful for providing feedback to improve the laser damage resistance of optics.

This chapter will discuss the various laser detection and analysis techniques. Subsequent sections will explain the evaluation methods and the implementation of the testing procedures, according to the ISO standard.

6.2   Detection and Analysis of Laser-Induced Damage

6.2.1   Laser Damage Detection Techniques

Laser damage is defined as any permanent laser-radiation-induced change in the characteristics of the surface or bulk of the specimen. These changes can be observed by a detection technique with a sensitivity related to the intended operation of the product concerned.[12] One of the important detection techniques recommended in ISO 21254 is the incident-light microscope with the Nomarski-type differential interference contrast (DIC). Other damage detection techniques, such as scattered light diagnostics,[19,20] plasma spark monitor,[21–23] transmission/reflection variation detection,[24,25] and acoustic shock wave detection,[35,36] are considered as alternatives. As a matter of fact, DIC microscopy is used as a standard to calibrate the other detection techniques.

6.2.1.1   Microscopy

As the standard damage detection technique defined in ISO 21254, Nomarski-type DIC microscope enhances the image contrast based on the principle of interferometry.[37] In principle, DIC works by passing a linearly polarized light through a Wollaston prism, where it is separated into two orthogonally polarized mutually coherent parts, which are then spatially displaced at the sample plane and recombined before the observation. Here, the interference of the two phases at the recombination is sensitive to their optical path difference. Therefore, when two such separated lights pass through a damaged site, one of their phases is delayed. Consequently, when the two phases are recomposed by using a Wollaston prism and polarizing filter, the interference effect enhances the contrast and carries some information, which otherwise is invisible to the human eye.

Some typical damage morphologies detected by the Nomarski-type DIC microscope are shown in Figure 6.1. Here, (a) illustrates plasma scald, (b) displays delamination on the surface of coatings, (c) reveals the pinpoint inside the KDP crystal, and (d) presents the crater damage on the rear surface of fused silica. It is seen that damage can be readily discriminated from the gray level difference between the damaged site and its background.

DIC microscopy is capable of detecting nearly all kinds of damages. Nevertheless, it is difficult to realize online detection, primarily owing to its complicated setup. An online microscopic detection setup for detecting laser damage usually comprises a telescope and a charge-coupled device (CCD) camera.[16] A He–Ne laser or white light is often used as the illumination source. In the typical procedure, images are captured before and after the laser irradiation, the comparison of which can be used to identify the damage.

The image comparison procedure can be accomplished either by human judgment or by specific image processing techniques. Although the former is simple and convenient, results are highly variable as they tend to depend on the operator. On the contrary, image processing techniques can automatically and rigorously detect the damages, although it involves the use of complex image processing algorithms.

Chapter 6

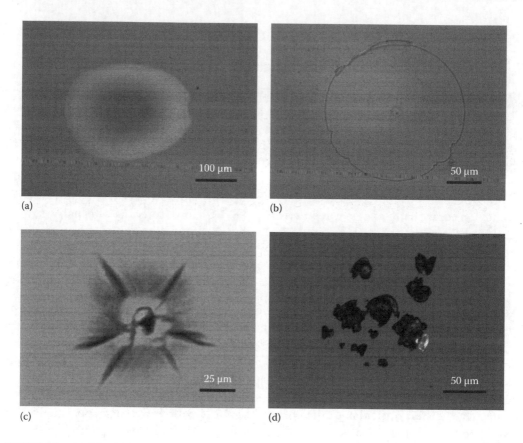

FIGURE 6.1   Surface damage morphologies on coatings (a) and (b), bulk damage in KDP crystal (c), and rear surface damage on fused silica (d).

The most straightforward method of image processing is direct comparison.[38] It is based on the pixel-to-pixel subtraction of images before and after laser shot, as illustrated in Figure 6.2a. The white pixels in the subtracted image indicate the occurrence of damage. However, this method has certain inherent disadvantages, including the difficulty associated with determining the pixel threshold level due to vignette or nonuniform illumination. Moreover, any misalignment, caused by sample motion or vibration, may lead to false positives. Subsequently, the defect comparison method[39] was developed to eliminate the false positives. In this method, each captured image, before and after laser irradiation, is subtracted from the background or reference image, which removes the effect of vignette or nonuniform illumination. Thenceforth, defect tables that record the position and size of damage sites can be created before and after the laser shot, as illustrated in Figure 6.2b. The number, location, and diameter in the two tables are compared using the image processing algorithm, taking into account the image displacement error due to vibration and misalignment. If all the three characteristics, namely, the number, location, and diameter, of the defects in the "before" and "after" tables are matched, then the defects in the "after" table are considered preexisting. In simple terms, matching characteristics in "before" and "after" table are not considered as damage, rather ignored as a preexisting condition. On the other hand, if the three characteristics do not match, then the defect is considered as a new one, and the presence of damage is determined.

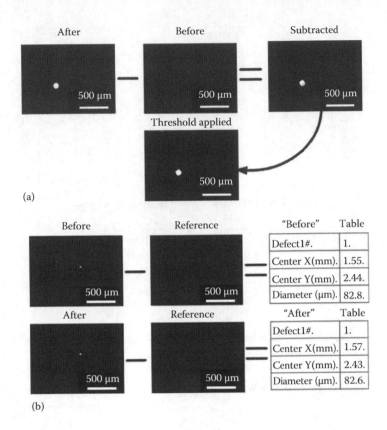

FIGURE 6.2   Protocol for image processing by (a) direct comparison method and (b) defect comparison method.

The online microscopy is capable of detecting most damages. In addition, it can also be used to precisely determine the dimension of the damaged sites.

6.2.1.2   Scattered Light Diagnostics

This is yet another popular detection technique that uses the light scattered from the damaged sites to determine the characteristics of the damage.[19,20] In this technique, the presence of damage is often determined by comparing the scattering signals of a probe beam, before and after the laser irradiation. Any signal difference that is greater than the background noise is indicative of the onset of damage (Figure 6.3).

Thus far, different scattering diagnostic techniques have been developed to detect the surface and bulk damages. Figure 6.4a shows the schematic diagram of scattering diagnostic technique used for detecting the surface damage. A He–Ne laser, which is used as the probe beam, is focused onto the target site, and the light scattered from the target site is collected and focused onto a detector. The probe beam is blocked from reaching the detector so as to collect only the light scattered from the target site. Figure 6.4b shows the schematic of the scatter diagnostic system used for detecting bulk damage. The aperture stop is used to control the amount of detected scattering light, while the field stop is used to eliminate the light scattered from the surface as it can severely interfere with the desired signal.[40–42]

Chapter 6

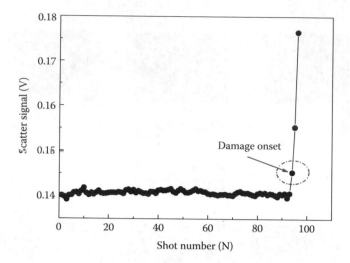

FIGURE 6.3 Change in the scattered light signal before and after the laser shots.

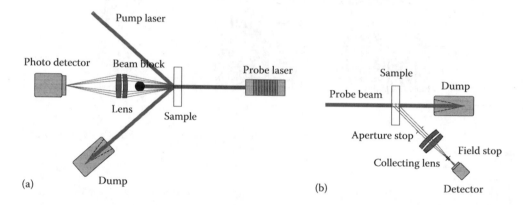

FIGURE 6.4 Layout of scatter diagnostic system used for detecting laser-induced damage (a) at the surface and (b) in the bulk.

Until now, scattering light diagnostics is one of the efficient detection techniques, and hence is applied to on-the-fly evaluation of laser-induced damage. The major limitation associated with this technique is the background noise that tends to lower its measurement sensitivity. The measurement precision is often improved by adopting the following three approaches: (1) determine multiple measurement results at the same test site and take the average value of measurements; (2) increase the gain of the detector; and (3) filter the background noises by means of other advanced techniques.[41]

6.2.1.3 Plasma Spark Monitor[21,22]

It is a well-known fact that many damage occurrences are often accompanied with a plasma spark.[21,22,43] Therefore, plasma scalds are always found at the damaged sites.[44–46] Figure 6.1a shows the morphology of the plasma scald, as observed by using DIC microscope. As can be seen from the photograph, the plasma scald is characterized by a relatively even surface when compared with other kinds of damages such as crater or pinpoint. Hence, it is hard to detect plasma scalds by online microscopy or scattered light diagnostics.

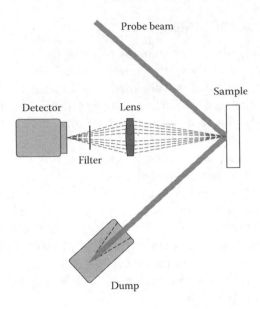

FIGURE 6.5   Schematic diagram of the plasma spark detection system.

Fortunately, the presence of plasma spark is an obvious symbol of damage occurrence, making it suitable for determination by using plasma spark detection method.[21]

The schematic diagram of the plasma spark detection method is shown in Figure 6.5. In the typical method, a collecting lens focuses the light from the plasma spark of the damaged site onto a detector. The response time of the detector should be smaller than the short duration of the plasma spark. The scattered test laser must be filtered so that it does not overwhelm the plasma spark.

The captured signal of plasma spark is shown in Figure 6.6. Typically, once the plasma is formed, the spark signal immediately grows to a maximum value in about 100 ns. The signal decreases slowly, with a fall time of approximately tens of microseconds. Experimental results show that the signal intensity is proportional to the energy of the plasma spark. However, the plasma spark caused by dust and air breakdown should be avoided in order to obtain precise results.

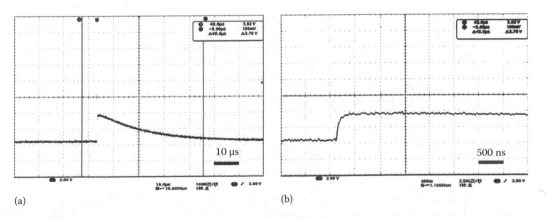

(a)   (b)

FIGURE 6.6   (a) The detected signal of plasma spark; (b) local enlarged view of the signals.

Chapter 6

6.2.2 Laser Damage Topography Analysis

The comprehensive analysis of the damage topography is indispensable for understanding the mechanism underlying the laser damage. Damage morphologies are often diverse owing to the complicated damage process associated with various optical components. The size and depth of the surface damage is critical for retrieving the damage precursors. Accordingly, the reconstruction of 3D topography is required to simulate the dynamics of the damage process.[47]

Topological analysis can be performed using various devices, including optical microscope, scanning electron microscope (SEM), atomic force microscope (AFM), step profiler, and white light interferometer (WLI). Each of these devices has their own features and application areas. However, there is no universal analysis device that is capable of analyzing all types of damage morphologies and that can simultaneously deal with both high-detection resolution and broad field of view. Therefore, it is necessary to choose the proper analysis technique or a combination of techniques, depending on the nature of the damage. This section presents flexible topography analysis methods for three typical damage morphologies.

The first typical damage morphology is a surface structure with a small aspect ratio, which is represented by the plasma scalds, as shown in Figure 6.7. The damage structure is characterized by a diameter of approximately 200 μm and shallow nanometer level depth. The step profiler is often preferred to precisely measure such plasma scalds.

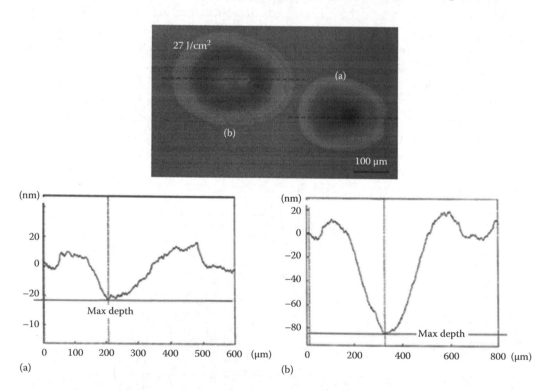

FIGURE 6.7 Typical surface profiler results of plasma scalds. (a, b) The surface profiler results of the plasma scalds marked by (a) and (b) in the top figure. The probe of the surface profiler moves along the dotted line shown in the top figure. (From Liu, X.F. et al., *Appl. Opt.*, 50, 4226, 2011. With permission.)

The damage site is readily scanned by the tip of the step profiler, over a considerable of maximum length of the order of tens of millimeters. Although some thickness measurement errors are unavoidably introduced due to the direct contact between the tip and the test site, the precision is sufficient enough to analyze a damaged site with a small aspect ratio.

The second typical damage morphology is surface structures with a large aspect ratio (~1), which is represented by nodular ejection pits that are typically generated in multilayer coatings. In such cases, conventional depth measurement tools such as AFM and step profiler are not suitable for obtaining information on the depth of the damage structure. When their tip runs through a nodular ejection pit with a vertical slope, it is hard to probe into the bottom. For such cases, focused ion beam (FIB) seems to be an effective tool to obtain depth information, although it is expensive and time consuming. Therefore, it still remains a challenge to effectively analyze the depth of nodular ejection pit using nondestructive tools. The simple and effective method is to count the layers from the captured SEM image, which is expected to give an estimate of the depth information.[48] The detailed process of this method is explained by the following example.

Figure 6.8 shows the cross-sectional SEM image of a nodular defect in HfO_2/SiO_2 multilayer with a design of $(HL)^{16}H2L$, where H and L represent HfO_2 and SiO_2, respectively. As is seen, the layers of HfO_2 and SiO_2 are observed as white and black colors, respectively, in the image. The analysis of the variations in white and black colors can provide a clue on the depth of the nodular ejection pit. As naked eyes are sensitive to the white HfO_2 layer, and HfO_2/SiO_2 appear in pairs, usually the white layers of the pit edge are counted to estimate the depth information. As shown in Figure 6.9a, 17 white layers in the damage site were counted out, which is equal to the total number of layer pairs in the coating. This implies that all the layers at the damaged site were destructed completely. This was further substantiated by the cross-sectional SEM image of the pit shown in Figure 6.9b.

The last typical damage morphology considered is the pinpoint structure inside the bulk. Conventional morphology analysis tools can only be applied to measure the damage morphology on the surface. Therefore, cleaving or etching is often used to expose

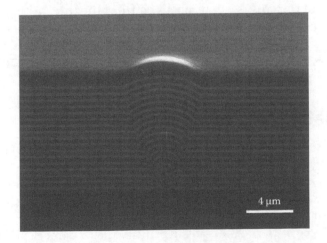

$4 \mu m$

FIGURE 6.8 Cross-sectional SEM image of a nodular defect.

Chapter 6

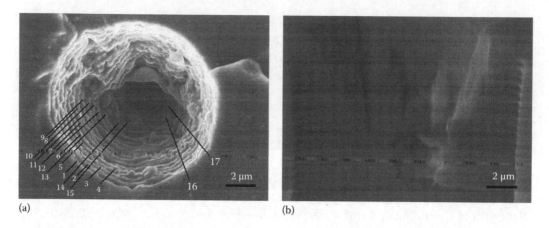

FIGURE 6.9 Damage depth information provided by SEM (a) layer number count from the pit edge and (b) cross section of the pit. (From Liu, X.F. et al., *Appl. Surf. Sci.*, 256, 3783, 2010. With Permission.)[49]

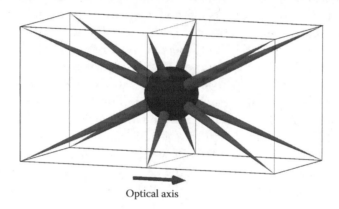

Optical axis

FIGURE 6.10 Reconstructed 3D morphology of a pinpoint in KDP crystal induced by 1064 nm laser pulses.

the bulk damage to the surface.[47] The 3D morphology of the bulk damage sites can be reconstructed from the cross sections obtained from the three orthogonal planes. The 2D surface morphology of a pinpoint, shown in Figure 6.1c, was obtained by etching KDP crystal to expose the bulk damage site in 60% distilled water and 40% ethanol solution. Figure 6.10 shows the 3D pattern of the pinpoint site in KDP crystal that was constructed from the three 2D surface morphologies.

6.3　Evaluation Methods and Statistics

In order to achieve high laser fluence, sufficient enough to execute laser damage testing for transparent optical materials, sampling with small laser spot is usually employed to evaluate the laser damage resistance in laboratories. The evaluation difficulties arise from the mismatch of small test spot with the full aperture of optics. Therefore, the damage probability, resulting from spot sizes and damage precursor concentration, should be introduced to clarify the issue. In principle, laser damage probability of optical components may behave differently with laser radiation modes. Consequently, several evaluation methods, including single-shot, multishot, ramp testing, and raster scan, have been developed to fully characterize the damage resistance and behaviors from different aspects. These four testing methods are helpful to gain insight on the laser damage process.

6.3.1 Definition of Laser-Induced Damage Threshold

Laser-induced damage threshold is defined as the highest quantity of laser radiation incident upon the optical component, for which the extrapolated probability of damage is zero. Here, the quantity of laser radiation may be expressed as energy density, power density, or linear power density.[12] In order to obtain a suitable power/energy density high enough to execute laser damage testing, the laser beam is usually focused onto a small spot on the target plane. Subsequently, the test sites are exposed to laser radiation of different radiation modes and power/energy densities so as to obtain a plot that depicts the damage probability as a function of power/energy density. Finally, the LIDT is obtained by extrapolating the damage probability data. Here, adopting a precise detection technique and considering sufficient testing sites on the sample are indispensable to improve the accuracy of the damage probability, especially for defects-driven cases. In addition, the choice of extrapolation methods such as linear[1,12] or exponential methods[49–51] could also affect the fitted LIDT values. The selection of the appropriate extrapolation method mainly depends on the defect characteristics involved in the laser beam spot.

6.3.2 Single-Shot Test (1–on–1 Test)[1,13]

Single-shot test, also known as the 1-on-1 test, is a testing procedure that uses one shot of laser radiation on each unexposed site on the specimen. As shown in Figure 6.11, 1-on-1 test requires at least 10 sites in a row on the specimen exposed to the same radiation fluence. The fraction of damaged sites is recorded as the damage probability for a given fluence. This procedure is repeated for other fluence levels until the range of fluences is sufficiently broad to include fluence values resulting in both zero and 100% damage probabilities. With these data, the damage probability is plotted as a function of fluence (Figure 6.12). Linear extrapolation of the damage probability data to zero damage probability yields the threshold value, as shown in Figure 6.12. As only one shot

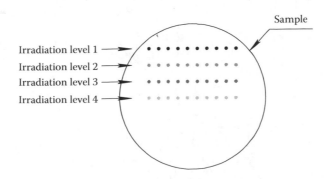

FIGURE 6.11   Schematic diagram of the single-shot testing procedure.

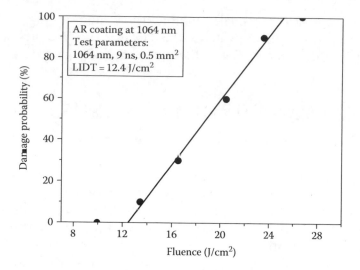

FIGURE 6.12   Representative example showing the damage probability plotted as a function of the fluence, which can be used to determine the damage threshold.

of laser radiation is exposed to each test site on the specimen, the measured LIDT is capable of characterizing the initial characteristics of the samples.

In case of nanosecond or long-pulse laser applications, most of the damage behaviors are defect driven and the damage probability data may be severely influenced by the sampling number, N_f, for each fluence, and the associated damage precursor density. Figure 6.13 shows two typical damage probability data for N_f corresponding to 10 and 40, respectively. As is seen, for N_f satisfying the basic requirement number defined in the ISO standard, the damage probability data in the plot do not show a good linear tendency. Thus, it is difficult to determine LIDT by the standard linear extrapolation.

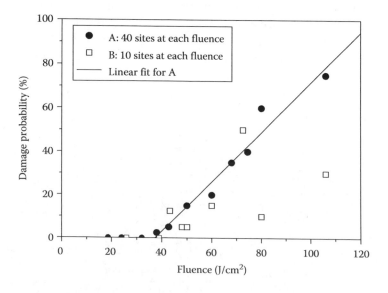

FIGURE 6.13   Damage probability versus laser fluence in 1-on-1 testing.

In contrast, as N_f is increased up to 40, the linearity of the damage probability data is observed to improve and hence the LIDT can be derived readily. Here, choosing a suitable value for N_f depends on the damage precursor density involved in the laser spot size as well as the operational experience.

The damage probability data obtained from the single-shot test not only reflects the current damage resistance (damage probability of zero) but also discloses the potential ability with respect to laser damage resistance (damage probability of 100%) as long as the damage precursors are eliminated by improvements of the manufacturing process. In addition, the density and categories of the precursors can also be derived from the slope angle of the fitted line.[51,53] All these information can be utilized to optimize the manufacturing process of the optical component.

6.3.3   Multishot Test (S–on–1 Test)[2,13]

Multishot test, also known as the S-on-1 test, is a testing procedure that uses a series of S-pulses irradiating each unexposed site on the specimen. Each pulses train has a constant fluence. Moreover, a short and constant time interval, given by the reciprocal of the pulse repetition rate, exists between two successive pulses. The measurement procedure of S-on-1 test is similar to that of the single-shot test, except that the control of pulse shots and record of the responsible shots leading to damage are required. For different values of S, a set of damage probability data can be obtained by the similar method described for 1-on-1 testing. Subsequent linear extrapolation of each data produces LIDTs for all the shots, as shown in Figure 6.14. Here, S is varied as 1, 10, 100, and 1000.

The S-on-1 test, in some sense, simulates the running status of the optics in the real laser systems and hence is of great interest for laser engineers. This technique is often employed to estimate the lifetime of the optics under the multishot laser radiation.

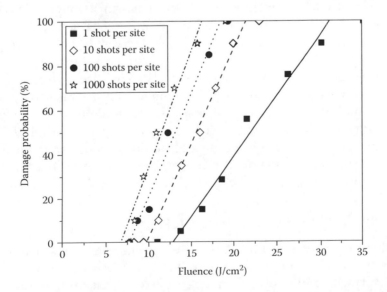

FIGURE 6.14   Damage probability versus fluence for S-on-1 testing, where S = 1, 10, 100, and 1000.

Chapter 6

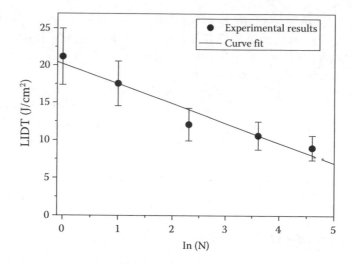

FIGURE 6.15 Extrapolation of the lifetime of coating samples.

Figure 6.15 gives a representative example of extrapolation of the lifetime of a coating sample using the damage threshold, plotted as a function of the number of laser shots.

The fitted curve of the experimental data of LIDT versus the number of laser shots can be expressed by Equation 6.1[54,55]:

$$LIDT(N) = A - B \cdot (\ln N) \tag{6.1}$$

where
    N is the number of laser shots
    A and B are the two calibration parameters

The lifetime, defined as the total number of laser shots, can be roughly derived from the specific reliable fluence of optics, according to the fitting equation.

Moreover, it was found that the laser damage resistance of the optics during the S-on-1 test process gradually degrades with increase in the number of laser radiation shots. This phenomenon is usually referred to as the accumulation effect, which can be used to explore the laser damage mechanisms.[56] The accumulation effect is mainly composed of two factors, namely, thermal accumulation and defects accumulation. The former could play a dominant role if the laser radiation trains have a high repetitive rate (>1 kHz), while the latter demonstrates a much more complex behavior, resulting from the deterioration of intrinsic defects and creation of regenerative defects. Almost all of them are mainly influenced by the number of laser shots, repetitive rate, as well as the relaxation time.

6.3.4   Ramp Test (R-on-1 and N-on-1 Tests)[31,32]

Laser conditioning by ramping the laser fluence is an effective means to improve the laser damage resistance of certain optical components, such as mirror coatings, polarizers, KDP/DKDP crystals, and so on.[57–60] In the past decades, the ramp test has been developed to evaluate the potential damage resistance of the optics by laser conditioning.[31,32]

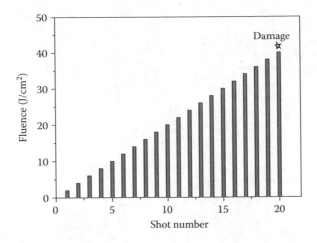

FIGURE 6.16   Sketch of the R-on-1 procedure.

In the R-on-1 test, the same site is irradiated and the fluence is ramped step by step until damage occurs. For each test site, the ramp should have the same initial fluence and fluence step, as illustrated in Figure 6.16. This procedure is repeated for a significant number of test sites, depending on the sample size and laser spot size. In fact, the improvement of laser damage resistance is achieved by the characteristic change of material exposed to lower fluences, depending on the concentration and damage initiation fluence of the damage precursors.[54] However, the R-on-1 test is time consuming because of the ramping procedure. According to the concentration fluences of precursors obtained from the damage probability curve, R-on-1 test can be simplified as N-on-1 test by decreasing the number of fluence steps.

The damage probability of each laser fluence can be obtained from the following two steps: (1) listing all the laser fluence data in a monotonous sequence and (2) computing the damage probability for the kth fluence according to Equation 6.2:

$$P_k = \frac{\sum_{i=1}^{k} N_i}{N}, \quad (k = 1, \ldots N)$$ (6.2)

where

N is the total number of test sites

N_i is the number of damaged sites for the ith laser fluence

$\sum_{i=1}^{k} N_i$ is the sum of the number of damaged sites from the first fluence (i = 1) to the kth fluence

The plot of the damage probability as a function of laser fluence, as obtained from the aforementioned analysis, can be employed to estimate the LIDT for ramp test. Figure 6.17a through c shows the measurement damage probability data for coatings, KDP crystals, and fused silica, respectively. There are no suitable fitting functions to extrapolate the fluence data to zero damage probability. Therefore, the largest fluence with zero damage probability is usually chosen as the LIDT. As long as the ramp step is small enough, the precision of measurement results is acceptable in practice.

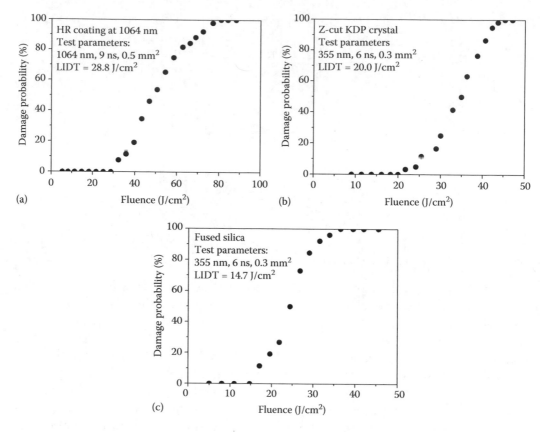

FIGURE 6.17 The R-on-1 laser damage probability curve for (a) coating; (b) KDP crystal; and (c) fused silica. (a: From Liu, X.F. et al., *Chin. J. Lasers*, 36, 1545, 2009. With permission.)

6.3.5   Raster Scan Test[33,34]

In the aforementioned three testing methods, sampling was performed on a number of random sites on the specimen. Therefore, the precision of this testing method is primarily influenced by the total number of sampling sites. However, some cases demand a more detailed evaluation of the damage behavior over the whole surface of the sample. This requirement can be satisfied by performing the raster scan test.

As illustrated in Figure 6.18, the raster scan test method is performed by zigzag movement of the testing optics through a stationary laser beam. The fluence of the laser beam is fixed during each scanning process. As the incident laser beam tends to have a circular shape in most cases, overlapping of the neighboring pulses is required to make a nearly homogeneous distribution of laser fluence over the scanned area. Usually, the incident laser has a Gaussian intensity distribution and hence the movement step is selected as the beam width at 90% of the peak intensity.

Each scan is performed at a constant fluence, and scans are repeated at specific increments, until the necessary fluence specification for the optics is reached. The output of the raster scan test is either a plot depicting the mapping of the damage initiation fluence or a curve showing the damage cumulative concentration. In Figure 6.19,

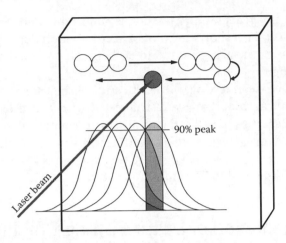

FIGURE 6.18   Schematic illustration of the raster scan testing procedure.

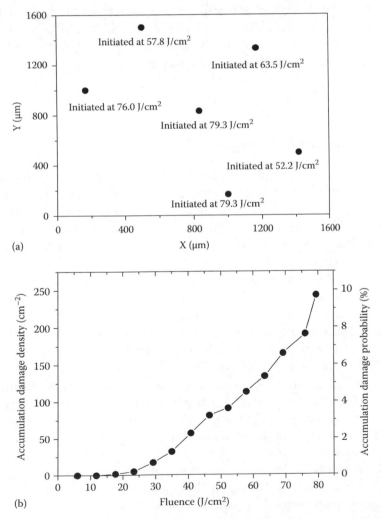

FIGURE 6.19   Raster scan results obtained from 14 scans. (a) Damage initiations during the scans; (b) accumulation density/probability versus fluence.

(a) illustrates the damaged sites and the corresponding fluence initiated in a portion of the scanned area and (b) shows the plot of accumulation density/probability versus fluence, for laser fluence ranging from 6 to 80 J/cm² (1064 nm, 9 ns) over ~2 cm² area of an optical coating.

In case of the raster scan testing, the laser conditioning effects should be taken into account for the evaluation of damage resistance. Nowadays, raster scan test has been developed as a full aperture posttreatment process for optical components in order to improve their laser damage resistance. Accordingly, a new evaluation method has been established, known as the functional damage threshold. It is beyond the content of this section and will be discussed in Section 6.5.

6.4 Implementation of Laser Damage Testing

With the advent of ISO 11254 standards, the accuracy of LIDT measurement has been improved remarkably. Nevertheless, the implementation of the principles defined in this standard is another important issue in reality. The basic requirement for LIDT measurement is the construction of an effective laser damage test system. The beam quality and stability are influenced by the delivery of laser beam in the test system. Therefore, the control of beam quality and measurement of laser parameters are essential for the determination of LIDT.

6.4.1 Laser Damage Test System

According to the LIDT evaluation methods discussed in Section 6.3, all the laser damage test procedures entail irradiation of a certain number of pulses with preselected laser fluence on several sampling sites, and subsequent detection of the damage. Here, the laser energy and the spatial and temporal profiles should be measured, in order to ensure that the laser provides reliable performance during the test and that all the laser pulses are of the right traceability.

During the past decades, lots of works have been dedicated to develop a reliable laser damage testing system.[17,20,62] The basic approach to laser damage testing is shown in Figure 6.20. In the typical process, the output power of a well-characterized laser is adjusted using a variable attenuator comprising of a half-wave plate and a fixed polarizer. Another half-wave plate is used to change the polarization of laser beam, according to the test specification for polarization. A focusing system is required to obtain laser fluence high enough to induce damage at the target plane. The laser beam is sampled with a beam splitter, which directs a portion of the beam to a diagnostic package that permits real-time measurement of the pulse radiation energy, and the spatial and temporal profiles. A manipulator, in which the sample is mounted, is used to position the different test sites and set the angle of incidence. The damage detection system monitoring the exposed area is used to determine the occurrence of damage.

The laser system, which is used as the light source for laser damage testing, should meet the requirements defined in the ISO standards, such as wavelength, pulse duration, output energy/power, stability, and beam quality. Commercially available lasers readily meet these requirements.

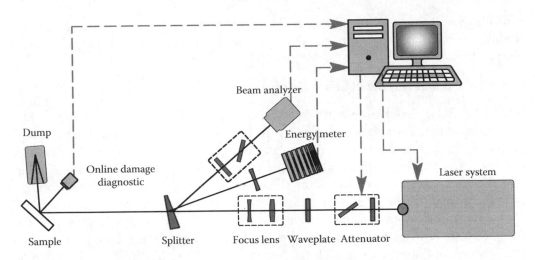

FIGURE 6.20 Schematic illustration of the basic approach adopted for laser damage testing.

When different levels of laser radiation are required for the process, a laser attenuator is often used to continuously adjust the energy/power of laser incident onto the target plane. At the same time, it is also equally important to maintain the beam quality. Continuously adjustable laser attenuation can be achieved either through absorption or through reflection. In the case of a low-power laser, both reflection and absorption can be applied, such as neutral density filter. On the other hand, in the case of a high-power laser, the main obstacles arise from the damage and heating instability. For such cases, reflection-type attenuator is often recommended, while the variable attenuator consisting of a half-wave plate and a polarizer is generally used for the linear polarized laser.

The focusing system is applied to meet the need of a desirable beam size at the target plane. The effective beam diameter at the target plane is determined by the focal length, wavelength, divergence, and aperture. In principle, the effective beam diameter should not be less than 0.8 mm, as described in ISO 21254. Otherwise, the spot size effect on the LIDT value will show up.[63–65] In most cases, a focused beam can be accomplished by a single positive lens with relatively long focal length, as larger spot size (>0.8 mm) demands a longer focal length. However, in some specific cases, the long focal length of the lens is limited by the dimension of test platform, and a folding mirror is also hard to be applied due to the high-energy density of the follow-up light path behind the focus lens. In such cases, a focus system, with a negative lens behind a positive one, is employed to achieve larger effective beam spot within a limited space.

6.4.2 Measurement of Laser Parameters

6.4.2.1 Laser Energy

As shown in Figure 6.20, the energy incident on the target is monitored in situ by the sampling light reflected from the splitter and can be computed by multiplication of the sampling beam energy and the splitting ratio. Prior to the LIDT testing, the splitting ratio can be obtained by simultaneously measuring the target and the sampling energy. The splitting ratio is influenced by the polarization orientation and the incident angle to the splitter and should be re-measured once these statuses are changed.

6.4.2.2   Spatial Profile of Laser Beam

6.4.2.2.1   Burn Pattern Method[66]

If a plate, such as burn paper or Polaroid film, positioned at the target plane is melted by the laser with a Gaussian profile, then a burn pattern is formed on the surface of the plate. This pattern can be employed to estimate the beam size. In this method, the relation between the diameter of the burn pattern and the laser energy is used to restore the Gaussian profile, from which the spot size of laser beam could be extracted. Usually, the dimension of the burn pattern can be conveniently measured by using a conventional microscope. Figure 6.21 shows the representative laser beam profile obtained from the burn pattern method. It is seen that the experimental data can be well fitted by a standard Gaussian function. Even though the burn pattern method is considered to be very primitive, it can provide a rough intensity profile of the test laser.

6.4.2.2.2   Knife-Edge Method[67]

Knife-edge method is a classic method based on the gradual eclipsing of laser beam by a sharp knife-edge. As illustrated in Figure 6.22, when the knife-edge intersects the beam in the direction perpendicular to the direction of the beam, the power/energy meter attached behind the knife-edge collects the intensity of the unmasked portion of the beam. If the laser beam has a Gaussian spatial shape, then the signal measured by the power/energy meter is equivalent to the integral of the Gaussian function. From this, the beam profile can be easily obtained by calculating the derivative of the measurement data. As the knife-edge can be driven in a small step, the measurement data well reflect the actual integral of Gaussian function.

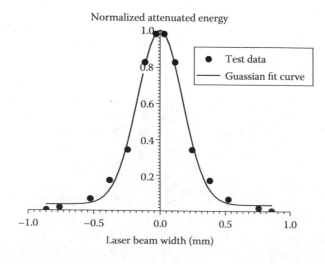

FIGURE 6.21   Laser profile showing the diameter of the laser, as determined from the burn pattern method.

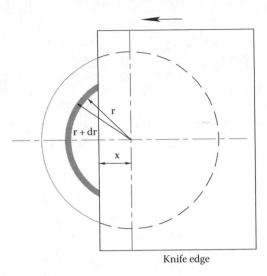

FIGURE 6.22   Edge displacement from the center of the laser beam.

For a Gaussian beam with the following intensity profile,

$$I(x,y) = \frac{2P}{\pi\omega} \exp\left[-\frac{2\left(x^2 + y^2\right)}{\omega^2} \right] \tag{6.3}$$

where
    P is the power of the beam
    ω is the $1/e^2$ intensity radius, the portion of laser beam transmitted through the knife-edge P(a) can be given as

$$P(a) = \int\limits_{-\infty}^{\infty} \int\limits_{a}^{\infty} I(x,y)\,dx\,dy = \frac{P}{2}\,\mathrm{erfc}\left(\frac{a\sqrt{2}}{\omega} \right) \tag{6.4}$$

where erfc(x) is the complementary error function,[68] erfc(x) = 1 − erf(x).

Using Equation 6.4, it is possible to obtain theoretical result of unmasked power versus knife movement distance, shown as the solid line in Figure 6.23. In addition, experimental data (dotted line) using knife-edge method are also shown for comparison in Figure 6.23. It is seen that both the theoretical and experimental data are in good agreement.

6.4.2.2.3   Camera Imaging Method[69]

CCD/CMOS camera imaging method is the most popular technique adopted for the measurement of spatial profile. The intensity distribution of the laser beam can be accurately digitalized by the array cells in the camera. Figure 6.24 shows a 2D contour image of laser beam captured by using a camera, from which the effective beam area can be obtained by image processing. The camera imaging method is applicable to measure the laser beam with arbitrary spatial profiles. Moreover, any light modulation can be easily realized from the measurement profiles.

In most cases, the spatial profile of the laser beam is obtained by positioning the camera normal to the laser beam. For the oblique incident laser, the actual spot size on the

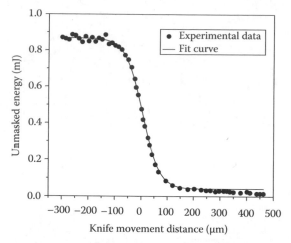

FIGURE 6.23   Portion of the laser beam transmitted through the knife-edge.

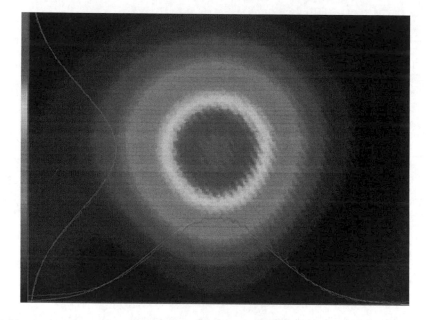

FIGURE 6.24   2D Contour image showing the spatial distribution of the laser beam.

target plane is equal to the measured value divided by the cosine of the incident angle. Moreover, this method can be employed for shot-to-shot real-time analysis.

All the aforementioned methods, namely, burn pattern method, knife-edge method, and camera imaging method, that are used for determining the spatial profile of the laser beam have their own advantages and disadvantages. Appropriate method should be chosen according to the requirements of the applications. Although the burn pattern method has a relatively low accuracy, it can provide a rough estimate of the size of the laser spot, without demanding any additional hardware. On the other hand, the other two methods are suitable for precise analysis. Of the three methods, the camera imaging method is the most popular measurement system because of the advantages of real-time visualization. In order to guarantee test accuracy, the plane of the camera should be located at the target plane or in an equivalent plane. This method is often recommended for moderately convergent beam. For highly convergent beam, it is very difficult to locate the camera cell array on the right position and hence may involve errors. On the other hand, the knife-edge method is much convenient in terms of finding the target plane, although the result is the average of lots of laser pulses.

6.4.2.3 Temporal Profile of Laser Beam

The temporal profile of laser beam is generally measured using a photodiode detector and a sampling oscilloscope. The photodiode transforms the laser intensity into a proportional voltage signal. Following that, the sampling oscilloscope analyzes the voltage signal and provides the temporal profile as the output (Figure 6.25). In principle, the response time of photodiode should be short enough to capture the pulse. In other terms, the response time of the photodiode detector and the sampling rate of the oscilloscope determine the precision and accuracy of laser temporal profile measurement. In the nanosecond range, the temporal profile of the laser beam can be directly determined by using commercially available photodiodes and oscilloscopes.

On the other hand, in the picosecond and femtosecond regimes, photodiodes cannot directly detect the ultrafast pulses. Alternatively, the picosecond pulse is characterized by using a streak camera, which transforms the temporal profile of a laser pulse into a spatial profile. In addition, autocorrelation technique is yet another popular method for measuring the temporal profiles of femtosecond pulses.[70]

The streak camera is an ultrafast detector that can capture the temporal phenomenon of light occurring in an extremely short time period (of the order of picoseconds). The general operating principle of streak camera is to transform the temporal profile of a

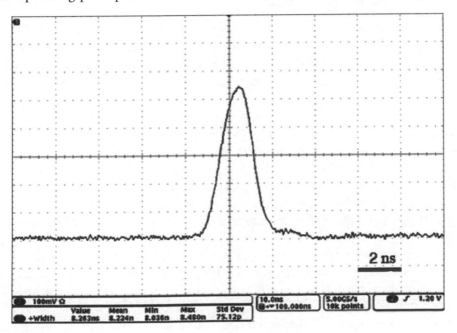

FIGURE 6.25 Measurement result of laser temporal profile.

light pulse into a spatial profile on a detector.[71] In particular, a light pulse enters the instrument through a narrow slit, along one direction. Subsequently, a line image is projected onto a photocathode that creates an electron replica inside the streak tube. As the electrons pass through the tube, the initial bias voltage on the deflection plates is reversed. The time-varying voltage causes the electrons to streak across a phosphor screen, located at the back of the tube. Subsequently, the optical intensity can be determined from the phosphor screen, and the time and incident light position can be obtained from its location in the phosphor screen. Now, the light pulse with 180 fs duration has been measured by using the steak camera.[72]

The intensity autocorrelation measurement is an indirect method for measuring the duration of light pulse. In principle, the pulse to be measured is split into two replicas. One replica passes through a delay layout, creating a time delay with respect to the other. Subsequently, the two replicas are spatially overlapped. Following that, a kind of nonlinear optical process occurs in a nonlinear material between the two replicating light pulses, such as the second harmonic generation (SHG) process in an SHG crystal. A detector placed next to the SHG crystal is used to collect the intensity of the SHG signal. For an ultrafast light pulse, the detector responds much slowly and measures the time integral. Then, the measured signal is calibrated by using the spatial delay layout. The light pulse duration can be obtained from the calibrated result and the pulse shape factor. Besides the SHG-autocorrelation measurement, methods such as PG-autocorrelation measurement are based on third-order nonlinear process. Furthermore, techniques such as frequency-resolved optical gating (FROG) have been developed to measure ultra-short laser pulses.[73,74]

6.5 Functional Damage

It can be realized that the laser damage test platform, along with different LIDT evaluation methods discussed in the previous sections, can effectively characterize the laser damage resistance of individual optical components. The measurement results not only provide a feedback for the manufacturing process but also reveal the mechanism underlying the damage of optical components at diverse operating conditions. When an optical component is installed in a laser system, any damage to the optical component more or less influences the output performance. For instance, a laser-induced damage site leads to beam modulation in near- or far-field,[75,76] and the concentration of damaged sites degrades the transmittance/reflectivity. Additionally, the roughness of the surface caused by damage increases the contrast of laser beam.[77] However, the tolerance for the performance degradation caused by the damages is different for various optical or laser systems. When the influence of the damage exceeds the tolerance of the system, the optical component is considered to be malfunctioning.

Given these considerations, a new concept of functional damage threshold (FDT) has been developed to assess the severity of damage permitted for a specific optical component in a laser system. FDT is defined as the minimum laser energy or power density at which a damaged optic degrades the performance of a laser system.[78] Laser fluence higher than FDT can lead to failure damage, which can be specified by the severity of laser damage. Failure damage specification varies according to the properties of the system and optics inside. For example, the Lawrence Livermore National Laboratory has established the following specifications of failure damage for small optical components used in the National Ignition Facility:

1. The number of damaged sites is more than 1% of the tested sites.
2. The size of damage is larger than 100 μm.
3. Damage grows upon further laser illumination.

The specification of FDT requires that the laser damage test procedure not only determines the occurrence of damage but also provides an estimate of the size, density, and growth characteristics of the damaged site. As mentioned in Section 6.3.5, the raster scan test method can be applied to accurately evaluate the size and density of initial damages, while the S-on-1 procedure can be used to confirm their stability under subsequent laser irradiations.

In fact, most damage sites with various morphologies can be regarded as imperfections of the optical components if they are stable under certain fluence. The surface damage can be classified as digs or scratches, while the pinpoint bulk damage can be ranked as bubbles. In ISO 10110[79] and MIL-PRF-13830B,[80] the imperfections in the optical components have been graded according to various applications. For example, 5/N × A is the expression used for indicating the surface imperfection, where N is the allowed number of imperfections with maximum permitted size and A is the square root of the maximum area allowed. Furthermore, ISO 10110 and MIL-PRF-13830B provide the fundamentals qualifications of optical components suitable for use in laser or optical systems. They do not distinguish the origin of the imperfections, rather are just concerned with the number and size of imperfections in the desired area. The imperfections that are allowed to exist in an optical component demonstrate that various damages, as one form of creating imperfections, are permitted in different systems. This implies that a more practical meaning of laser damage is its influence on the performance of the laser, instead of mere permanent physical changes. In other terms, laser damage practically refers to the degradation of output performances of the whole laser system rather than a simple physical change of the optical component.

References

1. ISO11254-1. 2000. Lasers and laser-related equipment-determination of laser-induced damage threshold of optical surfaces—Part 1: 1-on-1 test.
2. ISO11254-2. 2001. Lasers and laser-related equipment-determination of laser-induced damage threshold of optical surfaces—Part 2: S-on-1 test.
3. Lee H. A. 1969. Damage-threshold testing of laser glass at Owens-Illinois. *ASTM Spec. Tech. Publ.* 469:79–83.
4. Turner A. F. 1971. Ruby laser damage thresholds in evaporated thin films and multilayer coatings. *Natl. Bur. Stand. U.S. Spec. Publ.* 356:119–123.
5. Bliss E. S. and Milam D. 1972. Laser induced damage to mirrors at two pulse durations. *Natl. Bur. Stand. U.S. Spec. Publ.* 372:108–122.
6. Wood R. M., Taylor R. T., and Rouse R. L.1975. Laser damage in optical materials at 1.06 μm. *Opt. Laser Technol.* 7:105–111.
7. Walker T. W., Guenther A. H., and Nielsen P. E. 1981. Pulsed laser-induced damage to thin-film optical coatings—Part1: Experimental. *IEEE J. Quant. Electron.* 17:2041–2052.
8. Foltyn S. R. and Newnam B. E. 1981. Multiple-shot laser damage thresholds of ultraviolet reflectors at 248 and 308 nanometers. *Nat. Bur. Stand. U.S. Spec. Publ.* 620:265–276.

Chapter 6

9. Merkle L. D., Koumvakalis N., and Bass M. 1984. Laser-induced bulk damage in SiO$_2$ at 1.064 μm, 0.532 μm, and 0.355 μm. *J. Appl. Phys.* 55:772–775.

10. Guenther K. H., Humpherys T. W., Balmer J. et al. 1984. 1.06-μm laser damage of thin-film optical coatings: A round-robin experiment involving various pulse lengths and beam diameters. *Appl. Opt.* 23:3743–3752.

11. Guenther K. H. and Menningen R. G. 1984. 1.06-μm laser damage of optical coatings: Regression analyses of round-robin test-results. *Appl. Opt.* 23:3754–3758.

12. ISO 21254-1. 2011. Lasers and laser-related equipment—Test methods for laser-induced damage threshold—Part 1: Definitions and general principles.

13. ISO 21254-2. 2011. Lasers and laser-related equipment—Test methods for laser-induced damage threshold—Part 2: Threshold determination.

14. ISO 21254-3. 2011. Lasers and laser-related equipment—Test methods for laser-induced damage threshold—Part 3: Assurance of laser power (energy) handling capabilities.

15. ISO 21254-4. 2011. Lasers and laser-related equipment—Test methods for laser-induced damage threshold—Part 4: Inspection, detection and measurement (Technical Report).

16. Reichelt W. H. and Stark E. E. 1973. Radiation induced damage to NaCl by 10.6 μm fractional joule, nanosecond pulses. *Natl. Bur. Stand. U.S. Spec. Publ.* 387:175–180.

17. David W. M., Arenberg J. W., and Frink M. E. 1987. Production oriented laser damage testing at Hughes Aircraft Company. *Natl. Bur. Stand. U.S. Spec. Publ.* 752:201–210.

18. Natoli J. Y., Gallais L., Bertussi B., Commandre M., and Amra C. 2003. Toward an absolute measurement of LIDT. *Proc. SPIE* 4932:224–237.

19. Sheehan, L. and Kozlowski M. 1995. Detection of inherent and laser-induced scatter in optical material. *Proc. SPIE* 2541:113–122.

20. Hue J., Dijon J., and Lyan P. 1996. The CMO YAG laser damage test facility. *Proc. SPIE* 2714:102–113.

21. Henderson B. E., Getty R. R., and Leroy G. E. 1971. Plasma formation upon Laser irradiation of transparent dielectric materials. *Natl. Bur. Stand. U.S. Spec. Publ.* 356:31–37.

22. Giuliano C. R. 1972. The relation between surface damage and surface plasma formation. *Natl. Bur. Stand. U.S. Spec. Publ.* 372:46–54.

23. Liu X. F., Li D. W., Zhao Y. A., Li X., Ling X. L., and Shao J. D. 2010. Automated damage diagnostic system for laser damage threshold tests. *Chin. Opt. Lett.* 8:407–410.

24. Said A. A., Xia T., Dogariu A., Hagan D. J., Soileau M. J., Vanstryland E. W., and Mohebi M. 1995. Measurement of the optical-damage threshold in fused quartz. *Appl. Opt.* 34:3374–3376.

25. Singh A. P., Kapoor A., and Tripathi K. N. 2003. Recrystallization of germanium surfaces by femtosecond laser pulses. *Opt. Laser Technol.* 35:87–97.

26. Rosencwaig A. and Willis J. B. 1980. Photoacoustic study of laser damage in thin films. *Appl. Phys. Lett.* 36:667–669.

27. Sheehan L., Kozlowski M., and Rainer F. 1995. Diagnostics for the detection and evaluation of laser induced damage. *Proc. SPIE* 2428:13–22.

28. Mundy W. C. and Hughes R. S. 1984. Photothermal deflection microscopy of thin film optical coatings. *Natl. Bur. Stand. U.S. Spec. Publ.* 669:349–354.

29. Wu Z. L., Reichling M., Fan Z. X., and Wang Z. J. 1991. Applications of pulsed photothermal deflection technique in the study of laser-induced damage in optical coatings. *Proc. SPIE* 1441:214–227.

30. Hu H. Y., Fan Z. X., and Luo F. 2001. Laser-induced damage of a 1064-nm ZnS/MgF$_2$ narrow-band interference filter. *Appl. Opt.* 40:1950–1956.

31. Sheehan L., Kozlowski M., Stolz C., Genin F., Runkel M., Schwartz S., and Hue J. 1996. Large area damage testing of optics. *Proc. SPIE* 2775:357–369.

32. Hue J., Garrec P., Dijon J., and Lyan P. 1996. R-on-1 automatic mapping: A new tool for laser damage testing. *Proc. SPIE* 2714:90–101.

33. Chow R., Runkel M., and Taylor J. R. 2005. Laser damage testing of small optics for the National Ignition Facility. *Appl. Opt.* 44:3527–3531.

34. Borden M. R., Folta J. A., Stolz C. J. et al. 2005. Improved method for laser damage testing coated optics. *Proc. SPIE* 5991:59912A(1)–59912A(10).

35. Parks J. H. and Rockwell D. A. 1975. Surface studies with acoustic probe techniques. *Natl. Bur. Stand. U.S. Spec. Publ.* 435:157–163.

36. Lee K. C., Chan C. S., and Cheung N. H. 1996. Pulsed laser-induced damage threshold of thin aluminum films on quartz: Experimental and theoretical studies. *J. Appl. Phys.* 79:3900–3905.

37. Differential interference contrast microscopy, http://en.wikipedia.org/wiki/Nomarski_interference_contrast (last modified January 14, 2014).

38. Gallais L. and Natoli J. Y. 2003. Optimized metrology for laser-damage measurement: Application to multiparameter study. *Appl. Opt.* 42:960–971.

39. Wolfe J. E. and Schrauth S. E. 2007. Automated laser damage test system with real-time damage event imaging and detection. *Proc. SPIE* 6403:640328(1)–640328(9).

40. Franck J. B., Seitel S. C., and Hodgkin V. A. 1986. Automated pulsed testing using a scatter-probe damage monitor. *Natl. Bur. Stand. U.S. Spec. Publ.* 727:71–76.

41. Seitel S. C. and Babb M. T. 1987. Laser-induced damage detection and assessment by enhanced surface scattering. *Proc. SPIE* 752:83–88.

42. Sharp R. and Runkel M. 2000. Automated damage onset analysis techniques applied to KDP damage and the Zeus small area damage test facility. *Proc. SPIE* 3902:361–368.

43. Davit J. 1973. Damage threshold in 10.6 μm laser materials. *Natl. Bur. Stand. U.S. Spec. Publ.* 387:170–174.

44. Laurent L., Cavarro V., and Allais C. 2002. Time-resolved measurements of reflectivity plasma formation and damage of Hafnia/Silica multilayers mirrors at 1064 nm. *Proc. SPIE* 4679:410–419.

45. Génin F. Y. and Stolz C. J. 1997. Morphologies of laser-induced damage in hafnia-silica multilayer mirror and polarizer coatings. *Proc. SPIE* 2870:439–448.

46. Rigatti A. L. and Douglas J. S. 1997. Status of optics on the OMEGA laser after 18 months of operation. *Proc. SPIE* 2966:441–450.

47. Hu G. H., Qi, H. J., He, H. B., Li, D. W., Zhao, Y. A., Shao, J. D., and Fan, Z. X. 2010. 3D morphology of laser-induced bulk damage at 355 and 1064 nm in KDP crystal with different orientations. *Proc. SPIE* 7842:78420Y(1)–78420Y(8).

48. Liu, X. F., Zhao, Y. A., Li, D. W., Hu, G. H., Gao, Y. Q., Fan, Z. X., and Shao, J. D. 2011. Characteristics of plasma scalds in multilayer dielectric films. *Appl. Opt.* 50:4226–4231.

49. Liu X. F., Li D. W., Zhao Y. A., Li X., and Shao J. D. 2010. Characteristics of nodular defect in HfO_2/SiO_2 multilayer optical coatings. *Appl. Surf. Sci.* 256:3783–3788.

50. Krol H., Gallais L., Grezes-Besset C., Natoli J. Y., and Commandre M. 2005. Investigation of nanoprecursors threshold distribution in laser-damage testing. *Opt. Commun.* 256:184–189.

51. Li D. W., Shao J. D., Zhao Y. A., Yi K., and Qi H. J. 2007. The exponential fitting of laser damage threshold and analysis of testing errors. *Proc. SPIE* 6720:67200B(1)–67200B(7).

52. Porteus J. O. and Seitel S. C. 1984. Absolute onset of optical-surface damage using distributed defect ensembles. *Appl. Opt.* 23:3796–3805.

53. Capoulade J., Gallais L., Natoli J. Y., and Commandre M. 2008. Multiscale analysis of the laser-induced damage threshold in optical coatings. *Appl. Opt.* 47:5272–5280.

54. Natoli J. Y., Bertussi B., and Commandre M. 2005. Effect of multiple laser irradiations on silica at 1064 and 355 nm. *Opt. Lett.* 30:1315–1317.

55. Zhao Y. A., Shao J. D., He H. B., and Fan Z. X. 2005. Laser conditioning of high-reflective and anti-reflective coatings at 1064 nm. *Proc. SPIE* 5991:599117(1)–599117(6).

56. Maldutis E. K., Balickas S. K., and Kraujalis R. K. 1983. Accumulation and laser damage in optical glasses. *Natl. Bur. Stand. U.S. Spec. Publ.* 638:96–102.

57. Sheehan L., Kozlowski M., and Tench B. 1995. Full aperture laser conditioning of multilayer mirrors and polarizers. *Proc. SPIE* 2633:457–462.

58. Wolfe C. R., Kozlowski M. R., Campbell J. H., Rainer F., Morgan A. J., and Gonzales R. P. 1989. Laser conditioning of optical thin films. *Proc. SPIE* 1438:360–375.

59. Swain J. E., Stokowski S. E., Milam D., and Rainer F. 1982. Improving the bulk laser damage resistance of potassium dihydrogen phosphate crystals by pulsed laser radiation. *Appl. Phys. Lett.* 40:350–352.

60. Runkel M., Yoreo J. D., Sell W. D., and Milam D. 1997. Laser conditioning study of KDP on the optical sciences laser using large area beams. *Proc. SPIE* 3244:51–63.

61. Liu, X. F., Li, D. W., Li, X., Zhao, Y. A., and Shao, J. D. 2009. 1064 nm laser conditioning effect of HfO_2/SiO_2 high reflectors deposited by e-beam. *Chin. J. Lasers* 36:1545–1549.

62. Porteus J. O., Soileau M. J., Bennet H. E., and Bass M. 1975. Laser damage measurement at CO_2 and DF wavelength. *Natl. Bur. Stand. U.S. Spec. Publ.* 435:207–215.

63. Deshazer L. G., Newnam B. E., and Leung K. M. 1973. Role of coating defects in laser-induced damage to dielectric thin films. *Appl. Phys. Lett.* 23:607–609.

64. Picard R. H., Milam D., and Bradbury R. A. 1977. Statistical analysis of defect-caused laser damage in thin films. *Appl. Opt.* 16:1563–1571.

Chapter 6

65. Foltyn S. R. 1982. Spotsize effects in laser damage testing. *Natl. Bur. Stand. U.S. Spec. Publ.* 669:368–379.
66. Decker J. E., Xiong W., Yergeau F., and Chin S. L. 1992. Spot-size measurement of an intense CO_2 laser beam. *Appl. Opt.* 31:1912–1913.
67. Firester A. H., Heller M. E., and Sheng P. 1977. Knife-edge scanning measurements of subwavelength focused light beams. *Appl. Opt.* 16:1971–1974.
68. Olver F. W. J., Lozier B., and Clark C. W. 2010. *NIST Handbook of Mathematical Functions*. Cambridge University Publishers, New York.
69. Stewart A. F. and Guenther A. H. 1983. Preliminary experimental results of spot size scaling in laser induced damage to optical coatings. *Natl. Bur. Stand. U.S. Spec. Publ.* 638:517–531.
70. Liang X. Y., Leng Y. X., Wang C. et al. 2007. Parasitic lasing suppression in high gain femtosecond petawatt Ti:sapphire amplifier. *Opt. Express* 15:15335–15341.
71. Guide to streak camera, http://www.hamamatsu.com/resources/pdf/sys/e_streakh.pdf (accessed July 11, 2014).
72. Takahashi A., Nishizawa M., Inagaki Y., Koishi M., and Kinoshita K. 1994. New femtosecond streak camera with temporal resolution of 180fs. *Proc. SPIE* 2116:275–284.
73. Diels J. C. and Rudolph W. 2006. *Ultrashort Laser Pulse Phenomena* (2nd edn). Academic Impressed, London, U.K.
74. Trebino P. 2002. *Frequency-Resolved Optical Gating: The Measurement of Ultrashort Laser Pulses*. Kluwer Academic Publishers, Boston, MA.
75. Runkel M., Hawley-Fedder R., Widmayer C., Williams W., Weinzapfel C., and Roberts D. 2005. A system for measuring defect induced beam modulation on inertial confinement fusion-class laser optics. *Proc. SPIE* 5991:59912H(1)–59912H(9).
76. Hunt J. T., Manes K. R., and Renard P. A. 1993. Hot images from obscurations. *Appl. Opt.* 32:5973–5982.
77. Schmidt J. R., Runkel M. J., Martin K. E., and Stolz C. J. 2007. Scattering-induced downstream beam modulation by plasma scalded mirrors. *Proc. SPIE* 6720:67201H(1)–67201H(10).
78. Taniguchi J., LeBarron N. E., Howe J., and Smith D. J. 2001. Functional damage thresholds of hafnia/silica coating designs for the NIF laser. *Proc. SPIE* 4347:109–117.
79. ISO10110-7. 2008. Optics and photonics—Preparation of drawings for optical systems—Part 7: Surface imperfection tolerances.
80. MIL-PRF-13830B. 1997. Optical components for fire control instruments; general specification governing the manufacture, assembly, and inspection.

7. Statistics of Laser Damage Threshold Measurements

Jonathan W. Arenberg

Chapter 7

Laser-Induced Damage in Optical Materials. Edited by Detlev Ristau © 2015 CRC Press/Taylor & Francis Group, LLC. ISBN: 978-1-4398-7216-1.

7.1 Introduction

Laser damage threshold is a description of the ability of an optical component to maintain the integrity of performance under irradiation. Below this threshold, the component is expected to perform as designed and manufactured. Above this threshold, there is a risk of degraded performance and ultimately failure of the component and the system it belongs to. Viewed in this light, the damage threshold is the definition of the safe operating level for the component or system. Ideally, the value of this threshold would be certain and repeatable, like many other engineering parameters. This unfortunately is not the case for laser damage thresholds. Damage thresholds have been known from the earliest days of the laser to be far from repeatable. An early, but by no means unique, example is cited by Ready's comments (Ready 1971) on a set of results from Olness (1968), whose threshold values varied from 0.5 to 2.6 J/cm² for nominally identical samples. In the early days of the laser, the cause of this wide variance in observed thresholds was ascribed entirely to the variation in the intrinsic behavior of the samples. This chapter will demonstrate that the sample's intrinsic behavior, the type of measurement, and the manner in which it is made also contribute to poor repeatability of damage results.

The usual manner for dealing with such an uncertainty in system performance is twofold; tolerate the risk of degradation or include some manner of mitigation such as "derating" the operational level of the system relative to the laser damage threshold. Both of these coping strategies have costs; the larger the uncertainty that has to be accounted for, the greater the cost or programmatic risk. For high-reliability systems, such as space, medical, and military applications, mitigation of these uncertainties can be a major contributor to both the recurring and nonrecurring systems cost.

The accurate, standardized definition and measurement of laser damage thresholds is seminal to a robust global commerce in laser optics and key to the improvement of laser systems. The key role that measurements play has been known from the dawn of the laser age. The need for good widely comparable measurements was one of the reasons for the first meeting on laser damage in 1969 (Lee 1969, Stewart and Guenther 1984). The importance of good measurements continues in importance and is noted as a key area for research on the occasions of major reviews or retrospectives on the subject (Guenther 1999). Laser damage has long been viewed as the limiting technology for the advancement of the laser as a tool, for military, medical, industrial, and scientific applications (Seitel and Teppo 1983) and has been reported in the literature from the earliest days of laser development in the 1960s (Giuliano 1964, Hercher 1964, Bernal 1965).

Even in the early days of laser development, laser damage was a well-reported phenomenon. Unfortunately, definitions of what exactly damage was and what threshold meant were not common and infrequently, if ever, described. With many definitions of damage being used in the literature and inconsistent use of the term "threshold," direct comparison of results was difficult at best (Ready 1971). Few, if any, of these early works, however, discuss in detail exactly what data were collected and how it were processed to determine a "threshold." Even the meaning of threshold changed from report to report, especially in the early days, up to the emergence of the onset threshold in the 1980s (Porteus and Seitel 1984).

Laser manufacturers have used in-house testing methods and techniques to obtain acceptable optics and coatings for their systems for many years, especially in the early decades of laser development (Mordaunt et al. 1986). With the appearance of commercial testing houses in the mid-1980s, the field is converging on standardized tests for most applications with some unique applications defining specific tests, such as NIF in the United States and LMJ in Europe. Today, over 50 years after the invention of the laser, there is no universal catalog of damage performance available to guide the laser designer although some references give listings of results (Wood 2003).

International standards for making laser damage measurements have existed for over a decade (International Standards Organization 2000). Despite the large step forward the ISO standards represent, they are not perfect and their results cannot be used *carte blanche*. The hypothetical laser designer basing a design on published threshold data without detailed analysis and study is putting their design at substantial risk of failure. So, in terms of published data, or data from vendors it is *caveat emptor,* buyer beware. This chapter's goal is to give the reader a sufficient background in the statistics of laser damage threshold measurements to be an informed consumer and user of such data. It must be noted that there are in general differences between the damage threshold measured and the performance of a component in system. Determining the reduction in performance due to system effects is the job of the designer. However, the designer's task is impossible without a solid understanding of the laboratory measurement.

The word *statistic* is defined as "a value calculated from and observed sample with a view to characterizing the population from which is drawn" (Fisher 1944). The scope of this chapter is dedicated to the statistics of the onset threshold, T_{onset}, above which there is a finite chance of damage and its measurement. This chapter will show that the laser damage threshold is inherently statistical, that the observed or measured threshold, T_M, is dependent on how the data are collected and processed, the test procedure, and the intrinsic properties of the optical surface being measured. There are many reasons for the statistical behavior of the optic, material nonuniformities, voids, inclusions, contamination, and surface scratches, to name a few.

After this introduction, Section 7.2 will review the history of laser damage measurements, showing the evolution of the definition of the threshold and measurement techniques. Section 7.3 will introduce the analytic framework for analyzing damage thresholds and will derive the probability of damage (PoD) curve from Poisson statistics and models of the distribution of defects in space and with fluence. The behavior of the PoD curve under a few conditions of illumination and defect distribution will also be discussed. This section also derives the probability distribution for the weakest site on a sample to quantify the distribution of the onset threshold, T_{onset}. In Section 7.4, the main measurement techniques for laser damage threshold m are introduced. Each method's distribution of measured results, T_M, is derived. Section 7.5 shows that the observed damage threshold is a convolution of the effects of the measurement method and the properties of the optic. From this viewpoint, we have a complete statistical model of laser damage threshold observations. The chapter is summarized in Section 7.6.

Chapter 7

7.2 Selected History of Laser Damage Threshold Measurements

A selected timeline of events is given as Figure 7.1. It is easily seen from Figure 7.1 that shortly after the demonstration of the first laser in May of 1960 (Mainman 1960), reports of laser-induced damage in components began appearing. The earliest reports of laser-induced damage were made in 1964 in conference and refereed literature (Cullom and Waynant 1964, Giuliano 1964, Hercher 1964). In these early reports, no real attention was given to what data were collected and what processing was used to determine the reported threshold, this rather lax trend continued well into the 1970s (Giuliano et al. 1970). It was also becoming apparent that the damage resistance of the then contemporary state of the art of laser components was an impediment to the advancement of high-power systems (Glass and Guenther 1969, 1973). Many authors reported wide variations in thresholds and difficulty in defining exactly what was being measured and reported (Giuliano et al. 1970).

In June of 1969, a 1-day meeting was held under the auspices of the American Society for Testing and Materials (ASTM) at the Boulder, CO, laboratories of the US National Bureau of Standards, now the National Institute of Standards and Technology (Glass and

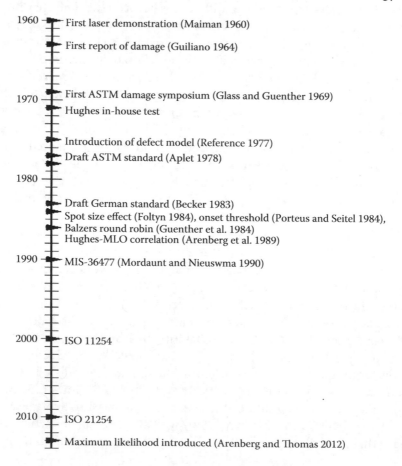

FIGURE 7.1 A timeline of selected events in the history of laser damage measurements.

Guenther 1969). The organizers reported an enthusiastic response (Glass and Guenther 1969) and held a subsequent meeting at the same location the following year (Glass and Guenther 1970). By 1971, this meeting became an annual meeting known colloquially as the Boulder Damage Symposium (BDS). The BDS continues to this day as an SPIE conference and is the main forum for workers from around the world to share results and collaborate. From its inception, the BDS has always contained a dedicated section of the meeting dealing explicitly with measurements (Glass and Guenther 1969).

Concurrent with the establishment of the annual Boulder meeting on damage, a number of panels assessed the state of the art in a series of reports in the early 1970s (Bloembergen 1970, 1972, Sahagran and Pitha 1971). This period saw the initial observations of defect-based damage (Milam et al. 1973) and development of defect (degenerate)-based models for analyzing damage (DeShazer et al. 1973, Picard et al. 1977). Interestingly, damage studies from the early 1970s defined the threshold as the median or 50% probability point, and many of these publications included analyses that show the probability of damage varies with spot size (Picard et al. 1977).

Also in the early 1970s, laser system reliability was becoming a major concern for the US Department of Defense as tactical systems such as the ruby-based range finder for the US M60A3 tank were going into production (Nieuwsma 2013). The need for a production-oriented, quality control–type test was noticed at Hughes Aircraft Company in the 1970s to enhance production and reliability of ruby-based systems for the US military. The earliest Hughes test involved placing a test optic within a laser resonator, designed to operate at higher than nominal flux, and firing the laser (Nieuwsma 2013). If the part was damaged, it was discarded, and if not damaged, it was sent on for production. Also in the late 1970s, the first attempt at documented laser damage test procedure was developed for the ASTM, Committee F-1, Subcommittee 2 (Aplet 1978).

In the 1980s, the Hughes test was revised to use a focused beam to test whether the witness sample from a production lot was damaged, and if the average intensity of damage on the witness piece was within 50% of a known good standard or reference optic, the lot was accepted; otherwise, it was rejected for use in production (Mordaunt et al. 1986). This test proved very successful, with this author knowing of no returns to Hughes for service due to laser damage (Nieuwsma 2013). This type of comparative test was also followed at other laboratories (Milam et al. 1981).

In 1983, a major collaborative effort to compare damage threshold measurements on identical samples at varying test conditions was conducted (Guenther et al. 1984). The samples were supplied by Balzers of Lichtenstein and sent to laboratories in the United States, Europe, and Asia. The threshold used in this landmark study was the average of the highest fluence that did not cause damage and the lowest fluence that does result in damage (Guenther et al. 1984).

Also in 1984, there was the emergence of a consensus away from the use of the median fluence as the definition of the laser damage threshold. The argument against the use of the median was made around several points (Foltyn 1984). Onset, or the zero percent probability of damage, has an analogy with accepted laser engineering nomenclature for the term threshold. Onset represents a safe operating level and was argued to be independent of spot size. Seitel and Porteus argued for the adoption of the "onset" and used a nondegenerate defect distribution to derive the probability of damage curve (Seitel and Porteus 1985). The works of Foltyn (1984) and Seitel and Porteus (1985) argued for a

Chapter 7

curve-fitting procedure to be used to extrapolate to the point of onset. It is at this point in the history of the field that the linear extrapolation enters the picture.

Hughes Aircraft was not the only manufacturer testing for laser damage. Every other major US government supplier performed some type of quality check on laser components, while these tests did differ, some commonality existed (Seitel et al. 1988). In the late 1980s, the first commercial testing house for laser damage, Montana Laser Optics (MLO and now Big Sky Laser) opened its doors. There was a major effort between Hughes and MLO to prove that viable interlaboratory comparisons of test results could be made. It was shown experimentally that it was possible to correlate the test results if the test procedures were the same (Arenberg et al. 1989). The US Army funded an effort to extend the Hughes–MLO correlation, which resulted in the issuance of MIS-36477 in 1990 (Mordaunt and Nieuwsma 1990).

In the late 1980s, an ISO working group began work and made its first public disclosure of the draft standard in 1993 (Becker and Bernhardt 1993, Seitel et al. 1993, Starke and Bernhardt 1993). The draft standard was developed in the 1990s and emerged as the ISO standard test, ISO 11254 Part 1, for 1-on-1 testing in 2001. A German round-robin experiment confirmed the ability to achieve reasonable levels of correlation using this test procedure (Riede et al. 1998). ISO 11254 has been reconfirmed, reorganized, and reissued in 2011 as ISO 21254 (International Standards Organization 2011).

7.3 Analytic Framework

7.3.1 Poisson Statistics

The two outcomes of a damage experiment, damage, D, and no damage, ND, are mutually exclusive and exhaustive, meaning

$$\Pr(D) + \Pr(ND) = 1, \tag{7.1}$$

where the notation $\Pr(X)$ means probability of event X occurring. Subtracting $\Pr(ND)$ from each side gives

$$\Pr(D) = 1 - \Pr(ND). \tag{7.2}$$

The ND event is where there are exactly zero defects illuminated above their thresholds in the test spot. The Poisson distribution is used to calculate the rate of observations of an event, given a mean rate (DeShazer et al. 1973). Since the defects that cause damage are assumed to be isotropically (uniformly) distributed on the optic's surface, Poisson statistics apply.

The Poisson distribution gives the probability of n defects being in the test area and illuminated above threshold, when the mean areal density of this occurring is $\langle n \rangle$ by the expression (Zwillinger and Kokoska 1999)

$$\Pr\left(n | \langle n \rangle\right) = \frac{\langle n \rangle^{n} \exp\left(-\langle n \rangle\right)}{n!}. \tag{7.3}$$

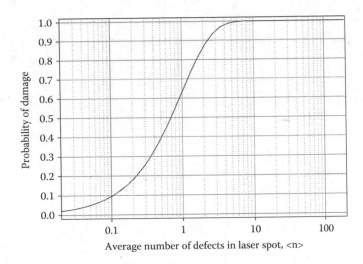

FIGURE 7.2   Plot of Equation 7.6 showing the behavior of the PoD curve.

The probability of event ND or no defects in the test area is when n = 0, namely,

$$Pr(ND) = \frac{\langle n \rangle^0 \exp(-\langle n \rangle)}{0!}. \tag{7.4}$$

Since $\langle n \rangle^0 = 1$ and $0! = 1$, Equation 7.4 reduces to

$$Pr(ND) = \exp(-\langle n \rangle). \tag{7.5}$$

Substitution of Equation 7.5 into Equation 7.2 gives

$$Pr(D) = 1 - \exp(-\langle n \rangle). \tag{7.6}$$

Equation 7.6 is plotted as Figure 7.2 for various values of $\langle n \rangle$. It can easily be seen from Equation 7.6 that for Pr(D) to be zero, $\langle n \rangle$ must be identically zero. Pr(D) increases very slowly for small values of $\langle n \rangle$, until $\langle n \rangle \sim 0.1$, grows rapidly until $\langle n \rangle \sim 3$, and saturates at approximately $\langle n \rangle \sim 4$. In Section 7.3.4, several important cases will illustrate how $\langle n \rangle$ can vary for different illuminations and defect distributions.

7.3.2   Irradiation Distributions

The irradiation distribution of the laser beam determines the local fluence at any point in the beam. Two kinds of distributions are introduced, the circular flat-top beam and the circular Gaussian profile. The circular flat-top beam simplifies the analysis while the Gaussian beam is the profile most commonly used.

Chapter 7

7.3.2.1   Flat-Top Beam

The fluence profile for a circular flat-top beam with peak fluence, ϕ_{max}, and a spot radius of w is

$$\phi(r) = \begin{cases} \phi_{max} & \text{if } r \le w \\ 0 & \text{if } r > w \end{cases}.$$

(7.7)

The area at or above a fluence level, ϕ, is found to be

$$a(\phi) = \begin{cases} \pi w^2 & \text{if } \phi \le \phi_{max} \\ 0 & \text{if } \phi > \phi_{max} \end{cases}.$$

(7.8)

7.3.2.2   Gaussian Beam

For a Gaussian beam, with peak axial fluence, ϕ_{max} and a 1/e radius of w, the local fluence is given by

$$\phi(r) = \phi_{max} \exp\left(-\frac{r^2}{w^2}\right).$$

(7.9)

Equation 7.9 can be manipulated to give the area at or above a given fluence, ϕ, $a(\phi)$,

$$a(\phi) = \begin{cases} \pi w^2 \ln\left(\dfrac{\phi_{max}}{\phi}\right) & \text{if } \phi \le \phi_{max} \\ 0 & \text{if } \phi > \phi_{max} \end{cases}.$$

(7.10)

For a Gaussian beam, it is worth noting that a $(\phi_{max}) = 0$, meaning that there is no area at the peak fluence. So, while the peak fluence is traditionally reported as the irradiation level for a Gaussian bean, no area is actually being tested at that level.

7.3.3   Distribution of Defects with Fluence

A defect-based model has long been used to predict, describe, and analyze laser damage; the first descriptions entering the literature in the early 1970s and is in wide use today (Sparks 1972, DeShazer et al. 1973, Fradin and Sua 1974, Porteus and Seitel 1984, Krol et al. 2005, Arenberg 2012). The defect model assumes that the surface has an intrinsic threshold of damage, T_i, and hosts a series of defects whose damage threshold is lower. The defects that fail on the fluence interval, $[\phi, \phi + \delta\phi]$, are assumed to be uniformly or isotropically distributed with a mean areal density, $\langle n(\phi) \rangle$.

Damage occurs when a laser illuminates a given site with a fluence exceeding the local threshold. Figure 7.3 shows a pictorial representation of a simple (degenerate) defect model and four different irradiation conditions. Test beam A shows no damage

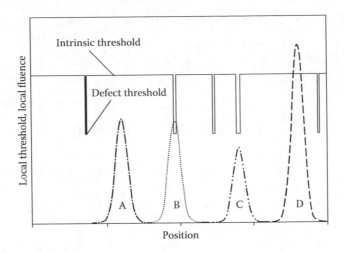

FIGURE 7.3   Pictorial representation of a simple (degenerate) defect model and four different irradiation conditions.

since the laser fluence is lower than the local threshold. Test beam B exceeds the local threshold as the beam is centered on a defect and damage would be observed. Since the fluence of test beams A and B are greater than the defect's threshold, damage will occur when a defect is located in the high-fluence portion of the beam. Test beam C, whose peak fluence is below that of both the intrinsic and defects, will show no damage, regardless of its location. Finally, test beam D, whose fluence exceeds the local (intrinsic) threshold, will show damage, regardless of location.

The defect distribution function, $f(\phi)$, is defined by

$$f(\phi) = D_T N g(\phi),$$ (7.11)

where
    N is a normalization constant
    D_T is the total areal density of defects
    $g(\phi)$ is the fluence distribution (shape) function

The normalization coefficient, N, is given by

$$\frac{1}{N} = \int_0^\infty g(\phi) d\phi.$$ (7.12)

Let $\int g(s) ds = G$, then Equation 7.12 can be written as

$$N = \frac{1}{G(\infty) - G(0)}.$$ (7.13)

Chapter 7

Substitution of Equation 7.13 into Equation 7.11 gives f, the defect distribution as

$$f(\phi) = \frac{D_T g(\phi)}{G(\infty) - G(0)}. \tag{7.14}$$

7.3.4   Derivation of the Probability of Damage Curve

The value of $\langle n(\psi) \rangle$ for a given irradiation will depend on the distribution of defects in space, their threshold distribution, $f(\phi)$, and the local fluence profile, $\phi(x, y)$. The defect distribution, $f(\phi)$, is the areal density of defects that fail on the interval $[\phi, \phi + d\phi]$. The relevant area is the area of the test beam that exceeds the threshold for the defects, t, $a(\phi)$. $\langle n \rangle$ is given by

$$\langle n(\phi_{max}) \rangle = \int_{Beam} da \int_0^{\phi_{max}} f(\phi') d\phi'. \tag{7.15}$$

The cumulative probability of damage curve (PoD curve) is given by substitution of Equation 7.15 into Equation 7.6 and evaluated for the specific conditions of the experiment. The literature contains cases for the PoD curve where $f(\phi)$, the defect distribution, has been a Dirac delta function, power law, or Gaussian (Picard et al. 1977, Porteus and Seitel 1984).

7.3.4.1   Degenerate Flat-Top Beam

The easiest case to analyze and understand is the so-called degenerate case, where all defects fail at a single threshold T irradiated by a circular flat-top intensity profile beam.

The distribution of defects, $f(\phi)$, is

$$f(\phi) = D_T \delta(\phi - T), \tag{7.16}$$

where

D_T is the density of defects

$\delta(\phi - T)$ is a Dirac delta function, with the usual properties (Weisstein 2014)

For this situation, $\langle n \rangle$ is written as

$$\langle n(\phi_{max}) \rangle = \pi w^2 \int_0^{\phi_{max}} D_T \delta(\phi' - T) d\phi'. \tag{7.17}$$

Using the properties of the Delta function, Equation 7.17 is evaluated to be

$$\langle n \rangle = \begin{cases} 0 & \text{if } \phi_{max} \leq T \\ \pi w^2 D_T & \text{if } \phi_{max} > T \end{cases}. \tag{7.18}$$

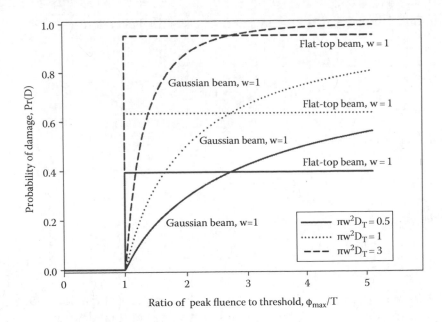

FIGURE 7.4   Probability of damage plotted versus peak fluence level for both flat-top and Gaussian profile beams with various defect distributions.

Substitution of Equation 7.18 into Equation 7.6 gives the PoD curve as

$$\Pr(D) = \begin{cases} 0 & \text{if } \phi_{max} \leq T \\ 1 - \exp(-\pi w^2 D_T) & \text{if } T_i > \phi_{max} > T. \\ 1 & \text{if } \phi_{max} \geq T_i \end{cases} \tag{7.19}$$

In Equation 7.19, the value of $\Pr(D)$ is zero until the threshold T is reached and, at higher fluences, $\Pr(D)$ is fixed at a value depending on the $\pi w^2 D_T$ product. When the intrinsic coating threshold, T_i is exceeded damage is certain. Equation 7.19 is plotted in Figure 7.4 for various values of D_T.

7.3.4.2   Degenerate Case Gaussian Beam

For the case of a Gaussian beam, the $\langle n \rangle$ is found by substitution of Equations 7.10 and 7.16 into Equation 7.15, which yields

$$\langle n \rangle = \pi w^2 \int_0^{\phi_{max}} D_T \delta(\phi' - T) \ln\left(\frac{\phi_{max}}{T}\right) d\phi'. \tag{7.20}$$

Evaluation of the integral gives $\langle n \rangle$ as

$$\langle n \rangle = \begin{cases} 0 & \text{if } \phi_{max} \leq T \\ \pi w^2 D_T \ln\left(\frac{\phi_{max}}{T}\right) & \text{if } \phi_{max} > T. \end{cases} \tag{7.21}$$

Chapter 7

Substitution of Equation 7.21 into Equation 7.6 gives the PoD curve as

$$Pr(D) = \begin{cases} 0 & \text{if } \phi_{max} \leq T \\ 1 - \exp\left(-\pi w^2 D_T \ln\left(\dfrac{\phi_{max}}{T}\right)\right) & \text{if } \phi_{max} > T \end{cases} \tag{7.22}$$

Equation 7.22 simplifies to

$$Pr(D) = \begin{cases} 0 & \text{if } \phi_{max} \leq T \\ 1 - \left(\dfrac{\phi_{max}}{T}\right)^{-\pi w^2 D_T} & \text{if } T_i > \phi_{max} > T \\ 1 & \text{if } \phi_{max} \geq T_i \end{cases} \tag{7.23}$$

The PoD curves are given in Figure 7.4 for the flat-top beam, Equation 7.19, and for a Gaussian beam, Equation 7.23, for three different values of total density of defects, D_T; $D_T = 0.5$, 1, and 3. The PoD curve for a flat-top beam has an instantaneous jump when $\phi_{max}/T = 1$, remaining constant thereafter. The PoD curve for the Gaussian beam is monotonically increasing after $\phi_{max}/T = 1$.

7.3.4.3 Multiple Degenerate Distributions

Consider a general distribution of defects that have a cumulative areal density of D, with a density distribution of $p(\phi)$, $f(\phi)$ is

$$f(\phi) = D_T p(\phi). \tag{7.24}$$

The probability distribution function $p(\phi)$, which is normalized to unity, can be discretized into a series of δ functions and the results of Equations 7.18 and 7.21 will be used to derive $\langle n \rangle$ for this more general case. The distribution p is written as a series

$$p(\phi) = \sum_i p(i\delta\phi)\delta(\phi - i\delta\phi). \tag{7.25}$$

Substitution of Equation 7.25 into Equation 7.24 gives $f(\phi)$ as

$$f(\phi) = D_T \sum_{i=0}^{m} p(i\delta\phi)\delta(\phi - i\delta\phi). \tag{7.26}$$

For the top-hat beam, the peak fluence is

$$\phi_{max} = m\delta\phi. \tag{7.27}$$

Using Equation 7.26 in Equation 7.18 results in

$$\langle n \rangle = \begin{cases} 0 & \text{if } \phi_{max} \leq T \\ \pi w^2 D_T \sum_{i=0}^{m} p(i\delta\phi) & \text{if } \phi_{max} > T \end{cases}. \tag{7.28}$$

For the case of a Gaussian beam, Equation 7.26 is inserted into Equation 7.21, giving

$$\langle n \rangle = \begin{cases} 0 & \text{if } \phi_{max} \leq T \\ \pi w^2 D_T \sum_{i=0}^{m} p(i\delta\phi) \ln\left(\dfrac{\phi_{max}}{i\delta\phi}\right) & \text{if } \phi_{max} < T \end{cases}. \tag{7.29}$$

The reader should note that the denominator of the natural logarithm term in Equation 7.29 is no longer a constant, but changes with i, since each value of i is a different class of degenerate defects. The results (Equations 7.28 and 7.29) are quite general as $\delta\phi$ can be as small as is desired.

7.3.4.4 Gaussian Distribution of Defects

Defects can have a Gaussian distribution in fluence, given by

$$f(\phi) = \frac{2D_T}{N} \exp\left(-\left(\frac{\phi - T}{2\sigma}\right)^2\right), \tag{7.30}$$

where

N and D_T have the same meaning as previously given

σ is the standard deviation of the Gaussian distribution

the role of T is now that of the mean fluence of failure not the onset threshold

The normalization, N, is given by

$$N = \frac{1}{2}\left(1 + \text{erf}\left(\frac{T}{\sigma\sqrt{2}}\right)\right), \tag{7.31}$$

where erf(x) is the well-known error function of argument x (Zwillinger and Kokoska 1999).

For a flat-top beam, the PoD is written by substitution of Equation 7.30 into Equation 7.15, yielding

$$\Pr(D(\phi)) = 1 - \exp\left(-\frac{D_T \pi w^2}{N} \int_0^{\phi_{max}} \exp\left[-\frac{1}{2}\left(\frac{\phi - T}{\sigma/2}\right)^2\right] d\phi\right). \tag{7.32}$$

Chapter 7

Equation 7.32 can be integrated to give

$$Pr\left(D(\phi)\right) = 1 - \exp\left(-\frac{D_T\pi w^2}{N}\left(\text{erf}\left(\frac{T\sqrt{2}}{\sigma}\right) - \text{erf}\left(\frac{(T-\phi_{max})\sqrt{2}}{\sigma}\right)\right)\right). \tag{7.33}$$

When the laser profile is a Gaussian beam and there is a Gaussian distribution of defects, the PoD curve is

$$P(\phi) = 1 - \exp\left(-\frac{D_T\pi w^2}{N}\int_0^{\phi_{max}}\exp\left(-\frac{1}{2}\left(\frac{\phi-T}{\sigma/2}\right)^2\right)\ln\left(\frac{\phi_{max}}{\phi}\right)d\phi\right). \tag{7.34}$$

Equation 7.34 does not have a closed-form solution and must be integrated numerically.

Figure 7.5 shows Equation 7.33 evaluate for $T = 30$ J/cm^2 and σ varying from 0.1 to 30 J/cm^2. For the small values of σ, 0.1 and 1 J/cm^2, most of the distribution has a similar failure level, and the curves look quite a bit like the degenerate distribution. As σ gets larger, $\sigma = 10$ J/cm^2, in the current example, there is finite probability of damage at all fluences. For the s less than 10 J/cm^2, the median (50% probability of damage) occurs at the same fluence, 30 J/cm^2. This perhaps gives some insight as to why the median was an early definition of damage threshold (see Section 7.2). When σ gets large, the median probability shifts to higher fluences. This shift is caused by the fact that when σ is large, $T < 3\sigma$, and therefore, at very small fluences, the value of the left error function term in Equation 7.33 is not zero in these cases and it moves the median to higher fluences.

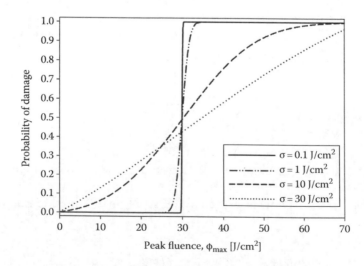

FIGURE 7.5   PoD curves for Gaussian distributed defects with a flat-top beam.

7.3.5   Behavior of the PoD Curve

This section examines the behavior of the PoD curve for two important cases. The first behavior is the probability of damage as a function of spot size. The second area of examination is how T_{onset}, the weakest site on an optic of area A, varies and with $f(\phi)$.

7.3.5.1   Spot Size Effect

For the purposes of this discussion, the spot size effect is defined to be a change in the PoD curve for a given defect distribution, $f(\phi)$, when the spot size (radius), w, is varied. Consider the simplest case of the PoD curve, which has been calculated for a flat-top beam and a degenerate distribution as given by Equation 7.19 and reproduced for the convenience of the reader as follows:

$$\Pr(D) = \begin{cases} 0 & \text{if } \phi_{max} \leq T \\ 1 - \exp\left(-\pi w^2 D_T\right) & \text{if } T_i > \phi_{max} > T. \\ 1 & \text{if } \phi_{max} \geq T_i \end{cases} \tag{7.35}$$

Equation 7.35 shows that PoD is entirely dependent on the spot size and then the fluence in the interval $[T, T_i)$ the PoD does not depend on the fluence, ϕ.

For a Gaussian beam, the PoD curve is given by Equation 7.23, repeated for the convenience of the reader as Equation 7.36:

$$\Pr(D) = \begin{cases} 0 & \text{if } \phi_{max} \leq T \\ 1 - \left(\dfrac{\phi_{max}}{T}\right)^{-\pi w^2 D_T} & \text{if } T_i > \phi_{max} > T. \\ 1 & \text{if } \phi_{max} \geq T_i \end{cases} \tag{7.36}$$

It is worth noting again, see Figure 7.4, that for both Equations 7.35 and 7.36, the onset, the maximum fluence where $\Pr(D) = 0$ is T_{onset}, and is independent of spot size, w. It is the independence of T_{onset} from w that is the root of the argument to use an onset, or 0% damage probability point as the definition of damage threshold. In the remainder of this chapter, we shall examine the extrapolation to T_{onset} under more generalized conditions.

7.3.5.2   Threshold (Onset)

In the analysis to develop PoD curves introduced in the previous section, the implicit assumption has been made that the sample (optic under test) areas are infinite. For an optic of area A, the distribution of defects is effectively filtered and the PoD curve goes to near zero when ϕ satisfies $f(\phi) \sim 1/A$. This presents a well-known problem when scaling from small areas to large optics (Feit et al. 1998, Arenberg 2000).

This section poses and then addresses the question, "What is the probability distribution of the weakest site, the true onset threshold on an optic of a finite area A with defect distribution $f(\phi)$?" The probability distribution of the onset threshold having a given value f, $\Pr(T_{onset} = \phi)$, for a sample of area A is determined by formulating the probabilities of two events. Event 1 occurs when there are no sites on the optic that can damage at fluence levels below ϕ. Event 2 occurs when there is at least one site that fails at exactly ϕ. Events 1 and 2 are independent events so the $\Pr(T_{onset} = \phi)$ can be written as

$$\Pr(T_{onset} = \phi) = \Pr(Event 1)\Pr(More than zero fail at \phi), \tag{7.37}$$

which becomes

$$\Pr(T_{onset} = \phi) = \Pr(No sites fail below \phi)\Pr(More than zero fail at \phi). \tag{7.38}$$

The probability that there are exactly no sites that can fail at level $\phi = 0$ is given using the defect distribution and the Poisson distribution as

$$\Pr(No sites that can fail at \phi = 0) = e^{-Af(0)}. \tag{7.39}$$

The probability that no sites can fail at $\delta\phi$

$$\Pr(No sites that can fail at \delta\phi) = e^{-Af(\delta\phi)}. \tag{7.40}$$

The probability that there are no sites that can possibly fail on the interval $[0, \delta\phi)$ is the product of the probabilities of no sites able to damage at 0 and none at $\delta\phi$:

$$\Pr(No sites that can fail below \delta\phi) = e^{-Af(0)}e^{-Af(\delta\phi)}. \tag{7.41}$$

By extension, the probability that there are no sites that fail below ϕ is given by

$$\Pr(Event 1) = \prod_{i=0}^{p-1} e^{-Af((\phi/p)i)}, \tag{7.42}$$

where $\delta\phi = \phi/p$. Equation 7.42 can be rewritten in more obvious form as

$$\Pr(Event 1) = \exp\left[-A\sum_i f\left(\frac{\phi}{p}i\right)\right]. \tag{7.43}$$

As $p \to \infty$, the summation in Equation 7.43 goes over to an integral

$$\Pr(Event 1) = \exp\left[-A\int_0^\phi f(\phi')d\phi'\right]. \tag{7.44}$$

The probability of Event 2 occurring is simply that at exactly ϕ, Event 1 does not occur, namely, there is a finite number (not zero) of sites that could fail at ϕ is

$$\Pr(\text{Event}\,2) = 1 - \exp\left[-Af(\phi)\right].$$ (7.45)

Substitution of Equations 7.44 and 7.45 into Equation 7.38 gives the final result

$$\Pr(T_{\text{onset}} = \phi) = \left(\exp\left[-A\int_0^\phi f(\phi')d\phi'\right]\right)\left(1 - \exp\left[-Af(\phi)\right]\right).$$ (7.46)

From inspection of Equation 7.46, it is clear that T_{onset} is distributed in ϕ and thus not generally constant. The distribution of T_{onset} depends on the manufacturing process and materials that determine $f(\phi)$, and the area, A, of the optic.

7.3.5.2.1 Onset Threshold Distribution-Degenerate Defects
In the following sections, the distribution of T_{onset} is calculated for a flat-top beam and degenerate defect distribution and linear defect distribution. The flat-top beam is chosen so that the integrals in Equation 7.46 are as simple as possible; this will allow the reader to gain insight. The linear defect model is intended as an approximation for the low-fluence wing of a more complex defect distribution.

In the case of the degenerate defects, the optic fails at either the level associated with the defects, T, or the intrinsic level associated with the coating, T_i. Since there are only two outcomes, this distribution is bimodal. The probability that there is a defect on the optic that can fail at T is

$$\Pr(T_{\text{onset}} = T) = 1 - \exp\left[-AD_T\right].$$ (7.47)

The probability that there are no defects at all, which will cause the onset threshold to be T_i, is the probability of no defects at all within optic of area A,

$$\Pr(T_{\text{onset}} = T_i) = \exp\left[-AD_T\right].$$ (7.48)

Figure 7.6 shows Equations 7.47 and 7.48 for various values of AD_T.

7.3.5.2.2 Onset Threshold Distribution-Linear Distribution of Defects
When the defect density varies linearly with fluence, $f(\phi)$ is given by

$$f(\phi) = \begin{cases} 0 & \phi \leq T \\ m(\phi - T) & \phi > T \end{cases}.$$ (7.49)

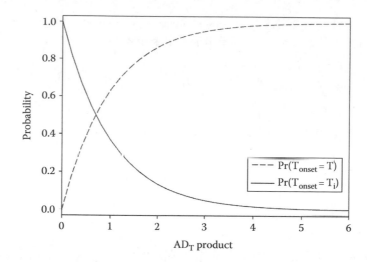

FIGURE 7.6 Plots of Equations 7.47 and 7.48 show that with increasing numbers of defects on the test optic (increasing AD_T product) the true onset is increasingly likely to be determined by the defects and not the intrinsic threshold.

Equation 7.49 is substituted into Equation 7.46 giving the distribution of the onset threshold for linear $f(\phi)$ as

$$\Pr\left(T_{onset} = \phi\right) = \begin{cases} 0 & \phi \leq T \\ \left(\exp\left[-A\int_T^\phi m\left(\phi' - T\right)d\phi'\right]\right)\left(1 - \exp\left[-Am\left(\phi - T\right)\right]\right) & \phi > T \end{cases} \quad (7.50)$$

Evaluation of the integral in Equation 7.50 gives

$$\Pr\left(T_{onset} = \phi\right) = \begin{cases} 0 & \phi \leq T \\ \left(\exp\left[-\dfrac{Am}{2}\left(\phi - T\right)^2\right]\right)\left(1 - \exp\left[-Am\left(\phi - T\right)\right]\right) & \phi > T \end{cases} \quad (7.51)$$

Figure 7.7 shows Equation 7.51 for various values of optic area A for m = 0.1. As the value of A increases, the distribution narrows and the mode decreases and begins to approach T. The result shown in Figure 7.7 clearly shows that onset threshold (weakest site) on an optic has an extended distribution and that distribution is dependent on optic area A. The result here is quite important. It means that for optics manufactured with the same process, yielding the same $f(\phi)$, that the underlying onset threshold (weakest site on the sample) should be expected to be constant in general.

7.4 Analysis of Measurements

The goal of this section is to introduce three different methods of determining laser damage threshold, meaning the onset threshold. Once each method is described, its distribution of measured values, T_M, is discussed. The distribution of T_M is generally a

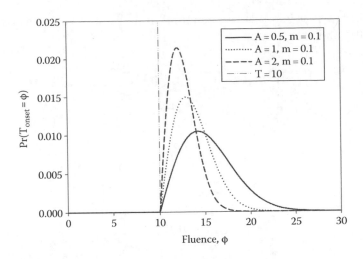

FIGURE 7.7   Probability distribution of onset thresholds, $\Pr(T_{onset} = \phi)$.

complicated dependence on the various test parameters and choices in the analysis of the collected data. In an ideal test method, the distribution of T_M would be immune to these choices. Frequently, the derivation of T_M for a measurement technique exposes a bias in the measurement as well as the techniques' ability to detect and report the true threshold.

This section analyzes the damage frequency method (DFM), which at the core of the ISO standards (International Standards Organization 2000, 2011), the binary search technique (BST) (Arenberg 2012), and the emerging maximum likelihood (ML) method (Arenberg and Thomas 2012). The DFM and BST have long been the subject of study analysis and have well-known systematic errors or biases (Arenberg 1995, Melninkaitis et al. 2009, Arenberg 2012). The ML method has only recently been applied to the problem of threshold measurement but shows good initial promise to be closer to an unbiased measurement technique (Arenberg and Thomas 2012).

7.4.1 Damage Frequency Method

The DFM is the recommended technique in ISO 21254-2, the international laser damage test standard (International Standards Organization 2011). In this test, a sample is exposed to a series of exposures at different fluence levels, ϕ_i. The number of damaged sites, d_i, at each fluence is divided by the number of sites exposed at each ϕ_i, n_i to give observed frequency of damage, F_i. The data (ϕ_i, F_i) are fit to a line

$$F(\phi) = m\phi + b. \tag{7.52}$$

The measured threshold, T_M, is given by

$$T_M = -\frac{b}{m}. \tag{7.53}$$

Chapter 7

The fitting formulas for b and m can be found in many standard texts (Bevington and Robinson 2003) and are given as

$$b = \frac{\sum_{i=1}^{r}\left(\phi_i/\sigma_i^2\right)\sum_{i=1}^{r}\left(F_i/\sigma_i^2\right) - \sum_{i=1}^{r}\left(\phi_i/\sigma_i^2\right)\sum_{i=1}^{r}\left(\phi_iF_i/\sigma_i^2\right)}{\sum_{i=1}^{r}\left(1/\sigma_i^2\right)\sum_{i=1}^{r}\left(\phi_i^2/\sigma_i^2\right) - \left(\sum_{i=1}^{r}\left(\phi_i/\sigma_i^2\right)\right)^2} \tag{7.54}$$

and

$$m = \frac{r\sum_{i=1}^{r}\left(\phi_iF_i/\sigma_i^2\right) - \sum_{i=1}^{r}\left(\phi_i/\sigma_i^2\right)\sum_{i=1}^{r}\left(F_i/\sigma_i^2\right)}{\sum_{i=1}^{r}\left(1/\sigma_i^2\right)\sum_{i=1}^{r}\left(\phi_i^2/\sigma_i^2\right) - \left(\sum_{i=1}^{r}\left(\phi_i/\sigma_i^2\right)\right)^2}, \tag{7.55}$$

where

σ_i is the standard deviation of the ith observed damage frequency
r is the number of observations used for the fit

The data to be fitted should not include any observations of $F_i = 0$ or $F_i = 1$. It has been shown analytically that the DFM yields the most accurate results on average when the maximum value of $F_i \sim 0.6$ (Arenberg 1994, Melninkaitis et al. 2009).

The individual substitution of Equations 7.55 and 7.54 into Equation 7.53 gives

$$T_M = -\frac{\sum_{i=1}^{r}\left(\phi_i/\sigma_i^2\right)\sum_{i=1}^{r}\left(F_i/\sigma_i^2\right) - \sum_{i=1}^{r}\left(\phi_i/\sigma_i^2\right)\sum_{i=1}^{r}\left(\phi_iF_i/\sigma_i^2\right)}{r\sum_{i=1}^{r}\left(\phi_iF_i/\sigma_i^2\right) - \sum_{i=1}^{r}\left(\phi_i/\sigma_i^2\right)\sum_{i=1}^{r}\left(F_i/\sigma_i^2\right)}. \tag{7.56}$$

Since $P_i = d_i/n_i$, Equation 7.56 can be written as

$$T_M = -\frac{\sum_{i=1}^{r}\left(\phi_i/\sigma_i^2\right)\sum_{i=1}^{r}\left(F_i/\sigma_i^2n_i\right) - \sum_{i=1}^{r}\left(\phi_i/\sigma_i^2\right)\sum_{i=1}^{r}\left(\phi_id_i/\sigma_i^2n_i\right)}{r\sum_{i=1}^{r}\left(\phi_id_i/\sigma_i^2\right) - \sum_{i=1}^{r}\left(\phi_i/\sigma_i^2\right)\sum_{i=1}^{r}\left(d_i/\sigma_i^2n_i\right)}. \tag{7.57}$$

It must be noted that a linear fitting form has emerged as a consensus simplification despite its known foibles (Arenberg 2012, Melninkaitis et al. 2009). Foltyn (1984) used Equation 7.23 in addition to linear function, and Porteus and Seitel (1984) used a power law model, but clearly noted its computational complexity. The arguments over computation complexity were well founded in the 1980s when high-performance computing power was scarce. This argument cannot be made today with the ubiquity of high-performance computing, even on mobile telephones!

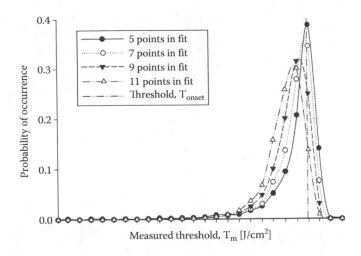

FIGURE 7.8 T_M distributions for the DFM. Note the long tail toward lower fluences and the short tail above true threshold.

7.4.1.1   Distribution of Measured Thresholds—DFM

The DFM has been analyzed by several authors using Monte Carlo methods (Arenberg 1994, 2012, Melninkaitis et al. 2009). It has been shown that the measured threshold, T_M, from the DFM technique has a mode near the true value, assuming that the maximum value of F used in the fitting is ~0.6 (Arenberg 1992). The distribution of T_M is wide and skewed with the precise details depending on the selection of the many parameters involved. The skewness shows a general trend to underestimate the true threshold.

An example of these T_M distributions, generated via Monte Carlo methods, is shown in Figure 7.8. Note the long tail toward fluences below the true threshold and the short tail above it. The means of these distributions are significantly below the true value, which is why it is said that the DFM tends to underestimate the threshold.

7.4.2   Binary Search Technique

The BST is a simple test. The test sample is exposed at different fluences and results plotted as damage (1) or no damage (0). A flowchart of a typical BST measurement is shown in Figure 7.9. The minimum damaging fluence (observed) is the measured threshold. The binary nature of the output gives this test its name as can be seen from Figure 7.10, the results of a typical BST measurement. A flowchart of a typical BST measurement is shown in Figure 7.9. The BST has a clear systematic bias because the reported threshold is the minimum observed damage, the PoD must be finite, therefore the reported threshold must be overestimate of the T_{onset}. Data gathering in such a binary manner can be found in the literature from the 1980s although the threshold definition used in this period was calculated quite differently (Guenther et al. 1984).

The BST has no standard rules for test fluence selection, which would guide the operator to make fluence selections after determination if damage did or did not occur. At least two authors have reported on several algorithms for determining such rules (Arenberg 1995, Melninkaitis et al. 2009). The observed threshold, T_M, depends on the fluence

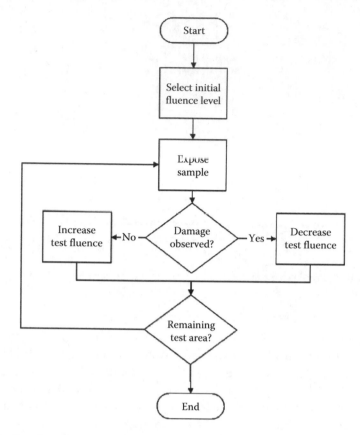

FIGURE 7.9 The flowchart for a typical BST.

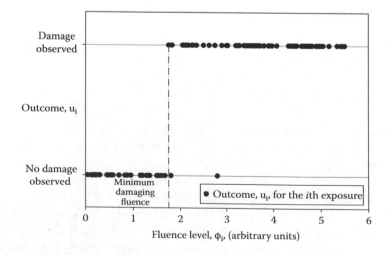

FIGURE 7.10 Test results for a typical BST.

selection rules and number of sites in the measurement (search length) (Arenberg 1995). The known strength of the BST is high repeatability of T_M.

7.4.2.1 Distribution of Measured Threshold—BST

The distribution of the BST method is derived in closed form using a probability argument. For a BST threshold to report a value of ϕ^* as the threshold, two events must occur:

Event 1 (E1): No damage occurs below ϕ^*.
Event 2 (E2): Damage occurs at ϕ^*.

These events are independent so the probability of both being true is the product of the probabilities, namely,

$$\Pr\left(T_M = \phi^*\right) = \Pr\left(E1\right)\Pr\left(E2\right). \tag{7.58}$$

For this analysis, the BST consists of a set of even fluence steps and that at each step i, m_i shots are taken at the same fluence. The probability of observing no damage at this ith fluence level is

$$\Pr\left(\text{No damage at } \phi_i\right) = \left(1 - P\left(\phi_i\right)\right)^{m_i}. \tag{7.59}$$

Event 1 is the case where at all levels it shows no damage and is given as the product of Equation 7.59 for all levels i = 1 to n − 1:

$$\Pr\left(E1\right) = \prod_{i=1}^{n-1}\left(1 - P\left(\phi_i\right)\right)^{m_i}. \tag{7.60}$$

Event 2 is when there is some damage at the nth fluence level. The easiest way to calculate when some damage occurs is to recognize that this event is the complement of the probability of no damage occurring at the nth level. This gives the expression that Event 2 is true as

$$\Pr\left(E2\right) = 1 - \left(1 - P\left(\phi_n\right)\right)^{m_n}. \tag{7.61}$$

Substitution of Equations 7.61 and 7.62 into Equation 7.58 gives the final expression

$$\Pr\left(T_M = \phi_m\right) = \left(1 - \left(1 - P\left(\phi_n\right)\right)^{m_n}\right)\prod_{i=1}^{n-1}\left(1 - P\left(\phi_i\right)\right)^{m_i}. \tag{7.62}$$

Figure 7.11 shows the results for Equation 7.62; note that there is no probability of underestimating the threshold with this technique and that the bias is toward overestimation of the *true* threshold. To understand the role of the many test parameters and choices that affect the BST results, the reader is referred to the literature (Arenberg 1995, Melninkaitis et al. 2009).

Chapter 7

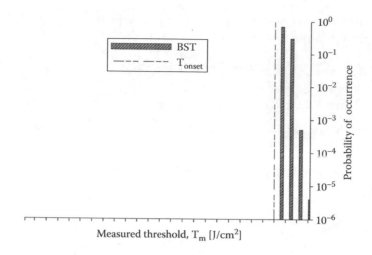

FIGURE 7.11   Distribution of results for a typical BST measurement; note the overestimation of the true threshold.

7.4.3   Maximum Likelihood Method

An emerging development in the field of laser damage threshold measurement is a revised method for determining the onset threshold from a BST measurement. Using maximum likelihood methods, an unbiased estimation of the threshold is derived, replacing the lowest observed damage as T_M with a direct estimation of the threshold. The maximum likelihood method has been used very successfully in many other fields, such as psychology, fungicide effectiveness, and explosives testing, where limited and quantal or binary data must be modeled for a prediction. Initial investigations applying ML to laser damage have been quite promising with the threshold estimate unbiased and quite accurate (Arenberg and Thomas 2012). More extensive theoretical investigations are planned along with experiments (Arenberg and Thomas 2014).

In the MLE method, the fundamental question asked of the data is, "What model gives the highest likelihood of observing the experiment?" This is in contrast to current LDT measurement techniques, which seek to fit observed data, usually via least-squares estimation to a model. Least squares asks, "Given a model, what is the best description of the data observed?" (Myung 2003).

The likelihood function, L, is defined as the probability that an event, U, which has already occurred, will yield a specific outcome. The MLE method formulates L in terms of a model, $P(\phi; \theta_1, \dots \theta_n)$ whose parameters, θ_i, are varied until L is maximized, which is interpreted as the most likely model.

A typical LDT test involves exposing the test specimen to various fluences, ϕ_i, and observing the outcome, u_i. The outcomes of the laser irradiation are assigned values, $u_i = 1$ is damage, $u_i = 0$ is no damage. The cumulative damage probability model is written, $P(\phi; \theta_1, \dots \theta_n)$. The likelihood of observing the outcome of the ith exposure, u_i, L_i, is

$$L_i = (1 - u_i)(1 - P(\phi_i, \theta_1, \dots \theta_n)) + u_i P(\phi_i, \theta_1, \dots \theta_n). \tag{7.63}$$

All of the individual observations are independent so the likelihood of observing the entire LDT test is simply the product of all the L_i

$$L = \prod_{i=1}^{n}\left(1 - u_i\right)\left(1 - P\left(\phi_i, \theta_1 \ldots, \theta_n\right)\right) + u_i P\left(\phi_i, \theta_1 \ldots, \theta_n\right). \qquad (7.64)$$

In MLE, the analyst usually does not deal directly with L, but rather its (natural) logarithm. This is done to simplify the algebra and making the values from calculations easier to handle. The (natural) logarithm of the likelihood function, $\mathscr{L}$ is

$$\mathscr{L} = \sum_{i=1}^{n} \ln\left(\left(1 - u_i\right)\left(1 - P\left(\phi_i, \theta_1 \ldots, \theta_n\right)\right) + u_i P\left(\phi_i, \theta_1 \ldots, \theta_n\right)\right). \qquad (7.65)$$

The values of the θ_i that maximize $\mathscr{L}$ give the most likely estimation of $P(\phi; \theta_1, \ldots \theta_n)$. The $P(\phi; \theta_1, \ldots \theta_n)$ is then analyzed to extract the threshold. The probability of damage models can take any form, consistent with being a probability distribution. Recent work has looked at two such options, $P(\phi)$ being derived from a degenerate defect distribution and having a clearly defined threshold and also taking the form of a logistic curve with finite probability of damage at all fluence levels (Arenberg and Thomas 2013).

7.4.3.1 Distribution of Measured Threshold—ML

A Gaussian distribution of T_M is an expected result of the ML technique. Several authors have solved the problem of Equation 7.65 analytically, usually using the equivalent of Gaussian distribution of strengths, and have also calculated the expected variance of that estimate (Langlie 1962). These results indicate unbiased estimates, which are distributed in a Gaussian fashion around the mean (accurate result). The fundamentally Gaussian nature of the uncertainties in model parameters is the reason that ML methods are a truly exciting area of research. But as this work has just begun, the ability to formulate the relationship between laser damage test parameters and the threshold distribution has not yet been accomplished and is an area of active research (Arenberg and Thomas 2014).

7.5 Distribution of Threshold Measurements

The results of the distributions of measured thresholds, given in Sections 7.4.1.1, 7.4.2.1, and 7.4.3.1 can be written as

$$\Pr\left(T_M = \phi | f(\phi), a(\phi), \vec{M}\right) = h(\phi). \qquad (7.66)$$

Equation 7.66 means that the probability of measuring a given result, given the defect distribution, $f(\phi)$, irradiation pattern $a(\phi)$ and the measurement technique and parameter values denoted as $\vec{M}$ can be described by a function h, depending on ϕ. For the DFM, the result is usually a beta function, for the BST h is given by Equation 7.62 and for the

ML method, it is a Gaussian. From Equation 7.66, we can derive the distribution of ΔT_M, the error in the measurement process. ΔT_M can be viewed as how well the measurement process identifies the true value of the threshold on the sample. ΔT_M can be written as

$$\Pr(\Delta T_M = T_{onset} - \phi) = h(\phi). \tag{7.67}$$

Section 7.3.5.2 showed that the true threshold of a sample can also be a random variable and depend on defect distribution f and optic area A.

We are now able to address the fundamental question posed at the beginning of this chapter, "What is the distribution of repeated threshold observations, T_{Obs}, on identical samples?" Identical samples are those that are made with the same processes and are described by the same defect distribution $f(\phi)$.

The probability of observing a specific value T_{Obs} is derived from recognizing that

$$T_{Obs} = \Delta T_M + T_{onset}. \tag{7.68}$$

Equation 7.68 means that the observed value of the threshold is the sum of the true value, T, and the error induced by the measurement technique applied, ΔT_M. The distribution of T_{Obs} is given by

$$\Pr(T_{Obs}) = \int_{-\infty}^{\infty} \Pr(\Delta T_M = x) \Pr(T_{onset} = T_{Obs} - x) dx. \tag{7.69}$$

Equation 7.69 is a convolution integral.

If the standard deviations of the measurement error and the weakest point on the surface are $\sigma_{\Delta T_M}$ and σ_T, respectively, the standard deviation of observation, $\sigma_{T_{onset}}$, is given by

$$\sigma_{T_{Obs}} = \sqrt{\sigma_{\Delta T_M}^2 + \sigma_{T_{onset}}^2}. \tag{7.70}$$

Equation 7.70 shows that both the behavior of the sample and the measurement techniques contribute to the width of the distribution of the observed threshold and shows that spending a great deal of effort to lower only one of the terms in the radical is likely to be an unrewarding exercise.

7.6 Summary/Review

This chapter introduced the central problem of laser damage threshold measurement and its uncertainties. Then a selected history was introduced to follow the development of threshold measurements from ad hoc observations to various kinds of statistical averages and finally to onset thresholds. An analytic framework using a defect-based model and developing the probability of damage curves followed. Analysis of the PoD curves showed why an onset definition for threshold is independent of spot size, assuming a finite extent of the distributions in fluence. If the defects are spread more widely as in a Gaussian case, it is not possible to define an onset that all observers could measure. The defect model was also used to show that the weakest point on an optic is not generally constant. The major test methods were introduced and described. In the final section, the various portions of the analysis were combined to derive an expression for the threshold observed over many identical samples. It is shown that this distribution is wider than either the distribution of measurement errors expected on a single sample or the intrinsic distribution of the threshold of a sample optic. In short, threshold observations do not often repeat in general because the statistics forbid it.

This chapter has just opened the door to this important and dynamic field, for the most current results, the reader is directed to the literature, in particular the *Proceedings of the Boulder Damage Symposium* published by SPIE annually.

Acknowledgments

I would like to express thanks to the reviewers of this manuscript, the two anonymous reviewers from among my fellow authors, and the colleagues I personally inflicted an early version of this chapter on. Thank you especially to Dr. Bernd Huettner and Dr. Wolfgang Reide, from DLR in Stuttgart Germany, for their time and comments. A special thank you goes to Dan Nieuwsma, recently retired as senior principal physicist at Raytheon in El Segundo, CA. Dan read the entire manuscript and helped insure historical accuracy, and started my path to writing this chapter, by assigning me to be responsible for running the Hughes Damage Test in the early 1980s where I got first-hand exposure to the problem of repeatable laser damage measurements.

References

Aplet, L.J. Method of test for relative optical damage thresholds of optical materials, ASTM Subcommittee II on Lasers and Laser Materials Document 2A05. 1978.

Arenberg, J.W. Investigation of the accuracy and precision of the damage-frequency method of measuring laser damage threshold. *Proceedings of the SPIE*, vol. 1848, *24th Annual Boulder Damage Symposium Proceedings—Laser-Induced Damage in Optical Materials: 1992*. Boulder, CO, 1992. p. 72.

Arenberg, J.W. Calculation of uncertainty in laser damage thresholds determined by use of the damage frequency method. *Proceedings of the SPIE*, vol. 2428, *Laser Induced Damage in Optical Materials: 1994*. Boulder, CO; Bellingham, WA: SPIE, 1994.

Arenberg, J.W. Accuracy and precision of laser damage measurements made via binary search techniques. *Proceedings of the SPIE*, vol. 2714, *Laser Induced Damage in Optical Materials: 1995*. Boulder, CO; Bellingham, WA: SPIE, 1995.

Arenberg, J.W. Extrapolation of the probability of survival of a large area optic based on a small sample. *Proceedings of the SPIE, Laser Induced Damage in Optical Materials: 2000*. Bellingham, WA: SPIE, 2000. p. 83.

Arenberg, J.W. Direct comparison of the damage frequency method and binary search technique. *Optical Engineering* 51(12) (2012): 121819.

Arenberg, J.W., M.E. Frink, D.W. Mordaunt, G. Lee, Seitel, and E.A. Teppo. Correlating laser damage tests. *Applied Optics* 28(1) (1989): 123–126.

Arenberg, J.W. and M.D. Thomas. Improving laser damage threshold measurements: An explosive analogy. *Proceedings of the SPIE*, vol. 8530, *Laser-Induced Damage in Optical Materials: 2012*. Bellingham, WA: SPIE, 2012.

Chapter 7

Arenberg, J.W. and M.D. Thomas. Laser damage threshold measurements via maximum likelihood estimation. *Proceedings of the SPIE*, vol. 8885, *Laser-Induced Damage in Optical Materials: 2013*. Bellingham, WA: SPIE, 2013.

Arenberg, J.W. and M.D. Thomas. Further investigations into use of the Maximum Likelihood method in the measurement of laser damage thresholds. In preparation.

Becker, J. and A. Bernhardt. ISO 11254: An international standard for the measurement of laser induced damage threshold. *SPIE Proceedings*, vol. 2114, *Laser Damage in Optical Materials*. Bellingham, WA: SPIE, 1993. pp. 703–713.

Bernal, E. Absorption of laser radiation by transparent crystalline solids. *Journal of the Optical Society of America* 55(5) (1965): 602.

Bevington, P.R. and D.K. Robinson. *Data Reduction and Error Analysis for the Physical Sciences*, 3rd edn. New York: McGraw Hill, 2003.

Bloembergen, N. *Fundamentals of Damage in Laser Glass*, National Materials Advisory Board Publication NMAB-271. Washington, DC: National Academy of Sciences, 1970.

Bloembergen, N. *High Power Infrared Laser Windows*, National Materials Advisory Board Publication NMAB-272. Washington, DC: National Academy of Sciences, 1972.

Cullom, J.H. and R.W. Waynant. Determination of laser damage thresholds for various glasses. *Applied Optics* 3 (1964): 989.

DeShazer, L.G., B.E. Newnam, and K.M. Leung. Role of coating defects in laser induced damage to dielectric thin films. *Applied Physics Letters* 23(11) (1973): 607–609.

Feit, M., M.R. Rubenchik, M.R. Kozlowski, F.Y. Genin, S. Schwartz, and L.M. Sheehan. Extrapolation of damage test data to predict performance of large area NIF optics at 355 nm. *Proceedings of the SPIE*, vol. 3578, *Laser Induced Damage in Optical Materials: 1998*. Bellingham, WA: SPIE, 1998. pp. 226–231.

Fisher, R.A. *Statistical Methods for Research Workers*, 9th edn. Edinburgh, U.K.: Oliver and Boyd, 1944.

Foltyn, S.R. Spot size effects in damage testing. National Bureau of Standards (US) Special Publication 669. Washington, DC: US Government Printing Office, 1984.

Fradin, D.W. and D.P. Sua. Laser-induced damage in ZnSe. *Applied Physics Letters* 24(11) 1974: 555.

Giuliano, C.R. Laser damage to transparent dielectrics. *Applied Physics Letters* 5 (1964): 137.

Giuliano, C.R., R.W. Hellwarth, L.D Hess, and G.R. Rickel. *Damage Threshold Studies in Laser Crystals*. Malibu, CA: Hughes Research Labs, 1970.

Glass, A.J. and A.E. Guenther. *Damage in Laser Glass*, ASTM Special Publication 469. Philadelphia, PA: American Society for Testing and Materials, 1969.

Glass, A.J. and A.E. Guenther. *Damage in Laser Materials*, NBS Special Publication 341. Washington, DC: US Government Printing Office, 1970.

Glass, A.J. and A.E. Guenther. Laser induced damage of optical elements—A status report. *Applied Optics* 12(4) (1973): 537–549.

Guenther, A.E. Symposium welcome. *Boulder Damage Symposium—Laser-Induced Damage In Optical Materials*. Collected papers. Bellingham, WA: SPIE, 1999. Disk 1.

Guenther, K.H. et al. 1.06-μm laser damage of thin film optical coatings: A round-robin experiment involving various pulse lengths and beam diameters. *Applied Optics* 23(21) (1984): 3743–3752.

Hercher, M. Laser-induced damage in transparent media. *Journal of the Optical Society of America* 54(4) (1964): 563.

International Standards Organization. ISO 11254-1:2000, Lasers and laser-related equipment—Determination of laser-induced damage threshold of optical surfaces—Part 1: 1-on-1 test. 2000. http://www.iso.org/iso/iso_catalogue/catalogue_ics/catalogue_detail_ics.htm?csnumber = 19232 (Accessed February 17, 2014).

International Standards Organization. ISO 21254-2:2011 Lasers and laser-related equipment—Test methods for laser-induced damage threshold—Part 2: Threshold determination. 2011. http://www.iso.org/iso/home/store/catalogue_tc/catalogue_detail.htm?csnumber = 43002 (Accessed February 17, 2014).

Krol, H., L. Gallais, C. Natoli, J.Y. Grezes-Besset, and M. Commandre. Investigation of nanoprecursors threshold distribution in laser-damage testing. *Optics Communications* 256(1–3) (2005), 184–189.

Langlie, H.J. A reliability test for "one-shot" items. *Eighth Conference on the Design of Experiments in Army Research Development and Testing*. Washington, DC: Defense Technical Information Center, 1962.

Lee, H.A. Damage threshold testing of laser glass at Owens-Illinois. *Damage in Laser Glass*, ASTM Special Technical Publication 469. Philadelphia, PA: ASTM, 1969, STP48614S.

Mainman, T.H. Stimulated optical radiation in ruby. *Nature* 187 (1960): 493–494.

Melninkaitis, A., G. Bataviciute, and V. Sirukaitis. Numerical analysis of laser-induced damage threshold search algorithms and their uncertainty. *Proceedings of SPIE*, vol. 7504, *Laser Induced Damage in Optical Materials: 2009*. Bellingham, WA: SPIE, 2009.

Milam, D., R.A. Bradbury, and M. Bass. Laser damage threshold for dielectric coatings as determined by inclusions. *Applied Physics Letters* 23(12) (1973): 654–657.

Milam, D., J.B. Willis, F. Rainer, and G.R. Wirtenson. Determination of laser damage by comparison with an absolute standard. *Applied Physics Letters* 38 (1981): 402.

Mordaunt, D.W., J.W. Arenberg, and M.E. Frink. Production oriented laser damage testing at Hughes aircraft. *Laser Induced Damage in Optical Materials, 1986*; NIST Special Publication, 752. Washington, DC: US Government Printing Office, 1986. p. 207.

Mordaunt, D.W. and D.E. Nieuwsma. Development and implementation of MIS-36477 laser damage certification of designator optical components. *SPIE Proceedings*, vol. 1441, *Laser Damage in Optical Materials, 1990*. Bellingham, WA: SPIE, 1990.

Myung, I.J. Tutorial on maximum likelihood estimation. *Journal of Mathematical Psychology* 47 (2003): 90–100.

Nieuwsma, D.E. Personal communication (2013).

Olness, D. Laser-induced breakdown in transparent dielectrics. *Applied Physics Letters* 39(6) (1968): 263.

Picard, R.H., D. Milam, and R.A. Bradbury. Statistical analysis of defect caused damage in thin films. *Applied Optics* 16(6) (1977): 1563.

Porteus, J.O and S.C. Seitel. Absolute onset of optical surface damage using distributed defect ensembles. *Applied Optics* 23(21) (1984): 3796–3805.

Ready, J.F. *Effect of High Power Laser Radiation*. New York: Academic Press, 1971.

Riede, W., U. Uwe Willamowski, M. Dieckmann, D. Ristau, U. Broulik, and B. Steiger. Laser-induced damage measurements according to ISO/DIS 11 254-1: Results of a national round robin experiment on Nd:YAG laser optic. *Proceedings of the SPIE*, vol. 3244, *Laser-Induced Damage in Optical Materials: 1997*. Bellingham, WA: SPIE, 1998.

Sahagran, G.C.S. and C.A. Pitha (eds.). *Conference on High Power Infrared Laser Materials*. Bedford, MA: Air Force Cambridge Research Laboratories, L.G. Hanscom Field, 1971.

Seitel, S.C., A. Giesen, and J. Becker. International standard test method for laser-induced damage threshold of optical surfaces. *Proceedings of the SPIE*, vol. 1848, *24th Annual Boulder Damage Symposium Proceedings—Laser-Induced Damage in Optical Materials: 1992*. Boulder, CO: SPIE, 1993.

Seitel, S.C. and J.O. Porteus. Toward improved accuracy in limited-scale pulsed laser damage testing via the onset method. *Damage in Laser Materials; 1983*, National Bureau of Standards (US) Special Publication 688. Washington, DC: US Government Printing Office, 1985.

Seitel, S.C. and E.A. Teppo. Damage testing for optics and procurement. *Laser Induced Damage in Optical Materials, 1983*: NIST Special Technical Publication. Washington, DC: US Government Printing Office, 1983. Abstract Only.

Seitel, S.C., E.A. Teppo, and J.W. Arenberg. Emergence of "consensus standards" in laser damage of production Nd:YAG laser optics. *Laser Damage in Optical Materials, 1988*; National Institute of Standards and Technology Special Publication 775. Washington, DC: US Government Printing Office, 1988.

Sparks, M. *Theoretical Studies of High-Power Infrared Window Materials*. Los Angeles, CA: Xonics, 1972.

Starke, K. and A. Bernhardt. Laser damage threshold according to ISO 11254: Experimental realization at 1064 nm. *Proceedings of the SPIE*, vol. 2114, *Laser Induced Damage in Optical Materials, 1993*. Bellingham, WA: SPIE, 1993. pp. 212–219.

Stewart, A.F. and A.E. Guenther. Laser-induced damage: An introduction. *Applied Optics* 23(21) (1984): 3741–3742.

Weisstein, E. Delta Function. From *MathWorld–A Wolfram Web Resource*. 2014. http://mathworld.wolfram.com/DeltaFunction.html (Last accessed July 22, 2014).

Wood, R.M. *Laser-Induced Damage of Optical Materials*. New York: Institute of Physics Press/Taylor & Francis Group, 2003.

Zwillinger, D. and S. Kokoska. *CRC Standard Probability and Statistics Tables and Formulae*. Boca Raton, FL: CRC Press, 1999.

Chapter 7

8. Measurement of Light Scattering, Transmittance, and Reflectance

Sven Schröder and Angela Duparré

Chapter 8

Laser-Induced Damage in Optical Materials. Edited by Detlev Ristau © 2015 CRC Press/Taylor & Francis Group, LLC. ISBN: 978-1-4398-7216-1.

8.1   Introduction

When a surface, thin film coating, or material is irradiated with light of a specific wavelength, certain parts of the radiation are reflected, transmitted, scattered, and absorbed, depending on the dielectric and structural properties. Many optical elements are specified in terms of their specular reflection and transmission properties, which usually are to be maximized. Aside from absorption, diffusely scattered radiation constitutes a loss mechanism and is connected with possible sources of laser-induced damage. It is, therefore, usually to be minimized.

On the one hand, accurate photometric measurement techniques are required to characterize optical properties such as specular reflectance, transmittance, and scattering, and a considerable amount of work has been dedicated to obtain measurements with the highest accuracy. On the other hand, these techniques together with theoretical models have become important tools to analyze loss mechanisms and to investigate the origins of laser-induced damage.

In particular, light scattering techniques have become widely recognized as a powerful tool for the inspection and characterization of optical and nonoptical surfaces, coatings, and materials. With light scattering techniques, a large variety of damage-relevant effects such as nanostructures, roughness, defects, as well as coating and material inhomogeneities can be advantageously investigated. Scattering methods are highly sensitive and at the same time noncontact and nondestructive, fast, flexible, and robust. This also offers opportunities to use light scattering for the in situ monitoring and investigation of irradiation-induced modifications and damage processes of surfaces, coatings, as well as in bulk materials.

This chapter is intended to give an overview of techniques for the measurement of light scattering, transmittance, and reflectance. In Section 8.2, the measurement techniques and the main types of instruments for measurements of angle resolved and total scattering as well as of specular reflectance and transmittance are discussed. The brief theoretical background given in Section 8.3 describes relationships between light scattering, and therewith reflectance and transmittance, and interface roughness. This particular case of scattering mechanism was selected for demonstration because it is of particular relevance and very instructive. In Section 8.4, a selection of examples is discussed to demonstrate the capabilities and limitations of the different measurement techniques for various applications of practical interest to the community dealing with high-end optical components and confronted with problems regarding laser-induced damage.

8.2   Measurement Techniques

The techniques for light scattering measurements of surfaces can be roughly divided into two major categories: angle-resolved scattering (ARS) and total scattering (TS). While ARS is mostly based on goniophotometers, instruments based on Coblentz (hemi) spheres or integrating spheres (Ulbricht spheres) are used for TS. For high-quality optical components, the terms reflectance and transmittance usually refer to the specular or directed quantities. Specular reflectance and transmittance are measured using either spectrophotometers or laser ratiometers with different advantages and disadvantages. In addition, cavity ring-down (CRD) techniques have become more and more established.

8.2.1   Angle–Resolved Scattering

8.2.1.1   Definitions and Standardization

A standard procedure for ARS measurements was defined in ASTM standard E 1392 (ASTM E 1392, 1990). This procedure was successfully implemented and verified in various round-robin experiments at different wavelengths (Leonard and Rudolph, 1993; Asmail et al., 1994) but is restricted to opaque samples. Recently, a new ASTM standard has been established (ASTM E 2387, 2011) that has a wider range of application. Currently, an ISO standard procedure for ARS measurements is being developed by the international working group TC172/SC9/WG 6 of the International Organization for Standardization to meet the increased demands concerning wavelength ranges, sensitivity, flexibility, and practicability.

ARS describes the power ΔP scattered into a given direction (θ_s, ϕ_s) in space:

$$\mathrm{ARS}(\theta_s, \phi_s) = \frac{\Delta P(\theta_s, \phi_s)}{P_i \Delta \Omega},$$
(8.1)

where
  $\Delta \Omega$ is the solid angle of collection
  P_i is the incident power
  θ_s and ϕ_s denote the polar and azimuthal angles of scattering with respect to the sample normal

ARS is thus identical with the normalized radiometric scattered intensity. It can be transformed into the bidirectional reflectance or transmittance distribution function, BRDF or BTDF, respectively, by dividing by $\cos \theta_s$.

8.2.1.2   Instrumentation

The measurement of ARS is mostly based on goniophotometers with special emphasis on high sensitivities and high dynamic ranges sometimes referred to as scatterometers. There are in-plane instruments, in which the range of detected scatter angles is confined to the plane of incidence, and 3D scatterometers that cover more or less the entire scattering sphere.

Chapter 8

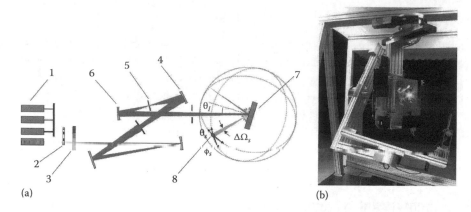

FIGURE 8.1 Instrument for ARS measurements in UV–VIS–IR spectral range. (a) Schematic setup. (b) Photograph of sample positioning and goniometer systems.

High-precision instrumentation for ARS measurements has been set up in a variety of laboratories (Elson et al. 1980; Bousquet et al., 1981; Orazio et al., 1983; Amra et al., 1989; Asmail et al., 1994; Neubert et al., 1994; Stover, 1995; Bennett et al., 1999; Germer et al., 1999). The majority of such systems is designed for operation at visible light wavelengths such as 632.8 nm (He–Ne laser). Although special instruments exist that are capable of measuring ARS even in the IR and UV spectral regions (Hogrefe et al., 1987; Amra et al., 1993; Schiff et al., 1993; Newell et al., 1997; Duparré et al., 2002; Schröder et al., 2005a; Schröder et al., 2010). Recent developments aim at spectral ARS measurements based on tunable laser sources (Zerrad at al., 2011; Schröder et al., 2014).

Figure 8.1 shows the schematic principle of a 3D scatterometer for laser-based angle-resolved light scattering, reflectance, and transmittance measurements in the UV–VIS–IR spectral regions. The instrument described in detail in Schröder et al. (2011a) is located in a clean room to suppress light scattering from particles in the air and to avoid sample contamination.

Several lasers are used as light sources (1) with wavelengths ranging from 325 nm to 10.6 μm. A chopper (2) allows for lock-in amplification and noise suppression. Neutral density filters (3) are used to adjust the incident power and to increase the dynamic range. The beam is focused by a spherical mirror (4) onto a pinhole (5), which acts as spatial filter. The pinhole is then imaged by spherical mirror (6) over the sample (7) onto the detector aperture (8). The sample (7) is located on a positioning system to adjust the irradiated position and the angle of incidence. The instrument allows plane or curved samples with diameters ranging from a few mm up to 670 mm to be investigated. Typical illumination spot diameters at the sample position are between 1 and 5 mm, although focusing to about 50 μm is also possible if required. Typical irradiances on the sample are about 500 mW/cm², which is well below the damage threshold of optical materials.

The detector (8), usually a photomultiplier tube (PMT), can be rotated within the plane of incidence (in-plane scatter measurement) or within the entire sphere around the sample (3D scatter measurement) to investigate the out-of-plane scattering of anisotropic samples. The diameter of the aperture in front of the PMT defines the detector solid angle $\Delta\Omega$. Aperture diameters between 0.1 and 5 mm are typically used depending on the specific requirements regarding sensitivity, speckle reduction, and near angle limit.

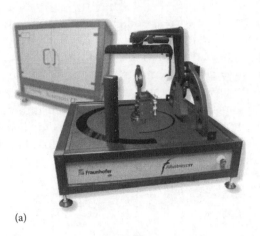

(a)   (b)

FIGURE 8.2   Compact instruments for ARS measurements. (a) Table-top 3D scatterometer. (b) CMOS-based light scattering sensor.

Up to 15 orders of magnitude dynamic range are achieved in the visible spectral range with the instrument shown. This is sufficient to investigate samples ranging from super-polished substrates, thin film coatings, and optical materials to nanostructured and technically rough surfaces.

Meanwhile, compact instruments are available for angle-resolved scatter measurements that maintain key parameters such as high-sensitivity and high dynamic range but are much smaller, easier to use, and can even be integrated into fabrication environments. A table-top 3D scatterometer developed recently at IOF (Finck et al., 2011, 2014) is shown in Figure 8.2a. Another tool developed at IOF that is based on a CMOS sensor rather than a scanning detector for rapid measurements of the 3D scattering distribution (Herffurth et al., 2011, 2013) is shown in Figure 8.2b.

Calibration: Careful calibration of scatter measurements is of crucial importance in order to compare results obtained with different instruments as well as to compare measurement with modeling results. Calibration of ARS measurements is performed either by measuring the incident power and the detector solid angle directly (absolute method) or by measuring the scattering of a diffuse reflectance standard with known ARS (relative method). Pressed PTFE powders, such as Spectralon™, are often used for this purpose in the UV–VIS range. They exhibit a Lambertian ARS proportional to the cosine of the scatter angle (constant BRDF) with a maximum of $1/\pi$ (Stover, 1995). The relative method has the advantage of being less sensitive to system nonlinearity.

Uncertainty of measurement: The estimation of the uncertainty of scatter measurements was performed according to the "Guide to the estimation of uncertainty in measurement" (GUM, 1995) using a Monte Carlo method (Herffurth et al., 2011). The dominating sources of uncertainties in ARS measurements are the effective size of the detector solid angle, fluctuations of the output power of the laser, the transmittances of the attenuation filters, as well as shot noise and excess noise of the PMT. In addition, considering the large dynamic range required, linearity of the detection system is of crucial importance. The relative uncertainty of ARS measurements following from error propagation is typically about 10% when sample inhomogeneities neglected. It is important

Chapter 8

to note that this rather large relative uncertainty corresponds to an exceptionally high absolute accuracy in determining scatter losses, in particular, for low-scattering samples such as superpolished surfaces and high-quality coatings. Another yet less-critical issue is the mechanical accuracy of goniometric instruments that can be in the range of 0.01°, although the actual reproducibility as well as gravity-induced distortions of the measurement arms has to be taken into consideration as well. Also, a certain part of the backscattering hemisphere is usually not accessible because the detector head will obstruct the incident beam. In the instruments described in Schröder (2011a) and von Finck (2014), this obstruction area is minimized by using a small mirror to direct the scattered light into the detector system.

8.2.2 Total Scattering

8.2.2.1 Definitions and Standardization

TS is defined in the international standard (ISO 13696, 2002) as the power P_s scattered into the forward or backward hemisphere normalized to the incident power P_i:

$$TS = \frac{P_s}{P_i},$$

(8.2)

where the range of scattered radiation to be detected (range of acceptance angles) should be $\theta_s = 2°, 85°$, and $\phi_s = 0°...360°$ in the corresponding backward or forward hemispheres. Total backscattering is denoted as TS_b and total forward scattering as TS_f. TS is hence equivalent to the scattering loss of the component. The well-defined range of acceptance angles ensures that the specular beam is excluded from the measurement.

Opaque as well as transparent surfaces and coatings can be measured both in the backward and forward directions. The influence of sample inhomogeneities is drastically reduced if the suggested procedure comprising a sample cross-scan, data reduction and averaging is performed. The ISO 13696 standard procedure was verified in an international round-robin experiment at 632.8 nm (Kadkhoda et al., 2000a).

Another quantity, total integrated scattering (TIS), is still in use but should not be confused with TS. TIS is defined as the ratio between the diffuse reflectance and the total reflectance (Davies, 1954; Bennett et al., 1961). A procedure for measuring TIS is prescribed in an ASTM standard (ASTM F 1048, 1987). The drawback of TIS is that it is defined for opaque not for transparent or semitransparent samples such as substrates, AR coatings, and beam splitters. Hence, TIS cannot be determined for such optical elements unless ambiguous additional assumptions are made. TS_b and TIS can be converted into each other if the reflectance of the surface is known.

8.2.2.2 Instrumentation

TS and TIS instruments based on the principle of either integrating spheres (Ulbricht spheres) or the Coblentz (collecting) hemispheres have been developed for more than two decades in a number of laboratories (Bennett, 1978; Guenther et al., 1984; Detrio et al., 1985; Kienzle et al., 1994; Rönnow et al., 1994, 1995; Duparré et al., 1997, 2002; Bennett et al., 1999; Balachninaite et al., 2005; Krasilnikova et al., 2005; Schröder et al., 2005a).

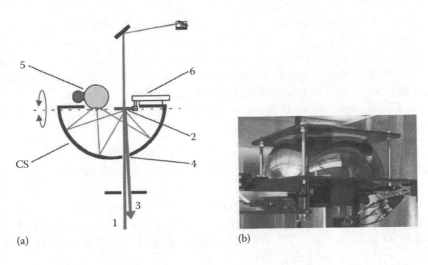

(a) (b)

FIGURE 8.3 The Coblentz (hemi)sphere arrangement for TS measurement in UV–VIS–IR spectral range. (a) In backscatter mode (schematic). (b) In forward scatter mode (photograph).

Integrating spheres are coated with a diffusely scattering material such as Spectralon™. The light scattered from a sample undergoes multiple scattering events inside the sphere leading to a homogeneous light distribution inside the sphere that is then measured by a PMT. In contrast, the Coblentz hemispheres are specularly reflective hemispherical mirrors that focus the light scattered from the sample onto the detector. Radiation propagating within 2° from the specular directions leaves the Coblentz hemisphere without being detected. Because of the smaller scattering volume and the collection of the scattered light, the Coblentz hemisphere–based instruments usually offer substantially lower background scatter levels and higher sensitivities.

Figure 8.3 displays a schematic picture of a TS arrangement based on a Coblentz (hemi)sphere (CS) in backscatter (TS_b) mode and a photograph of the setup in forward scatter (TS_f) mode. With the instrument described in detail in Duparré et al. (2002), total backward and forward scattering are detected according to the instructions in the international standard ISO 13696 at wavelengths between 325 nm and 10.6 μm using several laser sources. The laser beam is prepared in the illumination system similar to the instrument for ARS described in Section 8.2.1.2. The incident beam (1) hits the sample surface (2) at nearly 0°, and the specularly reflected beam (3) is guided back through the entrance/exit port of the Coblentz hemisphere (4). The detector unit (5) consists of a PMT for UV–VIS–NIR wavelengths or a cooled HgCdTe element for measurements in the IR located in a small integrating sphere for homogeneous illumination of the detector area. The positioning system (6) enables 1D and 2D scanning of the sample surface. A special arrangement allows switching between backscatter and forward-scatter modes by rotating the entire unit comprising the sample, the Coblentz hemisphere, and the detector. By carefully adjusting the beam preparation system to suppress parasitic stray light, extremely low background levels of as low as 5×10^{-8} at 632.8 nm can be achieved (Schröder et al., 2005a). Special instruments for TS measurements at VUV/DUV wavelengths below 200 nm were presented in Kadkhoda et al. (2000b), Gliech et al. (2002), Otani et al. (2002), and Schröder et al. (2005b) with background scatter levels down to 10^{-6} (Schröder et al., 2005b).

Chapter 8

Calibration of TS measurements is performed using a diffuse scattering standard with known total reflectance. For the UV–VIS spectral regions, pressed PTFE powders, such as Spectralon™, are commonly used while calibration in the IR is usually based on highly reflective diffuse gold. At wavelengths below 200 nm, no diffuse standards are available, so far. In Schröder et al. (2005b), a method is presented that allows for calibrating TS measurements at 193 nm and 157 nm using calcium fluoride diffusors and the energy balance.

Uncertainty of measurement: Similar considerations have to be made to estimate the uncertainty of TS measurements as for ARS measurements. However, since the specular components are excluded in TS, a lower dynamic range is usually required than for ARS, which leads to a smaller risk of systematic errors caused by system nonlinearities. Typical relative uncertainties are in the order of 10%. For a superpolished surface with a TS_b of about 10^{-5} (mirror with 0.2 nm rms roughness at 532 nm), this corresponds to an absolute uncertainty of as low as 1 ppm.

8.2.3 Specular Reflectance and Transmittance

8.2.3.1 Definitions and Standardization

Specular (or directed, or regular) reflectance ρ and transmittance τ are defined as the ratios of the specularly reflected or transmitted power, P_ρ or P_τ, to the incident power:

$$\rho = \frac{P_\rho}{P_i}; \quad \tau = \frac{P_\tau}{P_i}. \tag{8.3}$$

Two standards have been established for specular reflectance and transmittance measurements. The international standard (ISO 15368, 2001) describes measurements based on spectrophotometers. It applies to coated or uncoated plane optical elements in the spectral range between 190 nm and 25 µm. The spectral composition, polarization, and geometrical distribution of the radiation incident on the sample must be specified. Measurements according to ISO 15368 are considered to be at a medium accuracy level.

A measurement technique with higher precision is described in the international standard (ISO 13697, 2006). The technique measures the sample and reference signals merely simultaneously using a laser-ratiometric method by comparing the sample reflectance or transmittance with the reflectance of a highly reflective chopper mirror.

A different approach to measure specular reflectance with the highest accuracy, the CRD technique, is discussed as a work draft for a new international standard (ISO/WD 13142, 2010). CRD measures the temporal decay of a specular signal within a cavity containing the specimen. In contrast to the aforementioned techniques, it is, however, restricted to highly reflective samples with a specular reflectance close to unity (typically $\rho > 99.9\%$).

All three methods were evaluated in round-robin experiments at 1064 nm (Nickel et al., 2003; Duparré et al., 2008, 2010). The results will be discussed in Sections 8.4.2.1 and 8.4.2.2.

8.2.3.2 Instrumentation

There is a variety of spectrophotometers available for the measurement of specular reflectance and transmittance as a function of wavelength such as the Perkin Elmer Lambda series (Stemmler, 2005), Cary 6000i spectrophotometers, V-VASE spectroscopic, and M-2000 multiple-angle spectroscopic ellipsometers from Woollam. Special instruments were developed for the VUV/DUV (Blaschke et al., 2003) and EUV (Balasa et al., 2011) spectral ranges.

Spectrophotometers for the UV–VIS spectral range are usually dispersion-type instruments that are based on a broadband light source and a grating or a prism monochromator. The beam preparation system contains collector optics, a chopper, and a polarizer. In the so-called double beam spectrophotometers, the incident beam is split into a test and a reference beam to compensate fluctuations of the source. Photomultiplier tubes or photodiodes located in an integrating sphere or behind a diffusor disk are used as detectors.

Fourier-type spectrophotometers are mainly used in the IR spectral range. The method is based on recording interference fringes with an interferometer, often a Michelson interferometer. The measured interferograms are then Fourier transformed to obtain the reflectance or transmittance spectra.

Instruments for high-precision reflectance and transmittance measurements according to ISO 13697 are described in Voss et al. (1994), Jupé et al. (2003), Nickel et al. (2003), and ISO 13697 (2006). The measurement technique is based on a laser light source and a highly reflective chopper mirror. A typical setup is shown schematically in Figure 8.4. The light emitted by a laser (1) is spatially filtered (2) and reflected (3) to an optically smooth, highly reflective chopper mirror (4). The chopper divides the laser beam into a test beam (5) and a reference beam (6). The test beam is reflected by the chopper mirror and the sample (7), whereas the reference beam directly hits the rotating rear surface of an integrating sphere that contains the detector (8). Thus, both beams, alternating temporally, impinge upon the same spot on the detector system. A lock-in amplifier is used to measure even small differences in the reflectance of the chopper mirror and the sample with high sensitivity. A method with further enhanced accuracy is based on additional high-frequency

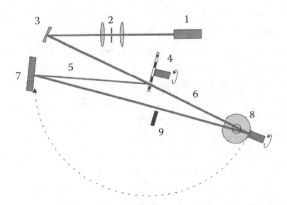

FIGURE 8.4 Instrument for high-precision reflectance and transmittance measurements according to ISO 13697.

amplitude modulation of the laser itself and the phase-sensitive amplification of the detector signals using the lock-in technique.

The CRD technique described in detail, for example, in Berden et al. (2000), Ren et al. (2006), and Schmidl et al. (2009), is based on measuring the decay of signal (specular power) after multiple reflections from the sample as a result of optical losses. For this purpose, the specimen under test is placed within a folded laser cavity consisting of two concave highly reflective mirrors and the specimen under test between them. Either a cw laser with an amplitude modulator or a pulsed laser light source is used. The pulses are reflected between the cavity mirrors and the specimen. After the laser source is turned off (or after the end of the laser pulse), the temporal behavior of the signal is measured and the data are evaluated by fitting the measured cavity energy to a generic exponential function. The mirror reflectance is then retrieved from the decay constant referred to as ring-down time.

Calibration of spectrophotometric measurements according to ISO 15368 is performed by successive measurements with and without the sample (direct method) or of the sample and a known reference sample (relative method) with different optical arrangements. The main challenge for calibration of measurements according to ISO 13697 is the determination of the reflectance of the chopper mirror that has to be known. This requires an additional measurement with a modified setup.

Uncertainty: Main sources of uncertainty in spectrophotometric measurements are the wavelength accuracy, spectral resolution, fluctuations of the light source, stray light, alignment errors, and linearity. Using two-beam instruments, the influence of source fluctuations can be reduced. Typical values of the accuracy are 0.1%–0.3% in the UV–VIS range and 0.5%–1% in the IR. Additional uncertainties are usually introduced in reflectance measurements because these require a modification of the setup leading to different positions on the sample and the detector being illuminated during measurement and calibration. One way to reduce the impact of alignment errors is discussed in Stentzel et al. (2009). In summary, while spectrophotometers are the most important tool for coating engineers because of the spectral information they provide, their application to accurately measure specular quantities close to unity is limited.

Significantly higher accuracies can be achieved using the method according to ISO 13697. The main source of uncertainty is noise in the laser intensity and in the lock-in system. Also scattering from the mirrors limits the accuracy. Yet because fluctuations and drifts of the laser power are compensated for by the measurement technique, substantially higher accuracies as compared to spectrophotometry can be achieved. Although the determination of the chopper reflectance might introduce additional systematic errors, nevertheless, absolute measurement uncertainties of as low as 5×10^{-5} at 1064 nm were reported (Jupé et al., 2003; Nickel et al., 2003).

Since CRD is based on the measurement of the decay rate rather than the measurement of changes in light power, the method is insensitive to fluctuations of the laser output power and highly sensitive to optical losses within the cavity. The accuracy depends on the resolution of the decay signal and thus on the value of the reflectance itself; the measurement accuracy increases as the specular reflectance approaches unity. A precision of up to 10^{-7} was achieved for highly reflective mirrors.

8.2.4   Relationship between ARS, TS, ρ, and τ

By comparing Equations 8.1 and 8.2, it becomes obvious that TS_b and TS_f can be derived from ARS by integration over the corresponding scattering hemispheres:

$$TS = 2\pi \int_{2°}^{85°} ARS(\theta_s) \sin \theta_s \, d\theta_s. \tag{8.4}$$

For a given sample irradiated with a certain incident power, the sum of all specular and diffuse components and the absorptance (A) should be equal to 1. This is usually referred to as the energy balance:

$$\rho + \tau + TS_b + TS_f + A = 1. \tag{8.5}$$

The energy balance is often used as an approach to determine losses from specular reflectance and transmittance measurements. However, separating the different loss mechanisms absorption and scattering requires additional assumptions. Even if the absorptance is assumed to be negligible, the determination of TS requires ρ and τ to be measured within a cone of exactly 2° around the specular directions. Otherwise, part of the radiation is missing or detected twice as will be discussed in the example of Section 8.4.1.5. In particular, for low-loss optical components, direct scatter (and absorption) measurements are both much more sensitive and accurate than indirect methods. Nevertheless, the energy balance offers various possibilities if the desired quantities cannot be accessed directly or, for example, for investigations of material properties for high-power applications considering that the loss mechanisms exhibit different nonlinear properties.

8.3   Theory

Theoretical models provide links between light scattering phenomena and their origins. They are thus extremely useful in order to interpret the results of experimental observations and to retrieve important information that might reveal the key parameters for further improvements. All sorts of surface and material imperfections give rise to light scattering, such as surface roughness, contaminations, defects, subsurface damage, and bulk inhomogeneities. The magnitude of the scattered light depends on the refractive index inhomogeneity, the size, and the density of the imperfections. As a result, the dominating scattering mechanism of high-quality optical surfaces is surface roughness. This scattering mechanism is also rather instructive, and the theoretical background is briefly resumed in this section starting with definitions of the relevant roughness quantities.

8.3.1   Relationship between Light Scattering and Surface Roughness

The following discussion of roughness-induced scattering from optical surfaces is based on a number of groundbreaking works presented in detail in Church et al. (1975, 1979),

Chapter 8

Harvey et al. (1978), Carniglia (1979), Elson et al. (1979), Bousquet et al. (1981), Stover (1995), and Bennett et al. (1999). The following discussion is confined to randomly and isotropically rough structures, which is the case for most surfaces that are generated by grinding, polishing, and coating. In the references, further theories can be found that also include nonisotropy, deterministic surface structures and defects (Church et al., 1979), as well as scatter effects from volume imperfections (Elson, 1984; Kassam et al., 1992; Duparré et al., 1993).

Any surface profile can be described as a Fourier series of sinusoidal waves with different periods, amplitudes, and phases. According to the grating equation, each single grating with spacing g causes scattering at the wavelength λ into the angle θ_s with $\sin\theta_s = \lambda/g$ for normally incident light. Then, $f = 1/g$ represents one single spatial frequency within this grating assembly. A stochastically rough surface contains a large diversity of components with different spatial frequencies and amplitudes. This is quantitatively described by the power spectral density (PSD), defined as the absolute square of the Fourier transform of the surface profile (Church et al., 1979; Elson et al., 1979; Stover, 1995; Bennett et al. 1999; Duparré et al., 2002):

$$\mathrm{PSD}\left(f_x, f_y\right) = \lim_{L \to \infty} \frac{1}{L^2} \left| \int_{-L/2}^{L/2} \int_{-L/2}^{L/2} z(x,y)\, e^{-j2\pi\left(f_x x + f_y y\right)} \mathrm{d}x \mathrm{d}y \right|^2 . \tag{8.6}$$

The PSD provides the relative strength of each roughness component as a function of spatial frequencies f_x and f_y. For isotropic surfaces, the two-dimensional PSD can be replaced by a one-dimensional function, the so-called 2D-isotropic PSD(f), which is calculated by averaging over all azimuthal directions after transformation into polar coordinates (Duparré et al., 2002). Integration of the PSD over a certain range of spatial frequencies leads to the band-limited rms roughness:

$$\sigma = \sqrt{2\pi \int_{f_{min}}^{f_{max}} \mathrm{PSD}(f)\, f \mathrm{d}f} . \tag{8.7}$$

The band limits f_{min} and f_{max}, and thus the relevant roughness, depend on the application at hand as will become clearer later.

Light scattering from surface roughness can be modeled through both scalar and vector theories. Scalar theories (Carniglia, 1979) are usually restricted to near-angle scattering with the exception of the model described in Krywonos et al. (2011) and Schröder et al. (2011c). Vector theories (Church et al., 1975, 1979; Elson et al., 1979; Bousquet et al., 1981) are more general and include the polarization properties of the incident and scattered light. For practical applications, the theory in Bousquet et al. (1981) was found particularly useful. All these theories are first-order perturbation models and are hence valid only if the roughness is small compared to the wavelength. A quantitative smooth surface criterion is $\sigma/\lambda \leq 0.02$ (Schröder et al., 2011c), which is clearly fulfilled by the optical surfaces of interest here. In this section, without the loss of generality, the formulas refer to the case of normal incidence. The overall formalism, however, allows the inclusion of all other cases including non-normal incidence and arbitrary polarization. For a single surface, the ARS caused by roughness is given by

$$\text{ARS}(\theta_s) = K(\lambda, n, \theta_s) \text{PSD}(f), \tag{8.8}$$

where n is the refractive index. The optical factor $K = 16\pi^2/\lambda^4 \cos^2\theta_s \cos\theta_i Q(\lambda, n, 0_s)$ contains the optical properties of the perfectly smooth surface (excluding the roughness) as well as the illumination and observation conditions. Q is the generalized reflectance and can be approximated by the Fresnel reflectance for s polarized light and sufficiently small angles (Stover, 1995). The simple relationship Equation 3.3 enables predicting the scattering of a surface with given roughness or, vice versa, to measure surface roughness by analyzing the scattered light.

The relationship between scattering angles and spatial frequencies f is given by the grating equation: $f = \sin \theta_s/\lambda$. This means that the roughness relevant for light scattering depends on the range of scatter angles and on the wavelength of light.

In cases where the correlation length, which represents the average lateral extension of the roughness, is much larger than λ, simple approximations for total backscattering and total forward scattering (for transparent materials) can be derived:

$$\text{TS}_b = \rho_0 \left(\frac{4\pi\sigma}{\lambda} \right)^2, \quad \text{TS}_f = \tau_0 \left(\frac{2\pi\sigma}{\lambda}(n-1) \right)^2, \tag{8.9}$$

where
   σ is the rms roughness
   ρ_0 and τ_0 are the specular reflectance and transmittance, respectively

Other equations have been derived for the case of correlations lengths that are small compared to the wavelength, a case that was discussed to be important for thin film coating structures (Elson et al., 1983; Duparré, 1995). Recently, it was shown, however, that Equations 8.9 hold in general if σ is identified as the band-limited roughness relevant for scattering at the given wavelength (Schröder et al., 2011b), rather than the total roughness used in the original paper (Elson et al., 1983).

8.3.2 Extension to Thin Film Coatings

Compared to single surface scattering, modeling light scattering from surfaces coated with dielectric films or multilayer stacks is substantially more complex. This is mainly because the amplitudes of the scattered fields from all interfaces add up to the overall scatter rather than just the intensities, and cross-correlation effects between the individual interface profiles have to be taken into account (Elson et al., 1980; Bousquet et al., 1981;

Chapter 8

Duparré et al., 1993, 1995; Amra, 1994a; Jacobs et al., 1998). The ARS of a thin film stack of M layers is given by (Bousquet et al., 1981)

$$\text{ARS}(\theta_s) = \sum_{i=0}^{M} \sum_{j=0}^{M} K_i K_j^* \ \text{PSD}_{ij}(f), \tag{8.10}$$

where
K_i denotes the optical factor at the ith interface
K_j^* denotes the conjugate complex optical factor at the jth interface

The optical factors now describe the properties of the perfectly smooth multilayer (refractive indices, film thicknesses) together with the illumination and observation conditions. The functions PSD_{ij} comprise the PSD of all interfaces (for $i = j$) as well as their cross-correlation properties (for $i \neq j$) (Duparré et al., 1988). For most thin film coatings, the roughness of the film interfaces and the substrate are the main sources of scattering. Nevertheless, volume scattering from microstructures within the film bulk may also occur and can then be described by models as given, for example, in Elson (1984), Kassam et al. (1992), and Amra et al. (1994b).

Equation 8.9 is the most general result for optical interference coatings and can be seen as an exact forward solution to first order of the scattering problem if the interface roughness is sufficiently small compared to the wavelength of light. However, the number of parameters required to model the ARS is proportional to the square of the number of layers. It is therefore practically impossible to predict the scattering of a coating or to retrieve information about the coating from scatter measurements except for single films (Duparré et al., 1988; Duparré, 1991; Rönnow, 1997). This can, however, be accomplished if additional assumptions are made to reduce the number of parameters. One approach is the following simplified model with two parameters:

1. The optical parameter δ describes the average deviation of the layer thickness from the perfect design expressed in quarter-wave optical thickness (real thickness = $(1 + \delta)$ design thickness).
2. The roughness parameter β describes the evolution of roughness from interface to interface according to a power law: (rms roughness (layer) $\propto$ layer number$^\beta$). For $\beta = 0$, all interfaces are assumed to be identical and fully correlated.

The parameters are then used to model the multilayer design and roughness evolution starting from a given design and top-surface PSD. The degrees of freedom of the problem can thus be drastically reduced, and the measured ARS can be analyzed by recursive modeling. The final result is a comprehensive model of the coating that provides important insight into the optical and structural properties. A detailed description of the method is given in Schröder et al. (2011a).

8.4   Examples of Applications

In this section, we present the results of investigations of optical surfaces, thin film coatings, and materials regarding angle-resolved and total light scattering, and specular reflectance based on the experimental methods described in Section 8.2. The interpretation of the results is based on the theoretical background given in Section 8.3. Examples have been selected that represent a cross section of topics relevant in particular for high-end applications facing the problem of possible laser-induced damage.

8.4.1   Light Scattering

8.4.1.1   Assessment of Surface Defects and Contaminations

Surface inhomogeneities such as local variations of surface roughness, surface contaminations, or subsurface defects play a critical role for both the scattering properties and the laser stability of optical surfaces. Local defects and particulate contaminations are known to play a critical role as damage precursors and can substantially reduce the damage threshold as a result of field enhancements (Bloembergen, 1973; Palmier et al., 2008). Surface cleaning is thus of crucial importance for high-power applications. Profilometric inspection methods, such as atomic force microscopy, offer high resolution but are limited to small sample areas. Light scattering methods have the potential of investigating larger sample areas with high sensitivity (Truckenbrodt et al., 1992; Duparré, 2006).

In Figure 8.5, the results of TS_b mappings at 532 nm of a polished BK7 surface before and after cleaning using denatured ethanol and cotton batting are shown. The average TS value before cleaning is $(8.7 \pm 29) \times 10^{-6}$, the large deviation being caused by the impact of defects—presumably particulate contaminations. After applying the data reduction algorithm prescribed in ISO 13696, a TS of $(7.3 \pm 0.4) \times 10^{-6}$ is obtained. This value now represents merely roughness-induced surface scattering as well as bulk scattering within the material. The TS value after cleaning is almost identical $(7.6 \pm 0.3) \times 10^{-6}$. This demonstrates that the data reduction procedure substantially enhances the robustness of TS measurements by reducing the impact of local defects onto the measurement result.

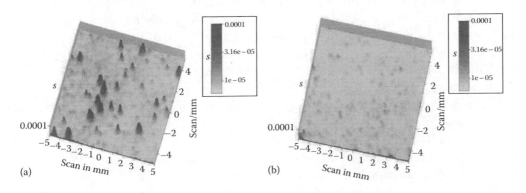

FIGURE 8.5 TS_b mapping at $\lambda = 532$ nm of a polished BK7 surface (a) before and (b) after cleaning.

Chapter 8

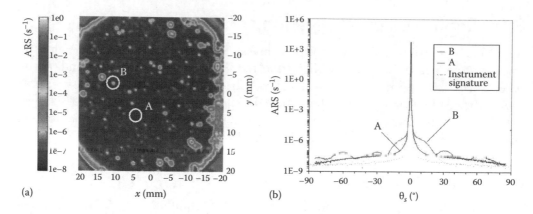

FIGURE 8.6 Results of ARS measurements at λ = 532 nm of superpolished silicon wafer. (a) Mapping at a fixed scatter angle of 15°. (b) ARS distribution at two distinct positions on sample revealing roughness-induced scattering (A) and scattering of a single particle (B).

The distribution of particles over the sample and the effect of cleaning are clearly visible by comparing the scatter maps before and after cleaning.

The TS method is very capable for performing fast and sensitive investigations of surface roughness using Equations 8.9, although this requires separation of the contribution from surface and bulk scattering in the case of transparent samples. It is also very useful for detecting local defects on larger areas with high sensitivity. Yet no detailed information can be obtained about the nature of the defects because of the integral measurement of scattering. This can be overcome by analyzing the angular scattering distribution.

In Figure 8.6, results of ARS measurements at λ = 532 nm of a superpolished silicon surface are shown (Schröder et al., 2011b). The sample surface was first scanned at a fixed scatter angle of 15° to gather information about the homogeneity and defects (Figure 8.6a). At certain positions (A and B), the ARS distribution was then measured to analyze the detected defects (Figure 8.6b). The roughness-induced scattering (A) is particularly low and close to but still above the sensitivity limit of the instrument. The TS_b calculated by integrating the ARS curve according to Equation 8.4 is 7.0×10^{-8}. This corresponds to a roughness of only 0.015 nm as calculated using Equation 8.9. This result is in excellent agreement with the value of 0.016 nm determined from atomic force microscopy in $10 \times 10 \ \mu m^2$ scan areas and integration of the PSD in the band between $f = 0.06 \ \mu m^{-1}$ and $1.9 \ \mu m^{-1}$ relevant for scattering at λ = 532 nm (Equation 8.7).

The scattering distribution at position B exhibits a fringe structure that is typical for the scattering from isolated particles that are larger than the wavelength used (van de Hulst, 1957). The particle diameter was estimated by evaluating the positions of the first minima (Maure et al., 1996) to be 1.4 μm. Although other evaluation methods are required for particles smaller than the wavelength (van de Hulst, 1957; Germer, 2001), it was estimated that single particles with a diameter of as small as 60 nm can be detected using UV light scattering with focused illumination. Other approaches for the detection of defects on large surfaces based on matrix detectors are described in Herffurth et al. (2013) and Zerrad et al. (2014).

8.4.1.2 Measurement of Surface Roughness through Light Scattering Analysis

Thorough assessment of surface roughness is essential for optical components for high-power applications. Roughness does not only lead to additional losses, but it is also correlated to other properties such as subsurface damage and the adherence of contaminations such as water and hydrocarbon layers or particles. In particular, for practical applications that involve complex sample geometries, surface finishing is still a challenging process and questions about both the roughness level achieved and the homogeneity often arise. This requires measurement techniques that are noncontact, sensitive, robust with respect to vibrations, and that are capable of covering large sample areas. It is interesting to note that only light scattering techniques meet all these demands.

As discussed in Section 8.3, for optically smooth surfaces, ARS is directly proportional to the surface PSD. Thus, the surface PSD and, consequently, the rms roughness can be measured by analyzing the scattered light. While the vertical resolution is given by the sensitivity of the instrument, the lateral resolution is defined by the wavelength and the range of scatter angles used.

In Figure 8.7, scatter-based roughness measurements of a 665 mm diameter collector mirror for EUV laser-produced plasma sources are shown (Trost et al., 2011). ARS measurements on the substrate before coating were performed at a wavelength of 442 nm using the instrument described in Schröder et al. (2011a). The PSD was then determined from ARS using Equation 8.8. Because of the wavelength used, the measured PSD does not cover the entire high-spatial frequency roughness (HSFR) range relevant for the application wavelength of 13.5 nm. However, polished substrate surfaces usually exhibit a fractal PSD—a straight line when plotted on a log–log scale (Church, 1988). This allows an analytic PSD function to be fitted to the measured data and extrapolation to the high-spatial frequency range.

A complete roughness map of the entire surface can then be retrieved by scanning the entire sample area while performing a scatter measurement at each point of interest. This provides useful information about the homogeneity and defects. In the example

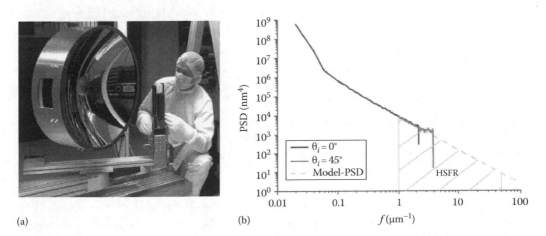

FIGURE 8.7 Surface roughness measurement of EUV collector before coating through light scattering analysis. (a) Mirror substrate under investigation with instrument for angle-resolved scatter measurements at λ = 405 nm. (b) Surface PSD function determined from ARS data and calculation of high spatial frequency roughness (HSFR).

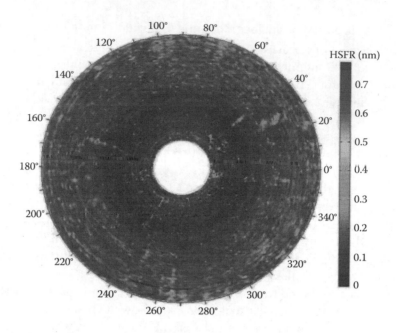

FIGURE 8.8   HSFR map of EUV collector substrate (diameter 660 mm) retrieved from scatter measurements and scanning the sample.

shown in Figure 8.8, the average HSFR is well below 0.2 nm in most areas. Using a roughness evolution model for the EUV coating (Stearns et al., 1998; Schröder et al., 2007) and the multilayer scattering theory described in Section 8.3.2, the scattering and reflectance properties of the multilayer-coated EUV mirror at λ = 13.5 nm were predicted. The results were proven to be in good agreement (within 1%) with the mirror reflectance measured at λ = 13.5 nm. The presented method has now become an internal standard procedure to characterize EUV collector mirrors at the Fraunhofer IOF.

8.4.1.3   Bulk Scattering of Fused Silica at the DUV Wavelength of 193 nm

Aside from the scattering induced by surface and interface roughness, bulk scattering can be a critical factor in transmissive optical elements. Bulk defects such as inclusions can be an important source of laser-induced damage. But even in high-quality optical materials, tiny fluctuations of the refractive index lead to enhanced scattering and absorption. In particular, polished amorphous materials, such as fused silica, may exhibit higher bulk scattering than interface scattering. Synthetic fused silica is one of the main materials for optical elements in DUV lithography systems. There exists a variety of synthetic fused silica manufactured by different process chains with different properties regarding losses and long-term stability under laser irradiation. Because measuring the bulk scattering is one way to accurately determine the bulk absorption from reflectance and transmittance measurements using the energy balance, detailed investigations of the bulk scattering properties of various materials were performed in Schröder et al. (2006).

The results of ARS measurements performed on a large disk (diameter 200 mm) illuminated through its lateral surface with s and p polarized incident radiation are shown in Figure 8.9. This approach enables surface and bulk scattering to be separated because

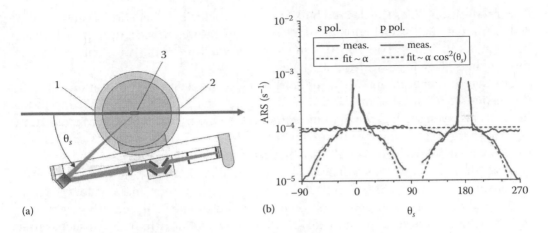

(a)

(b)

FIGURE 8.9   (a) Setup for the measurement of ARS at $\lambda = 193$ nm of fused silica disc. (b) ARS results for incident radiation polarized perpendicular (s pol.) and parallel (p pol.) to the plane of measurement. (Reproduced from Schröder, S. et al., *Opt. Expr.*, 14, 10537.)

only radiation emitted from the center of the sample is in the field of view of the detector except for the specular regions around 0° and 180° that contain scattering from the entry and exit surfaces. The flat ARS for s-polarized light and the cosine-squared-shaped ARS for p-polarized light indicate that the Rayleigh scattering from intrinsic index fluctuations in the amorphous silica structure is the main scattering mechanism. By fitting the Rayleigh model curves (van de Hulst, 1957) to the measurement data and evaluating the absolute scatter level, the bulk scattering coefficient can be determined if the field of view, hence the scattering volume, is known.

Another approach that can be used also for more convenient sample geometries is based on TS measurements. Assuming that the total scattering of a sample is the sum of the actual bulk scattering and surface effects and considering that only bulking scatter is a function of the sample thickness, the bulk scattering coefficient can be determined by investigating a thickness series. TS mappings of samples with thicknesses ranging from 2 to 20 mm are shown in Figure 8.10. Plotting TS versus the sample thickness and

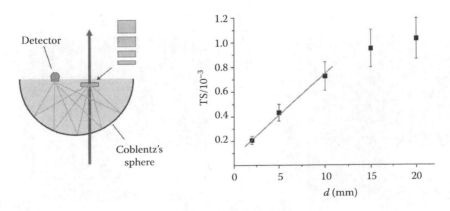

FIGURE 8.10 TS_b measurements at $\lambda = 193$ nm of fused silica samples with increasing thickness d (thickness series). The slope of TS versus d yields the bulk scattering coefficient.

Chapter 8

linear regression yields the desired bulk scattering coefficient. Depending on the material properties such as OH content and fictive temperature, bulk scattering coefficients between 6×10^{-4} cm^{-1} and 1.7×10^{-3} cm^{-1} were observed at $\lambda = 193$ nm. A more detailed discussion is given in Schröder et al. (2006).

For certain applications, it can be interesting to investigate modifications of bulk properties during irradiation. Since the Rayleigh scattering is directly linked to index fluctuations, even small changes in index induced by intense laser (or other) irradiation will cause significant changes in the scatter levels. Scatter measurements can thus be used for in situ investigations of laser-induced modifications in glasses and solids in general. Similar conclusions will be drawn in Section 8.4.1.6.

Subsurface damage (SSD) of optical surfaces is another important issue for the fabrication and the application of optical components and can be analyzed using light scattering methods. The separation of SSD scatter from roughness-induced surface scatter and bulk scatter is a particularly challenging task. Approaches for the characterization of SSD are discussed in Trost et al. (2013).

8.4.1.4   Metal Fluoride Highly Reflective Coatings at 193 nm

For applications at the DUV wavelength if 193 nm such as microlithography and material processing, coatings with low losses and high damage resistance are required. Metal fluorides offer low intrinsic absorption. However, the low energetic deposition techniques used in order to maintain sufficient stoichiometric properties lead to columnar growth and considerable amounts of interface roughness and scattering as well as adsorption of water in the porous film structure (Duparré et al., 1996; Jacobs et al., 1998; Protopapa et al., 2001; Ristau et al., 2002; Schröder et al., 2005c; Wang et al., 2007). Questions regarding the influence of substrate or thin film roughness or about optical thickness errors and their influence often arise. ARS measurements and modeling provide illustrative answers to these questions.

AFM surface profiles of an uncoated CaF_2 substrate and the same substrate coated with a highly reflective quarter wave stack with 20 periods of AlF_3/LaF_3 pairs are shown in Figure 8.11 and reveal a substantial increase in the surface roughness as a result of columnar growth.

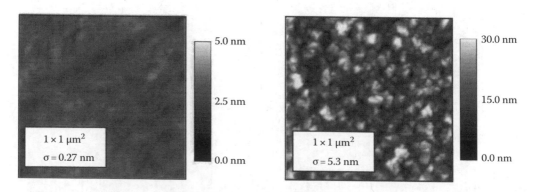

FIGURE 8.11   Atomic force microscopy images of the uncoated substrate (left) and the top surface of an all-fluoride high reflective (HR) coating for $\lambda = 193$ nm (right).

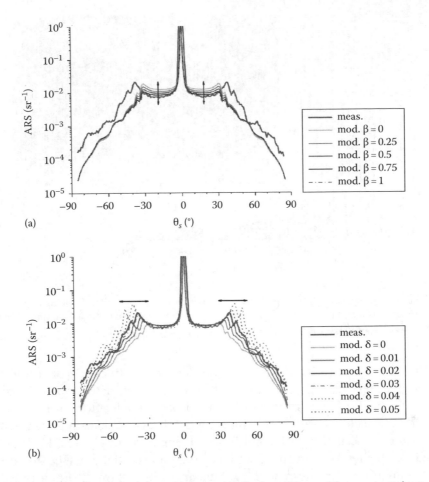

FIGURE 8.12   Angle-resolved scattering of HR coating for λ = 193 nm. Measurement (meas.) results obtained at 193 nm and modeling (mod.) results by varying (a) the roughness parameter and (b) the optical parameter.

The results of ARS measurements performed at λ = 193 nm using the instrumentation described in Schröder et al. (2008) are shown in Figure 8.12 together with modeling results using the simplified procedure described in Section 8.3.2.

The measured curves exhibit distinct peaks at 0° corresponding to the direction of specular reflection as well as typical shoulders and ripples caused by interference effects of waves scattered at different interfaces within the coating. The TS_b calculated using Equation 8.4 is 2.8%.

The modeling results shown in Figure 8.12a reveal that varying the roughness evolution parameter β but leaving the thickness parameter δ constant (δ = 0 corresponding to perfect design) merely influences the heights of the shoulders of the modeled curves. The best fit for β = 1 indicates that the coating exhibits a rapid roughening from the substrate through the multilayer in agreement with the AFM results. However, in contrast to AFM, the ARS method also provides quantitative information about the roughness of the inner interfaces.

Varying the optical thickness error δ but now leaving β = 1 constant (Figure 8.12b) basically shifts the angular position of the resonant scattering wings. The best fit for

Chapter 8

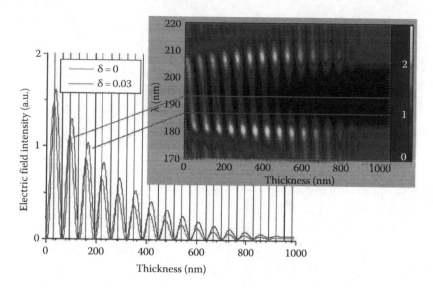

FIGURE 8.13   Standing wave electric field intensities with and without optical thickness errors. Vertical lines represent interfaces. For better comparability, δ has been converted into a variation of the incident wavelength: λ′ = 193 nm/(1 + δ).

δ = 0.03 indicates a deviation of 3% of the average optical thickness of each layer from the perfect quarter-wave design. This result is directly correlated to a spectral shift of the peak reflectance of the mirror, which is a well-known effect for porous coatings and indicates the inclusion of water. Even slight optical thickness deviations can lead to substantial field enhancements in the multilayer as illustrated in Figure 8.13. These induce additional scattering and absorption and may thus substantially reduce the laser stability. For the given coating, it can be shown that an accurate film thickness (δ = 0) would lead to a reduction of the scatter loss from 2.8% to 1.4%.

8.4.1.5   ARS of Low-Loss HR Coatings at 1064 nm

Optical components for high-power applications in general exhibit rather low losses corresponding to a throughput close to unity. Direct measurements of the optical losses are, therefore, indispensable for thorough investigations. Within an international round-robin experiment, highly reflective dielectric mirrors for 1064 nm had to be characterized in terms of specular reflectance and losses (Duparré et al., 2010). Sixteen layer pairs of SiO_2 and Ta_2O_5 were deposited on superpolished fused silica substrates using magnetron sputtering. The specular reflectance was measured using the methods described in Section 8.2.3 with significant deviations between the results of different participants as will be discussed in detail in Section 8.4.2.2. Additional scatter measurements were performed to verify the results of the reflectance measurements and to explain deviations of the results from different instruments by analyzing the influence of the angular distribution of scattering.

The results of ARS measurements performed at 1064 nm together with a curve modeled using the background described in Section 8.3 are shown in Figure 8.14. The results reveal a significant effect of replicated substrate roughness at small scatter angles as well as fringes corresponding to resonant scattering from the multilayer.

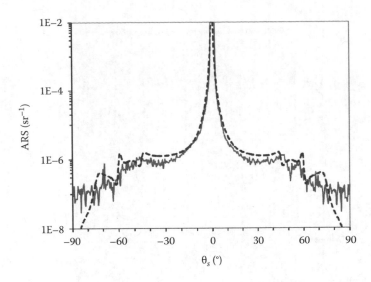

FIGURE 8.14 ARS at $\lambda = 1064$ nm of HR coating on fused silica. Measurement (solid line) and modeling (dashed line) results.

The total scatter loss calculated from ARS is as low as 7.2×10^{-6} when integrating from 2° to 85° according to the standard (ISO 13696, 2002). Though, varying the lower integration limit can lead to substantially different results. When integrating from 0.2°, the scatter loss increases to 38×10^{-6}. Considering the ranges of acceptance angles (or spatial frequency ranges) is thus of crucial importance when comparing results of reflectance measurements obtained by different techniques as will be discussed also in Section 8.4.2.2.

8.4.1.6 Investigation of Scattering Characteristics of Rugate Notch Filter after Laser Damage Test

Rugate films have been demonstrated to be a top candidate for optical coatings with substantially enhanced laser stability compared to standard stacks (Jupé et al., 2006). In contrast to conventional multilayer stacks, Rugate filters consist of material mixtures to achieve gradually changing index profiles. For highly reflective mirrors, it has been demonstrated that the damage threshold of Rugate filters can exceed that of a standard multilayer coating by a factor of 10. Investigations at 1064 nm revealed that laser-induced damage in Rugate films often occurs as alterations of the optical properties inside the films compared to the well-known ablation effects of standard stacks.

A Rugate notch filter for 355 nm was fabricated at the Laser Zentrum Hannover (LZH) by Ion Beam Sputtering. LID S-on-1 tests were performed at LZH at 355 nm by irradiating different sample positions on a regular matrix (effective beam diameter 250 μm, repetition rate 10 Hz, effective pulse duration 11 ns). Light scattering measurements were then performed at 325 nm after the damage tests (Schröder et al., 2011a). The results are shown in Figure 8.15. The scatter mapping performed by scanning the sample and measuring the scattering at a fixed scatter angle of 45° clearly reveals damaged and undamaged irradiation sites on the rather homogeneous intrinsic scatter of the non-irradiated areas. This illustrates the capability of scatter measurements for automatic

Chapter 8

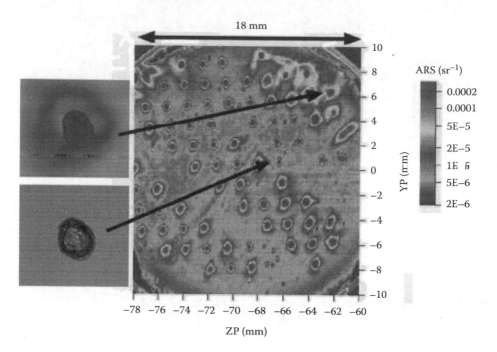

FIGURE 8.15 Scatter map at λ = 325 nm revealing different types of defects and DIC images of two different defects indicating surface and bulk effects.

postevaluation procedures of laser damage tests or for correlating sample properties before and after irradiation in order to check for possible damage precursors.

In addition, different types of defects can be identified: (1) sharp defects (small dots in scatter map) corresponding to pure surface defects and (2) defects surrounded by halos. The differential interference contrast images (DIC, field of view 0.85 × 0.85 mm²) shown also in Figure 8.15 indicate that the halos correspond to altered optical properties in the bulk of the film caused by intense laser irradiation.

ARS measurements at several positions near a damaged site reveal resonant scattering peaks near 15° from the specular direction that shift to larger angles as the position of investigation approaches the defect site (Figure 8.16a). This indicates that the halo in fact originates from the bulk of the film, and the optical thickness of the coating increases toward the defect site as a result of the high thermal load during irradiation.

In order to apply the ARS modeling procedure to gradient index films, the continuous index profile was approximated by a set of 700 single layers with about 200 different dielectric functions. The results of modeled and measured ARS in an undamaged region shown in Figure 8.16b demonstrate that the modeling exactly predicts the existence and the position of the observed resonant peaks near 15°. This indicates that the scattering of Rugate films is caused by similar mechanisms than that of standard stacks. Considering that an angular shift of the peak of 1.5° can be detected, the method is sensitive to laser-induced alterations of optical thickness of as low as 0.3%. The method might, therefore, be used to investigate laser-induced degradation even before ultimate damage events. More results on the scattering properties of Rugate films are presented in Herffurth et al. (2014).

Light scattering is sensitive to all sorts of structural inhomogeneities of surfaces, coatings, and bulk materials as well as to changes of these properties as a result of

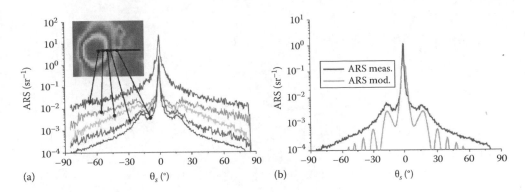

FIGURE 8.16 Angle-resolved scattering of a Rugate film. (a) Measured at $\lambda = 325$ nm at several positions near a defect site caused by laser-induced damage. (b) Results of ARS modeling compared to measured curve.

laser irradiation. Since scatter measurements are sensitive, fast, but nondestructive, this offers opportunities to use light scattering in general for the in situ characterization of damage processes during irradiation.

8.4.2   Transmittance and Reflectance

8.4.2.1   Spectrophotometry for Determining the Optical Constants of Single-Layer Coatings

The dielectric functions of thin films deposited using physical vapor deposition usually differ from the properties of the corresponding bulk materials depending on the deposition process parameters such as pressure, temperature, and ion assistance. As discussed in Section 8.4.1.4, even slight deviations of the spectral properties of the final coating from the design can lead to substantial field enhancements and reduced damage thresholds. The accurate determination of the optical constants of thin films is thus crucial. This is usually done by performing spectral reflectance and transmittance measurements and reverse engineering to determine the real and imaginary parts, n and k, of the index of refraction as a function of wavelength.

In an international round-robin experiment, the spectral characteristics of a Ta_2O_5 single layer on fused silica were to be determined from the specular reflectance and transmittance data measured at 45° angle of incidence (Duparré et al., 2008). Several spectrophotometric instruments based on the international standard ISO 15368 in different laboratories were used. Spectral transmittance measurements were performed by introducing the sample into the beam path of the instrument at the desired angle of incidence. The results are shown in Figure 8.17a. The interference pattern corresponds to the gradual change in the phase difference between the partial waves originating from the front and the rear interfaces of the single layer as the wavelength is increased. The transmittance for p-polarized incident light is higher than the s-polarized transmittance in agreement with the Fresnel equations. The deviations of the results with respect to the average values of all participants for selected wavelengths are shown in Figure 8.17b.

Spectral reflectance measurements involve modifications of the beam path and are, therefore, prone to additional measurement uncertainties from alignment errors as discussed in Section 8.2.3.2. This explains the rather large deviations of the reflectance spectra shown in Figure 8.18a. These issues become even more obvious when

Chapter 8

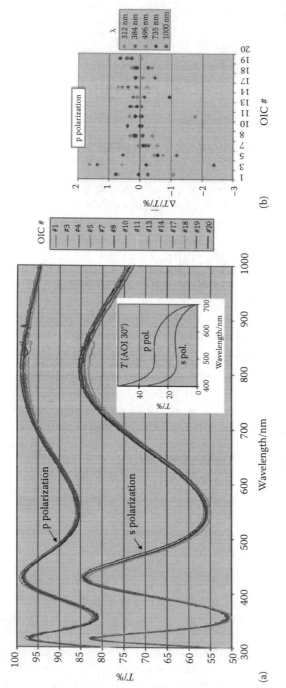

FIGURE 8.17 Spectral transmittance of a Ta_2O_5 thin film measured by different laboratories (OIC#). (a) Spectral transmittance curves. (b) Relative deviations of measurement results by different laboratories at specific wavelengths.

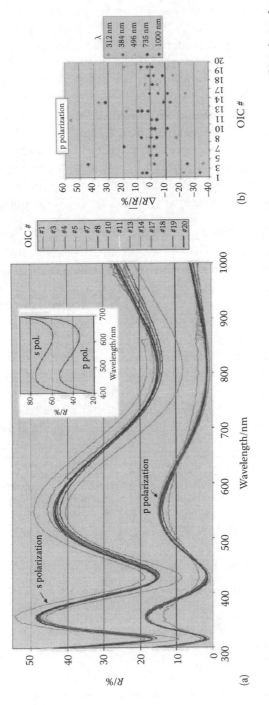

FIGURE 8.18 Spectral reflectance of a Ta_2O_5 thin film measured by different laboratories (OIC#). (a) Spectral transmittance curves. (b) Relative deviations of measurement results by different laboratories at specific wavelengths.

comparing the results at specific wavelengths shown in Figure 8.18b considering the different scale compared to Figure 8.17.

The optical constants were determined from the reflectance and transmittance curves using reverse engineering based on the background given, for instance, in Mcleod (2001), Stenzel (2005), and Stenzel et al. (2009). The results presented in detail in Duparré et al. (2008) illustrate that while n can be determined with an accuracy that represents the accuracy of the reflectance measurements, low *k* values are rather difficult to determine with sufficient accuracy using this approach. This is mainly because both the uncertainty of the reflectance and the transmittance measurements influence the result while *k* is determined indirectly using the energy balance. Since the *k* values are much smaller than the *n* values, they are much more affected by the uncertainties of *R* and *T*. In Stenzel et al. (2009), a reverse engineering procedure based on the Kramers–Kronig relations is presented that link *k* and *n* and thus enhances the accuracy of the presented method. Yet another issue is that both absorption and scattering contribute to the determined *k*. This once again demonstrates the necessity of direct measurements of losses.

8.4.2.2 Reflectance of Measurements of Low-Loss HR Coatings at 1064 nm

As a result of the challenging demands on the accuracy of reflectance measurements of low-loss coatings and recent developments in measurement capabilities, international round-robin experiments were created (Nickel et al., 2003; Duparré et al., 2010). The main task was to determine the reflectance of dielectric mirrors at 1064 nm as accurate as possible.

For the investigation presented in Nickel et al. (2003), highly reflective dielectric coatings with a nominal reflectance of 99.9% at normal incidence were deposited by e-beam evaporation. Spectrophotometry according to the standard for measurements at the medium accuracy level (ISO 15368, 2001) and measurements according to the standard for high-precision measurements (ISO 13697, 2006) as well as estimations based on simple transmission measurements were performed. The results depicted in Figure 8.19

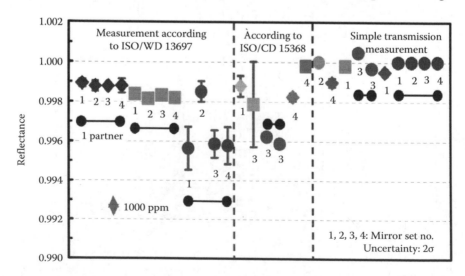

FIGURE 8.19 Results of reflectance measurements of HR coatings at λ = 1064 nm. (Reproduced from Nickel, D. et al., *Proc. SPIE*, 4932, 520, 2003.)

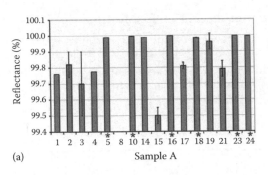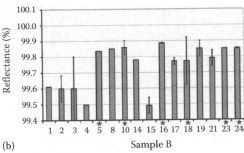

FIGURE 8.20   Results of reflectance measurements of dielectric HR coatings at $\lambda = 1064$ nm using spectrophotometry, laser ratiometry, and the CRD technique (*). (a) Type A samples (16 layer pairs). (b) Type B samples (10 layer pairs).

illustrate the higher precision of the method according to ISO 13697 but also reveal that the calibration based on measuring the chopper reflectance can lead to substantial systematic errors.

The round-robin experiment presented in Duparré et al. (2010) comprised spectrophotometry, laser ratiometry, as well as the CRD technique. The samples were HR dielectric laser mirror systems designed for normal incidence at 1064 nm. In addition to the design with 16 layer pairs of SiO_2 and Ta_2O_5 described in Section 8.4.1.5 (type A samples), coatings with 10 layer pairs (type B) were fabricated to generate a slightly lower reflectance. The results were again evaluated according to the regulations established in round-robin experiments in an anonymous manner.

The results for sample type A shown in Figure 8.20a reveal pronounced deviations between the two techniques. Reflectances above 99.95% were only retrieved from CRD methods marked by asterisks. One participant, who used spectrophotometry, deliberately refrained from providing a value for the reflectance $\rho = R$ of sample type A, because "it was not possible to determine R within acceptable confidence." He was, however, able to deduce from his measurements that "R is probably higher than 99.95%."

Figure 8.20b summarizes the results obtained for sample type B. Also for this sample type, significant differences are observed. Here, both CRD and spectrophotometry were found to render high ρ values, that is, $\rho > 99.8$ for type B samples.

As discussed in Section 8.4.1.5, scatter losses between 7.2 and 38 ppm were determined for sample type A depending on the range of scatter angles that contribute to the specular reflectance. In addition, absorption measurements were performed using laser calorimetry and photothermal methods with results between 3.7 and 19 ppm. The highest reflectance values reported from spectrophotometry was 99.987% leaving a rather large gap of more than 100 ppm in the energy balance assuming that the residual transmittance can be neglected for sample type A. In contrast, the highest result from CRD measurements was 99.9964%, which is in much better agreement with the losses determined directly from scatter and absorption measurements. The comparison of reflectance values obtained from spectrophotometry and CRD with scatter losses measured by ARS demonstrates that careful consideration of bandwidth effects is necessary. This issue is, however, usually not discussed in the current standards describing procedures for specular reflectance or transmittance measurements, and further work regarding homogenization of different measurement methods is required in the future.

Chapter 8

Another measurement task discussed within the 2013 OIC Measurement Problem was the measurement of the reflectance of AR coatings at near-normal incidence. Retrieving the front-surface reflectance by overcoming the rear-side problem turned out to be a particularly challenging task and led to substantial variations in the results. A more detailed discussion can be found in Duparré and Ristau (2013).

References

Amra, C. 1994a. Light scattering from multilayer optics. *J. Opt. Soc. Am. A* 11: 197–226.

Amra, C., Grèzes-Besset, C., Maure, S., and Torricini D. 1994b. Light scattering from localized and random interface or bulk irregularities in multilayer optics: The inverse problem. *Proc. SPIE* 2253: 1184–1200.

Amra, C., Grezes-Besset, C., Roche, P., and Pelletier E. 1989. Description of a scattering apparatus: Application to the problems of characterization of opaque surfaces. *Appl. Opt.* 28: 2723–2730.

Amra, C., Torricini, D., and Roche P. 1993. Multiwavelength (0.45–10.6 µm) angle-resolved scatterometer or how to extend the optical window. *Appl. Opt.* 32: 5462–5474.

Asmail, C. C., Cromer, C. L., Proctor, J. E., and Hsia J. J. 1994. Instrumentation at the national institute of standards and technology for bidirectional reflectance distribution function (BRDF) measurements. *Proc. SPIE* 2260: 52–61.

ASTM E 1392-90. 1990. *Standard Practice for Angle Resolved Optical Scatter Measurements on Specular or Diffuse Surfaces*. Philadelphia, PA: American Society for Testing and Materials.

ASTM E 2387-5. 2011. *Standard Practice for Goniometric Optical Scatter Measurements*. Philadelphia, PA: American Society for Testing and Materials.

ASTM F 1048-87. 1987. *Standard Test Method for Measuring the Effective Surface Roughness of Optical Components by Total Integrated Scattering*. Philadelphia, PA: American Society for Testing and Materials.

Balachninaite, O., Grigonis, R., Sirutkaitis, V., and Eckardt R. C. 2005. A coherent spectrophotometer based on a periodically poled lithium niobate optical parametric oscillator. *Opt. Commun.* 248: 15–25.

Balasa, I., Blaschke, H., and Ristau D. 2011. Broadband spectrophotometry on nonplanar EUV multilayer optics. *Proc. SPIE* 7969: 796928.

Bennett, H. E. 1978. Scattering characteristics of optical materials. *Opt. Eng.* 17: 480–488.

Bennett, H. E. and Porteus J. O. 1961. Relation between surface roughness and specular reflectance at normal incidence. *J. Opt. Soc. Am.* 51: 123–129.

Bennett, J. M. and Mattsson L. 1999. *Introduction to Surface Roughness and Scattering*, 2nd edn. Washington, DC: Optical Society of America.

Berden, G., Peeters, R., and Meijer G. 2000. Cavity ring-down spectroscopy: Experimental schemes and applications. *Int. Rev. Phys. Chem.* 19: 565–607.

Blaschke, H., Kohlhaas, J., Kadkhoda, P., and Ristau D. 2003. DUV/VUV spectrophotometry for high-precision spectral characterization. *Proc. SPIE* 4932: 536–542.

Bloembergen, N. 1973. Role of cracks, pores, and absorbing inclusions on laser induced damage threshold at surfaces of transparent dielectrics. *Appl. Opt.* 12: 661–664.

Bousquet, P., Flory, F., and Roche P. 1981. Scattering from multilayer thin films: Theory and experiment. *J. Opt. Soc. Am.* 71: 1115–1123.

Carniglia C. K. 1979. Scalar scattering theory for multilayer optical coatings. *Opt. Eng.* 18: 104–115.

Church, E. L. 1988. Fractal surface finish. *Appl. Opt.* 27: 1518–1526.

Church, E. L., Jenkinson, H. A., and Zavada J. M. 1979. Relationship between surface scattering and micro-topographic features. *Opt. Eng.* 18: 125–136.

Church, E. L. and Zavada J. M. 1975. Residual surface roughness of diamond-turned optics. *Appl. Opt.* 14: 1788–1795.

Davies, H. 1954. The reflection of electromagnetic waves from a rough surface. *Proc. IEE., Pt. III* 101: 209–214.

Detrio, J. A. and Miner S. M. 1985. Standardized total integrated scatter measurements of optical surfaces. *Opt. Eng.* 24: 419–422.

Duparré, A. 1991. Effect of film thickness and interface roughness correlation on the light scattering from amorphous and from columnar structured optical films. *J. Mod. Opt.* 38: 2413–2421.

Duparré, A. 1995. Light scattering of thin dielectric films. In *Thin Films for Optical Coatings*, eds. R. E. Hummel and K. H. Günther, Boca Raton, FL: CRC Press.

Duparré, A. 2006. Light Scattering techniques for the inspection of microcomponents and microstructures. In *Optical Methods for the Inspection of Microsystems*, ed. W. Osten, pp. 103–119. Boca Raton: Taylor & Francis.

Duparré, A., Ferré-Borrull, J., Gliech, S., Notni, G., Steinert, J., and Bennett J. M. 2002. Surface characterization techniques for determining the root-mean-square—Roughness and power spectral densities of optical components. *Appl. Opt.* 41: 154–171.

Duparré, A. and Gliech S. 1997. Non-contact testing of optical surfaces by multiple-wavelength light scattering measurement. *Proc. SPIE* 3110: 566–573.

Duparré, A. and Jakobs S. 1996. Combination of surface characterization techniques for investigating optical thin-film components. *Appl. Opt.* 35: 5052–5058.

Duparré, A. and Kassam S. 1993. Relation between light scattering and the microstructure of optical thin films. *Appl. Opt.* 32: 5475–5480.

Duparré, A. and Ristau D. 2008. Optical interference coatings 2007 measurement problem, *Appl. Opt.* 47: C179–C184.

Duparré, A. and Ristau D. 2011. Optical interference coatings 2010 measurement problem. *Appl. Opt.* 50: C172–C177.

Duparré, A. and Ristau D. 2014. Optical interference coatings 2013 measurement problem. *Appl. Opt.* 53: A281–A286.

Duparré, A. and Walther H. G. 1988. Surface smoothing and roughening by dielectric thin film deposition. *Appl. Opt.* 27: 1393–1395.

Elson, J. M. 1984. Theory of light scattering from a rough surface with an inhomogeneous dielectric permittivity. *Phys. Rev. B* 30: 5460–5480.

Elson, J. M. and Bennett J. M. 1979. Relation between the angular dependence of scattering and the statistical properties of optical surfaces. *J. Opt. Soc. Am.* 69: 31–47.

Elson, J. M., Rahn, J. P., and Bennett J. M. 1980. Light scattering from multilayer optics: Comparison of theory and experiment. *Appl. Opt.* 19: 669–679.

Elson, J. M., Rahn, J. P., and Bennett J. M. 1983. Relationship of the total integrated scattering from multilayer-coated optics to angle of incidence, polarization, correlation length, and roughness cross-correlation properties. *Appl. Opt.* 22: 3207–3219.

von Finck, A., Hauptvogel, M., and Duparré A. 2011. Instrument for close-to-process light scatter measurements of thin film coatings and substrates. *Appl. Opt.* 50: C321–C328.

von Finck, A., Herffurth, T., Schröder, S., Duparré, A., and Sinzinger, S. 2014. Characterization of optical coatings using a multisource table-top scatterometer. *Appl. Opt.* 53, A259–A269.

Germer, T. A. 2001. Polarized light scattering by microroughness and small defects in dielectric layers, *J. Opt. Soc. Am. A* 18: 1279–1288.

Germer, T. A. and Asmail C. C. 1999. Goniometric optical scatter instrument for out-of-plane ellipsometry measurements. *Rev. Sci. Instrum.* 70: 3688.

Gliech, S., Steinert, J., and Duparré A. 2002. Light-scattering measurements of optical thin-film components at 157 and 193 nm. *Appl. Opt.* 41: 3224–3235.

Guenther, K. H., Wierer, P. G., and Bennet, J. M. 1984. Surface roughness measurements of low-scatter mirrors and roughness standards. *Appl. Opt.* 23: 3820–3836.

Guide to the Expression of Uncertainty in Measurement (GUM). 1995. *International Organization for Standardization*, Geneva, Switzerland.

Harvey, J. E. and Shack R. V. 1978. Aberrations of diffracted wave fields. *Appl. Opt.* 17: 3003.

Herffurth, T., Coriand, L., Schröder, S., Duparré, A., and Tünnermann A. 2011. Nano-roughness assessment by light scattering techniques (in German). In *Neue Strategien der Mess- und Prüftechnik für die Produktion von Mikrosystemen und Nanostrukturen: Abschlussbericht DFG-Schwerpunktprogramm 1159 StraMNano*, ed. A. Weckenmann, Aachen, Germany: Shaker.

Herffurth, T., Schröder, S., Trost, M., Duparré, A., and Tünnermann, A. 2013. Comprehensive nanostructure and defect analysis using a simple 3D light-scatter sensor. *Appl. Opt.* 52: 3279–3287.

Herffurth, T., Trost, M., Schröder, S., Täschner, K., Bartzsch, H., Frach, P., Duparré, A., and Tünnermann, A. 2014. Roughness and optical losses of rugate coatings. *Appl. Opt.* 53: A351–A359.

Hogrefe, H. and Kunz C. 1987. Soft x-ray scattering from rough surfaces: Experimental and theoretical analysis. *Appl. Opt.* 26: 2851–2859.

International Organization for Standardization. 1995. *Guide to the Expression of Uncertainty in Measurement (GUM)*. Geneva, Switzerland: International Organization for Standardization.

Chapter 8

ISO 13696. 2002. *Optics and photonics—Lasers and Laser Related Equipment—Test Methods for Radiation Scattered by Optical Components*. Geneva, Switzerland: International Organization for Standardization.

ISO 13697. 2006. *Optics and Photonics—Lasers and Laser Related Equipment—Test Methods for Specular Reflectance and Regular Transmittance of Optical Laser Components*. Geneva, Switzerland: International Organization for Standardization.

ISO 15368. 2001. *Optics and Optical Instruments—Measurement of Reflectance of Plane Surfaces and Transmittance of Plane Parallel Elements*. Geneva, Switzerland: International Organization for Standardization.

ISO/WD 13142. 2010. *Optics and Photonics—Electro-Optical Systems—Cavity Ring-Down Technique for High Reflectance Measurement*. Geneva, Switzerland: International Organization for Standardization.

Jakobs, S., Duparré, A., and Truckenbrodt H. 1998. Interfacial roughness and related scatter in ultraviolet optical coatings: A systematic experimental approach. *Appl. Opt.* 37: 1180–1193.

Jupé, M., Grossmann, F., Starke, K., and Ristau D. 2003. High-precision reflectivity measurements: Improvements in the calibration procedure. *Proc. SPIE* 4932: 527.

Jupé, M., Lappschies, M., Jensen, L., Starke, K., and Ristau D. 2006. Laser-induced damage in gradual index layers and Rugate filters. *Proc. SPIE* 6403: 64031A.

Kadkhoda, P., Müller, A., Ristau, D. et al. 2000a. International Round-Robin experiment to test the International Organization for Standardization total-scattering draft standard. *Appl. Opt.* 39: 3321–3332.

Kadkhoda, P., Müller, A., and Ristau D. 2000b. Total scatter losses of optical components in the DUV/VUV spectral range. *Proc. SPIE* 3902: 118–127.

Kassam, S., Duparré, A., Hehl, K., Bussemer, P., and Neubert J. 1992. Light scattering from the volume of optical thin films: Theory and experiment. *Appl. Opt.* 31: 1304–1313.

Kienzle, O., Staub, J., and Tschudi T. 1994. Description of an integrated scatter instrument for measuring scatter losses of superpolished optical surfaces. *Meas. Sci. Technol.* 5: 747–752.

Krasilnikova, A. and Bulir J. 2005. A novel instrument for measurement of low-level scattering from optical components in the UV region. *Proc. SPIE* 5965: 59651I.

Krywonos, A., Harvey, J. E., and Choi N. 2011. Linear systems formulation of scattering theory for rough surfaces with arbitrary incident and scattering angles. *J. Opt. Soc. Am. A* 28: 1121–1138.

Leonhard, T. A. and Rudolph P. 1993. BRDF round robin test of ASTM E 1392. *Proc. SPIE* 1995: 285–293.

Macleod, H. A. 2001. *Thin-Film Optical Filters*, 3rd edn. Philadelphia, PA: Institute of Physics Publishing.

Maradudin, A. A. 2007. *Light Scattering and Nanoscale Surface Roughness*. New York: Springer.

Maure, S., Albrand, G., and Amra C. 1996. Low-level scattering and localized defects. *Appl. Opt.* 35: 5573–5582.

Neubert, J., Seifert, T., Czarnetzki, N., and Weigel T. 1994. Fully automated angle resolved scatterometer. *Proc. SPIE* 2210: 543–552.

Newell, M. P. and Keski-Kuha R. A. M. 1997. Extreme ultraviolet scatterometer: Design and capability. *Appl. Opt.* 36: 2897–2904.

Nickel, D., Fleig, C., Erhard, A. et al. 2003. Results of a round-robin experiment on reflectivity measurements at a wavelength of 1.06 μm. *Proc. SPIE* 4932: 520–526.

Orazio, F. D., Silva, R. M., and Stockwell W. K. 1983. Instrumentation for a variable angle scatterometer. *Proc. SPIE* 384: 123–132.

Otani, M., Biro, R., Ouchi, C. et al. 2002. Development of optical coatings for 157-nm lithography. II. Reflectance, absorption, and scatter measurement. *Appl. Opt.* 41: 3248–3255.

Palmier, S., Luc Rullier, J., Capoulade, J., and Natoli J. Y. 2008. Effect of laser irradiation on silica substrate contaminated by aluminum particles. *Appl. Opt.* 47: 1164–1170.

Protopapa, M. L., De Tomasi, F., Perrone, M. R. et al. 2001. Laser damage studies on MgF$_2$ thin films. *J. Vac. Sci. Technol. A* 19: 681.

Ren, G., Cai, B., Zhang, B., Xiong, S., Huang, W., and Gao, L. 2006. Study on precise measurement of high reflectivity by cavity ring-down spectroscopy. *Proc. SPIE* 6150: 61500O.

Ristau, D. Günster, S. Bosch S. et al. 2002. Ultraviolet optical and microstructural properties of MgF2 and LaF3 coatings deposited by ion-beam sputtering and boat and electron-beam evaporation. *Appl. Opt.* 41: 3196–3204.

Rönnow, D. 1997. Determination of interface roughness cross correlation of thin films from spectroscopic light scattering measurements. *J. Appl. Phys.* 81: 3627–3636.

Rönnow, D. and Roos, A. 1995. Correction factors for reflectance and transmittance measurements of scattering samples in focusing Coblentz spheres and integrating spheres. *Rev. Sci. Instrum.* 66: 2411–2422.

Rönnow, D. and Veszelei, E. 1994. Design review of an instrument for spectroscopic total integrated light scattering measurements in the visible wavelength region. *Rev. Sci. Instrum.* 65: 327–334.

Schiff, T. F., Knighton, M. W., Wilson, D. J., Cady, F. M., Stover, J. C., and Butler, J. J. 1993. Design review of a high accuracy UV to near IR scatterometer. *Proc. SPIE* 1995: 121–130.

Schmidl, G., Paa, W., Triebel, W., Schippel, S., and Heyer, H. 2009. Spectrally resolved cavity ring down measurement of high reflectivity mirrors using a supercontinuum laser source. *Appl. Opt.* 48: 6754–6759.

Schröder, S., Duparré, A., Coriand, L., Tünnermann, A., Penalver, D. H., and Harvey J. E. 2011c. Modeling of light scattering in different regimes of surface roughness. *Opt. Expr.* 19: 9820–9835.

Schröder, S., Duparré, A., and Tünnermann, A. 2008. Roughness evolution and scatter losses of multilayers for 193 nm optics. *Appl. Opt.* 47: C88–C97.

Schröder, S., Feigl, T., Duparré, A., and Tünnermann, A. 2007. EUV reflectance and scattering of Mo/Si multilayers on differently polished substrates. *Opt. Expr.* 15: 13997–14012.

Schröder, S., Gliech, S., and Duparré, A. 2005a. Sensitive and flexible light scatter techniques from the VUV to IR regions. *Proc. SPIE* 5965 1B: 1–9.

Schröder, S., Gliech, S., and Duparré, A. 2005b. Measurement system to determine the total and angle-resolved light scattering of optical components in the deep-ultraviolet and vacuum-ultraviolet spectral regions. *Appl. Opt.* 44: 6093–6107.

Schröder, S., Herffurth, T., Blaschke, H., and Duparré A. 2011a. Angle-resolved scattering: An effective method for characterizing thin-film coatings. *Appl. Opt.* 50: C164–C171.

Schröder, S., Herffurth, T., Duparré, A., and Harvey J. E. 2011b. Impact of surface roughness on the scatter losses and the scattering distribution of surfaces and thin film coatings. *Proc. SPIE* 8169: 81690R.

Schröder, S., Herffurth, T., Trost, M., and Duparré, A. 2010. Angle-resolved scattering and reflectance of extreme-ultraviolet multilayer coatings: Measurement and analysis. *Appl. Opt.* 49: 1503–1512.

Schröder, S., Kamprath, M., Duparré, A., Tünnermann, A., Kühn, B., and Klett, U. 2006. Bulk scattering properties of synthetic fused silica at 193 nm. *Opt. Expr.* 14: 10537–10549.

Schröder, S., Uhlig, H., Duparré, A., and Kaiser, N. 2005c. Nanostructure and optical properties of fluoride films for high-quality DUV/VUV optical components. *Proc. SPIE* 5963: 231–240.

Schröder, S., Unglaub, D., Trost, M., Cheng, X., Zhang, J., and Duparré, A. 2014. Spectral angle resolved scattering of thin film coatings. *Appl. Opt.* 53: A35–A41.

Stearns, D., G. Gaines, D. P., Sweeney, D. W., and Gullikson E. M. 1998. Nonspecular x-ray scattering in a multilayer-coated imaging system. *J. Appl. Phys.* 84: 1003–1028.

Stemmler, I. 2005. Angular dependent specular reflectance in UV/Vis/NIR. *Proc. SPIE* 5965: 468–478.

Stenzel, O. 2005. *The Physics of Thin Film Optical Spectra: An Introduction.* New York: Springer Series in Surface Sciences (Springer).

Stenzel, O., Wilbrandt, S., Friedrich, K., and Kaiser N. 2009. Realistische Modellierung der NIR/VIS/UV-Optischen Konstanten Dünner Optischer Schichten im Rahmen des Oszillatormodells. *Vak. Forsch. Prax.* 21: 15–23.

Stover, J. C. 1995. *Optical Scattering: Measurement and Analysis,* 2nd edn. Bellingham, WA: SPIE—The International Society for Optical Engineering. pp. 3–27, 133–175.

Trost, M., Herffurth, T., Schmitz, D., Schröder, S., Duparré, A., and Tünnermann, A. 2013. Evaluation of subsurface damage by light scattering techniques. *Appl. Opt.* 52: 6579–6588.

Trost, M. Schröder, S., Feigl, T., Duparré, A., and Tünnermann A. 2011. Influence of the substrate finish and thin film roughness on the optical performance of Mo/Si multilayers. *Appl. Opt.* 50: C148–C153.

Truckenbrodt, H., Duparré, A., and Schuhmann U. 1992. Roughness and defect characterization of optical surfaces by light scattering measurements. In *Specification and Measurement of Optical Systems,* ed. L. R. Baker, *Proc. SPIE* 1781: 139–151.

van de Hulst, H. C. 1957. *Light Scattering by Small Particles.* New York: John Wiley & Sons.

Voss, A., Plass, W., and Giesen A. 1994. Simple high-precision method for measuring the specular reflectance of optical components. *Appl. Opt.* 33: 8370–8374.

Wang, J., Maier, R., Dewa, P. G., Schreiber, H., Bellman, R. A., and Dawson Elli, D. 2007. Nanoporous structure of a GdF_3 thin film evaluated by variable angle spectroscopic ellipsometry. *Appl. Opt.* 46: 3221–3226.

Zerrad, M., Lequime, M., and Amra, C. 2011. Multimodal scattering facilities and modelization tools for a comprehensive investigation of optical coatings. *Proc. SPIE* 8169: 81690K.

Zerrad, M., Lequime, M., and Amra, C. 2014. Spatially resolved surface topography retrieved from far-field intensity scattering measurements. *Appl. Opt.* 53: A297–A304.

Chapter 8

Materials, Surfaces, Thin Films, and Contamination

9. Quartz and Glasses

Laurent Lamaignère

9.1 Introduction

Damage may occur either in the bulk or at the surface of an optical component due to intrinsic or extrinsic factors. Intrinsic processes include linear absorption, color-center formation, and nonlinear process such as self-focusing, multiphoton absorption, and electron avalanche breakdown. But undoubtedly the weakest parts in a laser system are the surfaces of the optical components. Surfaces have a lower damage threshold than the bulk.

Laser-Induced Damage in Optical Materials. Edited by Detlev Ristau © 2015 CRC Press/Taylor & Francis Group, LLC. ISBN: 978-1-4398-7216-1.

Chapter 9

One reason stems from the fact that after the polishing process of optical components, even good-quality polishing process, there are residual scratches, defects, and imperfections on the surface. In both cases, there are different ways of defining the damage phenomenon. One might define damage as the physical appearance of a defect in the material or by a degradation in the output performance of the laser system. From the point of view of the user, the performance deterioration and the lifetime of the optics are of more importance than the physical appearance of a defect in the material. The occurrence of a small defect does not systematically alter the laser performance. It is then crucial to determine whether the damage site remains constant or increases with time.

The aim of this chapter is to give the main features encountered during bulk and surface damages of quartz and glasses in general. Damage morphologies and parameters dependence (wavelength, pulse length) are presented. The following paragraph is devoted to bulk damage. Next, a more substantial paragraph deals with the surface damage. The latter is actively studied in the damage community, specifically in the framework of high-power lasers, where the maintenance cost is directly linked to the surface degradation and then to the performance decrease. Therefore, damage morphologies and growth phenomena are presented in detail. In the last paragraph, the occurrence of damage both in the bulk and on surfaces is discussed in the light of nonlinear effects in thick components.

9.2 Bulk Damage

9.2.1 Introduction

The damage phenomenon of an optical component is usually determined by the surface (subsurface polishing layer or coating) rather than the bulk material. Internal damage occurs relatively seldom, apart from the specific effect of self-focusing (Bercegol et al. 2003). The correlation between self-focusing and the increase in the rear surface damage is given in Section 9.4.2. Intrinsic processes that limit the laser resistance of materials include linear absorption, color-center formation, and a variety of nonlinear processes. Extrinsic factors include impurities and material defects (dislocations, voids). However, optical components exhibit few inhomogeneities, which generally form in the process of material production. The inhomogeneities consist of small bubbles, dielectric inclusions, and platinum particles in the case of laser glass.

9.2.2 Intrinsic Bulk Damage

The fundamental mechanism responsible for damage in an initially pure, nonabsorbing material is avalanche breakdown and/or electronic heating. It starts from an initial conduction band electron in the solid that is accelerated by the collisions with optical phonons and the electric field of the electronic wave. When the field is sufficiently intense, the electron may be accelerated, so that it can create a new conduction electron by ionization or by creation of an electron–hole pair. The electrons transmit a fraction of their energy to optical phonons that, in turn, lead to heating of the lattice. For glasses, the threshold for avalanche breakdown should lie above 10^3 J/cm^2 for a ns pulse. In the absence of inclusions, in a pure glass-host material, heating by multiphoton absorption and the Brillouin backscattering (SBS) are other possible mechanisms. But, the damage threshold for these processes appears to be higher than experimental thresholds.

9.2.3 Defects-Induced Bulk Damage

Most materials contain absorbing impurities. These may originate either from the raw materials or from the fabrication processes and can be minimized using appropriate fabrication technology. Although most fabrication processes aim to purify the melt, it is also possible to introduce even more damaging material into the lattice from the fabrication process itself. For example, many papers have been published on the presence and effect of platinum defects in melted Nd:glass (NMAB 1970). These microscopic Pt particles originate from the Pt crucible and are dispersed throughout the glass. Appropriate gas treatment has been shown to eliminate or melt these particles into even smaller particles. If the particle is large enough (>10λ of irradiation), the particle will heat up, add to the absorption coefficient, and act as scattering center. If the particle size is between 10λ and 0.1λ, the particle will absorb and strain the lattice. If the particle size is lower than 0.1λ, then it will only add to the absorption but not lead to either the scatter or the strain in the lattice. The problem of platinum inclusions has been solved by glass manufacturers either by using crucibles of ceramic and clay materials or by carefully controlling the oxygen content in the melt. The presence of inclusions is a statistical problem. Then, the bulk damage (Hopper and Uhlmann 1970) is characterized by the number of damage sites per unit volume as a function of incident-energy density.

In addition to Pt particles, nonmetallic inclusions are sometimes present: glassy and bubbles (gaseous). Gaseous inclusions arise from gas occluded during the melting process: they rarely cause damage in lasers. Glassy inclusions arise from emulsion formation and melting of crystals without complete solution. Results from literature indicate that metallic inclusions are more dangerous than nonmetallic inclusions. The general mechanism of damage (Bliss 1971) is due to the absorption of the energy from the beam by an inclusion with a high absorption coefficient. The inclusion becomes heated relative to the glass. The resulting tendency toward thermal dilatation stresses the surrounding glass, sufficiently for fracture to occur. The key parameters are the absorption coefficient for the incident radiation, the emissivity, the physical dimensions and orientation to the incident beam, the thermal conductivity, and heat capacity. The properties of the surrounding glass affecting damage resistance are thermal conductivity and heat capacity.

9.2.4 Wavelength Dependence

In the nanosecond regime, Kuzuu (1999) reports the laser-induced damage threshold (LIDT) of various synthetic fused silica at 1064, 532, 355, and 266 nm. In the range of 1064–355 nm, LIDT is proportional to the 0.43rd power of the wavelength:

$$I_{th} = 1.45\lambda^{0.43}. \tag{9.1}$$

At 266 nm, the LIDT is lower than expected by this power law. This difference is explained by the damage mechanism; at 266 nm, two-photon absorption-induced defects lower the LIDT as in the case of 355 nm laser-induced defects, whereas at longer wavelengths

Chapter 9

this process does not occur. At wavelengths shorter than half of the bandgap energy of amorphous SiO_2 ($\approx$9 eV), that is, 4.5 eV corresponding to λ = 280 nm, the laser damage occurs in the two-photon process. Furthermore, the bandgap of fused silica can strongly vary due to manufacturing process, for example, it will strongly depend on the OH content, any dopants like fluorine and stoichiometric imperfections resulting in strong absorption close to the intrinsic bandgap. Some differences are observed in the LIDT between fused silica and synthetic fused silica that are attributed to the effect and concentration of metallic impurities (like Al content in ppm). At 1064 nm, no difference is observed for several synthetic fused silica. This suggests that the LIDT is independent of impurities and defect structures in amorphous SiO_2.

9.2.5 Temporal Regime (from ns to fs)

Two regimes are generally distinguished. For pulses longer than about 100 ps, damage is caused by conventional heat deposition, resulting in melting and boiling. Energy is absorbed by conduction band electrons from the external field and transferred to the lattice. In this regime, the source of the initial seed electrons is local defects or impurities. The temperature near the local absorption centers is then sufficiently high to cause fracture and melting. Thermal conduction and diffusion into the lattice being the two limiting processes for temperature increase, the fluence required to raise the heated area to some critical temperature is proportional to $\tau^{0.5}$ (τ being the pulse duration). As the pulse width decreases below 100 ps, a gradual transition takes place from the long pulse, thermally dominated regime, to an ablative regime dominated by collisional and multiphoton ionization (Stuart et al. 1995). For pulses shorter than about 20 ps, electrons have no time to couple to the lattice during the laser pulse. The LIDT continues to decrease with decreasing pulse width, but at a rate slower than $\tau^{0.5}$. The damage morphology is then the characteristic of surface ablation. The damage site is limited to only a small area where the laser intensity is sufficient to produce a plasma without crack propagation.

9.3 Surface Damage

9.3.1 Damage Initiation

9.3.1.1 Front versus Rear Surface

Thermal damage to the front face of a component may either be in the form of a molten crater, line cracks, or catastrophic breaking. If the material melts before it cracks (short-pulse damage to an absorbing material), then it may protect the bulk of the component due to the plasma formation. Thermal damage inside a component is usually catastrophic unless the damage is initiated by small absorbing inclusions (e.g., Pt inclusion in laser glass, and misoriented crystallites in KDP). When damage is initiated on the rear surface of a component, it is usually far more evident than when it is initiated on the front face due to the fact that the damage occurs inside the sample and a hole is usually blasted out of the material.

The morphology of dielectric damage to the front face is usually minimal as any ejected material vaporizes in the form of a plasma, and this absorbs the incoming radiation. Damage in the bulk is usually catastrophic and is often in the form of self-focusing trails. Damage to the rear surfaces of materials and components with uncoated surfaces is lower than that of the front surfaces because of constructive interference between the incident and reflected beams. The peak power is enhanced (and the LIDT lowered) by the factor (n being the refractive index):

$$F = \frac{\text{Exit surface electric field}}{\text{Entrance surface electric field}} = \frac{4n^2}{(n+1)^2}. \tag{9.2}$$

As the electric field enhancement maximum is forward of the rear surface, the plasma formed is inside the material. In consequence of this, the rear surface damage is usually more catastrophic than the front surface damage. Moreover, the observation that an optical component illuminated by a collimated beam will usually damage at the exit surface at a lower fluence that at the entrance surface is also explained by the effect of the Fresnel reflection.

At normal incidence, there is a reflection at the air-to-sample boundary causing a phase shift of 180°, which results in a partially interference of the two light waves within a distance of $\lambda/4$ of the entrance surface. Then, the light intensity on the front surface (I_F) is related to the incident intensity I_0 by the relation

$$I_F = \frac{4}{(n+1)^2} I_0. \tag{9.3}$$

At the rear surface, there is no phase shift and the intensity (I_R) is related to I_0 by

$$I_R = \left(\frac{4n}{(n+1)^2}\right)^2 I_0. \tag{9.4}$$

Thus, the intensity on the rear face is larger than on the front face:

$$I_R = \left(\frac{4n^2}{(n+1)^2}\right) I_F. \tag{9.5}$$

For example, for $n = 1.55$, the ratio is 1.48.

Morphologies between front and rear damages are different following this concept. On the rear face, a plasma is formed and confined between the substrate and the defect. The density of the plasma and the pressure cause the particle to be ejected; then, the initial small pit leads to a large damage. On the front face, the plasma shields the incoming light and only a weak shockwave is launched into the substrate; the front surface is weakly damaged. Due to the occurrence of the rear damage event at lower fluences than those of front surface, the difference between front and rear surface damage thresholds is difficult to measure because of the destruction of the optic under test before the occurrence of front surface damage.

9.3.1.2 Defects and Mechanisms

Laser-induced damage in wide bandgap optical materials is the result of material modifications arising from extreme conditions occurring during this process. The material absorbs energy from the laser pulse and produces an ionized region that gives rise to broadband emission. Damage initiation from nanosecond pulses arises from defects located in nominally transparent material absorbing sufficient energy to produce an opaque plasma. The nature of such absorbers and the mechanism(s) by which they produce a plasma may vary among different materials. The appearance of surface damage on optical components is caused mainly by the presence of residual defects of submicron size in the glass, such as microscopic fractures stemming from polishing, bubbles, and metal or dielectric inclusions (Natoli et al. 2002). These defects absorb energy from the laser and serve as nuclei for damage, which appears at interfaces in the form of craters following the laser pulse (Bonneau et al. 2002).

It is essential to note that the size of the ionized region and resultant damaged region can be much larger than, and independent of, the size and nature of the precursors. Many studies have pushed forward the role of fractures in the generation of laser damage at the surface of silica optics (Salleo et al. 1998). Damage initiation was attributed to electromagnetic field enhancement (Génin et al. 2001) or to the presence of trapped contaminants in the crack (Feit and Rubenchik 2004). Recently, a 1D model of laser-microfracture interaction was developed by Bercegol (Bercegol et al. 2007) that takes into account the presence of a monomolecular absorbing layer inside microfracture. The induced absorption allows the local heating of silica up to the vaporization temperature. Close to the vaporization threshold, silica undergoes structural changes likely to induce important changes on electronic properties. Then, the pressure level reached in the material leads to irreversible densification associated with a systematic damage.

In the (D)UV, for rear surface damage, one should also consider long-term effects due to the effect of compaction in fused silica.

9.3.1.3 Damage Morphology

Damage sites include a crater (core region) from which most of the material has been ejected earlier in the damage process. The crater morphology is similar from 1 to 5 ns pulses, with evidence of resolidified molten material as a consequence of localized extreme conditions of temperature and pressure created during and shortly after each laser pulse in the growth sequence. The outside border of the sites contains mechanically damaged material extending out to tens and hundreds of microns (Negres et al. 2010).

The size and morphology of 351 nm damage sites in silica are strongly dependent on the laser fluence, number of pulses, and nature of the optic surface, for example, regular polish versus super polish. Qualitatively, the super polish surface is more resistant to damage in terms of fluence threshold and extent of the damage site. Quantitatively, the diameter of the damage crater increases monotonically with the number of pulses and is also a function of the laser fluence. The damage growth rate is also a function of the nature of the silica surface.

The morphology of damage regions in fused silica resulted from irradiation of a pulse of high fluence 3ω light consists of a collection of erosion pits (craters) spanning a

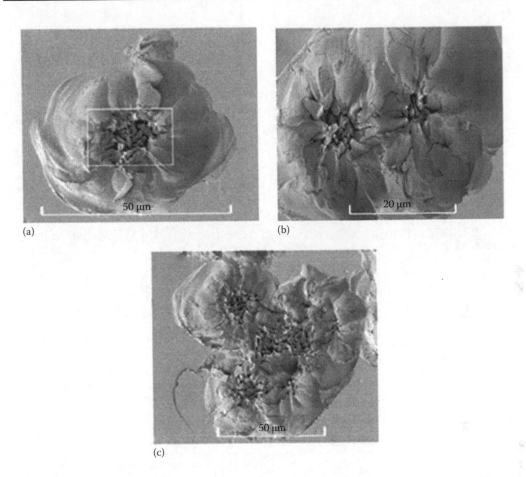

(a)

(b)

(c)

FIGURE 9.1 SEM micrograph of damage sites on a fused silica induced by a 3ω pulse. (a) Isolated crater, (b) doublet, and (c) multiplet. (From Wong, J. et al., *J. Non-Cryst. Solids*, 352, 255, 2006.)

fraction of the beam size on sample. The craters may be isolated as shown in Figure 9.1a, or a doublet shown in Figure 9.1b, or as a cluster of multiple craters shown in Figure 9.1c (Wong et al. 2006).

SEM examinations at higher magnification revealed that these damage craters consist of a core of molten nodules with a botryoidal morphology (Figure 9.2a) and fibers (Figure 9.2b) evident of a thermal explosion followed by quenching from the liquid state at this high-energy absorption region. The molten core is surrounded concentrically by an annulus of fractured material, indicative of mechanical damage accompanied by spallation of material as evident in Figure 9.2a. At a higher magnification, micron and submicron digs, pits, and surface cracks are also identified.

There is a general surface depression of ~10 nm with respect to the surface RMS value. There is a 5 nm pileup at a rim with a diameter of ~150 μm.

Moreover, nanocracks varying from 10 to 200 nm wide are observed at regions in the bulk material well below the bottom of the crater. Median cracks at the bottom of the damage crater and lateral cracks flanking on both sides are present (Figure 9.3).

Wong (Wong et al. 2006) correlates the current densification result with the equation of the state of fused silica from shock measurements: fused silica is highly compressible

Chapter 9

(a)   (b)

FIGURE 9.2   SEM micrograph of the center area of the damage crater shown in Figure 14.2a showing (a) a molten and (b) fibrous morphology. (From Wong, J. et al., *J. Non-Cryst. Solids*, 352, 255, 2006.)

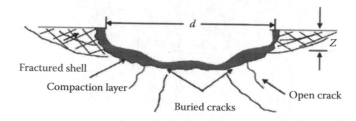

FIGURE 9.3   Schematics of damage crater model, morphology, microstructure, and point defects induced in fused silica by 3ω laser pulses at high fluences. (From Wong, J. et al., *J. Non-Cryst. Solids*, 352, 255, 2006.)

to ~30 GPa. Assuming negligible density relaxation after the passage of the shock wave, it may be inferred that the compaction of fused silica by 351 nm laser irradiation occurs at ~10 GPa. Results from Sugiura et al. (1997) show that permanent densification in fused silica starts at ~8.8 GPa and completes at 16 GPa.

The model proposed by Wong (Wong et al. 2006) postulates the absorption of laser energy at a subsurface nanoparticle in the glass and "instantaneous" energy release due to a possible thermal explosion. Three regions of interest are discernible. (1) An inner region in which material is subject to high pressure and high temperature to yield a resultant molten morphology. This region is characterized by less hydrogen and undergoes densification with a high concentration of oxygen-deficient center (Stevens-Kalceff 2002). (2) A second region, annular to the first, has spalled material due to the shock wave reflected off the free surface. This region has a fractured morphology. (3) An outermost region with high plastic deformation causing material cracking but no ejection.

The tomographic identification of a compaction layer at the bottom and wall of the damage crater substantiates the occurrence of a thermal explosion process that generates the shock wave needed for the compaction process. The work by Wong also renders a quantitative validation for the high compaction (20%) in fused silica (Wong et al. 2006).

In the case of repetitive illuminations at moderate fluences, cumulative effects of pulses numbers of 10^6–10^8 lead to compaction and/or rarefaction phenomena.

9.3.1.4 Pulse Length Dependence

The effect of pulse duration (in the ns regime) on damage initiation is well known and is often described by pulse scaling for both density and probabilistic measurements. Equation 14.6 gives the use of pulse scaling to damage density measurements:

$$\rho_2(\phi) = \rho_1 \left(\phi \left(\frac{\tau_1}{\tau_2} \right)^{\alpha} \right), \tag{9.6}$$

where ρ_1 and ρ_2 are the densities of damage sites produced by two pulses of the same fluence with durations of τ_1 and τ_2, respectively. The parameter α is the power or pulse scaling coefficient.

For pulses longer than a few tens of picoseconds, the generally accepted picture of damage to defect-free dielectrics involves the heating of conduction band electrons by the incident radiation and transfer of this energy to the lattice. Damage occurs via conventional heat deposition resulting in the melting and boiling of the dielectric material. Because the controlling rate is that of thermal conduction through the lattice, this model predicts a $\tau^{1/2}$ dependence of the threshold fluence upon pulse duration τ, in reasonably good agreement with numerous experiments that have observed a τ^{α} scaling with $0.4 < \alpha < 0.6$ (Stuart et al. 1995, Lamaignere et al. 2010) in a variety of dielectric materials from 100 ps to 10 ns (Figure 9.4).

A gradual transition from the long-pulse, thermally dominated regime to an ablative regime dominated by collisional and multiphoton ionization, and plasma formation is obtained while the laser damage threshold decreases. A general theoretical model of laser interaction with dielectrics, based on multiphoton ionization, Joule heating, and collisional (avalanche) ionization, is shown to be in good agreement with the data in the short-pulse regime and over a broad range of laser wavelength. This model is consistent with the observations of critical density plasmas produced

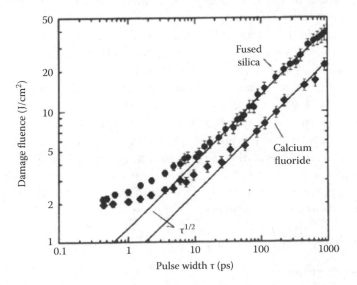

FIGURE 9.4 Observed values of damage threshold at 1053 nm for fuses silica (circles) and CaF_2 (diamonds). Solid lines are $\tau^{1/2}$ fits to long pulse results. (From Stuart, B.C. et al., *Phys. Rev. Lett.*, 74, 2248, 1995.)

Chapter 9

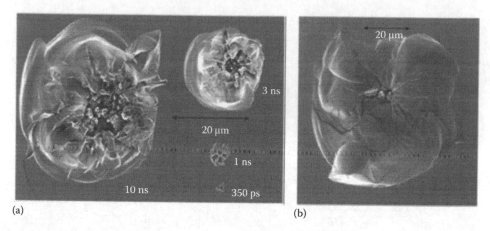

FIGURE 9.5 SEM observations of exit surface damage of fused silica using (a) Nd:Yag laser (short pulse length) and (b) excimer laser (long pulse duration). (From Carr, C.W. et al., *Opt. Expr.*, 19, A859, 2011.)

by ultrashort pulses (Figure 9.5). The damage threshold continues to decrease with decreasing pulse width, but at a rate slower than $\tau^{1/2}$ in the range of 0.1–20 ps. This deviation is accompanied by a qualitative change in the damage morphology indicative of rapid plasma formation and surface ablation. The damage site is limited to only a small region where the laser intensity is sufficient to produce a plasma with essentially no collateral damage. A theoretical model, in which initial electrons provided by multiphoton ionization are further heated resulting in collisional (avalanche) ionization, predicts short-pulse damage. For extremely short pulses ($\tau < 30$ fs), multiphoton ionization alone provides the critical density of electrons.

9.3.1.5 Damage Size versus Beam Parameters

Numerous observations lead to the following rule of thumb (Bertussi et al. 2009): laser damage sites have depth less than half the diameter of the damage site (Figure 9.6). Secondly, the distribution of damage sizes shifts to lower values when the pulse duration becomes shorter (Carr et al. 2007; Bertussi et al. 2009), see Figures 9.6 and 9.7, and when the fluence becomes smaller (Carr et al. 2007). These results are consistent with the model described by Rubenchik and Feit (2002).

The size of the damage is seen to increase with a power of pulse duration, the exponent being close to 1 (0.7). For a fixed pulse duration, the sizes of the sites are seen to increase with fluence. The approximately linear dependence of damage site size on pulse duration suggests that the physical processes that govern energy deposition and destructive heating are initiated at a precursor, a region of highly absorbing material grows outward from it with a nearly constant velocity. This growth may be driven by a shock wave or a process like laser-supported detonation in which there is a catastrophic increase in absorption at high temperature or pressure (Carr et al. 2007).

9.3.2 Damage Growth

9.3.2.1 Introduction

Steady growth is observed mainly on surfaces of optical components. Damage sites in the bulk of materials reach a limited size. Small micron-size zones created in KDP crystals

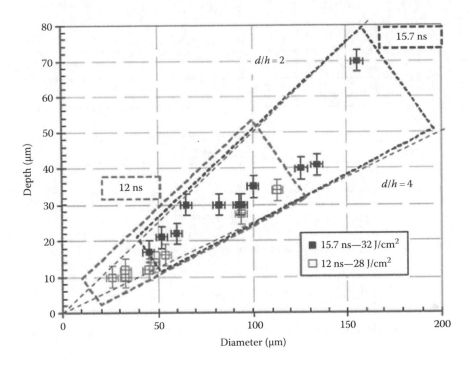

FIGURE 9.6 Damage depth including subsurface cracks versus diameter for two different nanosecond pulse durations ($F = 14$ J/cm² Equation 9.3 ns). (From Bertussi, B. et al., *Opt. Expr.*, 17, 11469, 2009.)

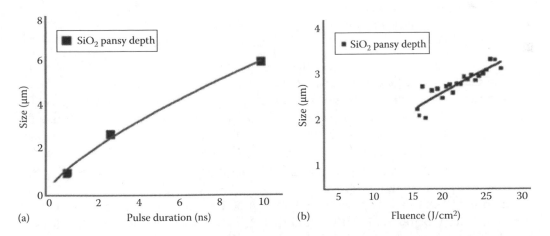

FIGURE 9.7 (a) Damage sites size versus pulse duration for SiO₂. (b) Damage site size versus fluence for 3 ns pulses. (From Carr, C.W. et al., *Proc. SPIE*, 6403, 64030K, 2007.)

keep almost the same size with following shots (Runkel et al. 1999). In Nd-doped laser glass, micron-size platinum inclusions generate damage sites that first grow but stabilize in size at several hundred microns in diameter (Razé et al. 2003).

The exit surface damage growth has been generally described by an exponential increase in diameter with the number of shots at fixed fluence (Figure 9.8):

$$d_N = d_0 \cdot \exp[\alpha_d(\phi) \cdot N] \tag{9.7}$$

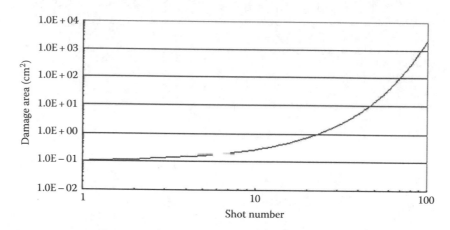

FIGURE 9.8 Exponential growth of the damage area following Equation 9.7.

or when considering the area (A) of the damage

$$A_N = A_0 \cdot \exp[\alpha_A(\phi) \cdot N] \tag{9.8}$$

with $\alpha_A = 2\alpha_d$

where
　　d is the site diameter (in μm) measured after the ith shot
　　N is the total number of shots at fixed fluence ϕ (in J/cm²)
　　α is the average exponential growth coefficient (dimensionless)

In practice, the α coefficient is found by plotting the measured site diameter during the growth sequence versus shot number and fitting the data to an exponential curve. For a given pulse duration, this procedure is repeated for many sites grown at discrete fluences to reveal the average growth trend, that is, the fluence dependence of the growth coefficient. For the case of long pulse durations, from 5 up to 15 ns, the exit surface growth exhibits the classic exponential behavior in most cases, in agreement with previous results by numerous groups.

In contrast, exit surface growth with shorter (1–2 ns) pulses is, in general, very different. Negres (Negres et al. 2010) found that site diameter increases linearly with shot number. A multiplicative growth factor as in Equation 9.7 is no longer appropriate to describe this linear dependence, but rather is described by an additive term to quantify the incremental growth in diameter with shot number as follows:

$$d_N = d_0 + g(\phi) \cdot N, \tag{9.9}$$

where g is the average linear growth coefficient with the same units as the diameter, that is, micrometer.

The multishot growth behavior, that is, average growth coefficients, as a function of fluence and pulse duration for the dominant behaviors, is then linear and exponential

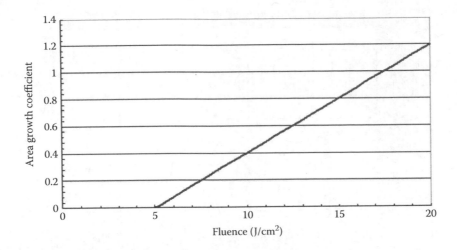

FIGURE 9.9 Growth coefficient versus fluence following Equation 14.9.

for short and long pulses, respectively. Previous growth studies at 351 nm were conducted primarily with longer pulses (i.e., 3–10 ns, near Gaussian temporal profiles) and established that the average exponential growth coefficient defined in Equations 19.7 and 19.8 increases linearly with fluence as follows (Figure 9.9):

$$\alpha(\phi) = A \cdot (\phi - B), \tag{9.10}$$

where B is the fluence growth threshold, and A is a constant.

9.3.2.2 Wavelength Dependence

Above the experimental growth threshold, the magnitude of the exponential growth coefficient increases linearly with fluence (Figure 9.10) up to 25 J/cm². Above a value, a saturation effect in the growth coefficients appears. For several wavelengths, it is considered that the total absorbed energy determines the damage growth rate. This is realistic if it is assumed that absorption at one or both wavelengths leads to plasma generation during the growth event and that all subsequent incident laser energy is absorbed. When several wavelengths are mixed, the threshold for growth is not altered significantly from the single harmonic threshold provided that it is the total fluence that is considered (Lamaignere et al. 2007; Norton et al. 2006). For example, for the combined 3w with 1w case, once growth is enabled, the growth rate is similar to the 3w only curve and the total energy determines the growth coefficient. This is in contrast to the combined 2w with 1w case, where again it is the total energy that determines the growth coefficient but now it corresponds to the 1w only curve. In both cases, the measured coefficient corresponds to that at the wavelength having the greatest damage growth rate.

9.3.2.3 Temporal Scaling Law

Multishot damage growth is largely exponential in nature for pulses longer than about 2 ns, while shorter pulses (1–2 ns) produce a dominant linear growth. More results from Negres et al. (2009) establish a significant dependence of exit damage growth parameters

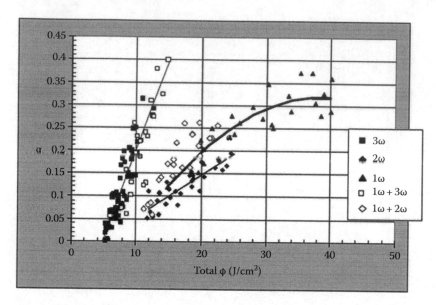

FIGURE 9.10   Growth coefficient for several wavelength and combination of wavelength (Norton, M.A. et al., *Proc. SPIE*, 5991, 599108, 2006.)

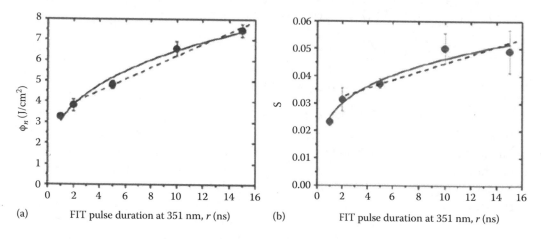

FIGURE 9.11   (a) Growth threshold *B* versus fluence. (b) Rate of increase in the growth coefficient *A* versus fluence. (From Negres, R.A. et al., *Proc. SPIE*, 7504, 750412, 2009.)

on pulse duration from 1 to 15 ns as $\tau^{0.3}$ for sites in the 50–100 μm size range for both the growth threshold *B* and the rate of increase coefficient *A* (see Figure 9.11).

These observations are well supported by a model developed by Bercegol et al. (2007) based on pressure. This model determines the pressure level that can be reached in the material as function of laser intensity, when the temperature at the fracture is high enough for all power laser to be absorbed. Growth with long pulses is primarily due to lateral and radial fractures; in contrast, growth with short pulses proceeds with initiations around the periphery, very similar to the input growth morphology. Sites grown with longer pulses (i.e., 5–15 ns in duration) appear to have most of the laser energy deposited in the central core, as suggested by the presence of molten material and flaking off large chips on the periphery (Bercegol et al. 2007). This type of growth results in an exponential increase in the site diameter (or surface area) with the number of shots. In contrast, for pulses as short as 1 or 2 ns, energy is deposited all across the damage site and growth proceeds by initiation on previously damaged material at the edge of the site.

9.4 Damage due to Nonlinear Effects

9.4.1 Introduction

The use of thick components makes the occurrence of nonlinear effects easier. With single longitudinal mode (SLM) pulses (corresponding to temporally Gaussian pulses), self-focusing occurs together with front surface damage, which can be attributed to a stimulated SBS wave. The use of multiple longitudinal modes (MLM) pulses (leading to temporal instabilities of the laser pulses) suppresses the occurrence of front surface damage and increases self-focusing (SF).

With SLM pulses, the observation of filaments seems coherent with the standard Kerr self-focusing effect and can be understood according to the treatment by Marburger (1975). However, when MLM pulses are used, filaments occur for much smaller peak intensities, by about a factor of 2. This second case is relevant to the situation of vacuum windows in high-power installation, where the laser spectrum is widened to get rid of SBS. For more details on the occurrence of nonlinear effects.

Self-focusing, SBS and front damage in relation to nonlinear optical effects are very complex processes, and there are different influencing parameters of the pulsed laser beam that have to be taken into account as, for example, divergence too. The aim of the following paragraphs is to give some clues about these nonlinear effects and their impacts on front, bulk and rear damage processes. For a detailed understanding of these phenomena, see the references given in the text.

9.4.2 Damage Density versus Thickness: Influence of Kerr Effects

Damage can also occur at points with no weak defects, when the intensity laser gets to very high values. Small-scale self-focusing is a possible cause of rear surface damage in thick optical components; this cause is enhanced by spatial modulation in the beam, caused by components defects or damage sites upstream, and amplified by nonlinear propagation. Figure 9.12 illustrates the rear surface damage of thick components due to the increase in fluence through the optic. Results are obtained for linearly and circularly polarized beams. These results are well fitted (Bercegol et al. 2004) with nonlinear calculations for which nonlinear refractive index ($n2$) values are taken from literature, with a 2/3 factor on the nonlinear index for circularly experiments.

9.4.3 Damage due to SBS

The Brillouin scattering is one of a number of characteristic scattering phenomena that occur when light interacts with solid, liquid, or gaseous media and corresponds to the scattering of light from thermally induced acoustical waves (propagating pressure/

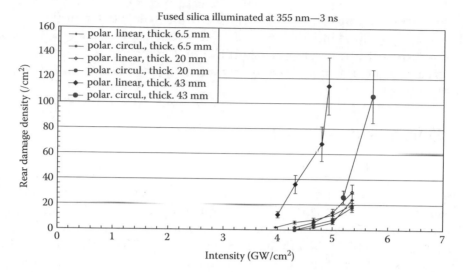

FIGURE 9.12 Rear surface damage density measured by rasterscan on samples of variable thickness, with linear or circular polarization. (From Bercegol, H. et al., *Proc. SPIE*, 5273, 136, 2004.)

density waves) present in media at all temperatures. At normal light levels, the amount of scattering is small. With intense coherent laser light, the rate of scattering can be so large that the acoustic wave amplitudes increase and the scattered light takes on an exponential growth. This regime corresponds to the phenomenon known as SBS. The conversion of incident light into backward scattered light is such that the transmission of even transparent media can be strongly reduced. The scattered light frequency is downshifted from the incident light by the acoustic frequency.

In materials with no optical absorption, SBS is mainly driven by the electrostriction strain produced by an intense laser pulse with long enough (nanosecond) durations. This strain excites acoustic waves on which a Stokes wave scatters a significant amount of energy, preferably in the direction opposite to the pump laser beam. For powerful pump beams, the counterpropagating Stokes wave can convey high enough fluence to cause severe damage.

Experiments performed at 355 nm on thick fused silica samples (thickness 5 cm) have shown front and rear surface damages (Sajer 2004). It is well established that self-focusing is involved in the rear surface damage. The front surface damage is due to an intense backscattered electromagnetic wave as a Brillouin stimulated wave. For parameters identical to the experimental ones, numerical evaluations indicate that the Brillouin gain is about $G_{Brillouin} \sim 30$. The fact that the stimulated Brillouin energy is as large as the incident laser energy explains the entrance face damage.

As example, Figure 9.13 illustrates this phenomenon: front surface damage is much more important as the thickness increases.

9.4.4 Competition between the SBS and the Kerr Effect

A way to suppress the SBS contribution to front surface damage consists to increase the laser bandwidth: a sufficiently large bandwidth reduces SBS to an indistinguishable level. Owing to this, the phase modulation will suppress front surface damage if

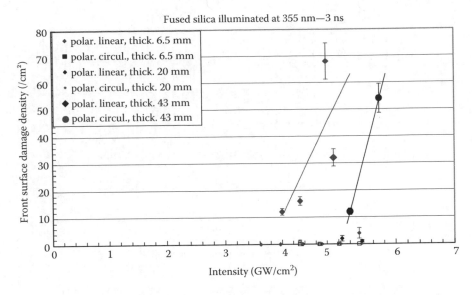

Fused silica illuminated at 355 nm—3 ns

FIGURE 9.13 Front surface damage density measured by rasterscan on samples of variable thickness with linear or circular polarization. (From Bercegol, H. et al., *Proc. SPIE*, 5273, 136, 2004.)

SBS is the only contributor but it will enhance the rear surface damage, the laser pulse being totally transmitted. Multimode broadband pumps, expected to suppress the Brillouin scattering, are observed to noticeably decrease the onset distance of filamentation, which is attributed to a higher Kerr nonlinear index (Bercegol et al. 2005).

For powerful pump beams, the counterpropagating Stokes wave can convey high enough fluence to cause severe damage and SBS appears as a harmful process that limits the pulse energy of high-power laser sources. SLM pulses in the nanosecond range initiated not only rear damage by the Kerr filamentation in fused silica but also front surface damage through the SBS. This competition is well described in Mauger et al. (2011).

An illustration of the competition between SBS and SF is given in Figure 9.14. The products of the intensity of the beam by the distance of self-focusing (Z_f) are reported as function of the peak power in the beam. Lower values are obtained with multimode configurations, meaning that all the intensity in the beam is transmitted through the optic and then the occurrence of self-focusing appears at shorter distances Z_f. At the opposite with the monomode configuration, a fraction of the intensity is converted into SBS wave reducing the intensity that propagates in the bulk: self-focusing effect appears at longer distance Z_f.

The two important conclusions are as follows:

1. Due to SBS, first, front surface damage is likely to happen for thick component with SLM pulses.
2. Second, all the entire beam is not transmitted, a part of the pump beam being converted into the Stokes wave. The quality of the beam at the exit of the optic is likely worse.

Chapter 9

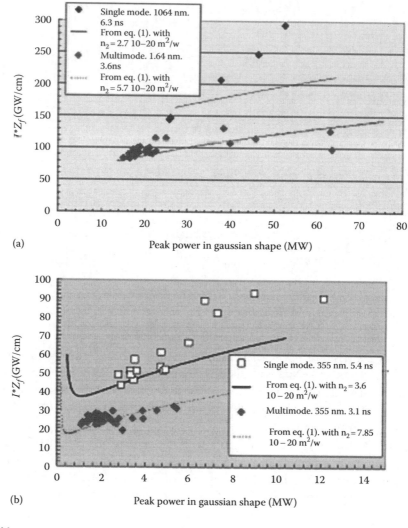

FIGURE 9.14   Intensity by length products $I \times Z_f$ at 1064 nm (a) and 355 nm (b). Theoretical curves are derived from theoretical work by Marburger with different values of nonlinear index. (From Marburger, J.H., *Prog. Quant. Electr.*, 4, 35, 1975; Bercegol, H. et al., *Proc. SPIE*, 5991, 59911Z, 2005.)

References

Bercegol, H., A. Boscheron, C. Lepage, E. Mazataud, T. Donval, L. Lamaignere, M. Loiseau, G. Raze, and C. Sudre. 2004. Self-focusing and surface damage in fused-silica windows of variable thickness with UV nanosecond pulses, *Proceedings of SPIE* 5273:136–144.

Bercegol, H., P. Grua, D. Hébert, and J.P. Morreeuw. 2007. Progress in the understanding of fracture related damage of fused silica, *Proceedings of SPIE* 6720: 672003.

Bercegol, H., L. Lamaignere, V. Cavaro and M. Loiseau. 2005. Filamentation and surface damage in fused silica with single-mode and multi-mode pulses. *Proceedings of SPIE* 5991:59911Z.

Bercegol, H., L. Lamaignere, B. Le Garrec, M. Loiseau, and P. Volto. 2003. Self-focusing and rear surface damage in a fused silica window at 1064 nm and 355 nm. *Proceedings of SPIE* 4932:276–285.

Bertussi, B., P. Cormont, S. Palmier, P. Legros, and J.L. Rullier. 2009. Initiation of laser-induced damage sites in fused silica optical components. *Optics Express* 17:11469.

Bliss E.S. 1971. Pulse duration dependence of laser damage mechanisms. *Opto-Electronics* 3:99.

Bonneau, F., P. Combis, J.L. Rullier, J. Vierne, M. Pellin, M. Savina, M. Broyer et al. 2002. Study of UV laser interaction with gold nanoparticles embedded in silica. *Applied Phyics. B* 75:803.

Carr, C.W., D.A. Cross, M.A. Norton, and R.A. Negres. 2011. The effect of laser pulse shape and duration on the size at which damage sites initiate and the implications to subsequent repair. *Optics Express* 19:A859.

Carr, C.W., M.J. Matthews, J.D. Bude, and M.L. Spaeth. 2007. The effect of laser pulse duration on laser-induced damage in KDP and SiO_2. *Proceedings of SPIE* 6403:64030K.

Feit, M.D. and A.M. Rubenchik. 2004. Influence of sub-surface cracks on laser induced surface damage. 2004. *Proceedings of SPIE* 5273:264–272.

Génin, F.Y., A. Salleo, T.V. Pistor, and L.L. Chase. 2001. Role of light intensification by cracks in optical breakdown on surfaces. *Journal of the Optical Society of America A* 18:2607–2616.

Hopper, R.W. and D.R. Uhlmann. 1970. Mechanism of inclusion damage in laser glass. *Journal of Applied Physics* 41:4023.

Kuzuu, N., K. Yoshida, H. Yoshida, T. Kamimura, and N. Kamisugi. 1999. Laser-induced bulk damage in various types of vitreous silica at 1064, 532, 355, and 266 nm: Evidence of different damage mechanisms between 266-nm and longer wavelengths. *Applied Optics* 38:2510.

Lamaignere L., M. Balas, R. Courchinoux, T. Donval, J.C. Poncetta, S. Reyné, B. Bertussi, and H. Bercegol. 2010. Parametric study of laser-induced surface damage density measurements: Toward reproducibility. *Journal of Applied Physics* 107:023105.

Lamaignere, L., S. Reyne., M. Loiseau, J.C. Poncetta, and H. Bercegol. 2007. Effects of wavelengths combination on initiation and growth of laser-induced surface damage in SiO_2. *Proceedings of SPIE* 6720:67200F.

Marburger, J.H. 1975. Self-focusing: Theory. *Progress in Quantum Electronics* 4:35–110.

Mauger S., L. Berge, and S. Skupin. 2011. Controlling the stimulated Brillouin scattering of self-focusing nanosecond laser pulses in silica glasses. *Physical Review A* 83:063829.

Natoli, J.Y., L. Gallais, H. Akhouayri, and C. Amra. 2002. Laser-induced damage of materials in bulk, thin-film, and liquid forms. *Applied Optics* 41:3156.

Negres, R.A., M.A. Norton, D.A. Cross, and C.W. Carr. 2010. Growth behavior of laser-induced damage on fused silica optics under UV, ns laser irradiation. *Optics Express* 18:19966.

Negres, R.A., M.A. Norton, Z.M. Liao, D.A. Cross, J.D. Bude and C.W. Carr. 2009. The effect of pulse duration on the growth rate of laser-induced damage sites at 351 nm on fused silica surfaces. *Proceedings of SPIE* 7504:750412.

NMAB National Materials Advisory Board. 1970. *Fundamentals of Damage in Laser Glass*. NMAB-271, Washington, DC.

Norton, M.A., E.E. Donohue, M.D. Feit, R.P. Hackel, W.G. Hollingsworth, A.M. Rubenchik, and M. Spaeth. 2006. Growth of laser damage in SiO_2 under multiple wavelength irradiation. *Proceedings of SPIE* 5991:599108–599111.

Razé, G., M. Loiseau, M. Josse, D. Taroux, and H. Bercegol. 2003. Growth of damage sites due to platinum inclusions in Nd-doped laser glass irradiated by the beam of a large scale Nd:glass laser. *Proceedings of SPIE* 4932:15–420.

Rubenchick, A.M. and M.D. Feit. 2002. Initiation, growth and mitigation of UV laser induced damage in fused silica. *Proceedings of SPIE* 4679:79–95.

Runkel, M., W. Williams, and J. DeYoreo. 1999. Predicting bulk damage in NIF triple harmonic generators. *Proceedings of SPIE* 3578:322–335.

Sajer, J. M. 2004. Stimulated Brillouin scattering and front surface damage. *Proceedings of* SPIE 5273:129.

Salleo, A., F.Y. Génin, J. Yoshiyama, C.J. Stolz, and M.R. Kozlowski. 1998. Laser-induced damage of fused silica at 355 nm initiated at scratches. *Proceedings of SPIE* 3244:341.

Stevens-Kalceff, M.A., A. Stesmans, and J. Wong. 2002. Defects induced in fused silica by high fluence ultraviolet laser pulses at 355 nm. *Applied Physics Letters* 80:758.

Stuart, B.C., M.D. Feit, A.M. Rubenchik, B.W. Shore, and M.D. Perry. 1995. Laser-induced damage in dielectrics with nanosecond to subpicosecond pulses. *Physics Review Letters* 74:2248.

Sugiura, H., R. Ikeda, K. Kondo, and T. Yamaskaya. 1997. Densified silica glass after shock compression. *Journal of Applied Physics* 81:1651.

Wong, J., J.L. Ferriera, E.F. Lindsey, D.L. Haupt, I.D. Hutcheon, and J.H. Kinney. 2006. Morphology and microstructure in fused silica induced by high fluence ultraviolet 3ω (355 nm) laser pulses. *Journal of Non-Crystalline Solids* 352:255–272.

Chapter 9

10. Crystalline Materials for UV Applications

Christian Mühlig and Wolfgang Triebel

10.1 Introduction

This chapter is dedicated to crystalline optical materials for applications in the UV (ultraviolet), DUV (deep ultraviolet), and VUV (vacuum ultraviolet) wavelength range. Furthermore, only bulk effects are considered and the authors refer to other chapters regarding surface, coating, and contamination effects.

In contrast to the previous chapters, the intention of the authors is lying on the material's absorption and absorption changes upon UV/DUV/VUV laser irradiation as part of the functional damage, that is, the undesired change of optical parameters or specifications. The prominently applied crystalline materials nowadays are of very high purity and therefore, the absorption is very low, even down to the VUV spectral region.

The authors like to dedicate this chapter solely to the standard crystalline optical materials. Therefore, nonlinear crystals (such as BBO, LBO, and KDP) as well as laser media (such as doped YAG and YVO$_4$) are not the subject herein. In addition, no effects of elevated or reduced sample temperature will be considered since this is practically only of minor interest for UV/DUV/VUV applications.

10.2 Crystalline Materials for UV/DUV/VUV Applications

When it comes to the selection of relevant materials for UV/DUV/VUV applications, one needs to define at least one criterion to be fulfilled, here with respect to the material's absorption. The authors decided to apply the criterion that the optical band gap of potential optical materials needs to significantly exceed the photon energy of the longest considered UV wavelength, for example, to realize a small signal absorption coefficient of equal or less than 10^{-3} cm^{-1}. The authors came to the decision that the wavelength of 355 nm, obtained

Laser-Induced Damage in Optical Materials. Edited by Detlev Ristau © 2015 CRC Press/Taylor & Francis Group, LLC. ISBN: 978-1-4398-7216-1.

Chapter 10

by third harmonic generation (3ω) of a Nd:YAG laser, is a reasonable value for the long wavelength edge of UV applications since most of the high-energy or high-intensity laser applications such as laser ignition (laser fusion) are covered.

The wavelength of 355 nm corresponds to the photon energy of 3.5 eV. Thus, the relevant optical materials should exhibit a band gap significantly larger than 3.5 eV in order to provide the desired small bulk absorption coefficient or, when even a two-photon absorption (TPA) process is virtually to be neglected, significantly larger than 7 eV (=twice the photon energy at the wavelength 355 nm). Obviously, the optical materials require in addition a very high purity in order to avoid absorbing energy states within the band gap. From the authors' point of view, however, this demand can be met for all the crystals mentioned later when choosing appropriate manufacturing processes and conditions.

With the exception of high-purity fused silica (a-SiO_2), only crystalline materials, mainly alkaline halides and alkaline earth halides, are able to fulfill these requirements for the optical band gap. A summary of those relevant crystalline materials is given in Table 10.1 with respect to their optical band gap values. Furthermore, it is stated whether these crystals are optically isotropic or not.

In the following, typical areas of application are given for the crystal materials from Table 10.1.

Lithium fluoride (LiF) is used for VUV optics and in x-ray monochromators. Although showing the largest band gap among all optical materials, its hygroscopic nature, softness (→difficult polishing), high sensitivity to thermal shock, and the fast high-energy radiation-induced color center formation (Becher et al. 1983; Liberman et al. 2002) prevent from using it broadly. Figure 10.1 shows the spectral position of the *F*-center absorption band in LiF compared with that in MgF_2.

Magnesium fluoride (MgF_2) is widely used from the VUV to the IR wavelength range like for space applications, in UV photomultiplier tubes, and in setups where extreme

Table 10.1 Overview of Appropriate Crystalline Materials for UV Applications, Their Optical Band Gaps, and Birefringence

Crystal	Optical Band Gap (eV)	Birefringence
Lithium fluoride (LiF)	13.6/14.2	No
Magnesium fluoride (MgF_2)	10.8/11.8	Yes
Calcium fluoride (CaF_2)	10.0/11/11.5/12.1	No[a]
Strontium fluoride (SrF_2)	9.4/10.9/11.2	No
Barium fluoride (BaF_2)	9.1/10.5/10.7	No
Sodium fluoride (NaF)	10.5	No
Sapphire (c-Al_2O_3)	8.3/9.9	Yes
Quartz crystal (c-SiO_2)	8.4/9	Yes

Sources: Optical bandgap data taken from different citations: Tomiki, T. and Miyata, T., *J. Phys. Soc. Japan*, 27, 658, 1969; Rubloff, G.W., *Phys. Rev. B*, 5, 662, 1972; Palik, E.D., *Handbook of Optical Constants*, Academic Press, London, U.K., 1998.

[a] At 157 nm, a weak birefringence CaF_2 has been revealed (Burnett et al. 2001; Letz et al. 2003).

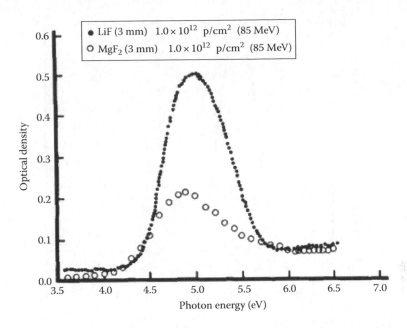

FIGURE 10.1 Shape of *F*-center absorption bands in LiF and MgF_2 irradiated with 85 MeV protons. (Reprinted from *J. Phys. Chem. Solids*, 44, Becher, J., Kernell, R.L., and Reft, C.S., Proton-induced *F*-Centers in LiF and MgF_2, 759–763, Copyright 1983, with permission from Elsevier.)

ruggedness and durability is required. Furthermore, it is an important low refractive index material for DUV-VUV coatings. Due to its birefringence, it is often used for UV-VUV polarization optics. Over many years, it has been applied as window material or mirror substrates for UV-VUV excimer lasers before being replaced by CaF_2 of high purity. The main reason for that change in the materials was the long lifetime of UV-VUV laser–induced color centers in MgF_2, namely, *F*-centers and their aggregates, and their broad absorption band(s) in the UV/DUV spectral region (Blunt and Cohen 1967; Sibley and Facey 1968; Facey and Sibley 1969; Davidson et al. 1993; Amolo et al. 2004) (Figure 10.2).

The use of *calcium fluoride* (CaF_2) in IR applications started almost a century ago and lasted for many decades. Nowadays, the abilities for growing it with extremely high purity and large dimensions made it the most applied crystal material for DUV-VUV optics. The most prominent example is the DUV laser lithography where it is the only material in use besides synthetic fused silica.

Strontium fluoride (SrF_2) is applied in UV-VUV spectroscopy and as DUV-VUV coating material. Compared with CaF_2, it has a much lower thermal conductivity and a much higher temperature-dependent refractive index dn/dT, resulting in a significantly stronger irradiation-induced thermal lens. Maybe this is a reason for its less importance for optical components in laser-based applications compared with CaF_2.

Barium fluoride (BaF_2) is the most resistant material against high energetic radiation and features high mechanical strength and temperature stability up to 500°C. It is mostly used in the IR spectral region in optics for astronomic applications or as window for CO_2 lasers. It is the fastest scintillating crystal and thus often used as scintillator material for x-rays and γ-radiation.

Chapter 10

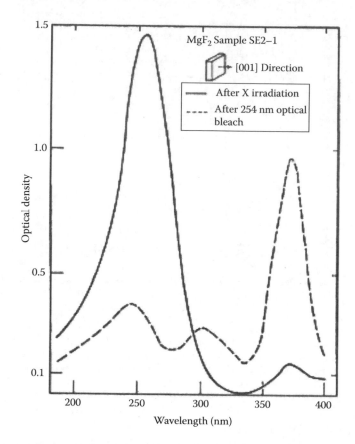

FIGURE 10.2   The optical absorption spectra of an MgF_2 crystal (0.05 cm thick) just after x-ray irradiation and just after subsequent 254 nm optical bleach. (Reprinted with permission from Blunt, R.F. and Cohen, M.I., *Phys. Rev.*, 153, 1031, 1967. Copyright 1967 by the American Physical Society.)

The only substantial application of *sodium fluoride* (NaF) is its use as the Cherenkov counter for particle detection.

Sapphire (c-Al_2O_3) is often used in spectroscopic instruments, especially when it comes to vacuum or high-/low-temperature applications. Here, the large Young's modulus allows for thinner window thickness than other materials. Furthermore, it has an extremely high thermal conductivity that prevents from stress-induced effects under intense light irradiation. There are some major absorption features in the UV/DUV wavelength range between 3.8 and 7 eV (Evans and Stapelbroek 1978; Draeger and Summers 1979; Crawford 1984; Innocenzi 1990), which can be assigned either to trace impurities or intrinsic defects (*F*- or *F*+-centers), for example, due to oxygen deficiency in the crystal. Light absorption into these absorption bands results in characteristic fluorescence centered in the range between 3 and 4 eV (Lee and Crawford 1979; Harutunyan et al. 1999). Figure 10.3 shows the emission and excitation spectra of the *F*-center in sapphire after neutron bombardment.

An important application for sapphire, doped with titanium, is its use as laser material especially as femtosecond or picosecond lasers.

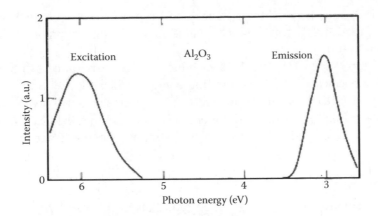

FIGURE 10.3 Excitation and emission spectra of the *F*-center from subtractively colored Al_2O_3 (neutron bombardment) measured at 300 K. (Reprinted with permission from Lee, K.H. and Crawford, J.H., *Phys. Rev. B*, 19, 3217, 1979. Copyright 1979 by the American Physical Society.)

Crystalline quartz (α-quartz) is used as window material in VUV and FIR (far infrared) applications. Furthermore, it is applied for piezoelectric components. Analogous to MgF_2, its birefringence makes it a very suitable material for polarization optics down to the DUV range.

Regarding absorbing defects in α-quartz, there are only little data available due to the heavy irradiation (neutrons, protons) required to create microscopic amorphous regions within the crystal lattice that could accommodate defects similar to those in fused silica (a-SiO_2). In those regions, the so-called oxygen deficiency centers (ODC) have been detected (Figure 10.4), for example, by their characteristic fluorescence bands

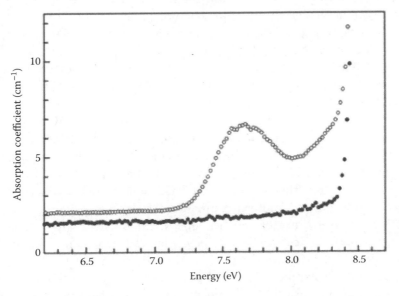

FIGURE 10.4 VUV absorption spectra, showing an absorption band of oxygen deficiency centers, detected at *T* = 290 K in samples of crystalline α-quartz, γ-irradiated with a dose of 10 MGy (empty symbol) and not irradiated (full symbol). (By permission from Cannas, M. et al., Luminescence of γ-radiation-induced defects in α-quartz, *J. Phys.: Condens. Matter.*, 16, 7931, 2004.)

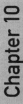

Chapter 10

(Brückner 1997; Cannas et al. 2004). Other defect centers common for a-SiO_2, for example, a free valance at a silicon atom (E'-center) or nonbridging oxygen holes (NBOH), have not been found.

In contrast to a-SiO_2, UV/DUV laser–induced defect generation is not expected in pure α-quartz because of the large binding energy of ideal Si–O bonds in the crystal network and the lack of distorted, low-energy Si–O bonds responsible for laser-induced absorption degradation in a-SiO_2. Therefore, it is assumed that the main origin for UV/DUV absorption in α-quartz under laser irradiation is related to trace impurities.

10.3   Lasers

Relevant laser sources are emitting pulsed radiation in the wavelength range between 150 nm and nearly 350 nm. Since damage processes are different for femtosecond pulses compared with longer pulses, we will only focus on the nanosecond and picosecond pulse durations.

To generate nanosecond pulses in the discussed spectral range, we can use solid-state bulk lasers, gas lasers, and the nonlinear frequency conversion of solid-state lasers emitting in the NIR spectral region.

Pulsed UV solid-state lasers are based, for example, on cerium-doped crystals such as Ce^{3+}:LiCAF (290 nm, 1 kHz, 1 mJ) or Ce^{3+}:$LiLuF_4$ (309 nm). These cerium doped lasers are pumped with nanosecond pulses from frequency-quadrupled Nd-lasers. Mode-locked operation of Ce:LiCAF lasers has been also realized (291 nm, 6 ps) (Fromzel et al. 2010).

Concerning the gas lasers, excimer lasers are the typical pulsed UV lasers. These lasers are very powerful UV sources emitting nanosecond pulses with repetition rates up to some kHz and average output powers between few watts and kilowatts. The generated laser wavelengths depend on the used laser gas (mostly rare gas halides) and range from 157 nm (F_2) to 351 nm (XeF).

Frequency-tripled or frequency-quadrupled Nd-lasers are important ns- and ps-laser sources at wavelengths around 355 and 266 nm. Using ytterbium (Yb)-doped material, the typical wavelengths are 343 and 257 nm.

The repetition rates of these lasers are determined by the pumping process of the active NIR emitting material. Flash lamp–pumped lasers are emitting in the some 10 Hz repetition rate regime, and diode-pumped lasers in the multikilohertz range.

Typical pulse energies of the third harmonic or the fourth harmonic radiation are in the μJ up to mJ region. The development of such solid-state lasers is a field with high dynamics including one- or multistage amplifier systems for very different applications like scientific purposes, medical applications, material processing, or laser-induced ignition. Parameters of selected UV high-power lasers are given in Table 10.2.

The world's largest laser system for 351 nm radiation is at the National Ignition Facility (NIF) of the Lawrence Livermore Laboratory under development. Each of the 192 power amplifier beams (Nd:laser) generates in KDP crystals with 40 × 40 cm interaction area third harmonics radiation at 351 nm. In 2012, the published record of the generated energy is 1.85 MJ, which corresponds to a peak power of 500 TW for the system (LLNL 2012). This means each beam delivers 9.8 kJ during the pulse duration of 22 ns.

Table 10.2 Parameters of Selected High-Power UV Lasers

Active Medium	Wavelength [nm]	Pulse Energy [mJ]	Pulse Duration [ns]	Repetition Rate [Hz]
Excimer laser				
F$_2$	157	50	20	200
ArF	193	600	20	200
KrF	248	1100	20	200
XeCl	308	2000	20	1200
XeF	351	400	20	200
Harmonics of solid-state lasers				
THG (Nd/Yb)	355/343	400	2.5–10	30
FHG (Nd/Yb)	266/257	200	15	20

10.4 Absorption Mechanisms/Origin of Absorption

To discuss absorption properties and absorption changes upon UV laser irradiation of the relevant crystalline materials with high transparency for UV applications, we want to include nonlinear absorption processes. Besides very weak single-photon absorption (SPA), the TPA is the main absorption mechanism inside the considered materials. TPA means the material absorbs a pair of laser photons simultaneously. The sum of energies of these photons is equal or higher than the band gap E_g of the material under consideration. The TPA is a nonlinear optical process yielding significant absorptions only at high laser intensities typical for pulsed lasers with nanosecond or shorter pulses. As a result of a TPA process, the transmission of a material would be significantly lower for pulsed laser excitation than for continuous laser irradiation of the same average laser intensity. The process of simultaneous absorption of two photons was first described theoretically by Marie Göppert-Mayer (1931). The first measurement of the TPA in earth alkaline materials was published in 1961 by W. Kaiser and C.G.B. Garrett for Eu-doped CaF$_2$ (Kaiser and Garrett 1961).

The intensity change in laser irradiation direction (z-direction) for laser radiation of intensity I inside the material with TPA cross section $\sigma^{(2)}$ is given by

$$\frac{dI}{dz} = -\sigma^{(2)} N_g I^2, \tag{10.1}$$

where N_g is the number of particles of the solid material per unit volume (cm^3) in the ground state. If the material also enables single photon absorption, for example, by spurious impurities, the propagation equation can be modified as

$$\frac{dI}{dz} = -\sigma^{(1)} N_g I - \sigma^{(2)} N_g I^2, \tag{10.2}$$

where $\sigma^{(1)}$ is the single photon absorption cross section.

Chapter 10

For a sample of length z, the transmitted laser intensity $I(z)$ for pure TPA is given by

$$I(z) = \frac{I_0}{1 + \sigma^{(2)} N_g I_0 z}, \tag{10.3}$$

where I_0 is the input intensity. This equation demonstrates the strong dependency of transmission $T = I(z)/I_0$ from the applied laser intensity.

Besides the discussed processes of TPA (or TPA combined with weak SPA), three photon absorption (ThPA) also is theoretically possible ($\hbar\omega_L \approx \frac{1}{3} E_g$, E_g...optical band gap). To estimate the probability of the ThPA process, we consider the probability W_n of an n-photon absorption process that is given by the relation

$$W_n \sim \left(\frac{E_{ph}}{E_0} \right)^{2n}, \tag{10.4}$$

where
E_{ph} is the amplitude of the electrical field strength of the applied laser radiation
E_0 characterizes the atomic field inside the crystal (Nathan et al. 1985)

Using a laser fluence of 50 mJ cm^{-2} and a pulse duration of 30 ns means $E_{ph} \approx 3 \cdot 10^6$ V m^{-1}. The atomic field E_0 inside the crystal has typical field strength in the range of 10^{11} V m^{-1}. From Equation 10.4, this would result in a quotient

$$\frac{W_3}{W_2} \approx 10^{-9} \tag{10.5}$$

for typical nanosecond laser pulses. Hence, the simultaneous absorption of three photons for such pulse durations is negligible.

We also want to discuss the possibility of a two-step absorption process in comparison with the simultaneous TPA process. Energy levels caused by defects inside the investigated crystals may induce a two-step absorption by single photon–induced excited state absorption. To distinguish between simultaneous TPA and two-step absorption is possible by application of equally shaped laser pulses with constant intensity but different pulse durations. The two-step absorption coefficient increases for pulses with longer pulse duration in contrast to the absorption coefficient of the simultaneous TPA process that will remain unchanged (Weber 1971; Görling et al. 2005).

Immediately after excitation of an electron from valence band to the conduction band, internal conversion processes take place in alkali halides and earth alkali halides and result in the formation of self-trapped excitons (STEs), that is, stabilized electron–hole pairs in the crystal lattice by exciton–lattice interaction. The characteristic time for these processes is in the 10^{-12} s range (Lindner 2000). The spectral properties of STE of the important materials were investigated in detail at low temperatures and at room temperature (RT) (Lindner et al. 2001; Song and Williams 1996). Besides transient absorption bands, the STEs also exhibit characteristic fluorescence bands. The final state of the fluorescence transition is the valence band of the material under investigation (Hayes 1974). The fluorescence lifetime is typically in the μs range (Mühlig et al. 2002). But not all generated STEs are deactivated by fluorescence. The radiationless decay of the STEs yields F- and H-centers in all crystalline materials. These centers are sources for the creation of more complex defect centers, namely, aggregates of multiple F-centers like M-center (two F-centers), R-center (three F-centers), etc., which can be responsible for laser-induced SPA of the applied UV laser radiation. The generation of STEs, their relaxation, and the generation of defect centers will be discussed in detail for CaF_2 in Section 10.5.

The processes of TPA were recently discussed by M. Rumi and J. W. Perry extensively and in detail (Rumi and Perry 2010). Their work describes comprehensively the measurement techniques and their limits especially to determine TPA cross sections of molecules. Rumi and Perry further consider the process of an SPA in molecules, starting from energy states excited by TPA for the applied laser wavelength. This effect causes additional absorption in molecules. In crystalline materials used as transparent UV components, this kind of two-step absorption process is not relevant because of the excitation of the conduction band by TPA. SPA for laser photon energy $\frac{1}{2} E_g \leq \hbar\omega_L \leq E_g$ can be created by current successive irradiation with laser pulses generating defects starting from the STE. As a result, differing from virgin materials, additional TPA and SPA processes can be present and the small signal transmission of the sample is lowered (Cramer et al. 2006; Mühlig et al. 2010). These facts result in the functional damage and furthermore in the lowering of the laser-induced damage threshold (LIDT).

10.5 CaF₂: The Most Applied Crystal in the UV-VUV Spectral Range

Since about the first century, CaF_2 has been of great interest for optical applications. Early in the twentieth century, CaF_2 became attractive because of its excellent infrared optical properties, and later it was one possible host for laser active materials due to its ability for high dopant concentrations. When the manufacturing process was changed to the use of synthetic raw materials instead of natural fluorite about 20 years ago, the enhanced purity enabled its application for optics down to the VUV spectral range. Nowadays, CaF_2 is, besides synthetic fused silica (a-SiO_2), the most applied optical material in the UV-VUV spectral range and often applied when a high durability against laser irradiation is required.

In contrast to synthetic fused silica, the initial absorption properties of CaF_2 in the UV-VUV spectral range are primarily determined by the residual amount of impurities within the crystal structure rather than by intrinsic defects. Furthermore, there is no permanent laser-induced refractive index change observed so far in CaF_2. The compaction (refractive index increase) or rarefaction (refractive index decrease) is a major functional damage effect in DUV-irradiated fused silica. In particular, the compaction can finally also yield catastrophic damage in the material (microchannel formation), especially for high-intensity applications.

Chapter 10

Over many decades, extensive studies have been made to investigate impurity-related absorption and fluorescence properties in CaF_2 using a wide variety of experimental techniques (Loh 1971; Staebler and Schnatterly 1971; Hayes and Lambourn 1973; Hayes 1974; Arkhangel'skaya et al. 1982; Rauch and Schwotzer 1982; Hamaïdia and Hachimi 1988; Mizuguchi et al. 1998, 1999; Mühlig et al. 2003). Besides, the impact of ionizing radiation on the absorption properties of CaF_2 crystals, with or without particular impurities, started to be of large interest. It was found that ionizing radiation (x-ray-, γ-, or intense DUV/VUV laser irradiation) can strongly increase the number of absorbing defects (in literature often referred to as "additionally colored"). In the case of laser irradiation, the additional defect generation is correlated with the radiation-induced generation of electron–hole pairs. Due to the large band gap of CaF_2 (≥ 10 eV, see Table 10.1), this is only likely via two- or three-photon absorption processes for the wavelength region considered within this chapter. Indeed, three-photon absorption in CaF_2 at 308 nm has been reported using intensities in the tens of MW cm^{-2} range (Petrocelli et al. 1991). However, the measurement results indicate color center formation that may affect the data due to multistep absorption processes via intermediate energy states in the band gap. At 248 nm where the photon energy of 5 eV is just about half of the lowest reported band gap values, there is a controversy in the reported literature. On the one hand, no two-photon but three-photon absorption has been observed (Tomie et al. 1989), whereas others report on two-photon-induced emission of STEs (Hata et al. 1990) or measured TPA coefficients (Taylor et al. 1988). The contrary results might be due to the strongly differing intensities—the I^2-ratio is $\geq 9 \times 10^4$—or related to a different purity of the investigated crystals since two-step absorption via intermediate energy states can strongly enhance an effective TPA. In contrast to the 248 nm wavelength, there is a general agreement in literature that TPA generates electron–hole pairs in CaF_2 at 193 and 157 nm since for both wavelength two times the corresponding photon energy of 6.4 and 7.9 eV, respectively, is higher than the largest optical band gap values from Table 10.1.

The ability of intense UV-VUV photons for bridging the optical band gap by nonlinear absorption processes and the resulting electron–hole pair (=free exciton) generation are the basic features that yield laser-induced changes in the optical properties (=functional damage) even of nowadays high-purity CaF_2 crystals. At 193 nm, these changes can either occur on a short-time (up to some 10,000 pulses) and/or a long-time (up to some 10^9 pulses) scales.

It is worth noticing that the intrinsic TPA is the primary process for both the short- and the long-time irradiation effects. Subsequent to the generation of free excitons, it is the strong exciton–lattice interaction in isolators possessing a large optical band gap that makes the exciton creating a lattice distortion and trapping itself therein (STE). This process of STE generation takes place on the picosecond time scale (Lindner 2000) and with a very high efficiency (Lindner 2000). The STE now can decay to its ground state by emitting a well-known fluorescence around 278 nm (Figure 10.5) with a lifetime (at room temperature) of about 1 μs (Williams et al. 1976; Song and Williams 1996; Mizuguchi et al. 1999; Mühlig et al. 2002; Görling et al. 2003).

The STE can also decay by the nonradiative energy transfer to phonons. The most important decay channel for the laser-induced optical property change, however, is

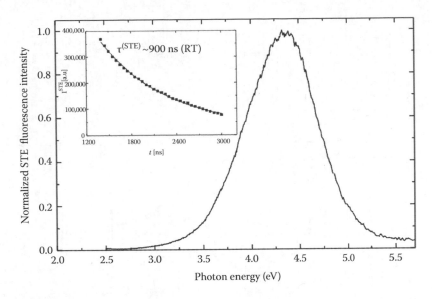

FIGURE 10.5 Normalized STE fluorescence at RT of a preirradiated CaF_2 sample detected 100 ns after laser excitation within a time gate of 100 μs. The inset shows the single exponential decay of the STE luminescence. (Reprinted from *Opt. Commun.*, 216, Görling, Ch., Leinhos, U., and Mann, K., Self-trapped exciton luminescence and repetition rate dependence of two-photon absorption in CaF_2 at 193 nm, 369–378, Copyright 2003, With permission from Elsevier.)

the formation of a stable pair of F- and H-centers where the F-center is an anion vacancy occupied by an electron and the H-center is a F_2^- molecule formed by a lattice fluorine ion and an interstitial fluorine atom. The H-center of the STE can diffuse away from the F-center by thermal activation with activation energy $E_a \sim 0.2$ eV (Mühlig et al. 2010; Rix 2011). The remaining F-center, that is, an electron, is now able to move within the crystal with a lifetime in the 1000 s range (Lindner 2000). It can be detected by a typical absorption band at 376 nm (Lindner 2000). On a short-time irradiation scale (up to some 10,000 pulses), the F-centers can be stabilized in the presence of impurities by forming the so-called F_A-centers or in the case of preexisting F_A-centers may form M_A-centers via $F + F_A \rightarrow M_A$ where the subscript A corresponds to monovalent impurities like sodium, potassium, lithium, or others. As a result, due to highly energetic irradiation and/or the presence of impurities, additional absorption bands as well as characteristic fluorescence can be observed (Hayes and Lambourn 1973; Arkhangel'skaya et al. 1982; Rauch and Schwotzer 1982; Hamaïdia and Hachimi 1988; Komine et al. 2000; Mühlig et al. 2003, 2010). In the DUV wavelength region, the effect of sodium impurity has been demonstrated to strongly increase the SPA and TPA in sodium-containing CaF_2 crystals (Komine et al. 2000; Mühlig et al. 2010). The origin of the sodium-related absorption increase was found to be a dynamic equilibrium between ArF laser–induced M_{Na}-center generation and annealing. A first simplified model for that interaction was given by Komine et al. (2000) based on their transmission measurements at 193 nm with sodium-doped CaF_2. This first model proposal, however, could not fully explain the multitude of further experimental findings for sodium-containing CaF_2 (Mühlig et al. 2010). Consequently, an extended rate

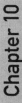

Chapter 10

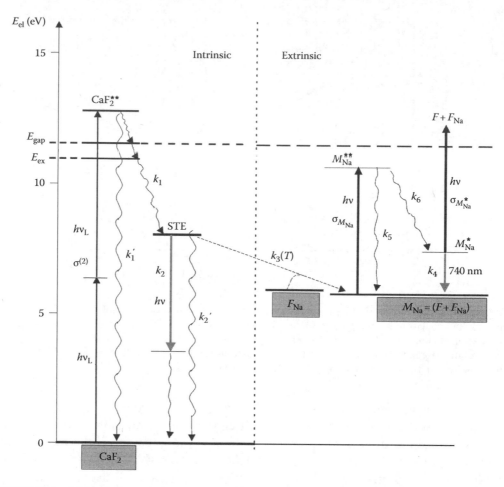

FIGURE 10.6 Scheme of ArF laser–induced generation and annealing of M_{Na}-centers in Na containing CaF_2. STE, self-trapped exciton; F_{Na}, sodium-stabilized F-center; M_{Na}, sodium-stabilized M-center; F, F-center; σ, absorption cross section; k, relaxation rate. (With kind permission from Springer Science+Bussiness Media: *Appl. Phys. B*, Influence of Na-related defects on ArF laser absorption in CaF_2, 99, 2010, Mühlig, C., Triebel, W., Stafast, H., and Letz, M., 525.)

equation–based model has been proposed (Figure 10.6) to explain the experimental findings related to ArF laser–induced absorption phenomena in sodium-containing CaF_2 crystals (Mühlig et al. 2010).

Table 10.3 shows values of the effective SPA and TPA coefficients α_{eff} and β_{eff}, obtained by Mühlig et al. (2010) in dependence on the sodium concentration represented by the intensity of the characteristic fluorescence emission of M_{Na}-centers around 740 nm.

On a long-time irradiation scale (up to some 10^9 pulses), CaF_2 was long believed to show no ArF laser–induced degradation. When, however, the applied laser pulse fluence strongly increased from just a few mJ cm^{-2} to tens of mJ cm^{-2}, for example, for laser windows, optical degradation has been detected in CaF_2 (Burkert et al. 2005). This degradation is assumed to result mainly from the increasing number of F-centers during intense long-time laser irradiation. Subsequent to the formation of M-centers (two F-centers) and R-centers (three F-centers), even large clusters of F-centers, that is, regions free of fluorine ions (=Ca clusters), are composed. Fortunately, the Ca

Table 10.3   Effective Single- and Two-Photon Absorption Coefficients α^{eff} and β^{eff} Obtained for CaF_2 Samples with Varying Sodium Concentrations, Represented by the Stationary Intensity $I_{fl}^{st}(740\ nm)$ of the M_{Na}-Center Fluorescence Emission in CaF_2

I_{fl}^{st} (740 nm) [a.u.]	α^{eff} (10^{-4} cm^{-1})	β^{eff} (10^{-9} cm · W^{-1})
0[a]	2.4 ± 1.4	1.7 ± 0.25
55[a]	5.7 ± 1.4	3.45 ± 0.4
65[a]	7.1 ± 1.3	3.8 ± 0.6
85	10 ± 0.6	4.2 ± 0.3
110[a]	6.0 ± 2.9	4.55 ± 0.7
150[a]	10 ± 5.0	4.8 ± 0.5
210[a]	9.3 ± 4.0	5.2 ± 0.5
240[a]	8.0 ± 0.7	4.8 ± 0.6
280[a]	9.8 ± 6.4	5.6 ± 0.5
430[a]	10.5 ± 2.5	6.7 ± 0.4
760	12.6 ± 0.8	7.8 ± 0.4
830	13.9 ± 0.6	8.3 ± 0.35
1110	16.8 ± 0.5	9.3 ± 0.5

Source: Mühlig, C. et al., *Appl. Phys. B*, 99, 525, 2010. With permission.

[a] Mean values of samples showing very similar I_{fl}^{st}.

lattice features the same symmetry and very similar lattice parameters compared to CaF_2. These Ca clusters, showing sizes in the 10–20 nm range at RT (Sanjuán et al. 1994; Bennewitz et al. 1995), act like metallic nanoparticles within the CaF_2 crystal, thus generating characteristic plasmonic resonances (=broad absorption bands). For Ca clusters in the CaF_2 crystal, two absorption bands around 500 nm and <250 nm have been calculated or measured (Creighton and Eadon 1991; Davidson et al. 2002; Cramer et al. 2005; Zeuner et al. 2011). Figure 10.7 shows the absorption band around 500 nm of Ca clusters in CaF_2 after electron irradiation as a function of the electron dose.

Recently, the ArF laser–induced VIS absorption bands in sodium-containing CaF_2 have been separated by fs-pump-probe spectroscopy and fluorescence excitation spectroscopy into contributions from Ca clusters (absorption band at 525 nm) and stable M_{Na}-centers (absorption band at 600 nm) (Zeuner et al. 2011) (Figure 10.8).

There are hints that the presence of impurities like sodium acts catalyzing to the Ca cluster formation (Shcheulin et al. 2007). Contrariwise, the Ca clusters seem to have a stabilizing effect on the M_{Na}-centers. It is, however, also found that formed Ca clusters are annealed by irradiation with wavelengths of their characteristic absorption bands (Cramer et al. 2005).

The effect of functional damage during intense long-term DUV/VUV laser irradiation results from the characteristic behavior of the metallic Ca clusters. For sophisticated optical systems, the strengthening of the thermal lens due to the composed absorption band(s) in the DUV/VUV region can become critical for the imaging quality. Furthermore, the thermally induced stress as well as the scattering-like behavior of

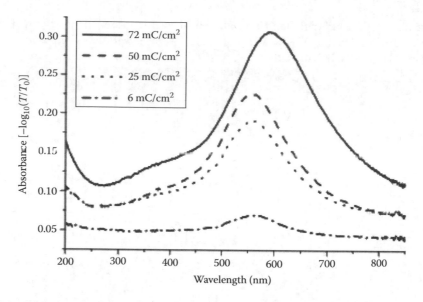

FIGURE 10.7   Absorption spectra of cleaved CaF_2 as a function of electron dose. The increase in absorption with increasing electron dose reflects increasing Ca nanoparticle densities. The shift of the main nanoparticle absorption to longer wavelengths at the highest electron dose reflects an increase in Ca nanoparticle diameter. (Reprinted with permission from Cramer, L.P., Schubert, B.E., Petite, P.S., Langford, S.C., and Dickinson, J.T., Laser interactions with embedded Ca metal nanoparticles in single crystal CaF_2, *J. Appl. Phys.*, 97(7), 74307. Copyright 2005, American Institute of Physics.)

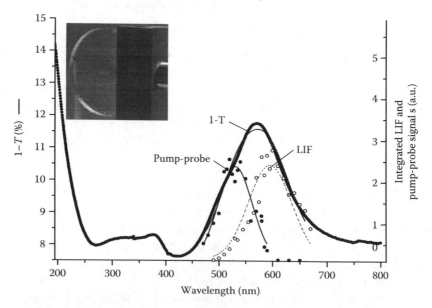

FIGURE 10.8   Effect of prolonged ArF laser irradiation on transmission properties of CaF_2; transient transparency spectrum from fs-pump-probe measurements (Ca-colloids, solid circles) and fluorescence excitation spectrum (M_{Na}-center, open circles) of the ArF laser preirradiated CaF_2 sample in comparison to the UV/VIS spectrometer record (dotted lines). Insight: rod-like luminescence and scatter trace upon 1 mm diameter laser illumination at 650 nm, visualizing the defects in the CaF_2 bulk material. (Reprinted with permission from Zeuner, T., Paa, W., Triebel, W., Mühlig, C., and Stafast, H., Picosecond kinetic confirmation of overlapping Ca cluster and M_{Na} absorption bands in UV grade CaF2, *J. Appl. Phys.*, 110, 056101. Copyright 2011, American Institute of Physics.)

the Ca clusters (Mie absorption) results in unwanted birefringence and depolarization. This can be in particular crucial for the polarized illumination in current polarization-enhanced optical lithography.

References

Amolo, G.O. et al., 2004. Visible and VUV optical absorption studies of Mg-colloids and colour centres in MgF$_2$ crystals implanted by 100 keV Mg-ions. *Nucl. Instrum. Methods B.* 218:244–248.

Arkhangel'skaya, V.A., V.M. Reiterov, L.M. Trofimova, and A.S. Shcheulin. 1982. Optical properties of fluorite-type crystals with M$_A$ color centers. *Zhurnal Prikladnoi Spektroskopii.* 37:644–648.

Becher, J., R.L. Kernell, and C.S. Reft. 1983. Proton-induced *F*-Centers in LiF and MgF$_2$. *J. Phys. Chem. Solids.* 44:759–763.

Bennewitz, R. et al., 1995. Size evolution of low energy electron generated Ca colloids in CaF2. *Appl. Phys. Lett.* 66:320–322.

Blunt, R.F and M.I. Cohen. 1967. Irradiation-induced color centers in magnesium fluoride. *Phys. Rev.* 153:1031–1038.

Brückner, R. 1997. *Encyclopedia of Applied Physics Vol. 18: Silicon Dioxide.* Weinheim, Germany: VHC.

Burkert, A. et al., 2005. Investigating the ArF laser stability of CaF$_2$ at elevated fluences. *Proc. SPIE.* 5878:58780E.

Burnett, J.H., Z.H. Levine, and E.L. Shirley. 2001. Intrinsic birefringence in calcium fluoride and barium fluoride. *Phys. Rev. B.* 64:241102.

Cannas, M. et al., 2004. Luminescence of γ-radiation-induced defects in α-quartz. *J. Phys.: Condens. Matter.* 16:7931–7939.

Cramer, L.P., S.C. Langford, and J.T. Dickinson. 2006. The formation of metallic nanoparticles in single crystal CaF$_2$ under 157 nm excimer laser irradiation. *J. Appl. Phys.* 99:054305.

Cramer, L.P., B.E. Schubert, P.S. Petite, S.C. Langford, and J.T. Dickinson. 2005. Laser interactions with embedded Ca metal nanoparticles in single crystal CaF$_2$. *J. Appl. Phys.* 97(7):74307.

Crawford, Jr., J.H. 1984. Defects and defect processes in ionic oxides: Where do we stand today? *Nucl. Instrum. Methods B.* 1:159–165.

Creighton, J.A. and D.G. Eadon. 1991. Ultraviolet-visible absorption spectra of the colloidal metallic elements. *J. Chem. Soc. Faraday Trans.* 87:3881–3891.

Davidson, A.T. et al., 2002. The colouration of CaF$_2$ crystals by keV and GeV ions. *Radiat. Eff. Defects Solids.* 157:637–641.

Davidson, A.T., J.D. Comins, T.E. Derry, and A.M.J. Raphuthi. 1993. Optical studies in the range 2–9 eV of ion-implanted MgF$_2$ crystals. *Phys. Rev. B.* 48:782–788.

Draeger, B.G. and G.P. Summers. 1979. Defects in unirradiated α-Al$_2$O$_3$. *Phys. Rev. B.* 19:1172–1177.

Evans, B.D. and M. Stapelbroek. 1978. Optical properties of the *F*+ center in crystalline Al$_2$O$_3$. *Phys. Rev. B.* 18:7089–7098.

Facey, O.E. and W.A. Sibley. 1969. Optical absorption and luminescence of irradiated MgF$_2$. *Phys. Rev.* 186:926–932.

Fromzel, V.A. et al., 2010. *Advances in Optical and Photonic Devices: Tunable, Narrow Linewidth, High Repetition Frequency Ce:LiCAF Lasers Pumped by the Fourth Harmonic of a Diode-Pumped Nd:YLF Laser for Ozone DIAL Measurements.* Rijeka, Croatia: InTech.

Göppert-Mayer, M. 1931. Über Elementarakte mit zwei Quantensprüngen. *Ann. Phys. (Leipzig).* 9:273–294.

Görling, Ch., U. Leinhos, and K. Mann. 2003. Self-trapped exciton luminescence and repetition rate dependence of two-photon absorption in CaF$_2$ at 193 nm. *Opt. Commun.* 216:369–378.

Görling, Ch., U. Leinhos, and K. Mann. 2005. Surface and bulk absorption in CaF$_2$ at 193 and 157 nm. *Opt. Commun.* 249:319–328.

Hamaïdia, A. and A. Hachimi. 1988. Colour centers in alkali metal doped CaF$_2$. *Phys. Stat. Sol. (b).* 149:711–724.

Harutunyan, V.V., V.A. Gevorkyan, and V.N. Makhov. 1999. Luminescence property studies of α-Al2O3 by means of nanosecond time-resolved VUV spectroscopy. *Eur. Phys. J. B.* 12:35–38.

Hata, K., M. Watanabe, and S. Watanabe. 1990. Nonlinear processes in UV optical materials at 248 nm. *Appl. Phys. B.* 50:55–59.

Hayes, W. 1974. *Crystals with the Fluoride Structure: Electronic, Vibrational and Defect Properties.* Oxford, U.K.: Clarendon Press.

Hayes, W. and R.F. Lambourn. 1973. Production of F and F-aggregates centres in CaF_2 and SrF_2 by irradiation. *Phys. Stat. Sol. (b).* 57:693–699.

Innocenzi, M.E. et.al., 1990. Room-temperature optical absorption in undoped α-Al_2O_3. *J. Appl. Phys.* 67:7542:7546.

Kaiser, W. and C.G.B. Garrett. 1961. Two-photon excitation in CaF_2:Eu^{2+}. *Phys. Rev. Lett.* 7:228–232.

Komine, N., S. Sakuma, M. Shiozawa, T. Mizugaki, and E. Sato. 2000. Influence of sodium impurities on ArF excimer-laser-induced absorption in CaF_2 crystals. *Appl. Opt.* 39:3925–3930.

Lee, K.H. and J.H. Crawford, Jr. 1979. Luminescence of the F-center in sapphire. *Phys. Rev. B.* 19:3217–3221.

Letz, M., L. Parthier, A. Gottwald, and M. Richter. 2003. Spatial anisotropy of the exciton level in CaF_2 at 11.1 eV and its relation to the weak optical anisotropy at 157 nm. *Phys. Rev. B.* 67:233101.

Liberman, V., M. Rothschild, P.G. Murphy, and S.T. Palmacci. 2002. Prospects for photolithography at 121 nm. *J. Vac. Sci. Technol. B.* 20:2567–2573.

Lindner, R. 2000. *Bildungs- und Relaxationsdynamik von Self-Trapped-Excitons in Erdalkalifluoriden.* PhD-thesis. Berlin, Germany: Freie Universität Berlin.

Lindner, R., R.T. Williams, and M. Reichling. 2001. Time-dependent luminescence of selftrapped excitons in alkaline-earth fluorides excited by femtosecond laser pulses. *Phys. Rev. B.* 63:075110.

LLNL. 2012. Lawrence Livermore's National Ignition Facility achieves record laser energy in pursuit of fusion ignition. LLNL press release, March 21.

Loh, E. 1971. Ultraviolet absorption spectra of photochromic centers in CaF_2 crystals. *Phys. Rev. B.* 4:2002–2006.

Mizuguchi, M., H. Hosono, H. Kawazoe, and T. Ogawa. 1998. Generation of optical absorption bands in CaF_2 single crystals by ArF excimer laser irradiation: Effect of yttrium impurity. *J. Vac. Sci. Technol. A.* 16:3052–3057.

Mizuguchi, M., H. Hosono, H. Kawazoe, and T. Ogawa. 1999. Time-resolved photoluminescence for diagnosis of resistance to ArF excimer laser damage to CaF_2 single crystals. *J. Opt. Soc. Am. B.* 16:1153–1159.

Mühlig, C., W. Triebel, H. Stafast, and M. Letz. 2010. Influence of Na-related defects on ArF laser absorption in CaF_2. *Appl. Phys. B.* 99:525–533.

Mühlig, Ch. et al., 2003. Laser induced fluorescence of calcium fluoride upon 193 nm and 157 nm excitation. *Proc. SPIE.* 5188:123–133.

Mühlig, Ch., W. Triebel, G. Töpfer, and A. Jordanov. 2002. Calcium fluoride for ArF laser lithography – characterization by in situ transmission and LIF measurements. *Proc. SPIE.* 4932:458–466.

Nathan, V., A.H. Guenther, and S.S. Mirra. 1985. Review of multiphoton absorption in crystalline solids. *J. Opt. Soc. Am. B.* 2:294–316.

Palik, E.D. 1998. *Handbook of Optical Constants.* London, U.K.: Academic Press.

Petrocelli, G., F. Scudieri, and S. Martellucci. 1991. Nonlinear absorption in ionic crystals determined by pulsed photothermal deflection. *Appl. Phys. B.* 52:123–128.

Rauch, R. and G. Schwotzer. 1982. Disturbed colour centres in oxygen- and alkali-doped alkaline Earth fluoride crystals after X-ray irradiation at 77 and 295K. *Phys. Stat. Sol. (a).* 74:123–131.

Rix, St. 2011. *Radiation-induced Defects in Calcium Fluoride and Their Influence on Material Properties under 193 nm Laser Irradiation.* PhD-thesis. Mainz, Germany: Johannes Gutenberg-Universität Mainz.

Rubloff, G.W. 1972. Far-ultraviolet reflectance spectra and the electronic structure of ionic crystals. *Phys. Rev. B.* 5:662–684.

Rumi, M. and J.W. Perry. 2010. Two-photon absorption: An overview of measurements and principles. *Adv. Opt. Photon.* 2:451–518.

Sanjuán, M.L., P.B. Oliete, and V.M. Orera. 1994. The enhanced Raman scattering of phonons in CaFz and MgO samples containing Ca and Li colloids. *J. Phys.: Condens. Matter.* 6:9647–9657.

Shcheulin, A.S., A.K. Kupchikov, and A.I. Ryskin. 2007. A high-stability medium based on CaF_2:Na crystals with colloidal color centers: I. Photothermochemical conversion of colloidal color centers in CaF_2:Na crystals. *Opt. Spectrosc.* 103:507–511.

Sibley, W.A. and O.E. Facey. 1968. Color centers in MgF_2. *Phys. Rev.* 174:1076–1082.

Song, K.S. and R.T. Williams. 1996. *Self-Trapped Excitons.* Springer Series in Solid-State-Science Vol. 95. Berlin, Germany: Springer.

Staebler, D.L. and S.E. Schnatterly. 1971. Optical studies of a photochromic color center in rare-earth-doped CaF_2. *Phys. Rev. B.* 3:516–526.

Taylor, A.J., R.B. Gibson, and J.P. Roberts. 1988. Two-photon absorption at 248 nm in ultraviolet window materials. *Opt. Lett.* 13:814–816.

Tomie, T., I. Okuda, and M. Yano. 1989. Three-photon absorption in CaF_2 at 248.5 nm. *Appl. Phys. Lett.* 55:325–327.

Tomiki, T. and T. Miyata. 1969. Optical studies of alkaline fluorides and alkaline Earth fluorides in VUV region. *J. Phys. Soc. Japan.* 27:658–678.

Weber, H.P. 1971. Two-photon-absorption laws for coherent and incoherent radiation. *J. Quantum Electr.* QE7:189–195.

Williams, R.T., M.N. Kabler, W. Hayes, and J. P. Stott. 1976. Time-resolved spectroscopy of self-trapped excitons in fluorite crystals. *Phys. Rev. B.* 14:725–740.

Zeuner, T., W. Paa, W. Triebel, C. Mühlig, and H. Stafast. 2011. Picosecond kinetic confirmation of overlapping Ca cluster and M_{Na} absorption bands in UV grade CaF_2. *J. Appl. Phys.* 110:056101.

Chapter 10

11. Materials for Lasers

Frequency Conversion, Q-Switching, and Active Materials

Anne Hildenbrand-Dhollande and Frank R. Wagner

Laser-Induced Damage in Optical Materials. Edited by Detlev Ristau © 2015 CRC Press/Taylor & Francis Group, LLC. ISBN: 978-1-4398-7216-1.

Chapter 11

11.1 Introduction

11.1.1 Motivation for Laser Damage Studies in Materials for Lasers

The development of more and more powerful and compact lasers raises the issue of laser damage of optical components in the laser sources. Active materials (like Nd:YAG crystals), frequency conversion crystals (like KDP crystals), and Q-switching components (such as RTP Pockels cells) are the main optical elements in a laser system. Hence, these materials are particularly concerned by laser damage phenomena due to intense laser irradiation. Moreover, these materials often have a lower laser damage threshold than glasses. In addition, nonlinear effects proper to crystals (such as second harmonic generation) also can reduce the laser damage threshold of these components. Thus, laser-active materials and nonlinear crystals are often a limitation for the generation or use of intense laser beams.

Generally, damage appears on the surface or the coatings of optical components. However, for nonlinear crystals and laser-active materials, damage often appears preferentially in the bulk, implying the necessity to study the laser damage of these materials. Indeed, fundamental laser damage studies aim to understand the damage mechanisms in order to improve the laser damage resistance of the material. Applied laser damage studies aim to determine the laser-induced damage threshold in given conditions in order to take it into account during the design of laser cavities or other high-power optical setups.

11.1.2 Difficulties upon Comparison of LIDT Literature Values

For a given material, laser-induced damage thresholds (LIDTs) from different publications can be very different. The dispersion of the literature values can be explained by many factors.

The main reason is certainly the variation of the test conditions between the different studies. The LIDT depends on many parameters aside the material itself: the wavelength, the pulse duration, the size of the irradiated surface, the number of irradiated sites, the number of pulses per site, the pulse repetition frequency, etc. [64,126]. Thus, the LIDT of a material has to be correlated with the test conditions. Note that in many publications the full set of test conditions is not given. In addition, the test method used (1-on-1 [1], S-on-1 [2], or R-on-1; cf. Section II, Measurements) can also induce different LIDT values.

Note that in 2011 a new standard (ISO 21254) was defined to describe test methods for laser-induced damage measurement [3–6]. This standard is divided into four parts. Part 1, *Definitions and general principles* [3], describes the fundamental principles to evaluate laser damage threshold of optical components in different conditions. Part 2, *Threshold determination* [4], details the methods of threshold determination. The 1-on-1 and S-on-1 modes are defined in this part. Part 3, *Assurance of laser power (energy) handling capabilities* [5], is focused on the possibilities of laser treatment of optical components. Finally, part 4, *Inspection, detection and measurement* [6], is a technical report on the damage detection methods and the inspection techniques of damage on optical surfaces and in the bulk of optical components.

Secondly, the tested material itself is not the same between the different publications. Different polishing and crystal growth processes result in different surface quality and different crystal quality, respectively. Generally, crystals are more resistant in more recent publications, a fact that is caused by improvements of the polishing and crystal growth processes. Besides, the publications do not always specify if the reported damages are located on the surface, on the coating, or in the bulk of the material.

Another cause for differing LIDT values can be different definitions of "laser damage." "Damage" is generally defined by an *irreversible* modification of the material. However, the sensitivity of the detection of material modifications depends strongly on the employed diagnostics method. Possible diagnostic methods are optical microscopy, plasma or sound wave detection, light diffusion or transmission, wave front deformation, scanning electron microscopy (SEM), atomic force microscopy (AFM), z-scan measurements, and many more. Indeed, a functional damage threshold, for example, transmission of more than 95%, and a physical damage threshold as observed by optical microscopy will be reached at different fluences and thus lead to different threshold values.

Furthermore, LIDT values are not always expressed with the same quantity, which may complicate comparisons. The commonly used quantity is the peak value of the surface density of pulse energy, also called fluence, and is expressed in J/cm^2. Finally, the definition of the LIDT can also be different. The most commonly used LIDT is the low threshold, or 0% threshold, given by the highest fluence under which no damage appears. Some publications also used the high threshold defined by the lowest fluence above which damage systematically appears, or the 50% threshold defined by the fluence for which damage appears with a 50% probability.

Thus, careful precautions have to be taken concerning the comparison of LIDT data for a given material from different publications, and simultaneous publication of the LIDT of a reference material like synthetic fused silica is recommended.

11.1.3 Specificities of Laser Damage Measurements in Crystals

Nonlinear and laser-active crystals are usually expensive and of small dimensions, making it difficult to perform detailed laser damage studies. Moreover, due to several effects proper to crystals, laser damage studies are more complex in nonlinear and active crystals than in isotropic materials. Laser damage measurements can be modified by effects due to birefringence of crystals (focusing aberrations, walk-off) and nonlinear phenomena (self-focusing, frequency conversion) [48]:

- Due to birefringence, that is, the anisotropy of the linear refractive index, aberrations can appear in focused beams and cause a strongly astigmatic beam profile in the crystal [48]. Modeling of the effect has been done for flat-top beams [55], but an experimental check is recommended. Usually, the problem can be avoided by working with only slightly focused beams and checking the absence of significant aberrations by characterizing the beam profile at the focus through a large thickness of sample material. In order to quantitatively take the aberrations into account, the beam profile of the test beam has to be characterized through a sample whose

Chapter 11

thickness corresponds to the distance from the entrance side to the focus point during the damage test. In both cases, quantitative measurement or qualitative check, the same angles and polarizations as during the actual damage test have to be used.

- Birefringence can also induce walk-off, which tends to separate the different waves propagating in the crystal. Thus, the test beam diameter should be significantly larger than the walk-off in the test situation.
- Self-focusing can appear at relatively low beam power in nonlinear crystals [72] as compared to glasses since the nonlinear refractive index is generally higher. Laser damage measurements can thus be influenced by unexpected moderate self-focusing appearing at beam powers below the critical value (cf. Chapter 4).
- Frequency conversion is a common application of nonlinear crystals. However, during laser damage measurements, the unwanted generation of new wavelengths can influence the measured laser damage threshold. Indeed, laser damage in nonlinear crystals often is a multiple wavelengths issue, and cooperative damage mechanisms have been evidenced (cf. Section 11.3.3).

Finally, in crystals, contrary to glasses, the anisotropy of several different physical parameters can induce anisotropy of the laser damage resistance. In consequence, the laser damage threshold can depend on the polarization state and/or the propagation direction of the laser in the crystal (see Section 11.3.2).

11.1.4 Aim of This Chapter

A large number of crystals for frequency conversion, electro-optic switching, and laser-active materials exist [84]. The choice of a material for an application depends on many criteria. For instance, in the case of frequency doubling of a high-power Q-switched Nd:YAG laser, the important parameters are a high conversion efficiency, a large phase-matching bandwidth (large thermal, spectral, and angular acceptance), good chemical and physical properties (thermal conductivity, nonhygroscopic crystal, etc.), an easy crystalline growth, and a high laser damage resistance. Each application defines a set of parameters, determining the optimal choice of the crystal.

The scope of this chapter is not to make an exhaustive list of all nonlinear crystals and laser-active materials. In the first part, this chapter gives an overview of laser damage studies found in the literature of the frequently used nonlinear crystals for frequency conversion and Q-switching, as well as the commonly used active materials. The second part deals with the specificities of the mechanisms of laser damage in nonlinear crystals.

11.2 Frequently Used Crystals for Lasers: Main Properties and Damage Thresholds

This section summarizes literature data concerning the frequently used nonlinear crystals for frequency conversion and Q-switching, as well as the commonly used active materials. For the different applications, the main properties and the laser damage thresholds are given.

11.2.1   Frequency Conversion

Even though dye lasers allow direct generation of practically all interesting wavelengths, all-solid-state lasers are preferred for many applications. In order to access, a given wavelength with an all-solid-state laser light of the fundamental wave is converted to the final wavelength by nonlinear frequency conversion in crystals. The most common nonlinear conversion processes are frequency doubling (usually abbreviated as SHG for second harmonic generation); sum (SFG) and difference frequency generation (DFG); and optical parametric oscillation (OPO), amplification (OPA), and generation (OPG). Nonlinear frequency conversion is efficient only at sufficiently high optical intensities. However, the higher the intensity, the higher the risk that the laser damages the crystal.

A large number of nonlinear crystals can be used depending on the application. Figure 11.1 gives a list of several commonly used nonlinear crystals with their transparency ranges. Typical crystals for SHG of Nd:YAG lasers are KDP, KTP, LBO, and BBO. A comparison of these crystals is reported in [12].

11.2.1.1   Niobates

11.2.1.1.1   Lithium Niobate (LiNbO₃, LN)   Thanks to excellent electro-optical, acousto-optical, nonlinear optical properties, and a wide transparency range (between 0.33 and 5.5 μm), lithium niobate ($LiNbO_3$, LN) is a very attractive crystal for many applications. It is often used for SHG of infrared (IR) wavelengths of more than 1 μm and OPOs

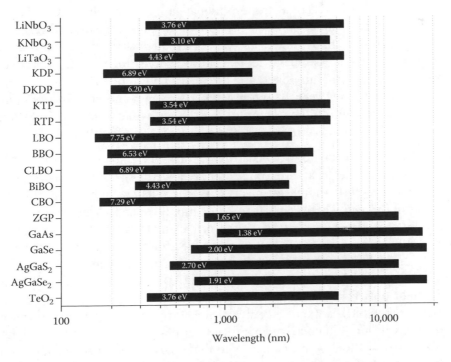

FIGURE 11.1   Transparency ranges for commonly used nonlinear crystals. Approximate values of the band gap ΔE_g are also given (white writing on the bars) and have been obtained from the short-wavelength transparency limit λ_{UV} by $\Delta E_g = 1240$ eV nm/λ_{UV}.

Chapter 11

pumped by Nd:YAG lasers. However, its use in high optical power devices is limited due to a photorefractive effect causing strong beam distortions at relatively weak intensities (150 W/cm^2 at 532 nm [115]). A large number of articles can be found concerning the photorefractive effect in LN (often named "optical damage," although the effect is usually reversible).

In lithium niobate crystals, this phenomenon is mainly due to Fe impurities and Nb_{Li} intrinsic defects (i.e., Nb^{5+} in Li^+ sites, also called antisites) [24,40]. Therefore, the photorefractive threshold can be improved by decreasing the Fe concentration and also by doping LiNbO$_3$ with ions such as Mg^{2+}, Zn^{2+}, Sc^{3+}, and In^{3+}, which can incorporate the Li sites and thus reduce the Nb antisites [67,110,116,129]. It was reported that the photorefractive threshold can be increased by a factor of 2 by doping with 4.6 mol.% magnesium oxide (MgO) [139]. However, Mg^{2+}-doped LN crystals of high optical quality are difficult to grow. Recently, tetravalent ions such as Hf^{4+}, Zr^{4+}, and Sn^{4+} were reported to be a good choice, as illustrated by high photorefractive thresholds, low doping thresholds, and a very good dopant distribution [65,66,123]. Another solution to prevent the photorefractive effect is to reduce the concentration of photoexcitable charge carriers by oxidizing the crystal. Recently, a thermoelectric oxidization technique applied to undoped LN crystals led to an increase in the photorefractive threshold by one order of magnitude [38]. Note also that raising the crystal temperature above room temperature substantially increases the photorefractive threshold, probably explained by higher carrier mobility [96].

However, Furukawa et al. showed that the photorefractive threshold and the bulk laser damage threshold (LIDT) of lithium niobate are not correlated [43]. High MgO doping levels (typically about 4.5 mol.%) improve the photorefractive threshold [139], whereas they reduce the LIDT, due to a lower crystal quality. LN doped with more than 3 mol.% MgO contained scattering centers (local aggregations of Mg) that decreased the transparency. Note that undoped and 1 mol.% MgO-doped LiNbO$_3$ have approximately the same LIDT: between 10 and 14 J/cm^2 at 1053 nm and 1 ns. In 1993, Sun et al. obtained similar data with 12 J/cm^2 for bulk and 22 J/cm^2 for surface damage thresholds of undoped LiNbO$_3$ crystals at 1064 nm, 10 ns, in 1-on-1 mode [108].

11.2.1.1.2 Potassium Niobate (KNbO$_3$, KN)

Potassium niobate (KNbO$_3$, KN) has high nonlinear coefficients, a wide transparency range (0.4–4.5 µm), and good phase-matching properties. It is an interesting crystal for frequency doubling for fundamental wavelengths in the region between 850 and 1064 nm. It is also well suited for OPO applications using pump laser wavelengths in the visible and near IR.

In 1992, Ellenberger et al. [37] measured single-shot damage thresholds of KNbO$_3$ for various crystal cuts (propagation directions) and polarizations in uncoated samples. At 1054 nm and 700 ps pulse duration, the LIDT varied from 8 to 26 J/cm^2. The LIDT for light polarized parallel to the c-axis is more than twice as high as for light polarized parallel to the a-axis. In contrast, at 527 nm and 500 ps pulse duration, the damage threshold for light polarized parallel to the c-axis is lower by 30% and the LIDT values varied from 4 to 7 J/cm^2. Contrary to the polarization anisotropy, no influence of the propagation direction on the LIDT was observed. Note that an Nd:YAG laser irradiation of KNbO$_3$ crystal at a fluence level of 10 J/cm^2, with 100 ps pulse duration at a repetition rate of 20 Hz during 1 h, did not produce any observable damage.

Similarly, no damage occurred after 55 h at a fluence of 1 J/cm^2 in a frequency doubling experiment pumped by a Q-switched Nd:YLF laser operated at 1 kHz and exhibiting 11 ns pulse duration [104].

11.2.1.2 Phosphates

11.2.1.2.1 Potassium Dihydrogen Phosphate (KH$_2$PO$_4$, KDP)

Potassium dihydrogen phosphate (KH$_2$PO$_4$, KDP) and deuterated potassium dihydrogen phosphate (KD$_x$H$_{2-x}$PO$_4$, DKDP or KD*P) are among the most commonly used commercial nonlinear crystals. With a transparency range from about 200 to 1500 nm, KDP and DKDP are widely used as second, third, and fourth harmonic generators for Nd:YAG and Nd:YLF lasers. KDP can be grown up to very large size with a good optical quality due to the rapid growth method (a $50 \times 50 \times 50$ cm^3 crystal is obtained in 2 months) [134]. Thus, large sizes and quite low-cost components are available. However, KDP has a low nonlinearity and is hygroscopic.

Due to its unique large size, KDP crystals are used as frequency converters in high-power lasers such as the National Ignition Facility (NIF) in the United States and the Laser Mégajoule (LMJ) in France developed for inertial confinement fusion. The power of these laser lines is limited by laser-induced damage in KDP. Thus, as far as laser-induced damage is concerned, KDP is the most studied nonlinear crystal with a very large number of publications.

As in many nonlinear crystals, nanosecond laser–induced damages in KDP generally appear in the bulk rather than on the surface at fluences well below the dielectric breakdown of the material [30]. Although the damage mechanism is still not well understood, it is admitted that bulk damage is initiated by localized absorption by either foreign nanoparticles incorporated during growth or intrinsic defect clusters formed during crystal growth. However, the nature and chemical composition of these defects called "damage precursors" are unknown. Nanosecond laser–induced damage typically manifests as small microcavities in the bulk surrounded by a zone showing melting and re-crystallization, which indicates highly localized absorption [23]. Once formed, damage sites in the bulk generally do not grow under laser irradiation with moderate fluences (below 20 J/cm^2). However, they affect the laser beam quality due to scattering. If damages are formed near the surface, they can erupt onto the surface and form a damage site with the potential to grow.

Due to the very large size of the crystals and the laser beam in inertial confinement fusion lasers, the presence of defects generating bulk damages is unavoidable. Thus, in recent publications concerning KDP, laser damage is often represented by the damage density for a given fluence instead of the damage probability generally used (cf. Chapter 7). Assessment of the laser damage behavior in terms of damage density allows defining application-dependent functional damage thresholds and helps, under certain conditions, to overcome variations of LIDT values caused by varying sampling volumes [70] (see also Section 11.1.2).

Concerning the growth conditions, laser-induced damage is not correlated with the impurity concentration for an optimized growth process with an acceptable level of absorbing impurities. However, laser damage depends on the growth temperature and is independent of growth rate variations at constant temperature. Note also that growth boundaries formed due to abrupt changes in growth conditions lead to dramatically enhanced damage densities [82] (cf. Section 11.3.1).

The existence of two distinct populations of damage precursors has also been demonstrated [28,80]. The first population initiates damage at longer wavelengths (such as at the fundamental of an Nd:YAG laser at 1064 nm) while the second population initiates at shorter wavelengths (second and third harmonics at 532 and 355 nm, respectively). The latter population of precursors is limiting the damage resistance of the crystal since it damages at lower fluences.

Natoli et al. measured 1-on-1 laser-induced damage in a Z-cut KDP crystal with an Nd:YAG laser at fundamental, second, and third harmonic wavelengths with 12 ns pulse duration and a spot size of several hundred micrometers. They reported LIDTs of 49 J/cm² at 1064 nm, 22 J/cm² at 532 nm, and 18 J/cm² at 355 nm [80]. Note that Natoli et al. also measured 1-on-1 LIDT in the bulk of synthetic fused silica (Suprasil) in the nanosecond regime, and they obtained 100 J/cm² at 1064 nm and 47 J/cm² at 355 nm [79]. However, recent measurements show that the LIDT of KDP is dependent on the polarization of the used light [91,97,130] (cf. Section 11.3.2). Furthermore, a cooperative damage mechanism is found in KDP under simultaneous exposure to 532 and 355 nm irradiation [29] (cf. Section 11.3.3). Pulse length (τ) scaling of the LIDT in the nanosecond range is somewhat particular in KDP where LIDT $\propto \tau^{0.35}$ was found [10,20], whereas commonly thermal processes lead to a $\tau^{0.5}$ dependency [106].

To conclude, KDP and DKDP laser damage resistance can be increased by preexposure of the crystal to subdamage threshold laser pulses, known as laser conditioning or laser annealing, which is employed for large-aperture laser systems [11,27]. These two last observations gave rise to a couple of theoretical simulations with the aim to better identify the damage precursors [33,34,73].

11.2.1.2.2 Potassium Titanyl Phosphate (KTiOPO₄, KTP) and Rubidium Titanyl Phosphate (RbTiOPO₄, RTP)

Potassium titanyl phosphate (KTiOPO₄, KTP) is an excellent nonlinear crystal that exhibits large optical nonlinearity, wide angular, thermal and spectral bandwidths, small walk-off, and a broad transparency range from 0.35 to 4.5 μm. KTP is among the most commonly used materials for frequency doubling of Nd:YAG and other Nd³⁺-doped lasers. It is also employed as OPO for near-IR generation up to 4 μm. However, it is limited to low- or moderate-power applications due to its susceptibility to photochromic damage, called gray tracking. The gray tracking threshold for 532 nm light in standard KTP was found to be 80 MW/cm² [17]. This low-power effect is at least partially caused by defects and impurities [103]. The so-called gray track resistant KTP crystals are available today from different providers, which allow using approximately twice the power compared with standard crystals.

Another well-known crystal of the same family is the rubidium titanyl phosphate (RbTiOPO₄, RTP) crystal. It has very similar optical and physical properties as KTP. However, with its higher effective electro-optic coefficient, RTP is particularly used for electro-optic applications (cf. Section 11.2.2.2). RTP can also be used for SHG at 1 μm and for OPOs generating 3 μm radiation.

Gray tracking in KTP is the subject of a large number of publications [138]. Photochromic damage consists in the generation of color centers during frequency doubling of 1064 nm light above a given intensity and number of shots [17]. The color centers absorb light of wavelengths covering the full visible spectrum and part of the near IR. The effect is also sometimes named GRIIRA (green-induced infrared absorption) [124]. Usually gray tracking is reversible (faster at higher temperatures), but it can become permanent at high fluences and above a given number of shots [54]. The color centers consist of an electron trap ($Ti^{4+} \rightarrow Ti^{3+}$) stabilized by interstitial potassium ions [124] or oxygen vacancies [99,103] and a hole trap that can be an impurity ion like ($Fe^{3+} \rightarrow Fe^{4+}$) [103] or an oxygen atom next to a K^+ vacancy [36]. The absorption of the laser beam by these color centers decreases the 532 nm intensity and can lead to damage. Gray track could also be obtained by the application of a continuous electric field [102]. This phenomenon is responsible for the gray track in RTP Pockels cells at high voltage.

Like for KDP, in KTP and RTP crystals, damages generally appear in the bulk rather than on the surface [46]. The comparison between KTP and RTP revealed no significant difference of the laser damage resistance [49]. In 1-on-1 mode, the obtained LIDTs are 18 J/cm^2 at 1064 nm and 7 J/cm^2 at 532 nm, for 7 ns pulse duration [49,120]. In S-on-1 mode at 1064 nm the LIDTs in both materials decrease with shot number to approximately 50% of the 1-on-1 LIDT, whereas no effect is observed at 532 nm [49,120,117]. Yoshida et al. measured a polarization-dependent LIDT anisotropy in KTP at 1064 nm [130,131]. KTP and RTP LIDTs are dependent on the polarization of the incident beam, with a significantly higher threshold for z-polarized light, whereas no effect of the propagation direction has been observed [49] (see Section 11.3.2). It has also been evidenced that gray tracking is polarization dependent [19]. The laser damage anisotropy has been attributed to a color center–based laser damage mechanism [122] where the lifetime of the color centers depends on the polarization of the laser light.

During SHG in KTP, the conversion of 1064–532 nm leads to a lower damage threshold than for pure 1064 or pure 532 nm irradiation [120]. The damage mechanism for simultaneous 1064 and 532 nm irradiations is strongly cooperative between these two wavelengths (see Section 11.3.3). The laser damage mechanism in KTP is based on preferential damage precursor (color center) generation by 532 nm light, followed by the preferential damage precursor activation by 1064 nm light [39,122]. Besides, a strong LIDT difference at 1064 nm was also observed for x-cut and y-cut RTP crystals in the Pockels cell configuration, with a two times higher LIDT for x-cut RTP [121]. This result can be explained by the influence of unwanted frequency doubling in RTP on the LIDT with a higher unwanted second harmonic intensity in the case of the y-cut RTP Pockels cell.

Laser damage in RTP was found to be independent on significant variations in crystal quality (ionic conductivity, IR absorption, structural defects) as long as optical quality and bulk absorption values are good [50] (see Section 11.3.1).

To conclude, the laser damage resistance of KTP and RTP in the nanosecond regime seems to be limited by inherent material properties due to an important role of the color centers in the damage mechanism [122].

11.2.1.3 Borates

Borate crystals are usually used for high-power ultraviolet (UV) generation because of their adequate transparency in the UV region, their large nonlinear optical coefficients, and their relatively high tolerance to laser damage. The most commonly used borates are lithium triborate (LBO), β-barium borate (BBO), cesium lithium borate (CLBO), bismuth triborate (BiBO), and cesium borate (CBO).

11.2.1.3.1 Lithium Triborate (LiB₃O₅, LBO)

11.2.1.3.1 Lithium Triborate (LiB$_3$O$_5$, LBO)   Lithium triborate (LiB$_3$O$_5$, LBO) is an excellent nonlinear optical crystal for UV applications. It exhibits high optical homogeneity, low absorption in a very broad transmission range (from 160 nm to 2.6 μm), good chemical and mechanical properties, and a very high laser damage threshold. However, LBO is slightly hygroscopic and has a high thermal dilatation coefficient. LBO has clear advantages in high-power UV applications. It is used for frequency doubling and tripling of high peak power pulsed Nd^{3+}-doped lasers and Ti:sapphire lasers, and for optical parametric processes.

In 1994, Nikogosyan listed LBO LIDT data from literature, underlining an extraordinary high LIDT [83]. The very weak presence of impurities in LBO, due to its crystal structure and the resulting regularity of the lattice, seems to contribute to its very high damage resistance [25] (see also Section 11.3.1). Bulk LIDT at 1064 nm has been reported to be around 1.5 times higher than the one of fused silica [16,42,119,130]. In 1-on-1 mode, bulk LIDTs measured were 155 J/cm^2 at 1064 nm, 74 J/cm^2 at 532 nm, and 22 J/cm^2 at 355 nm, with a 7 ns pulse duration [46,118,119]. In addition, multiple pulse measurements revealed a strong fatigue effect at 1064 nm. A weaker fatigue effect was observed at 532 nm, and no bulk fatigue effect was found at 355 nm [119]. Note that surface damage under UV irradiation can appear during long-term 355 nm generation (>500 h at 20 kHz) in the nanosecond regime due to the interaction of UV laser light with the LBO surface and foreign atoms in the ambient atmosphere [75]. Laser-induced optical damage upon sum-frequency generation was also investigated by simultaneous exposure to intense 1064, 532, and 355 nm pulsed laser light. Multiphoton absorption processes were considered to be important for possible damage mechanisms [42] (cf. Section 11.3.3).

Similarly to KTP, RTP, and KN, LBO exhibits polarization anisotropy of the damage threshold. A higher LIDT for an incident beam polarized along the z-axis was evidenced [119] (see Section 11.3.2). Some of these observations are similar to the ones in KTP, and possibly color centers that are stabilized by nearby Li$^+$ vacancies play an important role in the damage mechanism. Color centers causing transient optical absorption upon high-power UV irradiation have been identified by Hong et al. [52]. These trapped-hole centers due to their intrinsic nature cannot be eliminated, but they can be limited by minimizing the oxygen vacancies. In 2001, Kim et al. [63] showed that the incorporation of NaCl in the flux, in order to help the growth by decreasing the flux viscosity, leads to crystals that are more susceptible to cracking. A significant reduction (about 30%) in optical damage resistance was observed when the level of Na$^+$ ions incorporated into the lattice exceeded approximately 250 ppm.

11.2.1.3.2 β-Barium Borate (β-BaB₂O₄, BBO)

11.2.1.3.2 β-Barium Borate (β-BaB$_2$O$_4$, BBO)   β-Barium borate (β-BaB$_2$O$_4$, BBO) has a very good optical quality, a wide transmission range from 190 nm in the UV to beyond 3.5 μm in the IR, and a broad phase matching range. However, BBO is hygroscopic and has a high birefringence. Compared with LBO, BBO has a larger nonlinear coefficient and a wider thermal acceptance bandwidth, but a lower laser damage threshold. Like LBO, BBO is particularly well adapted for high-power applications in the deep UV. It is used for harmonics generation (up to the fifth) of Nd:YAG and Nd:YLF lasers, for frequency doubling and tripling of ultrashort pulses from Ti:sapphire lasers, and also for optical parametric conversions for high-power broadly tunable sources.

In 1988, Nakatani et al. measured single- and multiple-shot LIDT at 1064, 532, and 355 nm, and obtained lower LIDT under multiple irradiations revealing a fatigue effect of the crystal [78]. In 1999, Kouta et al. [68] reported bulk LIDTs under multiple irradiations (10^4 shots at a repetition rate of 10 Hz) of 45 J/cm^2 at 1064 nm, 10 ns pulse width; 20.7 J/cm^2 at 532 nm, 9 ns pulse width; 7.2 J/cm^2 at 355 nm, 8 ns pulse width; and 2.4 J/cm^2 at 266 nm, 8 ns pulse width. The surface LIDTs measured were approximately half of these values. Yoshida et al. [130] found that the damage threshold of BBO depends on the direction of the laser irradiation, with a nearly two times higher threshold for an incident beam along the c-axis at 1064 nm. For an irradiation along the a- or b-axis, the 1-on-1 bulk LIDTs were 12 J/cm^2 at 1064 nm, 1.1 ns pulse duration and 4 J/cm^2 at 355 nm, 0.75 ns pulse duration. Note that using the empirical scaling law in the nanosecond regime ($\tau_{0.5}$ dependency) [126], these values yield 36 J/cm^2 at 1064 nm, 10 ns and 13 J/cm^2 at 355 nm, 8 ns, which is quite close to the values obtained by Kouta et al. in multiple pulse tests [68]. Besides, the damage thresholds at 1064 and 532 nm are correlated with some types of inclusions and impurities present in BBO [14,111], such as Na$^+$-ion inclusions [13].

11.2.1.3.3 Cesium Lithium Borate (CsLiB$_6$O$_{10}$, CLBO)

Cesium lithium borate (CsLiB$_6$O$_{10}$, CLBO) is transparent from 180 nm to beyond 2700 nm, has low optical losses, and a wide phase matching range. A major disadvantage of CLBO is that it is severely hygroscopic. In comparison to BBO, CLBO exhibits larger angular and spectral bandwidths, a smaller walk-off angle, but also a lower nonlinear coefficient. It is particularly well suited for the generation of the fourth and fifth harmonics of Nd:YAG laser sources and is an excellent choice for sum frequency mixing applications.

Like for BBO crystals, the damage threshold of CLBO is around two times higher for an incident beam along the c-axis at 1064 nm than for the a- or b-axis. In these directions, the 1-on-1 bulk LIDTs were 16 J/cm^2 at 1064 nm (1.1 ns pulse duration) and 7 J/cm^2 at 266 nm (0.75 ns pulse duration) [130,131]. High-quality growth CLBO crystals with lower defect concentration have lower UV absorption and higher bulk laser damage threshold, which is approximately two times higher at 266 nm (11–15 J/cm^2, for 0.75 ns pulse duration) than for CLBO grown by the conventional method and for fused silica [85,86]. Moreover, the surface LIDT of CLBO can be enhanced by removing residual surface-polishing compounds using ion-beam etching [57]. Since CLBO is hygroscopic, moisture can degrade the crystal quality. Surface degradation could be suppressed by increasing the crystal temperature, but a water impurity is also present in the crystal. The bulk LIDT of CLBO in the UV region can be improved by eliminating water using a heat treatment of the crystal (150°C in ambient atmosphere and subsequently in dry atmosphere) [60].

11.2.1.3.4 Bismuth Triborate (BiB$_3$O$_6$, BiBO)

Bismuth triborate (BiB$_3$O$_6$, BiBO) is a recently developed nonlinear optical crystal. It possesses large effective nonlinear coefficients (3.5–4 times higher than that of LBO and 1.5–2 times higher than that of BBO). It has a broad transparency range from 280 to 2500 nm, a wide temperature bandwidth, and is not hygroscopic. Moreover, BiBO has a high damage threshold comparable to LBO [45]. It is well adapted for SHG and THG of middle- and high-power Nd^{3+}-doped lasers. It has been used for producing blue light through SHG of Nd^{3+} lasers at 914 and 946 nm or Ti:sapphire lasers.

Chapter 11

In 2010, Petrov et al. reviewed physical properties, linear and nonlinear optical characteristics of BiBO [90]. They reported some damage threshold values from literature. In 2011, Yun et al. reported that BiBO crystals are susceptible to optical damage by continuous waves (CW) UV and green lasers at high intensities and by a CW argon laser of 476, 488, and 515 nm wavelengths [133]. Investigation of the phenomenon showed that the beam deformation was induced by the photorefractive effect and could be repaired completely by annealing (300°C during 24 h). BiBO crystals are more susceptible to photorefractive effects in visible than in IR light [56].

11.2.1.3.5 Cesium Triborate (CsB_3O_5, CBO)

Cesium triborate (CsB_3O_5, CBO) is transparent far into the UV region (170–3000 nm). CBO has nearly the same nonlinear characteristics as LBO [127]. It is mainly used for third harmonic generation of Nd:YAG lasers. CBO has a high damage threshold of 26 J/cm² at 1053 nm and 1 ns pulse duration [127,128]. At 355 nm and 6 ns pulse, the 1-on-1 bulk LIDT was found to be about two times higher than fused quartz [95].

CBO can be grown by top-seeded solution growth (TSSG) from self-flux solutions. However, the presence of scattering centers in the as-grown crystal decreases its LIDT. Recently, a new postgrowth quenching method, providing fast cooling, allowed to decrease the concentration of scattering centers and improved its LIDT by a factor of 1.5 compared with the conventional cooling process [125].

11.2.1.4 Infrared Materials

Compared with nonlinear crystals for other wavelength regions, IR crystals have often lower laser damage thresholds. In the last decade, a lot of work was done to improve middle and deep IR nonlinear materials [140]. The most commonly used materials are zinc germanium phosphide (ZGP, $ZnGeP_2$), silver gallium selenide and sulfide ($AgGaSe_2$ and $AgGaS_2$), gallium selenide (GaSe), and orientation-patterned gallium arsenide (OP-GaAs).

11.2.1.4.1 Zinc Germanium Phosphide ($ZnGeP_2$, ZGP)

Zinc germanium phosphide ($ZnGeP_2$, ZGP) is a highly effective nonlinear optical crystal for the mid-IR. It has a wide transparency range from 0.75 to 12.0 μm, and its useful transmission ranges from 1.9 to 10.6 μm. It has a high nonlinear coefficient and a good thermal conductivity. ZGP crystals are successfully used in diverse applications like SHG of pulsed CO or CO_2 lasers and OPO of mid-IR wavelengths pumped by holmium lasers. This latter application motivated several laser damage studies of ZGP at 2 μm.

In the early development stage of this crystal, Peterson et al. [89] studied laser damage of ZGP from different boules with different polish and different AR coatings. Damage always occurred at the sample surface rather than in the bulk. The R-on-1 damage thresholds measured at 2.08 μm and 70 ns pulse duration ranged from 2 to 20 J/cm². The LIDTs varied between sites on a given sample and between samples, indicating a lack of uniformity and repeatability of the ZGP growth.

More recently, laser damage studies were performed in S-on-1 mode. Damage of the test sites was detected with an optical microscope after an exposure of 30 s [105,135,136]. The surface LIDT was found to be significantly lower for uncoated ZGP than for the AR-coated samples. ZGP seems to be sensitive to the fatigue effect with an LIDT dependency on repetition rates (LIDT = 2 J/cm² at 2.09 μm, 2 Hz and LIDT = 1 J/cm² at

2.05 µm, 10 kHz) [105]. As usual, the LIDT also depends on wavelength with better laser damage resistance at longer wavelengths (at 10 kHz: LIDT = 1 J/cm² at 2.05 µm, LIDT = 1.9 J/cm² at 3 µm, and LIDT = 2.6 J/cm² at 4 µm) [105]. However, no temperature dependency was observed over the range of −5°C to 55°C [105].

Zawilski et al. measured an LIDT of 2 J/cm² for AR-coated ZGP at 2.05 µm, 10 kHz, 15 ns pulses duration, and 130 µm beam diameter for an improved surface polish [135,136] compared with 1 J/cm² obtained in their previous work [105]. They observed that the removal of more material during polishing significantly improves the LIDT, indicating the importance of subsurface damage, which is more known in silica [81].

11.2.1.4.2 Silver Gallium Selenide (AgGaSe₂, AGSe) and Silver Gallium Sulfide (AgGaS₂, AGS) Silver gallium selenide (AgGaSe₂, AGSe) and silver gallium sulfide (AgGaS₂, AGS) have wide transmission ranges in the IR region (from 0.65 to 18 µm for AGSe and from 0.46 to 12 µm for AGS). Their wide phase-matching abilities make them excellent materials for OPO applications, producing tunable radiation over a wide range of wavelengths in the IR. For example, an AGSe OPO pumped by an Ho:YLF laser at 2.05 µm generates output from 2.5 to 12 µm. AGSe and AGS have also been demonstrated to be efficient frequency doubling crystals for IR radiation such as the CO_2 laser at 10.6 µm.

AGSe laser damage thresholds were measured at 2 and 10 µm. In 1991, Ziegler et al. [141] measured a damage threshold of 3.5 J/cm² for an AGSe crystal with AR-coated surfaces at 2.1 µm, 1 Hz, and 180 ns pulse duration in R-on-1 mode (with 100 shots per fluence step). Acharekar et al. [8] obtained a lower threshold at 2.09 µm for uncoated surfaces. The LIDT of AGSe was also measured with a continuous wave CO_2 laser at 9.042 µm in R-on-1 mode (with 5 min at each power level). A threshold of 60 kW/cm² was obtained for the AR coating [9]. Kildal et al. measured damage thresholds at 10.6 µm and 150 ns pulse duration. They obtained between 1.5 and 3 J/cm² and between 3 and 3.5 J/cm² for pit formation on the surface of AGSe and AGS, respectively [62].

11.2.1.4.3 Gallium Selenide (GaSe) Gallium selenide (GaSe) combines a large nonlinear coefficient and a wide transparency range (from 0.62 to 18 µm). It is very suitable for efficient SHG of CO_2 lasers and optical parametric generation within the 3.5–18 µm wavelength range.

In 2010, Telminov et al. obtained a laser damage threshold of 16 MW/cm² on the surface at 10.6 µm and 50 ns pulse duration [112].

11.2.1.4.4 Gallium Arsenide (GaAs) Gallium arsenide (GaAs) has a large nonlinear susceptibility, a good thermal conductivity, excellent mechanical properties, low absorption losses, and a wide transparency range (from 0.9 to 17 µm). Use of GaAs as an efficient nonlinear optical material has been enabled by the development of fabrication and growth technologies to achieve quasi-phase-matching (QPM) in orientation-patterned GaAs. Orientation-patterned GaAs (OP-GaAs) is a type of QPM GaAs in which periodic inversions of the crystallographic orientation are generated during crystal growth. OP-GaAs crystals were shown to be well adapted for efficient mid-IR OPOs and also permitted the demonstration of efficient SHG and difference frequency generation.

To our knowledge, no systematic laser damage study has been published on this material for the moment. The only available data are due to the use of OP-GaAs in OPO

Chapter 11

cavities pumped at 2.09 with a 200 µm spot size. Damages occurred on the coatings of the crystal at a fluence of about 2 J/cm² at 50 kHz, which is comparable to ZGP crystals in the same configuration [61]. In another OPO cavity, also pumped by an Ho:YAG laser, damages occurred at fluences of 1.9 and 1.5 J/cm² at 40 and 70 kHz, respectively [47]. Note that these values correspond to a single test and not to a systematic study.

11.2.2 Q-Switching

Q-switching is a method for obtaining typically nanosecond pulses from lasers by modulating the intracavity losses. Q-switching can be achieved with acousto-optic (AO) or electro-optic (EO) devices.

11.2.2.1 Acousto-Optic Crystals

Materials used for AO devices like Q-switches in lasers include fused silica (SiO_2), lithium niobate ($LiNbO_3$, LN), and tellurium dioxide (TeO_2).

Fused silica has the highest LIDT of the mentioned AO materials (cf. Chapter 9). Lithium niobate presented previously in Section 11.2.1.1 is widely used in AO devices. Tellurium dioxide is also commonly used for AO modulators. This material exhibits high optical homogeneity, low light absorption and scattering, and a transmission ranging from 330 nm to 5 µm, which is similar to the one of lithium niobate. In 1993, Acharekar et al. [7] published LIDT data for AO Q-switch materials. At 1 µm, TeO_2 and LN have a similar LIDT (327 and 302 MW/cm², respectively). At 2 µm, TeO_2 has higher LIDT than LN (454 and 240 MW/cm², respectively).

11.2.2.2 Electro-Optic Crystals

Several crystals are used for their electro-optical properties [100]. Deuterated potassium dihydrogen phosphate (KD_2PO_4, DKDP or KD*P), lithium niobate ($LiNbO_3$, LN), lithium tantalite ($LiTaO_3$), rubidium titanyl phosphate ($RbTiOPO_4$, RTP), and β-barium borate (β-BaB_2O_4, BBO) are the most frequently used. A comparison of these crystals for the Pockels cell applications is reported in Table 11.1.

Table 11.1 Comparison of Electro-Optic (E-O) Crystals for Pockels Cells Applications

Crystal	Transmission Range (µm)	Effective E-O Coefficient (pm/V)	Insertion Losses (%)	Extinction Ratio at 0.63 µm	LIDT (MW/cm²) at 1.06 µm. 10 ns
DKDP	0.2–2.1	$r_{63} = 23.6$	4	>1000:1	600
LiNbO₃	0.33–5.5	$r_{22} = 6.8$	<1	>400:1	100
LiTaO₃	0.28–5.5	$r_e = 21.5$	<1	>200:1	500 (at 1.06 µm, 1 ns)
RTP	0.35–4.5	$r_{e1} = 30.2$ $r_{e2} = 23.6$	<1	800:1	1800
BBO	0.19–3.5	$r_{22} = 2.2$	<2	>500:1	5000

Source: Roth, M. et al., *Glass Phys. Chem.*, 31(1), 86, 2005.

Note: The extinction ratio gives the relative minimum intensity that can be achieved for a Pockels cell and a polarizer using linearly polarized light at the input.

DKDP crystals are widely used for electro-optic applications. Thanks to a high degree of optical homogeneity, DKDP has a very good extinction ratio, and large apertures are possible. DKDP has a laser damage threshold higher than $LiNbO_3$. Moreover, as DKDP has a high effective electro-optic coefficient, the operating voltage is as low as for RTP crystals. However, the insertion losses of DKDP cells are 4% in the high transparency optical range of 0.3–1.2 μm. The main disadvantage of these crystals is that they are hygroscopic and so require protection against moisture. For more detailed laser damage information, refer to Section 11.2.1.2.

$LiNbO_3$ is a popular material used in Q-switched lasers. It is a nonhygroscopic crystal that has low insertion losses that do not exceed 1% in modern devices. Nevertheless, it has a low laser damage threshold (see Section 11.2.1.1), a mediocre extinction ratio, and it is very sensitive to piezoelectric ringing at frequencies above 1 kHz.

$LiTaO_3$ is also widely used due to its optical and electro-optical properties similar to those of $LiNbO_3$ crystals. Compared with $LiNbO_3$, the main advantage is its higher damage threshold [131].

RTP belongs to the same crystal family as KTP (cf. Section 11.2.1.2). Thus, RTP crystals have similar nonlinear optical properties as KTP. However, with a higher effective electro-optic coefficient, RTP is more used for electro-optic applications, whereas KTP is popular for SHG of Nd:YAG lasers. RTP can be operated at low voltage without piezoelectric ringing at high repetition rates. Moreover, RTP is a nonhygroscopic crystal and has low insertion losses. It has a high laser damage threshold and a good extinction ratio, which is better than 20 dB. However, thermal birefringence needs to be compensated by the use of two crystals turned by 90° around the light propagation axis.

BBO is the electro-optic material of choice for the high-power Pockels cell applications. BBO crystals are transparent down to 0.19 μm. It has a good extinction ratio, relatively low losses, and a very high laser damage threshold (more laser damage data are given in Section 11.2.1.3). No piezoelectric ringing was observed in BBO Pockels cells up to 6 kHz. However, with a very small effective electro-optic coefficient, the BBO operating voltage is very high. Thus, the aperture sizes have to be kept as small as possible. Moreover, BBO is a hygroscopic material.

11.2.3 Solid–State Laser-Active Materials

A solid-state laser-active material is defined by a solid-state host material doped with activator ions. A very large number of solid-state laser-active materials exists [59]. Solid-state host materials may be separated into two groups: glasses and crystalline materials. Glasses can be produced in large dimensions with very good optical quality. However, crystals usually have higher transition cross sections, higher thermal conductivity, and smaller absorption and emission bandwidths. In some cases, single crystals may be replaced by ceramic gain media, which constitute an attractive alternative because of their ease of manufacture, low cost, and scalability. These host materials can be doped with laser-active rare earth ions (neodymium [Nd^{3+}], erbium [Er^{3+}], ytterbium [Yb^{3+}], holmium [Ho^{3+}], thulium [Tm^{3+}], cerium [Ce^{3+}]) or with transition metal ions (titanium [Ti^{3+}], chromium [Cr^{4+}]).

Due to the development of more and more powerful lasers, solid-state laser materials are now also exposed to the laser damage issue. However, very few data can be found

Chapter 11

in literature. We can suppose that coatings are generally less resistant than the bulk of these materials. In this section, we present the most frequently used laser materials with their laser damage thresholds.

11.2.3.1 Glasses

Glasses form an important class of host materials. They have the advantages to be easily fabricated, available in large size and excellent optical quality. The most common optical glasses are oxide glasses, mainly silicates (SiO_2) and phosphates (P_2O_5). Glasses can be doped at very high concentration with excellent uniformity. They are often doped with Nd^{3+} (emitting at 1062 nm for SiO_2 and 1054 nm for P_2O_5) or also with Er^{3+} (emitting between 1.53 and 1.56 μm) and Yb^{3+} (emitting at 1030 nm).

In general, laser damage thresholds for glass lasers have been found to run higher than their crystalline counterparts [64]. The laser damage properties of glasses are detailed in Chapter 9.

11.2.3.2 Garnets

Yttrium aluminum garnet (YAG, with chemical formula $Y_3Al_5O_{12}$) is commonly used as a host material in various solid-state lasers. YAG is mechanically robust, optically isotropic, and transparent from 300 nm to 4 μm. YAG accepts laser active ions from both rare earth elements and transition metal elements. YAG is generally doped with neodymium. Nd^{3+}:YAG is the most commonly used active laser medium in solid-state lasers, producing IR light at a wavelength of 1064 nm. YAG can also be doped with erbium (emitting at 2.94 μm), ytterbium (emitting at 1.03 μm), holmium (emitting at 2.1 μm), thulium (emitting between 1.9 and 2.1 μm), or chromium (emitting between 1.35 and 1.65 μm).

Several papers studied laser damage in single crystalline and ceramic YAG. High-quality ceramic YAG and crystalline YAG have the same damage threshold [15,58]. In 2003, Bisson et al. [15] measured single-pulse damage at 1064 nm and 4 ns pulses duration. They obtained a bulk damage threshold of about 100 J/cm², which is nearly three times lower than the LIDT of synthetic fused silica (Suprasil) that measures 280 J/cm². In 2008, Kamimura et al. [58] obtained a threshold in YAG approximately half that of silica at 1064 nm and 8 ns. Low-level doping seems to have no influence on the LIDT. Bisson et al. reported no difference between 0.7%-Nd^{3+}-doped and undoped YAGs [15]. Zelmon et al. [137] observed no influence for doping levels up to 4%-Nd^{3+}. However, at 8% doping, the crystal transparency was diminished, and the damage threshold was a factor of 5 lower. Do et al. [32] found substantial variation of thresholds for different doping ions: 0.7%-Nd^{3+}, 8%-Yb^{3+}, and 0.2%-Cr^{3+}.

11.2.3.3 Sapphire

Titanium-doped sapphire (Ti^{3+}:Al_2O_3) is widely used for tunable lasers emitting in the range of 650–1100 nm, in particular at 800 nm, and to generate ultrashort pulses (femtosecond regime). Different papers studied the laser damage in Ti:sapphire, including a recent thesis on this subject [21].

In 1994, Sun et al. [109] published Ti:sapphire damage thresholds measured at 532 nm, 10 ns, conditions that correspond to a typical pump laser for a Ti:sapphire laser. They found that the surface damage threshold (varying from 6 to 23 J/cm²) is much lower than the bulk damage threshold (varying from 57 to 103 J/cm^2). Note that the thresholds were defined at 50% damage probability. In their study, bulk damage was mainly caused by self-focusing in the crystal. Recently, in 2010, Bussière et al. [21] found that bulk and uncoated surfaces, prepared with an optimized polishing process, have the same threshold in the nanosecond, picosecond, and femtosecond regimes. The surface damage mechanism is due to the material itself when the surface is exempt of defects like scratches. Note that using the same polishing process for undoped and doped sapphires results in a better surface quality of undoped sapphire. The surface LIDTs of undoped, uncoated, and superpolished sapphire disks are the following: 11 J/cm^2 at 532 nm, 8 ns; 18.5 J/cm^2 at 1064 nm, 50 ps; and 210 J/cm^2 at 800 nm, 100 fs [114]. These results reflect a weak LIDT dependency on wavelength in the range of 532–1064 nm, which can be understood comparing the band gap of sapphire (8.8 eV) to the photon energy range (1.2–2.3 eV). The damage threshold of sapphire is strongly correlated with the Ti^{3+} ion doping concentration: titanium-doped sapphire exhibits a higher threshold than undoped sapphire whatever the wavelengths with a more pronounced effect in nanosecond than picosecond (the effect is negligible in femtosecond) [22]. The authors attribute this effect to the possibility of another relaxation pathway of electronic excitation that has been introduced to the material by doping.

11.2.3.4   Fluorides

Yttrium lithium fluoride ($YLiF_4$) crystals are naturally birefringent and transparent within the spectral band of 0.12–7.5 μm. YLF is often doped with Nd^{3+}. The commonly used laser transitions occur at 1047 and 1053 nm. Nd:YLF offers an alternative to Nd:YAG for near IR operation thanks to a weak thermal lensing and a long fluorescence lifetime, which allow higher energy storage. However, the Nd:YLF crystal fractures easily and is slightly hygroscopic. No data on laser damage of YLF were found.

11.2.3.5   Vanadates

The vanadate (YVO_4) crystal is naturally birefringent, and laser output is linearly polarized, which avoids undesirable thermally induced birefringence. The most popular vanadate crystal is neodymium-doped yttrium orthovanadate ($Nd:YVO_4$), which typically emits at 914, 1064, or 1342 nm. $Nd:YVO_4$ has a large stimulated emission cross section that is five times higher than the one of Nd:YAG and a strong broadband absorption at 809 nm.

In 1974, Leung et al. measured the surface damage threshold of YVO_4 using a Q-switched ruby laser at 0.69 μm and a spot size of 190 μm [71]. They obtained an LIDT of 7 J/cm^2 and concluded that inclusions were the cause of damage. We can assume that nowadays the surface damage threshold is higher since the polishing processes have been optimized, and YVO_4 lasers are widely used now.

11.2.3.6   Tungstates

Neodymium-doped potassium gadolinium tungstate crystals (Nd:KGW, with chemical formula $Nd^{3+}:KGd(WO_4)_2$) are a highly efficient laser medium (three to five times higher compared with Nd:YAG) with a low lasing threshold. It is transparent from 0.35 to 5.5 μm and commonly used laser transitions occur at 911, 1067, and 1351 nm.

Chapter 11

The laser damage threshold of Nd:KGW was measured at 1064 nm, 8 ns with a focused beam of 14 μm diameter. The bulk LIDT obtained was 59 J/cm², and the antireflection coating LIDT was 29 J/cm² [46].

11.3 Specificities of Laser Damage in Nonlinear Crystals

In this section, we briefly discuss some of the physical mechanisms involved in laser-induced damage initiation in nonlinear crystals as far as they are understood today. Extensive experimental studies that allow to conclude on the physical phenomena involved in the initiation of laser-induced damage are available for only a few crystals: the $KTiOPO_4$ (KTP) family (including KTP and RTP) and KH_2PO_4 (KDP) and its deuterated form $KD_xH_{2-x}PO_4$ (DKDP) where the latter two are used for the inertial confinement fusion class lasers. Some studies are also available on LiB_3O_5 (LBO) or other borates like β-BaB_2O_4 (BBO). Practically, all of these studies use nanosecond lasers, the most popular being the Q-switched Nd:YAG laser at 1064 nm and its harmonics at 532, 355, and 266 nm. Unlike in fused silica or glass, a parallel beam traveling through a nonlinear crystal of KDP or the KTP family usually induces damage in the bulk and not on the uncoated surfaces [49,94]. For uncoated LBO, the weak point of the crystal depends on the used wavelength [119]. In order to avoid being influenced by polishing and coating technology, this section concentrates on the results obtained in the bulk material of the crystals.

11.3.1 Effects of Impurities

During early stages of crystal growth optimization, the influence of large-scale defects on the damage threshold can be observed [92]. For optically homogeneous crystals too, atomic scale or nanometer-sized impurity sites in crystals often cause enhanced absorption and reduced optical quality related to small refractive index variations. The question whether or not impurities cause reduced laser damage resistance has most extensively been studied for rapid growth KDP. Rapid growth KDP develops two growth sectors, pyramidal and prismatic sectors, where the prismatic growth sector contains in general higher impurity concentrations than the pyramidal one [41]. In particular, the iron impurities induce an absorption band at about 280 nm, which reduces the laser damage threshold for 355 nm radiation [41]. However, once the growth process has been optimized in order to reduce absorbing impurities to an acceptable level, the volume density of laser-induced damages at a given fluence no longer depends on the impurity level [91], but rather correlates with the growth temperature and the stability of the growth process [82]. The present understanding of the experimental data is that irregularities in the crystal structure enhance clustering of atomic defects (intrinsic or impurity related) that would not induce damage if they were isolated [30]. Structure irregularities (like dislocations) may also cause local charge compensation, thus stabilizing atomic defects that are usually only observed at low temperature. Thermal annealing [41,44,107] and laser annealing [27,33,73] have been shown to increase the laser damage resistance of KDP at 351 nm, which is probably related to a size or density reduction of defect clusters [73].

For high-quality $RbTiOPO_4$ (RTP) too, laser damage tests have been carried out in two different growth sectors and no significant difference in the laser damage probability curves at 1064 nm was observed despite a factor of 2 in the absorption values of the different growth sectors [50]. The absorption in the bulk material of this crystal has been determined by a photothermal technique [26] to be in the range of 100–300 ppm/cm. Similar results have also been presented for KTP [16] and LBO [42]. Further, RTP can be grown in a standard $Rb_6P_4O_{13}$ flux [53] or a self-flux [113], the latter resulting in crystals with better stoichiometry exhibiting much lower ionic conductivity (along the z-axis). Laser damage measurements with multimode 1064 nm, 6 ns laser pulses in both types of crystals coming from different providers are practically identical. Additionally, laser damage measurements in experimental growth RTP using different fluxes that influence ionic conductivity, absorption, and optical quality also resulted in only minor variations that can be ascribed to the variations in optical quality [50].

For the KTP family, where these results are corroborated by other studies [117,122], the laser damage mechanism thus involves light-induced laser damage precursors, in contrast to fabrication-induced laser damage precursors, as they can be found on fused silica surfaces [81] or in KDP, for example.

The sensitivity of KTP to gray tracking however still depends on the growth method or the applied posttreatment. The intensity where gray tracking begins to appear is too low for efficient light-induced damage precursor generation. Gray tracking, also named "photochromic damage," is in fact color center generation at defect sites in the crystal (see Section 11.2.1.2). This kind of damage, which appears typically during frequency conversion of 1064 nm nanosecond pulses to 532 nm, is caused by Ti ions in the Ti^{3+} oxidation state and may depend on Fe impurities in the crystals [103] but also on other influences. The topic has been studied in detail and a number of patents protect "gray track resistant" types of KTP [74,76]. During SHG of 1064 nm, the color centers are mainly created by two-photon absorption of 532 nm photons [18]. As the color centers absorb also 1064 nm radiation, some authors also speak about GRIIRA or BLIIRA (Blue-Induced InfraRed Absorption) [51,124]. At high intensities, the color centers may, however, be generated also in the absence of growth-related defects as the light-induced Frenkel pairs can act as charge traps in KTP [35,77,93,124].

Lithium triborate, LiB_3O_5 (LBO), is a high-quality material that exhibits a very high laser damage resistance for 1064 nm nanosecond pulses, even higher than the bulk of synthetic fused silica [42,83,119]. The crystalline structure of LBO is very compact, and the material thus presents only few sites that can accept impurity ions during the growth [25]. Remembering the findings for KDP and the KTP family, it is probably less the absence of impurities in the material that causes this high laser damage resistance, but more the very good regularity of the crystalline structure and the large band gap (see Figure 11.1). A large band gap is always an advantage in terms of laser damage resistance as the conditions for creating electron hole pairs are more difficult to achieve in this case. (Short wavelengths or very high intensities are needed.)

In summary, impurities are important for the laser damage threshold if they cause strong absorption, but once the growth procedure or the posttreatment of the crystals is optimized, impurities usually are no longer the limiting factor for the laser damage resistance of nonlinear crystals.

Chapter 11

11.3.2   Anisotropy of Laser Damage

Crystals are anisotropic materials and as such exhibit also anisotropic optical and mechanical properties. KDP, for example, is a uniaxial crystal while the KTP family crystals and LBO are biaxial crystals. The linear refractive index of these materials depends on polarization and propagation direction of the laser beam [101]. The absorption properties (especially in laser-active materials [69]) and the nonlinear optical properties of the crystals are anisotropic too [31].

Thus, one expects the laser damage behavior of crystals to be also anisotropic, and experimental studies on the anisotropy of laser damage have been published for a number of crystals. When speaking about the propagation direction or the polarization of the light, we will refer to the coordinate system x, y, and z where z is the axis with the highest refractive index for biaxial crystals or the direction of the optical axis for uniaxial crystals.

For KDP, Reyne et al. carried out a detailed study using a type II THG-cut KDP crystal for 1053 nm [97]. They rotated the crystal around the propagation direction of the light and thus varied the orientation of the linearly polarized 1064 nm test beam from y-polarization (ordinary polarization) to a polarization in the x–z plane (extraordinary polarization). They found a clearly enhanced (140%) laser damage resistance for the extraordinary polarization (e-polarization). Considering that the authors showed that unwanted frequency SHG was very low for o- and e-polarizations, this damage threshold difference has to be attributed to the orientation of the polarization of the pump beam. The experimental observations have successfully been described by a model postulating oblate-type ellipsoid-shaped damage precursors oriented along the z-axis (optical axis) of the crystal [97]. Other recent publications also report an enhanced laser damage threshold for z-polarized or e-polarized light [91,130], whereas earlier studies report that the damage threshold is more dependent on the light propagation direction than on the polarization [20,132]. As laser damage precursors in KDP are growth condition dependent, the differences probably result from evolutions in the KDP growth process.

Crystals of the KTP family and LBO show a clearly higher damage resistance with respect to z-polarized light [49,119–121]. This polarization-dependent anisotropy in the laser damage resistance is stronger at 1064 nm compared with 532 and 355 nm [49,119,120]. No significant differences were observed between x- and y-polarized lights, and also the propagation direction of the light has no influence on the laser damage properties as long as frequency conversion can be neglected.

Despite a factor of 10 in the absolute values of the LIDTs of KTP and LBO, similarities exist between these crystals: In KTP, the color centers that cause gray tracking are stabilized by K^+ vacancies [36]. Similar color centers, stabilized by Li^+ vacancies, exist in LBO [52]. The physical mechanism for the laser damage anisotropy, at least in KTP, is probably a lifetime reduction of the light-induced color centers by certain phonons that are created by the Raman scattering. As mentioned in Section 11.3.1, nanosecond laser damage in the KTP family crystals is caused by light-induced color centers. Boulanger et al. found that z-polarized light causes higher color center concentrations as x- or y-polarized light [19]. At intense z-polarized 1064 nm irradiation however, stimulated Raman scattering (SRS) at 269 cm^{-1} is observed and according to several observations these 269 cm^{-1} phonons interact with the color centers decreasing their lifetime [87,88,124].

The laser damage anisotropy is thus generated by the anisotropy of the SRS threshold. This tentative model also explains that the laser damage anisotropy is only observed at certain irradiation conditions, those where the z-polarized SRS threshold is lower than the damage threshold.

For LBO, available data on the properties of the color centers and their interaction with phonons are less detailed, but the similarity of the laser damage observations and the types of color centers may indicate an analogous explanation of the polarization-dependent laser damage anisotropy in this material.

11.3.3   Multiwavelength Irradiation

Nonlinear crystals are often used for frequency conversion tasks. In this kind of application, usually the second-order nonlinear susceptibility of the crystal is exploited, and the crystal is thus irradiated simultaneously by two or three different wavelengths. Literature reports mention at least two cases where laser damage of double wavelength irradiation cannot be predicted by single-wavelength experiments alone: irradiation of KDP with 532 and 355 nm [29,98] and irradiation of KTP with 1064 and 532 nm [120,122].

Figure 11.2 schematically shows the effect of cooperative laser damage mechanisms on the LIDT observed during simultaneous irradiation with two wavelengths λ_1 and λ_2. Typically, the thresholds for the single-wavelength situations T_1 and T_2 are different, reflecting the differing capacities of each of the wavelengths to couple energy to the sample. In the case of a noncooperative damage mechanism, where the interaction of wavelength λ_1 with the sample is not influenced by the presence of light of wavelength λ_2, a linear dependence of the LIDT (in terms of total fluence $F(\lambda_1) + F(\lambda_2)$) on any of the fluences $F(\lambda_1)$ or $F(\lambda_2)$ is expected (see dashed black line in Figure 11.2). For a cooperative

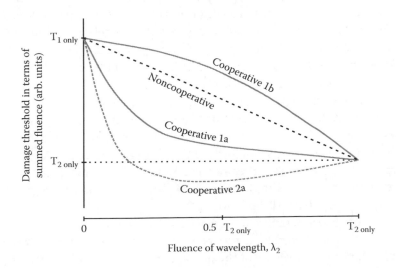

FIGURE 11.2   Dependency of the damage threshold in terms of total fluence $F(\lambda_1) + F(\lambda_2)$ on the fluence $F(\lambda_2)$. The dependency is sketched for three types of laser damage mechanisms: a noncooperative mechanism (dashed black line), two single-sided cooperative damage mechanisms (cooperatives 1a and b, solid gray lines), and a bidirectional cooperative damage mechanism (cooperative 2a, short-dashed gray line). The "a"-labeled mechanisms correspond to enhanced coupling efficiency of λ_1 light in the presence of λ_2 light.

damage mechanism, the presence of light of wavelength λ_2 influences the coupling efficiency of wavelength λ_1 to the sample. The coupling efficiency of λ_1 light may increase or decrease, but in any case the LIDT under simultaneous irradiation deviates from the linear behavior (see solid gray lines in Figure 11.2, "cooperative 1a" and "cooperative 1b," respectively). It is also possible that the cooperativeness of the mechanism is bidirectional: the presence of light of wavelength λ_2 enhances the coupling efficiency of wavelength λ_1 to the sample, and the presence of light of wavelength λ_1 enhances the coupling efficiency of wavelength λ_2 to the sample (see short-dashed gray line "cooperative 2a" in Figure 11.2).

In order to quantify the cooperatives of a laser damage mechanism under simultaneous double-wavelength irradiation, it is useful to define the "relative toxicity" γ between both wavelengths. γ is obtained by comparing fluence combinations that have the same damaging effect, for example, a certain damage density [29] or the threshold of damage probability curves obtained in similar experimental conditions [120]. In the latter case, the relative toxicity γ is defined by $T_{short\,only} = T_{short} + \gamma T_{long}$, where $T_{short\,only}$ is the threshold fluence for short-wavelength irradiation only and (T_{short}, T_{long}) is a combination of fluences used to reach the threshold upon simultaneous irradiation. A more detailed explanation of the definition of γ for the case of damage density measurements is given in R [29], where it is also shown that for KDP γ is independent of the damage density that is chosen as a reference. The relative toxicity γ is thus calculated by comparing the damaging effect of a particular simultaneous irradiation situation to the situation where only the shorter wavelength is used. γ may hence be considered as a function of the long wavelength fluence T_{long} used in the wavelength mixture.

For a noncooperative damage mechanism, $\gamma(T_{long})$ is constant and can be calculated by the two single-wavelength experiments ($\gamma = T_{short\,only}/T_{long\,only}$). This case has been observed upon simultaneous irradiation of KDP with 1064 and 355 nm, where $\gamma = 0.24 \pm 0.06$ [29]. This is in agreement with the fact that two different types of damage precursors have been observed at these wavelengths, one being responsible for damage at 1064 nm and another one being responsible for damage at 355 nm [28,80]. As a consequence, independent damage pathways exist for each of these wavelengths so that any cooperativeness of these wavelengths is excluded.

For simultaneous irradiation with 532 and 355 nm, the situation is different as γ increases from 0.3 to 0.85 with increasing 532 nm fluence [29]. (This case roughly corresponds to the case cooperative 1b in Figure 11.2.) In fact, both wavelengths act on the same type of damage precursor and more detailed measurements of this effect revealed insight into the electronic properties of the atomic defects that make up this type of damage precursors by clustering. The proposed electronic structure of the point defects involves two states in the band gap [30].

A case of a strongly cooperative damage mechanism is found in KTP with simultaneous 1064 and 532 nm irradiation, where γ decreases from 10 to 0.25 with increasing 1064 nm fluence [120]. The cooperativeness of the damage mechanism in KTP is probably caused by the fact that the generation of the light-induced damage precursors (color centers) is more efficient with 532 nm irradiation, whereas the catastrophic absorption by these damage precursors is driven more efficiently by 1064 nm irradiation. This damage mechanism quantitatively describes the damage threshold development for different SHG conversion efficiencies in KTP [122] and corresponds to the case cooperative 2a in Figure 11.2.

The strongly collaborative damage mechanism is also present in RTP with one of the standard applications of RTP being the Pockels cells for high repetition Q-switching at 1064 nm. In RTP Pockels cells, the electric field is applied in z-direction and the crystals may be y-cut or x-cut, but the x-cut crystals need an approximately 30% higher driving voltage. The polarization of the incoming light has to be oriented at 45° with respect to the z-axis so that unwanted type II SHG can appear. The corresponding phase mismatch is 300 cm^{-1} for y-cut crystals and 710 cm^{-1} for x-cut crystals. This means that the intensity of the unwanted green light is 5.8 times higher in the y-cut crystals compared with the x-cut crystals. This is probably the reason why x-cut RTP Pockels cells have a nearly two times higher damage resistance than y-cut cells [121]. Testing the same crystals with pure x- or y-polarization yields only a 15% difference in damage threshold.

Measurements in LBO also indicate collaborative effects, although detailed data as in KTP or in KDP are not available. As all nonlinear optical materials that have been studied in detail show to some degree and at certain wavelengths a collaborative damage mechanism, it is possible that this is a general feature of laser damage in nonlinear optical crystals under simultaneous multiwavelength irradiation. However, the practical importance of the effect seems to be strongest for the KTP family crystals.

11.4   Conclusion

We summarized laser damage data available for materials for lasers and gave an overview of the physical phenomena involved. The number of systematic studies is most important for nonlinear optical crystals. Further research is needed to reach a quantitative understanding of the mechanisms of laser-induced damage for all materials and at different wavelengths. Laser-induced damage in materials for lasers remains an up-to-date research topic laying the foundation for material scientists to further improve the crystals that we use in modern laser systems.

References

1. Determination of laser induced threshold on optical surfaces—Part 1: 1-on-1. ISO Standard 11254-1, 2000.
2. Determination of laser induced threshold on optical surfaces—Part 2: S-on-1. ISO Standard 11254-2, 2001.
3. Lasers and laser-related equipment. test methods for laser-induced damage threshold. Part 1: Definitions and general principles. ISO Standard 21254-1, (ISO/TR 21254-1:2011(E)), 2011.
4. Lasers and laser-related equipment. Test methods for laser-induced damage threshold. Part 2: Threshold determination. ISO Standard 21254-2, (ISO/TR 21254–2:2011(E)), 2011.
5. Lasers and laser-related equipment. Test methods for laser-induced damage threshold. Part 3: Assurance of laser power (energy) handling capabilities. ISO Standard 21254-3, (ISO/TR 21254-3:2011(E)), 2011.
6. Lasers and laser-related equipment. Test methods for laser-induced damage threshold. Part 4: Inspection, detection and measurement. ISO Standard 21254-4, (ISO/TR 21254-4:2011(E)), 2011.
7. M. Acharekar and J. Montgomery. Laser damage threshold measurements in Q-switch materials. *Proceedings of the SPIE*, 1848:35–45, 1993.
8. M. A. Acharekar, L. H. Morton Jr., and E. W. Van Stryland. 2 μm laser damage and 3–6 μm optical parametric oscillation in AgGaSe$_2$. *Proceedings of the SPIE*, 2114:69–81, 1994.
9. M. A. Acharekar, J. L. Montgomery, and R. J. Rapp. Laser damage threshold measurements of AgGaSe$_2$ crystal at 9 μm. *Proceedings of the SPIE*, 1624:46–54, 1991.

10. J. J. Adams, J. R. Bruere, M. Bolourchi, C. W. Carr, M. D. Feit, R. P. Hackel, D. E. Hahn et al. Wavelength and pulselength dependence of laser conditioning and bulk damage in doubler-cut KH_2PO_4. *Proceedings of the SPIE*, 5991:59911R, 2005.

11. B. Bertussi, H. Piombini, D. Damiani, M. Pommies, X. L. Borgne, and D. Plessis. SOCRATE: An optical bench dedicated to the understanding and improvement of a laser conditioning process. *Applied Optics*, 45:8506–8516, 2006.

12. G. C. Bhar, A. M. Rudra, P. K. Datta, U. N. Roy, V. K. Wadhawan, and T. Sasaki. A comparative study of laser second harmonic generation in some crystals. *Pramana Journal of Physics*, 44(1):45–53, 1995.

13. G. C. Bhar, A. K. Chaudhary, and P. Kumbhakar. Study of laser induced damage threshold and effect of inclusions in some nonlinear crystals. *Applied Surface Science*, 161:155–162, 2000.

14. G. C. Bhar, A. K. Chaudhary, P. Kumbhakar, A. M. Rudra, and S. C. Sabarwal. A comparative study of laser-induced surface damage thresholds in BBO crystals and effect of impurities. *Optical Materials*, 27:119–123, 2004.

15. J.-F. Bisson, Y. Feng, A. Shirakawa, H. Yoneda, J. Lu, H. Yagi, T. Yanagitani, and K.-I. Ueda. Laser damage threshold of ceramic YAG. *Japanese Journal of Applied Physics*, 42:1025–1027, 2003.

16. R. J. Bolt and M. van der Mooren. Single shot bulk damage threshold and conversion efficiency measurements on flux grown $KTiOPO_4$ (KTP). *Optics Communications*, 100:399–410, 1993.

17. B. Boulanger, M. M. Fejer, R. Blachman, and P. F. Bordui. Study of KTP gray-track at 1064, 532 and 355 nm. *Applied Physics Letters*, 65(19):2401–2403, 1994.

18. B. Boulanger, J. P. Feve, and Y. Guillien. Thermo-optical effect and saturation of nonlinear absorption induced by gray tracking in a 532-nm-pumped KTP optical parametric oscillator. *Optics Letters*, 25(7):484–486, 2000.

19. B. Boulanger, I. Rousseau, J. P. Feve, M. Maglione, B. Ménaert, and G. Marnier. Optical studies of laser induced gray-tracking in KTP. *IEEE Journal of Quantum Electronics*, 35(3):281–286, 1999.

20. A. K. Burnham, M. Runkel, M. D. Feit, A. M. Rubenchik, R. L. Floyd, T. A. Land, W. J. Siekhaus, and R. A. Hawley-Fedder. Laser-induced damage in deuterated potassium dihydrogen phosphate. *Applied Optics*, 42:5483–5495, 2003.

21. B. Bussiere. Etude des mécanismes l'endommagement par laser impulsionnel des cristaux de Saphir dopé Titane. PhD thesis, Universite Aix-Marseille, France, 2010.

22. B. Bussiere, O. Utéza, N. Sanner, M. Sentis, G. Riboulet, L. Vigroux, M. Commandré, F. Wagner, J.-Y. Natoli, and J.-P. Chambaret. Laser induced damage of sapphire and titanium doped sapphire crystals under femtosecond to nanosecond laser irradiation. *Proceedings of the SPIE*, 7504:75041N, 2009.

23. C. W. Carr, M. D. Feit, M. C. Nostrand, and J. J. Adams. Techniques for qualitative and quantitative measurement of aspects of laser-induced damage important for laser beam propagation. *Measurement Science and Technology*, 17:1958–1962, 2006.

24. M. Carrascosa, J. Villarroel, J. Carnicero, A. Garcia-Cabanes, and J. M. Cabrera. Understanding light intensity thresholds for catastrophic optical damage in $LiNbO_3$. *Optics Express*, 16(1):115–120, 2008.

25. C. T. Chen, Y. C. Wu, A. D. Jiang, B. C. Wu, G. M. You, R. K. Li, and S. J. Lin. New nonlinear-optical crystal—LiB_3O_5. *Journal of the Optical Society of America B-Optical Physics*, 6(4):616–621, 1989.

26. M. Commandre and P. Roche. Characterization of optical coatings by photothermal deflection. *Applied Optics*, 35(25):5021–5034, 1996.

27. P. DeMange, C. W. Carr, R. A. Negres, H. B. Radousky, and S. G. Demos. Laser annealing characteristics of multiple bulk defect populations within DKDP crystals. *Journal of Applied Physics*, 104:103103, 2008.

28. P. DeMange, R. A. Negres, H. B. Radousky, and S. G. Demos. Differentiation of defect populations responsible for bulk laser-induced damage in potassium dihydrogen phosphate crystals. *Optical Engineering*, 45(10):104205, 2006.

29. P. DeMange, R. A. Negres, A. M. Rubenchik, H. B. Radousky, M. D. Feit, and S. G. Demos. The energy coupling efficiency of multiwavelength laser pulses to damage initiating defects in deuterated KH_2PO_4 nonlinear crystals. *Journal of Applied Physics*, 103:083122, 2008.

30. S. G. Demos, P. DeMange, R. A. Negres, and M. D. Feit. Investigation of the electronic and physical properties of defect structures responsible for laser-induced damage in DKDP crystals. *Optics Express*, 18(13):13788–13804, 2010.

31. R. DeSalvo, M. Sheikbahae, A. A. Said, D. J. Hagan, and E. W. Vanstryland. Z-scan measurements of the anisotropy of nonlinear refraction and absorption in crystals. *Optics Letters*, 18(3):194–196, 1993.

32. B. T. Do and A. V. Smith. Bulk optical damage thresholds for doped and undoped, crystalline and ceramic yttrium aluminum garnet. *Applied Optics*, 48(18):3509–3514, 2009.

33. G. Duchateau. Simple models for laser-induced damage and conditioning of potassium dihydrogen phosphate crystals by nanosecond pulses. *Optics Express*, 17(13):10434–10456, 2009.

34. G. Duchateau and A. Dyan. Coupling statistics and heat transfer to study laser-induced crystal damage by nanosecond pulses. *Optics Express*, 15(8):4557–4576, 2007.

35. A. Dudelzak, P. P. Proulx, V. Denks, V. Murk, and V. Nagirnyi. Anisotropic fundamental absorption edge of KTiOPO$_4$ crystals. *Journal of Applied Physics*, 87(5):2110–2113, 2000.

36. G. J. Edwards, M. P. Scripsick, L. E. Halliburton, and R. F. Belt. Identification of a radiation-induced hole center in KTiOPO$_4$. *Physical Review B*, 48(10):6884–6891, 1993.

37. U. Ellenberger, R. Weber, J. E. Balmer, B. Zysset, D. Eligehausen, and G. J. Mizell. Pulsed optical damage threshold of potassium niobate. *Applied Optics*, 31(36):7563–7569, 1992.

38. M. Falk, Th. Woike, and K. Buse. Reduction of optical damage in lithium niobate crystals by thermo-electric oxidization. *Applied Physics Letters*, 90(25):251912, 2007.

39. S. Favre, T. C. Sidler, and R. P. Salathe. High-power long pulse second harmonic generation and optical damage with free-running Nd:YAG laser. *IEEE Journal of the Quantum Electronics*, 39(6):733–740, 2003.

40. M. Fontana, K. Chah, M. Aillerie, R. Mouras, and P. Bourson. Optical damage resistance in undoped LiNbO$_3$ crystals. *Optical Materials*, 16:111–117, 2001.

41. K. Fujioka, S. Matsuo, T. Kanabe, H. Fujita, and M. Nakatsuka. Optical properties of rapidly grown KDP crystal improved by thermal conditioning. *Journal of Crystal Growth*, 181(3):265–271, 1997.

42. Y. Furukawa, S. A. Markgraf, M. Sato, H. Yoshida, T. Sasaki, H. Fujita, T. Yamanaka, and S. Nakai. Investigation of the bulk laser damage of lithium triborate, LiB$_3$O$_5$, single crystals. *Applied Physics Letters*, 65(12):1480–1482, 1994.

43. Y. Furukawa, A. Yokotani, T. Sasaki, H. Yoshida, K. Yoshida, F. Nitanda, and M. Sato. Investigation of bulk laser damage threshold of lithium-niobate single-crystals by Q–switched pulse laser. *Journal of Applied Physics*, 69(5):3372–3374, 1991.

44. F. Guillet, B. Bertussi, L. Lamaignere, and C. Maunier. Effects of thermal annealing on KDP and DKDP on laser damage resistance at 3 omega. *Proceedings of the SPIE*, 7842:78421T, 2010.

45. H. Hellwig, J. Liebertz, and L. Bohaty. Linear optical properties of the monoclinic bismuth borate BiB$_3$O$_6$. *Journal of Applied Physics*, 88:240–244, 2000.

46. A. Hildenbrand. Étude de l'endommagement laser dans les cristaux non linéaires en régime nanoseconde. PhD thesis, Université Aix-Marseille, France, 2008.

47. A. Hildenbrand, C. Kieleck, E. Lallier, D. Faye, A. Grisard, B. Gérard, and M. Eichhorn. Compact efficient mid-infrared laser source: OP-GaAs OPO pumped by Ho:YAG laser. *Proceedings of the SPIE*, 8187:81870H, 2011.

48. A. Hildenbrand, F. R. Wagner, H. Akhouayri, J.-Y. Natoli, and M. Commandré. Accurate metrology for laser damage measurements in nonlinear crystals. *Optical Engineering*, 47(8):083603, 2008.

49. A. Hildenbrand, F. R. Wagner, H. Akhouayri, J.-Y. Natoli, M. Commandré, F. Théodore, and H. Albrecht. Laser-induced damage investigation at 1064 nm in KTiOPO$_4$ crystals and its analogy with RbTiOPO$_4$. *Applied Optics*, 48(21):4263–4269, 2009.

50. A. Hildenbrand, F. R. Wagner, J.-Y. Natoli, and M. Commandré. Nanosecond laser induced damage in RbTiOPO$_4$: The missing influence of crystal quality. *Optics Express*, 17(20):18263–18270, 2009.

51. J. Hirohashi, V. Pasiskevicius, S. Wang, and F. Laurell. Picosecond blue-light-induced infrared absorption in single-domain and periodically poled ferroelectrics. *Journal of Applied Physics*, 101(3):033105, 2007.

52. W. Hong, M. M. Chirila, N. Y. Garces, L. E. Halliburton, D. Lupinski, and P. Villeval. Electron paramagnetic resonance and electron-nuclear double resonance study of trapped-hole centers in LiB$_3$O$_5$ crystals. *Physical Review B*, 68:094111, 2003.

53. J. C. Jacco, G. M. Loiacono, M. Jaso, G. Mizell, and B. Greenberg. Flux growth and properties of KTiOPO$_4$. *Journal of Crystal Growth*, 70(1–2):484–488, 1984.

54. J. C. Jacco, D. R. Rockafellow, and E. A. Teppo. Bulk darkening threshold of flux grown KTP. *Optics Letters*, 16(17):1307, 1991.

55. M. Jain, J. K. Lotsberg, J. J. Stamnes, and O. Frette. Effects of aperture size on focusing of electromagnetic waves into a biaxial crystal. *Optics Communications*, 266:438–447, 2006.

56. J. H. Jang, I. H. Yoon, and C. S. Yoon. Cause and repair of optical damage in nonlinear optical crystals of BiB$_3$O$_6$. *Optical Materials*, 31:781–783, 2009.

57. T. Kamimura, S. Fukumoto, R. Ono, Y. K. Yap, M. Yoshimura, Y. Mori, T. Sasaki, and K. Yoshida. Enhancement of CsLiB$_6$O$_{10}$ surface-damage resistance by improved crystallinity and ion-beam etching. *Optics Letters*, 27(8):616–618, 2002.

58. T. Kamimura, Y. Kawaguchi, T. Arii, W. Shirai, T. Mikami, T. Okamoto, Y. L. Aung, and A. Ikesue. Investigation of bulk laser damage in transparent YAG ceramics controlled with microstructural refinement. *Proceedings of the SPIE*, 7132:713215, 2008.

59. A. A. Kaminskii. Laser crystals and ceramics: Recent advances. *Laser and Photonics Reviews*, 1(2):93–177, 2007.

60. T. Kawamura, M. Yoshimura, Y. Honda, M. Nishioka, Y. Shimizu, Y. Kitaoka, Y. Mori, and T. Sasaki. Effect of water impurity in $CsLiB_6O_{10}$ crystals on bulk laser-induced damage threshold and transmittance in the ultraviolet region. *Applied Optics*, 48(9):1658–1662, 2009.

61. C. Kieleck, M. Eichhorn, A. Hirth, D. Faye, and E. Lallier. High-efficiency 20–50 kHz mid-infrared orientation-patterned GaAs optical parametric oscillator pumped by a 2 μm holmium laser. *Optics Letters*, 34(3):262–264, 2009.

62. H. Kildal and G. W. Iseler. Laser-induced surface damage of infrared nonlinear materials. *Applied Optics*, 15(12):3062–3065, 1976.

63. J. W. Kim, C. S. Yoona, and H. G. Gallagherb. The effect of NaCl melt-additive on the growth and morphology of LiB_3O_5 (LBO) crystals. *Journal of Crystal Growth*, 222:760–766, 2001.

64. W. Koechner. *Solid-State Laser Engineering*. New York: Springer, 2006.

65. E. P. Kokanyan, L. Razzari, I. Cristiani, V. Degiorgio, and J. B. Gruber. Reduced photorefraction in hafnium-doped single-domain and periodically poled lithium niobate crystals. *Applied Physics Letters*, 84:1880, 2004.

66. Y. Kong, S. Liu, Y. Zhao, H. Liu, S. Chen, and J. Xu. Highly optical damage resistant crystal: Zirconium-oxide-doped lithium niobate. *Applied Physics Letters*, 91:081908, 2007.

67. Y. Kong, J. Wen, and H. Wang. New doped lithium niobate crystal with high resistance to photorefraction: $LiNbO_3$: In. *Applied Physics Letters*, 66:280, 1995.

68. H. Kouta. Wavelength dependence of repetitive-pulse laser-induced damage threshold in beta-BaB_2O_4. *Applied Optics*, 38(3):545–547, 1999.

69. A. A. Lagatsky, N. V. Kuleshov, and V. P. Mikhailov. Diode-pumped CW lasing of Yb:KYW and Yb:KGW. *Optics Communications*, 165(1–3):71–75, 1999.

70. L. Lamaignere, T. Donval, M. Loiseau, J. C. Poncetta, G. Razé, C. Meslin, B. Bertussi, and H. Bercegol. Accurate measurements of laser-induced bulk damage density. *Measurement Science and Technology*, 20:095701, 2009.

71. K. M. Leung and L. G. DeShazer. Surface defects on crystals of TiO_2 and YVO_4 studied by laser-induced damage effects. *Proceedings of the SPIE*, 414:193–199, 1974.

72. H. Li, F. Zhou, X. Zhang, and W. Ji. Bound electronic kerr effect and self-focusing induced damage in second-harmonic-generation crystals. *Optics Communications*, 144:75–81, 1997.

73. Z. M. Liao, M. L. Spaeth, K. Manes, J. J. Adams, and C. W. Carr. Predicting laser-induced bulk damage and conditioning for deuterated potassium dihydrogen phosphate crystals using an absorption distribution model. *Optics Letters*, 35(15):2538–2540, 2010.

74. C. Miner. Periodically poled potassium titanyl phosphate crystal. U.S. Patent No. 7,101,431, 2006.

75. S. Moller, A. Andresen, C. Merschjann, B. Zimmermann, M. Prinz, and M. Imlau. Insight to UV-induced formation of laser damage on LiB_3O_5 optical surfaces during long-term sum-frequency generation. *Optics Express*, 15(12):7351–7356, 2007.

76. P. A. Morris. Process for reducing the damage susceptibility in optical quality crystals. U.S. Patent No. 5,411,723, 1995.

77. V. Murk, V. Denks, A. Dudelzak, P. P. Proulx, and V. Vassiltsenko. Gray tracks in $KTiOPO_4$: Mechanism of creation and bleaching. *Nuclear Instruments and Methods in Physics Research Section B-Beam Interactions with Materials and Atoms*, 141(1–4):472–476, 1998.

78. H. Nakatani, W. R. Bosenberg, L. K. Cheng, and C. L. Tang. Laser-induced damage in beta-barium metaborate. *Applied Physics Letters*, 53(26):2587–2589, 1988.

79. J.-Y. Natoli, B. Bertussi, and M. Commandré. Effect of multiple laser irradiations on silica at 1064 and 355 nm. *Optics Letters*, 30(10):1315–1317, 2005.

80. J.-Y. Natoli, J. Capoulade, H. Piombini, and B. Bertussi. Influence of laser beam size and wavelength in the determination of LIDT and associated laser damage precursor densities in KH_2PO_4. *Proceedings of the SPIE*, 6720:672016, 2007.

81. J. Neauport, P. Cormont, P. Legros, C. Ambard, and J. Destribats. Imaging subsurface damage of grinded fused silica optics by confocal fluorescence microscopy. *Optics Express*, 17(5):3543–3554, 2009.

82. R. A. Negres, N. P. Zaitseva, P. DeMange, and S. G. Demos. Expedited laser damage profiling of $KD_xH_{2-x}PO_4$ with respect to crystal growth parameters. *Optics Letters*, 31(21):3110–3112, 2006.

83. D. N. Nikogosyan. Lithium triborate LBO: A review of its properties and applications. *Applied Physics A*, 58:181–190, 1994.

84. D. N. Nikogosyan. *Nonlinear Optical Crystals: A Complete Survey.* New York, Springer, 2005.

85. M. Nishioka, A. Kanoh, M. Yoshimura, Y. Mori, and T. Sasaki. Growth of $CsLiB_6O_{10}$ crystals with high laser-damage tolerance. *Journal of Crystal Growth*, 279:7681, 2005.

86. R. Ono, T. Kamimura, S. Fukumoto, Y. K. Yap, M. Yoshimura, Y. Mori, T. Sasaki, and K. Yoshida. Effect of crystallinity on the bulk laser damage and UV absorption of CLBO crystals. *Journal of Crystal Growth*, 237–239:645–648, 2002.

87. V. Pasiskevicius, A. Fragemann, F. Laurell, R. Butkus, V. Smilgevicius, and A. Piskarskas. Enhanced stimulated Raman scattering in optical parametric oscillators from periodically poled $KTiOPO_4$. *Applied Physics Letters*, 82(3):325–327, 2003.

88. V. Pasiskevicius, H. Karlsson, F. Laurell, R. Butkus, V. Smilgevicius, and A. Piskarskas. High-efficiency parametric oscillation and spectral control in the red spectral region with periodically poled $KTiOPO_4$. *Optics Letters*, 26(10):710–712, 2001.

89. R. D. Peterson, K. L. Schepler, J. L. Brown, and P. G. Schunemann. Damage properties of $ZnGeP_2$ at 2 μm. *Journal of the Optical Society of America B*, 12(11):2142–2146, 1995.

90. V. Petrov, M. Ghotbi, O. Kokabee, A. Esteban-Martin, F. Noack, A. Gaydardzhiev, I. Nikolov, P. Tzankov et al. Femtosecond nonlinear frequency conversion based on BiB_3O_6. *Laser and Photonics Reviews*, 4(1):53–98, 2010.

91. M. Pommies, D. Damiani, B. Bertussi, J. Capoulade, H. Piombini, J. Y. Natoli, and H. Mathis. Detection and characterization of absorption heterogeneities in KH_2PO_4 crystals. *Optics Communications*, 267:154–161, 2006.

92. I. M. Pritula, M. I. Kolybayeva, V. I. Salo, and V. M. Puzikov. Defects of large-size KDP single crystals and their influence on degradation of the optical properties. *Optical Materials*, 30(1):98–100, 2007.

93. P. P. Proulx, V. Denks, A. Dudelzak, V. Murk, and V. Nagirnyi. Intrinsic electron excitations of $KTiOPO_4$ crystals. *Nuclear Instruments and Methods in Physics Research Section B-Beam Interactions with Materials and Atoms*, 141(1–4):477–480, 1998.

94. F. Rainer, L. J. Atherton, and J. J. De Yoreo. Laser damage to production- and research-grade KDP crystals. *Proceedings of the SPIE*, 1848:46–58, 1992.

95. D. Rajesh, M. Yoshimura, T. Eiro, Y. Mori, T. Sasaki, R. Jayavel, T. Kamimura et al. UV laser-induced damage tolerance measurements of CsB_3O_5 crystals and its application for UV light generation. *Optical Materials*, 31(2):461–463, 2008.

96. J. Rams, A. Alcazar de Velasco, M. Carrascosa, J.M. Cabrera, and F. Agullo-Lopez. Optical damage inhibition and thresholding effects in lithium niobate above room temperature. *Optics Communications*, 178:211–216, 2000.

97. S. Reyné, G. Duchateau, J.-Y. Natoli, and L. Lamaignere. Laser-induced damage of KDP crystals by 1ω nanosecond pulses: Influence of crystal orientation. *Optics Express*, 17(24):21652–21665, 2009.

98. S. Reyné, G. Duchateau, J.-Y. Natoli, and L. Lamaignere. Pump-pump experiment in KH_2PO_4 crystals: Coupling two different wavelengths to identify the laser-induced damage mechanisms in the nanosecond regime. *Applied Physics Letters*, 96:121102, 2010.

99. M. G. Roelofs. Identification of Ti^{3+} in potassium titanyl phosphate and its possible role in laser damage. *Journal of Applied Physics*, 65(12):4976–4982, 1989.

100. M. Roth, M. Tseitlin, and N. Angert. Oxide crystals for electro-optic Q–switching of lasers. *Glass Physics and Chemistry*, 31(1):86–95, 2005.

101. B. E. A. Saleh and M. C. Teich. Optics of anisotropic media. In *Fundamentals of Photonics*, pp. 210–223. New York: John Wiley & Sons, Inc., 1991.

102. M. N. Satiyanarayan, H. L. Bhat, M. R. Srinivasan, and P. Ayyub. Evidence for the presence of remnant strain in grey-tracked $KTiOPO_4$. *Applied Physics Letters*, 67(19):2810–2812, 1995.

103. M. P. Scripsick, D. N. Loiacono, J. Rottenberg, S. H. Goellner, L. E. Halliburton, and F. K. Hopkins. Defects reponsible for gray tracks in flux grown KTP. *Applied Physics Letters*, 66(25):3428, 1995.

104. W. Seelert, P. Kortz, D. Rytz, B. Zysset, D. Ellgehausen, and G. Mizell. Second harmonic generation and degradation in critically phase-matched $KNbO_3$ with a diode pumped Q-switched Nd:YLF laser. *Optical Letters*, 17(20):1432–1434, 1992.

105. S. D. Setzler, P. G. Schunemann, L. A. Pomeranz, and T. M. Pollak. $ZnGeP_2$ laser damage threshold enhancement. *Proceedings of the 14th American Conference on Crystal Growth and Epitaxy*, Seattle, WA, 2002.

106. B. C. Stuart, M. D. Feit, A. M. Rubenchik, B. W. Shore, and M. D. Perry. Laser-induced damage in dielectrics with nanosecond to subpicosecond pulses. *Physical Review Letters*, 74(12):2248–2251, 1994.

Chapter 11

107. S.-T. Sun, L.-L. Ji, Z.-P. Wang, M.-X. Xu, X.-L. Liang, B. Liu, X.-M. Mu, X. Sun, and X.-G. Xu. The effect of material, thermal and laser conditioning on the damage threshold of type II tripler-cut DKDP crystals. *Crystal Research and Technology*, 43(7):773–777, 2008.

108. Y. Sun, C. Li, Z. Li, and F. Gan. Investigation of nonlinear absorption and laser damage on lithium niobate single crystals. *Proceedings of the SPIE*, 1848:594–598, 1993.

109. Y. Sun, Q. Zhang, and H. Gong. Polarized 10-ns frequency-doubled Nd:YAG laser-induced damage to titanium-doped sapphire. *Proceedings of the SPIE*, 2114:166, 1994.

110. B. Maximov T. Volk and, S. Sulyanov, N. Rubinina, and M. Wohlecke. Relation of the photorefraction and optical-damage resistance to the intrinsic defect structure in $LiNbO_3$ crystals. *Optical Materials*, 23:229–233, 2003.

111. Q. Tan. Laser damage from harmful impurities in meta-barium borate crystal. *Journal of Crystal Growth*, 209:861–866, 2000.

112. A. E. Telminov, A. G. Sitnikov, A. N. Panchenko, D. E. Genin, S. Yu. Sarkisov, S. A. Bereznaya, Z. V. Korotchenko, and E. V. Vavilin. Damage threshold of modified GaSe crystals under irradiation of pulsed CO_2 laser with inductive energy storage and SOS-diodes. *Modification of Material Properties*, 322–323, 2010.

113. M. Tseitlin, E. Mojaev, and M. Roth. Growth of high resistivity $RbTiOPO_4$ crystals. *Journal of Crystal Growth*, 310(7–9):1929–1933, 2008.

114. O. Uteza, B. Bussiere, F. Canova, J.-P. Chambaret, P. Delaporte, T. Itina, and M. Sentis. Laser-induced damage threshold of sapphire in nanosecond, picosecond and femtosecond regimes. *Applied Surface Science*, 254(4):799–803, 2007.

115. J. Villarroel, O. Caballero-Calero, B. Ramiro, A. Alcazar, A. Garcia-Cabanes, and M. Carrascosa. Photorefractive non-linear single beam propagation in $LiNbO_3$ waveguides above the optical damage threshold. *Optical Materials*, 33:103–106, 2010.

116. T. R. Volk, V. J. Pryalkin, and M. M. Rubinina. Optical-damage-resistant $LiNbO_3$: Zn crystal. *Optics Letters*, 15(18):996–998, 1990.

117. F. R. Wagner, A. Hildenbrand, H. Akhouayri, C. Gouldieff, L. Gallais, M. Commandré, and J.-Y. Natoli. Multipulse laser damage in potassium titanyl phosphate: Statistical interpretation of measurements and the damage initiation mechanism. *Optical Engineering*, 51:121806, 2012.

118. F. R. Wagner, A. Hildenbrand, J.-Y. Natoli, and M. Commandré. Nanosecond-laser induced damage at 1064 nm, 532 nm, and 355 nm in LiB_3O_5. *Proceedings of the SPIE*, 7504:75041M, 2009.

119. F. R. Wagner, A. Hildenbrand, J.-Y. Natoli, and M. Commandré. Multiple pulse nanosecond laser induced damage study in LiB_3O_5 crystals. *Optics Express*, 18:26791–26798, 2010.

120. F. R. Wagner, A. Hildenbrand, J.-Y. Natoli, and M. Commandré. Nanosecond-laser-induced damage in potassium titanyl phosphate: Pure 532 nm pumping and frequency conversion situations. *Applied Optics*, 50(22):4509–4515, 2011.

121. F. R. Wagner, A. Hildenbrand, J.-Y. Natoli, M. Commandré, F. Théodore, and H. Albrecht. Laser damage resistance of $RbTiOPO_4$: Evidence of polarization dependent anisotropy. *Optics Express*, 15(21):13849–13857, 2007.

122. F. R. Wagner, G. Duchateau, A. Hildenbrand, J.-Y. Natoli, and M. Commandré. Model for nanosecond laser induced damage in potassium titanyl phosphate crystals. *Applied Physics Letters*, 99:231111, 2011.

123. L. Wang, S. Liu, Y. Kong, S. Chen, Z. Huang, L. Wu, R. Rupp, and J. Xu. Increased optical-damage resistance in tin-doped lithium niobate. *Optics Letters*, 35(6):883–885, 2010.

124. S. Wang, V. Pasiskevicius, and F. Laurell. Dynamics of green light-induced infrared absorption in $KTiOPO_4$ and periodically poled $KTiOPO_4$. *Journal of Applied Physics*, 96(4):2023–2028, 2004.

125. Z. Wang, D. Rajesh, M. Yoshimura, H. Shimatani, Y. Kitaoka, Y. Mori, and T. Sasaki. Enhancement of the CsB_3O_5(CBO) crystal quality by fast cooling after crystal growth. *Journal of Crystal Growth*, 318:625–628, 2011.

126. R. M. Wood. *Laser-Induced Damage of Optical Materials*. Bristol, U.K.: Institute of Physics Publishing, 2003.

127. Y. Wu, P. Fu, J. Wang, Z. Xu, L. Zhang, Y. Kong, and C. Chen. Characterization of CsB_3O_5 crystal for ultraviolet generation. *Optics Letters*, 22(24):1840–1842, 1997.

128. Y. Wu, T. Sasaki, S. Nakai, A. Yokotani, H. Tang, and C. Chen. CsB_3O_5: A new nonlinear optical crystal. *Applied Physics Letters*, 62(21):2614–2615, 1993.

129. J. K. Yamamoto, K. Kitamura, N. Iyi, S. Kimura, Y. Furukawa, and M. Sato. Increased optical damage resistance in Sc_2O_3 doped $LiNbO_3$. *Applied Physics Letters*, 61:2156, 1992.

130. H. Yoshida, H. Fujita, M. Nakatsuka, M. Yoshimura, T. Sasaki, T. Kamimura, and K. Yoshida. Dependences of laser-induced bulk damage threshold and crack patterns in several nonlinear crystals on irradiation direction. *Japanese Journal of Applied Physics*, 45(2A):766–769, 2006.

131. H. Yoshida, T. Jitsuno, H. Fujita, M. Nakatsuka, T. Kamimura, M. Yoshimura, T. Sasaki, A. Miyamoto, and K. Yoshida. Laser-induced damage in nonlinear crystals on irradiation direction and polarization. *Proceedings of the SPIE*, 3902:418–422, 2000.

132. H. Yoshida, T. Jitsuno, H. Fujita, M. Nakatsuka, M. Yoshimura, T. Sasaki, and K. Yoshida. Investigation of bulk laser damage in KDP crystal as a function of laser irradiation direction, polarization, and wavelength. *Applied Physics B*, 70(2):195–201, 2000.

133. I. H. Yun, S.-K. Hong, H. J. Kim, C. Lim, and C. S. Yoon. Highly efficient optical parametric chirped-pulse amplification based on BiB_3O_6 and beta-BaB_2O_4 crystals at low pump intensity. *Optics Communications*, 284:2341–2344, 2011.

134. N. P. Zaitseva, J. J. De Yoreo, M. R. Dehaven, R. L. Vital, L. M. Carman, and H. R. Spears. Rapid growth of large-scale (40–55 cm) KDP crystals. *Journal of Crystal Growth*, 180:255–262, 2001.

135. K. T. Zawilski, P. G. Schunemann, S. D. Setzler, and T. M. Pollak. Large aperture single crystal $ZnGeP_2$ for high-energy applications. *Journal of Crystal Growth*, 310(7–9):1891–1896, 2008.

136. K. T. Zawilski, S. D. Setzler, P. G. Schunemann, and T. M. Pollak. Increasing the laser-induced damage threshold of single-crystal $ZnGeP_2$. *Journal of the Optical Society of America B*, 23(11):2310–2316, 2006.

137. D. E. Zelmon, K. L. Schepler, S. Guha, D. Rush, S. M. Hegde, L. P. Gonzales, and J. Lee. Optical properties of Nd-doped ceramic yttrium aluminum garnet. *Proceedings of the SPIE*, 5647:255–264, 2005.

138. Q. Zhang, G. Feng, J. Han, B. Li, Q. Zhu, and X. Xie. High repetition rate laser pulse induced damage in KTP crystal: Gray-tracking and catastrophic damage. *Optik*, 122:1313–1318, 2011.

139. G. Zhong, J. Jin, and Z. Wu. Measurements of optically induced refractive-index damage of lithium niobate doped with different concentrations of MgO. *Journal of the Optical Society of America*, 70:631, 1980.

140. T. Zhu, X. Chen, and J. Qin. Research progress on mid- IR nonlinear optical crystals with high laser damage threshold in China. *Frontiers of Chemistry in China*, 6:1–8, 2011.

141. B. C. Ziegler and K. L. Schepler. Transmission and damage threshold measurements in $AgGaSe_2$ at 2.1 μm. *Applied Optics*, 30:5077–5080, 1991.

Chapter 11

12. Surface Manufacturing and Treatment

Jérôme Néauport and Philippe Cormont

12.1 Introduction

There are a wide variety of optic components varying widely in shape and material. The wavelength, the end purpose, and the system constraints will induce the laser system designer to use one or other of these components. More generally, an optical component consists of a material the surface or surfaces of which are shaped to meet a given optic function: a spherical polished glass surface for light focusing, diamond-turned metal to reflect incident beams, and a structured surface for light diffraction. The main objective guiding the manufacturer in the choice of his type of fabrication is generally meeting the manufacturing dimensional and optical specifications. For example, the reflected or transmitted wave surface quality, the surface roughness, and the scratch and dig specifications or diffraction efficiency (in the special case of a diffractive component) are criteria, which more often than not determine the manufacturer's choice. Manufacturing operations consisting of transforming the initial blank of material into a component with the required

Laser-Induced Damage in Optical Materials. Edited by Detlev Ristau © 2015 CRC Press/Taylor & Francis Group, LLC. ISBN: 978-1-4398-7216-1.

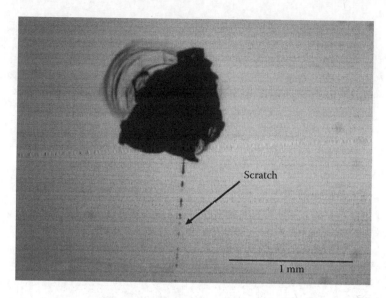

FIGURE 12.1   Laser damage initiated on a scratch present on a polished silica substrate after a 14 J/cm² laser shot (wavelength of 351 nm, pulsetime 2.5 ns).

optical function can affect the laser damage performance of the final component. In a paper on the laser-induced damage (LID) performance of various colored glasses at 1.064 μm in the nanosecond regime, Hack and Neuroth show that while the volume LID threshold depends on the type of glass tested, the surface threshold remains constant and is determined by the polishing process adopted (Hack and Neuroth 1982). The surface condition, cosmetically speaking (pits or scratches), can also have an impact. An example of such a phenomenon is given in Figure 12.1. A residual scratch made during the polishing of a silica part is the precursor of LID (wavelength of 351 nm, pulse duration of 2.5 ns, fluence of 14 J/cm²).

The aim of this chapter is to detail the link between the manufacturing and surface texturing processes and LID resistance. After a brief overview of a few basic notions of optical component polishing or finishing processes, a series of examples will be detailed where the polishing process proves to be a limiting factor on the LID performance. It will subsequently be shown that when fine characterization of the interface is possible, certain precursor defects of the interface can be related to laser damage. The case of surface defects induced by such processes (roughness, scratches, and digs) and of subsurface defects will also be addressed. A few methods limiting the impact of such defects will also be detailed. Finally, the final section of this chapter will examine diffractive components and the link between the diffractive structure and behavior under a high-fluence laser beam.

This chapter reviews phenomena related to the surface or near to the surface and will not present intrinsic material defects such as bubbles, inclusions, or lattice defects for crystals. Problems specific to surface treatments for production of antireflection, mirrors, or optic filters will be addressed in a dedicated chapter (cf. Chapters 13 through 15).

12.2   Surface Quality and LID

12.2.1   Manufacturing Optical Surfaces

There are many polishing techniques, the choice of which depends not only on the material but also on the shape of the surface to be produced and on the precision required. The basics of glass surfacing will be detailed in the following text since that covers the vast majority of the components used. The particular case of metals and crystals will also be briefly addressed.

The majority of techniques for the finishing or polishing of glass are by no means new and are based on an essentially oral know-how. It is interesting to note that major projects such as the laser megajoule (LMJ), the National Ignition Facility (NIF) (see Moses 2013), or the programs for the development of multisegmented telescopes (OWL, EURO50, TMT) are today pushing polishing processes to new limits (Dierickx 2007), requiring a more exhaustive scientific understanding of the phenomena involved. The surfacing of glass includes the basic stages illustrated in Figure 12.2. A complete description can be found in the work of H. H. Karow (Karow 1993).

Shaping consists of producing from a shapeless blank material a part of the lateral dimensions of which are those of the final optic and the thickness of which is greater than that of the final dimension. This enables implementation of the subsequent stages of fine grinding and polishing. Shaping involves two main operations: cutting (with diamond-tipped saws) and grinding with a diamond abrasive wheel. Shaping can also be performed with coarse abrasives (approximately 100 μm) in a suspension of water or another liquid, depending on the nature of the material. Fine grinding, used to reduce the roughness of the part and to improve its flatness, is generally performed using a medium-coarse abrasive (10 μm) in a water suspension with friction of the

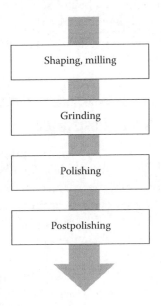

FIGURE 12.2   Main stages of a surfacing process.

Table 12.1 Type of Action according to the Surfacing Stages

	Type of Abrasive (Approx. Mean Diameter)	Type of Action
Shaping	Fixed (100 μm)	Mechanical
Grinding, fine grinding	Fixed or loose (10 μm)	Mechanical
Polishing	Loose (1 μm)	Mechanical and chemical
Postpolishing	Miscellaneous	Mechanical/mechanical and chemical

face to be ground against a brass, cast-iron, or ceramic plate. Polishing gives the part its transparency and the all-but final quality of flatness. As with fine grinding, a fine abrasive is used (1 μm) with friction of the part against a synthetic or pitch plate. Postpolishing can then be performed to further improve the flatness of the part and/or its LID resistance. Numerous processes can be employed: local robotized polishing, local ion beam polishing, local or total chemical etching or magnetorheological finishing (MRF). Each processing stage differs by the type of action employed to remove material (a mechanical or mechanical and chemical action) and by the type of abrasive: a loose abrasive, suspended in a liquid, or a fixed abrasive, held in the binder of a tool. Table 12.1 gives the combinations that exist in the majority of cases.

In the case of crystals used for optics, the main operations as described for glass are generally applied. Polishing compound, solvent, and pads used must of course be adapted, particularly in the case of certain hygroscopic crystals where all water-based solutions are prohibited. Furthermore, in certain cases, the techniques of single-point diamond turning (SPDT) have been successfully used as polishing process (semifinishing, finishing) instead of traditional loose abrasive polishing (Fuchs et al. 1986). While silicon carbide, a very hard material, can be polished with methods adapted from those used for glass, soft substrates such as metals can be polished by diamond machining using the SPDT method. A traditional postpolishing stage is occasionally added to eliminate the cutting marks specific to SPDT.

12.2.2 Overview of Certain Cases Where the Interface Degrades Laser Damage Resistance

Regardless of the material and process chosen, an optical component shaped by a polishing operation contains a certain number of imperfections and defects near or on its surface. These imperfections may be surface defects (roughness, scratches, pits, digs, etc.) or subsurface defects (microcracks, impurities, and dislocations for crystalline or polycrystalline materials). It will be seen that depending on the laser applications considered, the disrupted interface can considerably reduce the performance of the component under an intense laser beam.

In an exhaustive study of the LID resistance of monocrystalline metals (aluminum, copper, and nickel) at 1064 nm in the nanosecond regime, the surfaces generated by SPDT are postpolished by different methods and tested in a single-shot or a multishot regime at 10 Hz. The residual surface defects substantially impact the damage threshold of copper mirrors because of its high boiling point. On the other hand, the phenomenon is limited

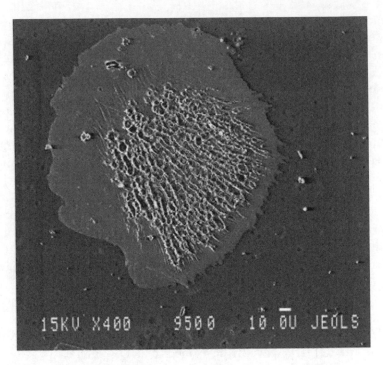

FIGURE 12.3   Laser damage in an SPDT copper mirror (1064 nm, 10 ns, single shot). (Reprinted from Gorshkov, A., Fusion Eng. Design, 865, Gorshkov, A., Bel'bas, I., Sannikov, V., and Vukolov, K., Frequency laser damage of Mo mirrors, 66–68, Copyright 2003, with permission from Elsevier.)

and even eliminated in the case of nickel or aluminum, because the intrinsic absorption of the material governs the damage mechanism (Jee et al. 1988). More recently, a study was made on the surface preparation of copper mirrors to be used as diagnostic mirrors of the International Thermonuclear Experimental Reactor (ITER) magnetic confinement fusion reactor (Gorshkov et al. 2005). The test conditions were similar (1064 nm, 10 ns, single shot and multishots). A mirror polished by SPDT was compared with a mirror polished by SPDT and then coated by a 1 μm layer of copper deposited by evaporation. In the single-shot regime, the threshold of the mirror obtained by SPDT was five times higher than that of the mirror coated with a copper layer (cf. Figure 12.3).

Gorshkov explains the difference by a densification of the surfaced introduced during the machining process without, however, supporting that explanation by a characterization. The difference in density would also be the cause of the difference in behavior of the samples in the multishot regime. Molybdenum was also the subject of similar work (Gorshkov et al. 2003) in identical test conditions (1064 nm, 10 ns, single shot and multi shot). The findings were similar: the threshold depends on the preparation of the substrate. An impact of the surface quality is one of the hypotheses put forward to explain these differences. All this work was performed in the nanosecond regime.

In the subpicosecond regime, in view of the physical phenomena involved in the damage (cf. Chapter 5 and Nolte et al. 1997), it is generally accepted that the role of the surface preparation is relatively limited if abstraction is made of ripple type damage that will be addressed in Section 12.2.3.3.

Chapter 12

The quality of surface in the case of crystals is also linked to LID performance. In the case of the titanium-doped sapphire used in the Ti:Sa amplifiers, the surface threshold in the nanosecond regime at 532 nm is lower than that of the bulk material (Sun et al. 1994, Bussiere et al. 2009). Sun evidences that the damage is induced by a collection of micropits that he correlates to the presence of a nanometric surface contaminant. Bussiere also stresses the difficulty of polishing the Ti:Sa compared to the nondoped sapphire, although in this case it is not supported by any measurement (Bussiere et al. 2009). Similar behavior was recorded on the nondoped sapphire (Uteza et al. 2007). In that case, the test was performed in the femtosecond regime at a wavelength of 800 nm and the damage threshold measured with quality polishing (superpolished) was approximately 1.5–2 times greater than that obtained with standard polishing. On the other hand, Bussiere or Uteza does not relate these differences to any particular precursor induced by the polishing process. This importance of the surface condition also applies to potassium dihydrogen phosphate (KDP), a material that is the only crystal available in large dimensions for the doubling or tripling of frequency and that is, therefore, used in facilities such as the NIF and LMJ (cf. Section 5.4). As previously stated, this type of crystal is polished by SPDT. Figure 12.4 shows typical surface damage of KDP.

Recent studies (Menapace et al. 2009) have evidenced that additional reworking with MRF using an MRF slurry specially developed for KDP (KDP is hygroscopic; hence, classic water-based MRF slurry cannot be used) produced a twofold gain in the damage threshold at 1064 nm, 10 ns, and a gain of over 1.5 for the threshold at 532 nm, 7.5 ns. The probable reason for this substantial improvement lies in the tangential action without any normal mechanical stress with this type of polishing process. This makes MRF an adequate process for subsurface damage (SSD) removal (Menapace et al. 2007).

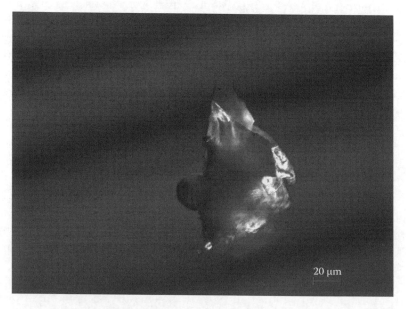

FIGURE 12.4 Surface damage on a KDP crystal polished by SPDT with antireflection sol-gel treatment, tested at 351 nm, 3 ns.

12.2.3   Working toward Greater Knowledge of Damage Precursors

It has been seen that in many cases, on metals, crystals, or glass, the finishing process can impact the LID performance. The following section will show that, using a suitable metrology, it is possible to gain knowledge of the damage precursor, which then makes it possible to adjust the polishing process to reduce their impacts. Surface defects will be addressed first, followed by the subsurface defects.

12.2.3.1   Surface Defects

The defects on the surface of optical components contribute substantially to a deterioration of their LID performance, especially if the defects are fractures (Bloembergen 1973). Such deterioration has been observed on both metal surfaces (Jee et al. 1988) and on dielectric surfaces (Salleo et al. 1998 or Génin et al. 2001). Use of minimally abrasive techniques, such as diamond turning (SPDT) or MRF, distinctly improves the LID performance without, however, totally eliminating the surface defects (Hurt 1985). The physical model enabling prediction of the damage according to the type of defect is still being developed, but recent work makes it possible to envisage the damage scenarios on silica in the nanosecond regime as from fractures in the silica (Bercegol et al. 2008, Miller et al. 2010). In the subpicosecond regime, given that the damage in most cases is governed by the intrinsic properties of the material, the problem of appearance defects is less sensitive.

The correlation between the defects and the manufacturing tools will be examined here. Surface defects are characteristics of the manufacturing tools used and processed. In what follows, examples mainly obtained on silica optics will be presented where considerable work has been undertaken on that material, given its wide use in major laser facilities such as the NIF or the LMJ. The observations on defects and results from silica optics can be nonetheless extended to glass and to crystals where polishing of the surfaces is performed in a traditional manner via abrasive slurry.

The defect found most frequently, and easy to identify on optics is a discontinuous scratch. This type of scratch generally appears during polishing and is due to the friction of a rogue particle, larger than the particles used for the polishing. There are many sources for such rogues; they can be contaminant or be present in the polishing liquid, with sedimentation of the polishing product (Suratwala et al. 2008). Rogues can be naturally present in abrasives with a wide particle size distribution, or even be created by the machine. The characteristics of such scratches (width, length, and depth) very much depend on the size and nature of the particles but also on the polishing conditions (polishing pads, pressures, speeds applied, etc.). A micrograph of a typical example of a discontinuous scratch is given in Figure 12.5.

Discontinuous scratches are very different from the scratches created in handling operations on optics during their life cycle (scratches illustrated by the example given in Figure 12.6). These handling scratches are less frequent but still present with certain manufacturers.

Polishing using MRF technology that has been developed over the last few years especially for the polishing of surfaces with complex shapes also creates very typical surface defects such as those illustrated in Figure 12.7. The defects resemble a comet, oriented according to the direction of the rotation of the MRF wheel. As explained by Menapace

Chapter 12

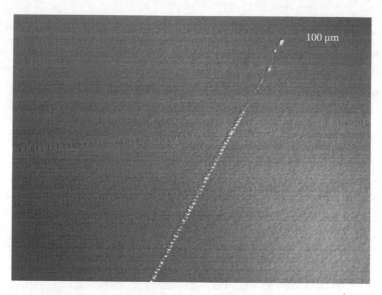

FIGURE 12.5 Discontinuous polishing scratch.

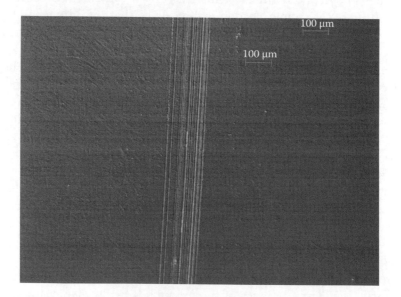

FIGURE 12.6 Typical continuous scratch from handling.

(Menapace et al. 2007), these defects are due to surface imperfections present prior to MRF and can be eliminated by also ensuring the absence of scratches prior to MRF and by a proper, prior preparation of the surface.

Hollow chips (see Figures 12.8 and 12.9) are defects that are difficult to see with a classic visual inspection, which means normal illumination, and inspection but that appear much more clearly with lighting of the edges of the component. With this type of defect, a very deep fracture produces a very wide diffusion of the light in the material. These defects are generated during shaping or rough grinding with a diamond grinding wheel prior to the stages of fine grinding and polishing. As the optical

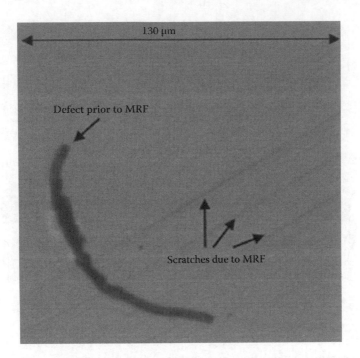

FIGURE 12.7   Scratches created by MRF due to a defect previously present on the surface image width = 130 μm.

FIGURE 12.8   Hollow chip in a neodymium glass laser plate.

component is not transparent at the roughing stage, hollow chips are identified only by painstaking visual inspections after polishing.

The defects found on the surface of crystals, such as KDP crystals, are generally pits caused by diamond machining. Fine scratches can also be encountered on this type of optic related to cleaning after machining.

FIGURE 12.9 Hollow chip in a silica optic.

Surface defects have different impacts on the use of the optical component concerned. They must, therefore, be classified, and their acceptability must be determined according to the end use of the component. The ISO 10110-7 standard offers a classification based on their dimensions measured on the surface. Depending on requirements, these specifications on the number and size of acceptable surface imperfections can be completed by indications on other aspects of the defect. For the requirements of LID resistance, for example, a distinction must be made between the so-called plastic scratches and scratches presenting fractures. The former have a higher LID resistance than the latter. That has been evidenced at the wavelength of 351 nm in the nanosecond regime (Salleo et al. 1998). The scratches caused by MRF illustrated in Figure 12.7 are typical of the plastic type, whereas discontinuous polishing scratches generally present fractures.

12.2.3.2 Treatments to Reduce Surface Defects

To reduce surface defects during manufacture, precautions must be taken in the choice of the abrasives, in the maintenance and the environment of the machine tools and in the handling of the optical components. Despite all these precautions, surface defects may occur, especially when the optics components are manufactured in large numbers and when they have a large surface area. To eliminate the remaining surface defects or at least to reduce their deleterious impact, several postpolishing techniques can be employed. For silica components, annealing consists of superficially remelting the silica, which eliminates certain defects. Annealing may be very localized, thanks to the use of the CO_2 laser that emits a wavelength of 10.6 μm absorbed by the silica, enabling heating of the silica over a thickness of a few millimeters (Cormont et al. 2010).

Despite being effect at mitigating scratches as shown in Figure 12.10, localized CO_2 laser annealing technique nonetheless poses problems because of stress and surface distortion (Cormont et al. 2010).

Hydrofluoric acid is also used on silica to remove the contaminants and to eliminate the fine fractures, as can be seen in Figure 12.11 (Suratwala et al. 2011).

FIGURE 12.10 Handling scratch partially re-melted in the center by firing of the CO_2 laser.

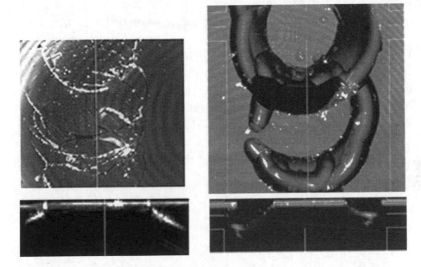

FIGURE 12.11 Scratches before and after acid treatment. On the left, the detail of a 20 μm wide scratch. On the right, detail of a scratch after the acid treatment. The lower images are in-depth cross sections obtained by confocal microscopy. These images make it possible to note that the acid spreads into the depth of the fractures and widens them. In this case, the scratches are 12 μm deep.

Postpolishing treatments to eliminate surface defects can be applied locally or more generally over the entire optic component. Local application typically gives rise to modifications in the other qualities of the component, such as its flatness, and is not widely used by manufacturers.

12.2.3.3 Roughness

The previous section presented a few examples of damage precursors (subsurface defects), which are extremely localized and often related to manufacturing incidents, the frequency of which varies depending on the techniques used and their implementation. Nevertheless, the manufacturing technique also generates more global characteristics on the surface, some of which will severely impact the damage. In the nanosecond

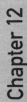

regime, numerous studies relate LID resistance to surface roughness. Over a very wide range of materials and machining techniques, the damage threshold increases when the roughness decreases (Wood et al. 1985). However, surface roughness is not sufficient to determine the behavior of an optical component under a high fluence. By working solely on the improvement of surface roughness, polishers can obtain optic components with excellent final surface roughness but a rather bad subsurface quality. This is, for example, the case when material removal during polishing is not gradual enough to remove defects created by early processing stages. As will be seen in the following paragraph, it is in fact the presence of fractures beneath the surface, which will limit the LID resistance of the optic component in the nanosecond regime.

Surface imperfections, such as roughness, may nonetheless create types of damage with highly characteristic morphologies, one of the most frequently encountered and the most recognizable being the ripple pattern that can be observed after laser exposure. Ripples have been observed on metals (Birnbaum 1965), semiconductors (Young et al. 1983), and dielectrics (Temple and Soileau 1981) and with both continuous and pulsed lasers over wavelengths ranging from the UV to the IR. The ripple pattern is attributed to the interference between the incident wave and a surface wave generated by surface imperfections, for example, roughness (Gorshkov et al. 2003, 2005). The surface imperfections causing the creation of ripples may also present a variety of aspects (Jee et al. 1988): pits, holes, craters, grooves, scratches, pores, embedded holes, heterogeneous surface layers, and surface irregularities. The morphology of the ripple pattern is characteristic, with a period of $d = \lambda/(1 \pm \sin \theta)$, where λ is the laser wavelength and θ the angle of incidence (Young et al. 1983). Moreover, the orientation of the ripples is perpendicular to the plane of incidence. An example of a ripple pattern on dielectric is given in Figure 12.12 (ripple pattern on the top layer of a multilayer dielectric pulse compression grating).

FIGURE 12.12   Ripple pattern observed on an MLD pulse compression grating after a burst of 100 laser shots at 1.053 μm, 500 fs.

In addition to these ripples with a period close to the wavelength, another family of ripples with a greater period is observed more especially in the femtosecond regime, both on metals and on semiconductors (Bonse et al. 2002) or glass (Shimotsuma et al. 2003). Considerable work is being pursed on the understanding and control of the formation of these nanogratings in the hope of being able to easily create sub-λ patterns on a variety of materials to change their optic properties.

12.2.3.4 Subsurface Defects

After a polishing operation, a component includes a disrupted zone between the bulk material and the outside environment (Rayleigh 1903). This configuration applies to glass and most other substrates.

This transition zone can be divided into two main areas: a so-called polishing layer (Beilby 1903) created by the physicochemical polishing process, extending over a few dozen nanometers (Rayleigh 1903) and measurable, for example, by ellipsometry (Sakata 1973); a fractured area (closed fractures) created during the cutting and shaping operations, propagated or partially occluded by subsequent lapping, grinding, and polishing operations, the extension of which ranges from a few microns to a few dozen microns (Parks 1990). The extent and nature of these two areas vary considerably according to the surfacing process and the characterization method chosen. Results in the literature give SSD depths ranging from a few microns (Dettman 1990) to over 100 μm (Camp et al. 1998). Local absorbing pollutants coming from the polishing process (tooling, polishing compounds, etc.) can also be embedded in the SSD (Kozlowski et al. 1998, Neauport et al. 2005b). Pollutants from the environment and the immediate surroundings can also be present. Their impact is addressed in Chapter 16.

12.2.3.5 Pollution Induced by the Manufacturing Process

Polishing can employ many tools and products including, for example, diamond abrasive wheels, the grains of which are held by a metal or resinoid binder, alumina, boron carbide, or silicon used in a water suspension in the grinding operations performed against a cast-iron, brass, or other plate, cerium or zirconium for polishing, etc. As the polishing process consists of a succession of abrasive stages with removal by mechanical action (shaping, grinding, or fine grinding) or physicochemical action (polishing), the polished glass interface is contaminated with these materials (Figure 12.13).

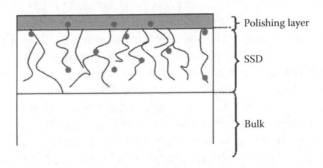

FIGURE 12.13   Schematic structure of the glass interface after polishing (dots = pollutants specific to the process).

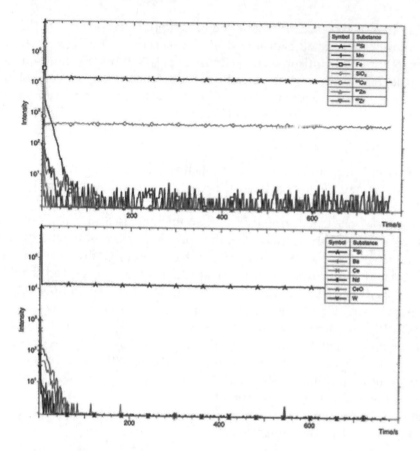

FIGURE 12.14   TOF-SIMS analysis of a cerium-polished sample. A polluted layer approximately 50 nm thick is visible (1 s = 1 nm).

Figure 12.14 illustrates this contamination, presenting a measurement performed by time of flight—secondary ion mass spectroscopy (TOF-SIMS) made on a cerium-polished silica sample. The existence of a zone some 50 nanometers deep contaminated not only by cerium (from the polishing slurry) but also by iron (from a cast-iron plate), barium, and neodymium (laser glass blank polished with the same slurry prior to polishing of the silica sample) can be seen.

If such pollutants exist, it can therefore be thought that they may play a role in the initiation of LID (Kozlwoski et al. 1998). This issue has been substantially addressed in the particular case of silica damage at 351 nm in the nanosecond regime. The impact of these pollutants is in fact supported by various authors who consider that the damage precursors are of nanometric size: bubbles, metal or dielectric inclusions, etc. (Bloembergen 1973, Dijon et al. 1997). Furthermore, attempts at characterizing the silica surface by various means (absorption, luminescence, etc.) and correlation with laser damage in the nanosecond regime have rarely proved to be fruitful. This can be due to the fact that the size of initiating defects are below the detection level of methods used. Various groups have also used engineered defects (e.g., nanoparticles embedded under silica vacuum deposited layer) to evidence that initiation may occur from nanoparticle type defects (see Section 1.3)

Relating the concentration of a pollutant to an LID threshold or density requires a method for analyzing the polished glass interface. It has been seen above that SIMS can be used for an elementary quantification of the pollutants (Kozlowski et al. 1998). Acid etching coupled with inductively coupled plasma mass spectroscopy (ICP-MS) has also proved to be pertinent (Neauport et al. 2005b). Synchrotron-based x-ray fluorescence spectrometry (Li et al. 2010) has been shown to be effective. Employing these diagnostics becomes possible to manufacture parts with a given combination of polishing products, to measure the pollutants in the interface of the glass polished and to attempt to correlate the level of pollutants with the threshold measured. Several authors have undertaken such an experimental program. Cerium, in particular, has been accused of being the potential damage initiator at 351 nm in the nanosecond regime given its potential absorption at that wavelength (Kozlwoski et al. 1998). The tests of D. W. Camp evidence that although the best sample is polished with zirconium, the correlation between the level of cerium and the damage threshold remains limited (Camp et al. 1998). Camp's result has been contradicted by the evidencing of a close cerium content/damage threshold correlation (Neauport et al. 2005b). That trend was confirmed, again in the nanosecond regime but at 1064 nm, by A. V. Smith (Smith and Do 2008). An absence of correlation was further-more evidenced in another series of tests with higher LID resistance samples (Neauport et al. 2008). There is therefore a certain degree of disagreement between the results that can be explained by the wide variety of surfacing processes used by the different authors. Moreover, modification of the polishing product often goes in hand with a modifica-tion of the lapping plate or pad material and more generally the polishing parameters (pressure, speed, etc.). It is therefore probably not just the level of pollutant that changes from one sample to another but more generally the nature of the interface and in par-ticular the quantity of micro-subsurface fractures, the role of which will be addressed in the following part. It nevertheless seems that the quantity of metal species plays a poten-tially important role in damage to the component in the nanosecond regime, especially when the quantity of pollutant is high (Li 2010). In that respect, removal of 10–20 nm, a thickness similar to the depth of the polishing layer contaminated with metal species, by ion beam figuring (IBF) on a polished silica surface pre-irradiated at 355 nm enables a practically twofold gain in the LID threshold at 355 nm in the nanosecond regime (Xu et al. 2008). The benefit of IBF in the elimination of contaminants brought in by the polishing process has also been recorded on other materials such as the CLBO crystal. A 30% improvement in the LID threshold at a wavelength of 266 nm and a pulse time of 0.75 ns was, for example, evidenced (Kamimura et al. 1998).

12.2.3.6 Subsurface Cracks and Dislocations

It has been seen that in the milling, grinding and polishing of an optic component hard abrasive grains (diamond wheel, abrasive powder, etc.) come into contact with the surface of the component. Material removal then occurs by successive fracturing, particularly during initial stages of shaping and grinding (Figure 12.13). An optical component, therefore, has residual quantity of that SSD, the geometry, density, and depth of which depend on the parameters of the process adopted. The existence of these fractures has been known for more than a century (Rayleigh 1903), and numerous destructive and nondestructive measurement means have been successfully employed (Wang et al. 2011).

Chapter 12

Generally speaking, the principle of a destructive measurement consists of performing local or total wear on the part to open up or reveal the microcracks, the depth of which can then be measured. Different types of wear methods have been studied. For example, Carr soaks the parts in successive HF baths and used a microscope observation to measure SSD (Carr et al. 1999). Other methods are polishing a taper (Hed et al. 1989) or a ball dimple (Zhou et al. 1994) in the part to be qualified using a suitable tool. If the shape of the local wear produced is known, it is possible to determine the SSD depth using conventional microscopic observation. The ball dimpling method can be improved by replacing mechanical polishing by polishing on an MRF machine (Randi et al. 2005). Randi's method can be extended over a larger surface area by generating a prism shape on the whole sample surface, again using the MRF process, to reveal the fractures (Suratwala et al. 2006). Finally, a method consisting of quantifying the evolution of the surface roughness during iterative acid HF treatments has also proved to be effective (Neauport et al. 2009). Most of these methods are effective on a very wide range of materials: silica, BK7, zerodur, silicon, LibNO$_3$, CaF$_2$, or MgF$_2$. While the vast majority of the measurement methods are destructive, there are a few nondestructive methods, including optical coherent tomography (Wuttig et al. 1999), total internal microscopy (Temple 1981), or more recently confocal fluorescence microscopy (see Figure 12.15, Neauport et al. 2009, Catrin et al. 2013) or light scattering (Trost et al. 2013).

The value of these measurement methods is that they provide tools for the development of polishing methods enabling the manufacture of optical components with the smallest possible number of SSD type defects. Application of such an approach makes it possible to study the comparative effect of roughing or grinding conditions on SSD

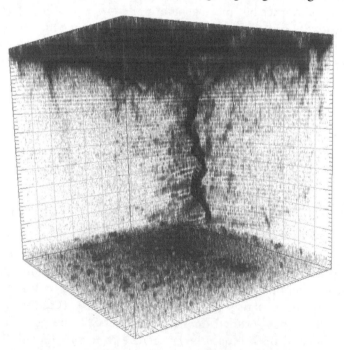

FIGURE 12.15 Microfracture on a silica part after a diamond wheel grinding step by confocal fluorescence microscopy with excitation at 405 nm, acquisition on the 435–660 nm band. The fracture is 150 μm deep.

(Suratwala et al. 2006, Wang et al. 2008, Neauport et al. 2009). Elimination of the polished silica interface defect by a combination of MRF, acid treatment, and laser conditioning also makes it possible to obtain a very marked reduction in the damage density at 351 nm in the nanosecond regime (Menapace et al. 2002). That result is attributed to the capability of MRF to remove material by tangential rather than by normal stress in a standard polishing process (LLE Review 117). Dissolution of SSD from an acid bath also improves LID resistance. Such a result has been reported mainly on silica (Kamimura et al. 2004, Suratwala et al. 2011) but should extend to other materials. Reactive ion machining processes have also been successfully implemented to eliminate SSD, such as the plasma jet–assisted chemical evaporation (PJCE) process, where, as in acid treatment, the material is removed without any abrasive, thanks to the combined effect of the plasma and its reactive gases, resulting in a considerable reduction of SSD (Schindler et al. 2001).

12.3   Diffractive Optics and LID

A diffractive component uses diffraction to control the light. There are many types of diffractive components (O'Shea et al. 2004): binary grating, phase volume grating, multilevel binary optic, volume phase hologram, kinoform phase plate, etc. Diffractive components in a laser can perform a very wide range of functions, such as beam sampling, pulse compression, beam focusing, or spatial shaping, to mention but a few. The majority of the studies published deal with the case of diffraction gratings, whether volume holographic or surface relief. In most cases, these results extend to multiple binary optics. As far as kinoform components are concerned, with a continuous shape variation, reference must be made to the previous section on laser damage and polishing processes. The mechanisms governing their damage are frequently similar, because such components are obtained by methods derived from polishing techniques (IBF, MRF, etc.).

A diffraction grating is an optical component that has been in use for over a century. Its wide dispersion makes it a component of choice in spectroscopy. It is used in a host of fields, from the analysis of chemical compounds to optic telecommunications. The use of large-size gratings in high-power lasers is less well known and more recent (Rouyer et al. 1993). Nevertheless, such gratings have been successfully installed in various facilities (cf. Chapter 19). The use of gratings in nanosecond pulsed power laser chains has benefited from considerable technological progress. For ultrashort laser pulses in the femtosecond regime, the chirped amplification techniques (linear variations of the frequency during the pulse) introduced in the mid-1980s led to the production of intensities 1,000–10,000 times higher than nanosecond regime pulses. A full description of the use of diffraction gratings in high-intensity laser applications is also available in the article by J. Britten in the work edited by K. Barat (Britten 2009). Finally, the manufacturing processes for different types of gratings are described in the book of E. G. Loewen and E. Popov (Loewen and Popov 1997).

12.3.1   Phase Volume Gratings

A phase volume holographic grating is formed by a variation of the index or the absorption in a solid material. Several types of material can be used, including dichromated gelatins (DCG) and polymer matrices. In addition, more recent developments have

Chapter 12

also enabled the use of photothermal refractive (PTR) glass (Efomiv et al. 1999, Glebov 2007). To obtain a grating, the photosensitive material is exposed in a field of interferences at several wavelengths in order to obtain the required periodic variations. Chemical or heat treatment, depending on the photosensitive material used, makes it possible to modify the index locally, thus producing the volume hologram. Comparative damage tests have been carried out on DCG and polymer gratings in transmission in the nanosecond regime (Loiseaux et al. 1997). It has been found that the thresholds of both types of material are relatively high (close to 50 J/cm² at the 800 nm wavelength). The threshold in the UV is low, probably induced by the intrinsic absorption of the photosensitive materials. For short pulses, the results are widespread with thresholds ranging from the order of 1 J/cm² to a fraction of a J/cm² depending on the references (Loiseaux et al. 1997, Reichart et al. 2001) for DCG and for polymer gratings. Similar findings were observed more recently with DCG gratings deposited on a dielectric mirror (Rambo et al. 2006). It must be noted that the problems of wave surface quality in the DCG and polymer gratings, due to the thickness of the photosensitive material in the order of a dozen microns, and the hygroscopic character of the DCG layers are such that their use in high LID resistance applications is uncommon. Surface relief gratings are generally preferred.

Finally, as far as the PTR glasses are concerned, the inorganic and vitreous nature of the photosensitive material provides a threshold at 1054 nm, ranging from the order of 20 J/cm² for 1 ns pulse width (Liao et al. 2007) to 40 J/cm², 8 ns pulse width (Glebov 2007, 2008), that is, similar to ordinary BK7 type glass. On the other hand, the thresholds in the femtosecond regime appear to be lower (Hernández-Garay et al. 2011).

12.3.2 Surface Relief Gratings

There are many manufacturing techniques for the production of surface relief gratings: direct engraving with a diamond-tipped tool, holography, electron beam engraving, lithography, etc. The choice of the method depends on the application in question and on the characteristics of the pattern required (pitch, depth, etc.). Whatever the process adopted, the end result is either a grating engraved directly on the substrate or a grating engraved in a layer deposited on a substrate (photopolymer or a replica layer in the case of a replicated grating). A distinction is made between transmission and reflection gratings.

The majority of transmission gratings used in a high LID resistance context are gratings engraved in a silica substrate. This is due to two main reasons: photosensitive resins offer limited thresholds and silica is the material that offers the highest threshold from the UV to the IR in both nanosecond and femtosecond regimes. These silica transmission gratings have many applications, with, for example, beam sampling (Britten and Summer 1998), beam deviation (Nguyen et al. 1997, Neauport et al 2005a), the production of a polarizing function (Clausnitzer et al. 2007, Wang et al. 2011), focusing (Neauport et al 2005a), and even pulse compression (Clausnitzer et al. 2003), outside of the context of ultrahigh intensities given the nonlinear index of silica. As far as the LID resistance behavior is concerned, while it has been occasionally reported that the engraving of the gratings maintains the threshold of the substrate in 1/1 mode tests at 355 nm, 3 ns (Nguyen et al. 1997, Neauport et al. 2003), deterioration of the LID resistance resulting from the engraving of the grating has sometimes been recorded at 1064 nm, 3 ns (Neauport et al. 2007). There are two potential sources for these differences in behavior: the manufacturing process, similar to what has been illustrated in the case of polishing processes (cf. Section 12.2.3), and the form of the periodic pattern itself. The periodic pattern engraved on the surface does in fact create intensity enhancement at the grating interface, the place and value of which depend on the characteristics of the grating (shape, depth, and period). Simple observation of the effect of the polarization on the damage morphology of a transmission grating would indicate that it contributes to the threshold of the final grating. This is illustrated in Figure 12.16. A 2400 l/mm grating engraved in silica was LID tested at 355 nm, 3 ns in TE (S) polarization or in TM (P) polarization.

The damage morphology can be seen to change on moving from one polarization to the other. That change matches the displacement of the peak of the electric field. When the field enhancement occurs in the line (TE polarization), the damage appears in the form of a molten line. When the field enhancement is at the interface (TM polarization), the line is broken. The field distribution is calculated here for the simple lamellar profile of a transmission grating; it can be calculated for a much more complex shape and therefore be extended to any type of diffractive component (Demesy et al. 2011).

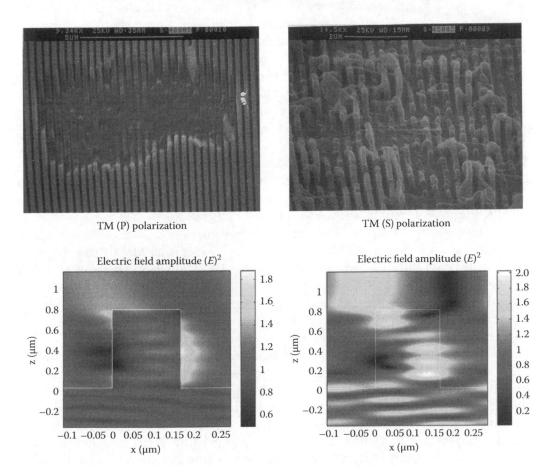

FIGURE 12.16 2400 l/mm grating engraved in silica. Effect of the position of field enhancement on the damage morphology. Test at 25°, 355 nm, 3 ns in both polarizations.

Considerable work has been undertaken on the study and optimization of the LID resistance of surface relief gratings, driven by the development of chirped pulse amplification systems that use the so-called pulse compression gratings operating in pairs. A compression grating by reflection must therefore be seen as a system with two distinct functions: a dispersion function obtained with the periodic structure and a reflection function. The grating must diffract the light in the required order, the order −1R (i.e., [−1] order in reflection) at an incidence close to Littrow. Numerous combinations can be imagined: a surface relief grating on or under the structure for the reflection function, a phase variation grating (volume holographic grating) either alone or combined with a high-reflection mirror (see the preceding text), and a grating engraved on a surface in total reflection.

The gold surface relief compression grating is, historically speaking, the first type of grating used for laser pulse compression. A gold surface relief grating comes in the form of a periodically modulated surface (the dispersion function) coated with a layer of gold (the reflection function). Considerable work has been undertaken in the LID of gold gratings in the nanosecond and femtosecond regimes, in particular, the work of B.C. Stuart, who turned his attention to this subject as early as 1994. He evidences (Stuart et al. 1996) that in short pulses (< a few 100 ps) on metals and dielectrics, the damage process at 1053 and 526 nm occurs by a multiphotonic absorption phenomenon. The damage threshold of a gold grating (in this case, a 1480 l/mm tested at 51°) remains constant, close to 0.5 J/cm^2 over a timescale from 100 fs to 300 ps. Moreover, the threshold is independent of the thickness of the gold coating for any thickness above approximately 100 nm. Above 200 ps, the threshold varies in $\tau^{1/2}$, where τ is the ratio of the pulse times considered. In his work, Stuart used metrology in the S/1 mode with approximately 600 shots per site at 10 Hz. Figure 12.17 shows the result of a damage test performed on a 1740 l/mm gold grating at an incidence of 72° in TM (P) polarization at a wavelength of 1053 nm and a pulse time of 500 fs. The test mode is S/1 with 100 shots per site and 200 sites tested. The test beam is Gaussian with a diameter of 180 μm at 1/e^2. The damage threshold is 0.67 J/cm^2. That value is comparable to Stuart's data after an angular correction to take account of the differences of test incidences between the 1480 l/mm (51°) gratings and the 1740 l/mm (72°) gratings.

To overcome the limitations of the gold compression grating, use must be made of materials offering higher thresholds in short pulses. That is what M. D. Perry proposed when he invented the multilayer dielectric grating (MLD) in 1995 (Perry et al. 1995). Instead of coating the surface relief grating (dispersion function) with a gold layer (reflection function), he inverted the structure by placing the mirror under the grating. Schematically, the grating emits two orders—0T and −1T—and the mirror reflects those two orders to recombine them constructively toward the order −1R.

A multidielectric mirror consisting of alternating high-index layers (HfO$_2$) and low-index layers (SiO$_2$) deposited by vacuum evaporation is chosen. The grating is then engraved in the top layer of the mirror, which is in silica to maximize the damage threshold. The choice of the HfO$_2$/SiO$_2$ in tandem meets three main constraints: obtaining the highest possible contrast index to minimize the number of layers to be deposited, using vacuum evaporation, and obtaining the highest possible LID resistance. The HfO$_2$/SiO$_2$ combination was rapidly adopted for laser requirements in nanosecond pulses and is now widely used (see Chapters 13 through 15). It has therefore logically been adopted here. The threshold of silica in short pulses is 2–2.5 J/cm^2 for a polished silica substrate tested

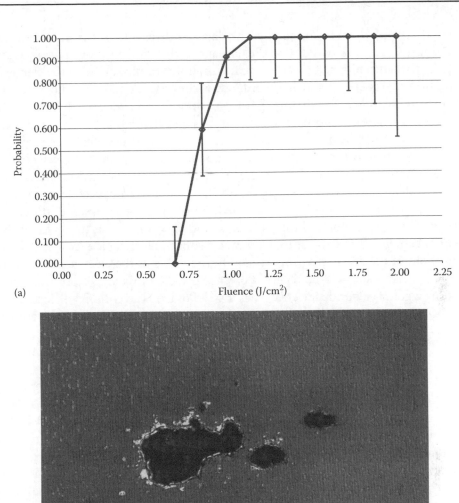

(a)

(b)

FIGURE 12.17   (a) Probability of damage to a HJY1740 l/mm grating, TM, incidence of 72°, 500 fs and (b) typical damage morphology (microscope observation).

at 500 fs (Stuart et al. 1996) with normal incidence and 1.053 μm. As far as deposited thin layers are concerned, the work of M. Mero gives an overview (Mero et al. 2005) of the comparative levels of performance of various dielectric monolayers deposited by ion beam sputtering and tested for their resistance at a wavelength of 800 nm and with different pulse times. HfO_2 appeared to be the most pertinent choice. This new design gives a gain in the order of 2 for the laser damage threshold compared to the conventional gold grating with a threshold of 2.5 J/cm^2 at 1053 nm, 1 ps for a MLD 1780 l/mm grating at 77.2° (Barty et al. 2004). That is equivalent to ~2 J/cm^2 at 500 fs using a time law in $\tau^{1/3}$, such as established by Mero on the short pulse dielectrics (Mero et al. 2005).

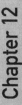

What are the phenomena impacting the damage threshold of these pulse compression gratings? C. P. J. Barty finds that the damage to the MLD grating appears to start at the edge of the silica pillars, that is, where the modulus of the square of the electric field $|E|^2$ is highest (Barty et al. 2004), in accordance with the work quoted earlier (Stuart et al. 1996). The test of an MLD 1800 l/mm grating at different incidences also evidenced that the threshold variation is approached better by a law in $1/|E|^2$ than by a law in the inverse of the cosine of the angle of incidence (Britten et al. 2004). The differences recorded are unfortunately small, especially if account is taken of a typical uncertainty on the test fluence of ±10%. He also noted that a grating with *thin* pillars tends to minimize the value of the electric field enhancement $|E|^2$ in the structure compared to a grating with *thick* pillars. A more systematic study carried out on MLD gratings with different line density profiles 1780 l/mm and the patterns of which vary so that the value of the $|E|^2$ field enhancement in the structure also varies subsequently evidences that the damage threshold of the grating is proportional to the value of the electric field (Neauport et al. 2007). That result will be repeated on the so-called mixed compression gratings, combining a gold layer and dielectric pairs instead of the conventional dielectric mirror (Neauport et al. 2010). Finally, scanning electron microscopy (SEM) observations also evidence that the damage occurs on the edge of the line opposite the incident beam in accordance with the field calculation and that a ripple is present around the damage and plays a part in initiating the damage (Hocquet et al. 2011).

In addition to the effect of the electric field, the final cleaning of the grating also appears to have a specific impact on the damage threshold of pulse compression gratings. A study has been devoted to the effect of various cleaning processes (PIRANHA at different temperatures, Nanostrip, etc.) on the threshold at 10 ps, 1053 nm. It would appear that the efficiency of the process and its positive impact on damage are linked to the reduction of the traces of photopolymers after processing (LLE Review 2005, 2007). Outside contaminants such as those from the outgassing of materials from the compression chamber can also noticeably lower the thresholds (Ashe et al. 2008).

References

Ashe, B., K. L. Marshall, D. Mastrosimone, and C. McAtee 2008. Minimizing contamination to multilayer dielectric diffraction gratings within a large vacuum system. *Paper Presented at the Meeting of the Boulder Damage Symposium, SPIE 7069.*

Barty, C. P. J., M. Key, J. Britten, R. Beach, G. Beer, C. Brown, S. Bryan et al. 2004. An overview of LLNL high-energy short-pulse technology for advanced radiography of laser fusion experiments. *Nuclear Fusion* 44 (12): PII S0029-5515(04)88687-3.

Beilby, 1903. Surface flow in crystalline solids under mechanical disturbance. *Proceedings of the Royal Society*, 72: 218–225.

Bercegol, H., P. Grua, D. Hebert, and J. P. Morreeuw 2008. Progress in the understanding of fracture related laser damage of fused silica—Art. no. 672003. *Laser-Induced Damage in Optical Materials:* 2007 6720: 72003–72003.

Birnbaum, 1965. Semiconductor surface damage produced by ruby lasers. *Journal of Applied Physics* 36 (11): 3688.

Bloembergen, N. 1973. Role of cracks, pores, and absorbing inclusion on laser induced damage threshold at surface of transparent dielectrics. *Applied Optics* 12: 661.

Bonse, J., S. Baudach, J. Kruger, W. Kautek, and M. Lenzner 2002. Femtosecond laser ablation of silicon-modification thresholds and morphology. *Applied Physics A-Materials Science & Processing* 74 (1): 19–25.

Britten J. A. 2009. *Diffraction Gratings for High Intensity Laser Applications*, pp 99 from Barat K. *Laser Safety, Tools and Training*, CRC Press, New York.

Britten, J. A., W. Molander, A. M. Komashko, and C. P. J. Barty 2004. Multilayer dielectric gratings for petawatt-class laser systems. *Proceedings of the SPIE—The International Society for Optical Engineering* 5273, pp. 1–7.

Britten, J. A. and L. J. Summers 1998. Multiscale, multifunction diffractive structures wet etched into fused silica for high-laser damage threshold applications. *Applied Optics* 37 (30): 7049–7054.

Bussiere, B., O. Uteza, N. Sanner, M. Sentis, G. Riboulet, L. Vigroux, M. Commandre, F. Wagner, J.-Y. Natoli, and J.-P. Chambaret 2009. Laser induced damage of sapphire and titanium doped sapphire crystals under femtosecond to nanosecond laser irradiation. *Paper Presented at the Meeting of the Boulder Damage Symposium*, SPIE 7504, pp. 75041N-1–75041N-9.

Camp, D. W., M. R. Kozlowski, L. M. Sheehan, M. Nichols, M. Dovik, and R. Raether, I. Thomas 1998. Subsurface damage and polishing compound affect the 355 nm laser damage threshold of fused silica surfaces. *Paper Presented at the Meeting of the Boulder Damage Symposium, SPIE* 3244, pp. 356–364.

Carr, W., E. Fearon, L. J. Summers, and I. D. Hutcheon, 1999. Subsurface damage assessment with atomic force microscopy. In *Proceedings of the First International Conference and General Meeting if the European Society of Precision Engineering and Nanotechnology,* Bremen, Germany.

Catrin, R., J. Neauport, P. Legros, D. Taroux, T. Corbineau, P. Cormont, and C. Maunier 2013. Using STED and ELSM confocal microscopy for a better knowledge of fused silica polished glass interface. *Optical Express* 21 (24): 29769–29779.

Clausnitzer, T., T. Kaempfe, E.-B. Kley, A. Tuennermann, A. Tishchenko, and O. Parriaux 2007. Investigation of the polarization-dependent diffraction of deep dielectric rectangular transmission gratings illuminated in Littrow mounting. *Applied Optics* 46 (6): 819–826.

Clausnitzer, T., J. Limpert, K. Zollner, H. Zellmer, H. J. Fuchs, E. B. Kley, A. Tunnermann, M. Jupe, and D. Ristau 2003. Highly efficient transmission gratings in fused silica for chirped-pulse amplification systems. *Applied Optics* 42 (34): 6934–6938.

Cormont, P., L. Gallais, L. Lamaignere, J. L. Rullier, P. Combis, and D. Hebert 2010. Impact of two CO_2 laser heatings for damage repairing on fused silica surface. *Optics Express* 18 (25): 26068–26076.

Demésy, G., L. Gallais, and M. Commandré 2011. Tridimensional multiphysics model for the study of photo-induced thermal effects in arbitrary nano-structures. *Journal of the European Optical Society. Rapid Publications* 6: 11037.

Dettman, L. 1990. *Optical Sciences Center*, Tucson, AZ, Estimation of subsurface damage depth by dimpling, technical digest, post conference edition, OSA.

Dierickx, P. 2007. *The European Extremely Large Telescope*, Highlights of Spanish Astrophysics IV, Springer, 15–28.

Dijon, J., T. Poiroux, and C. Desrumaux 1997. Nano absorbing centers: A key point in laser damage of thin film in Laser-Induced damage in Optical Materials. *Paper Presented at the Meeting of the Boulder Damage Symposium SPIE* 2966: 315.

Efimov, O. M., L. B. Glebov, S. Papernov, and A. W. Schmid 1999. Laser-induced damage of photo-thermo-refractive glasses for optical-holographic-element writing. *Proceedings of SPIE* 3578: 554–575.

Fuchs, B. A., P. Hed, and P. C. Becker 1986. Fine diamond turning of KDP crystals. *Applied Optics*, 25: 1733–1735.

Genin, F. Y., A. Salleo, T. V. Pistor, and L. L. Chase 2001. Role of light intensification by cracks in optical breakdown on surfaces. *Journal of the Optical Society of America A-Optics Image Science and Vision* 18 (10): 2607–2616.

Glebov, L. B. 2007. Fluorinated silicate glass for conventional and holographic optical elements. *Proceedings of SPIE* 6545: 654507.

Glebov, L. B. 2008. Volume holographic elements in a photo-thermo-refractive glass. *Journal of Holography Speckle* 5 (1): 1546.

Gorshkov, A., I. Bel'bas, M. Maslov, V. Sannikov, and K. Vukolov 2005. Laser damage investigations of Cu mirrors. *Fusion Engineering and Design* 74 (1–4): 859–863.

Gorshkov, A., I. Bel'bas, V. Sannikov, and K. Vukolov 2003. Frequency laser damage of Mo mirrors. *Fusion Engineering and Design* 66–68: 865–869.

Hack, H. and N. Neuroth 1982. Resistance of optical and colored glasses to 3-nsec laser pulses. *Applied Optics* 21 (18): 3239–3248.

Hed, P. P., D. F. Edwards, and J. B. Davis 1989. Subsurface damage in optical materials: Origin, measurements and removal, in *Collected Papers from ASPE Spring Conference on Subsurface Damage in Glass*, Tucson, AZ.

Chapter 12

Hernández-Garay, M. P., O. Martínez-Matos, J. G. Izquierdo, M. L. Calvo, P. Vaveliuk, P. Cheben, and L. Bañares 2011. Femtosecond spectral pulse shaping with holographic gratings recorded in photopolymerizable glasses. *Optical Express* 19 (2): 1516–1527.

Hocquet, S., J. Neauport, and N. Bonod 2011. The role of electric field polarization of the laser beam in the short pulse damage mechanism of pulse compression gratings. *Applied Physics Letters* 99 (6), 061101.

Hurt, H. H. 1985. Defects induced in optical-surfaces by the diamond-turning process. *Proceedings of the Society of Photo-Optical Instrumentation Engineers* 525: 16–21.

Jee, Y., M. F. Becker, and R. M. Walser 1988. Laser-induced damage on single-crystal metal-surfaces. *Journal of the Optical Society of America B-Optical Physics* 5 (3): 648–659.

Kamimura, T., S. Akamatsu, M. Yamamoto, I. Yamato, H. Shiba, S. Motokoshi, T. Sakamoto, T. Jitsuno, T. Okamoto, and K. Yoshida 2004. Enhancement of surface-damage resistance by removing subsurface damage in fused silica. *Paper Presented at the Meeting of the Boulder Damage Symposium* SPIE 5273, pp. 244–249.

Kamimura, T., M. Yoshimura, T. Inoue, Y. Mori, T. Sasaki, H. Yoshida, M. Nakatsuka, K. Yoshida, K. Deki, and M. Horiguchi 1998. Improvement of laser-induced surface damage in CsLiB6O10 crystal by ion etching. *Paper Presented at the Meeting of the Boulder Damage Symposium* SPIE 3244, pp. 695–701.

Karow, H. H. 1993. *Fabrication Methods for Precision Optics*, A Wiley series in pure and applied optics. Wiley Interscience Publication, New York.

Kozlowski, M. R., J. Carr, I. Hutcheon, R. Torres, L. Sheehan, D. Camp, and M. Yan, 1998. Depth profiling of polishing-induced contamination on fused silica surface. *Proceedings of Laser-Induced Damage in Optical Materials*, G. J. Exarhos, A. H. Guenther, M. R. Kozlowki, M. J. Soileau Eds, *Proceedings of SPIE* 3244: 365–375.

Li, C., X. Ju, W. Wu, X. Jiang, J. Huang, W. Zheng, and X. Yu 2010. Synchrotron micro-XRF study of metal inclusions distribution and variation in fused silica induced by ultraviolet laser pulses. *Nuclear Instruments and Methods in Physics Research Section B: Beam Interactions with Materials and Atoms* 268 (9): 1502–1507.

Liao, K.-H., M.-Y. Cheng, E. Flecher, V. I. Smirnov, L. B. Glebov, and A. Galvanauskas 2007. Large-aperture chirped volume Bragg grating based fiber CPA system. *Optical Express* 15 (8): 4876–4882.

LLE Review 2005, Vol 108, 194.

LLE Review 2007, Vol 112, 228.

Loewen E. G. and Popov E. 1997. *Diffraction Gratings and Applications*. Boca Raton, FL: CRC Press.

Loiseaux, B., D. Anne, H. Jean-Pierre, and E. Frederic 1997. Phase volume holographic grating for high-energy lasers. *Proceedings of SPIE* 3047: 957–962.

Menapace, J. A., P. J. Davis, W. A. Steele, M. R. Hachkowski, A. Nelson, and K. Xin 2007. MRF applications: On the road to making large-aperture ultraviolet laser resistant continuous phase plates for high-power lasers–art. no. 64030N. *Laser-Induced Damage in Optical Materials*: 2006 6403: 64030N.

Menapace, J. A., P. R. Ehrmann, and R. C. Bickel 2009. Magnetorheological finishing (MRF) of potassium dihydrogen phosphate (KDP) crystals: Nonaqueous fluids development, optical finish, and laser damage performance at 1064 nm and 532 nm. *Paper Presented at the Meeting of the of the Boulder Damage Symposium*, SPIE 7504, pp. 750414-1–750414-12.

Menapace, J. A., B. Penetrante, D. Golini, A. Slomba, P. E. Miller, T. Parham, M. Nichols, and J. Peterson 2002. Combined advanced finishing and UV-Laser conditioning for producing UV-damage-resistant fused silica optics. *Laser-Induced Damage In Optical Materials, SPIE Proceedings* 4679: 56–68.

Mero, M., J. Liu, W. Rudolph, D. Ristau, and K. Starke 2005. Scaling laws of femtosecond laser pulse induced breakdown in oxide films. *Physical Review B* 71 (11): 115109.

Miller, P. E., J. D. Bude, T. I. Suratwala, N. Shen, T. A. Laurence, W. A. Steele, J. Menapace, M. D. Feit, and L. L. Wong 2010. Fracture-induced subbandgap absorption as a precursor to optical damage on fused silica surfaces RID E-4791-2011. *Optics Letters* 35 (16): 2702–2704.

Moses, E. I. 2013. The National ignition campaign: Status and progress. *Nuclear Fusion* 53: 104020.

Neauport, J., C. Ambard, P. Cormont, N. Darbois, J. Destribats, C. Luitot, and O. Rondeau 2009. Subsurface damage measurement of ground fused silica parts by HF etching techniques. *Optical Express* 17 (22): 20448–20456.

Neauport, J., N. Bonod, S. Hocquet, S. Palmier, and G. Dupuy 2010. Mixed metal dielectric gratings for pulse compression. *Optical Express* 18 (23): 23776–23783.

Neauport, J., P. Cormont, L. Lamaignere, C. Ambard, F. Pilon, and H. Bercegol 2008. Concerning the impact of polishing induced contamination of fused silica optics on the laser-induced damage density at 351 nm. *Optics Communications* 281 (14): 3802–3805.

Neauport, J., P. Cormont, P. Legros, C. Ambard, and J. Destribats 2009. Imaging subsurface damage of grinded fused silica optics by confocal fluorescence microscopy. *Optical Express* 17 (5): 3543–3554.

Neauport, A., E. Journot, G. Gaborit, and P. Bouchut 2005a. Design, optical characterization, and operation of large transmission gratings for the laser integration line and laser megajoule facilities. *Applied Optics* 44 (16): 3143–3152.

Neauport, J., L. Lamaignere, H. Bercegol, F. Pilon, and J. C. Birolleau 2005b. Polishing-induced contamination of fused silica optics and laser induced damage density at 351 nm, *Optics Express* 13 (25): 10163–10171.

Neauport, J., E. Lavastre, G. Razé, G. Dupuy, N. Bonod, M. Balas, G. de Villele, J. Flamand, S. Kaladgew, and F. Desserouer 2007. Effect of electric field on laser induced damage threshold of multilayer dielectric gratings. *Optical Express* 15 (19): 12508–12522.

Nguyen, H. T., B. W. Shore, S. J. Bryan, J. A. Britten, R. D. Boyd, and M. D. Perry 1997. High-efficiency fused-silica transmission gratings. *Optics Letters* 22 (3): 142–144.

Nolte, S., C. Momma, H. Jacobs, A. Tünnermann, B. N. Chichkov, B. Wellegehausen, and H. Welling 1997. Ablation of metals by ultrashort laser pulses. *Journal of Optical Society of America B* 14 (10): 2716–2722.

O'Shea D., T. J. Suleski, A. D. Kathman, and D. W. Parther 2004. *Diffractive Optics, Design, Fabrication and Test*. Bellingham, WA: SPIE Press.

Parks, R. E. 1990. Subsurface damage in optically worked glass. *Technical Digest, post conference edition*, OSA.

Perry, M. D., R. D. Boyd, J. A. Britten, D. Decker, B. W. Shore, C. Shannon, and E. Shults 1995. High-efficiency multilayer dielectric diffraction gratings. *Optics Letters* 20 (8): 940–942.

Rambo, P., J. Schwarz, and I. Smith 2006, Development of a mirror backed volume phase grating with potential for large aperture and high damage threshold. *Optics Communications* 260: 403–414.

Randi, J. A., J. C. Lambropoulos, and S. D. Jacobs 2005. Subsurface damage in some single crystalline optical materials. *Applied Optics* 44 (12), 2241–2249.

Rayleigh J. W. S. 1903. *Scientific Papers*. Cambridge, U.K.: University Press, Vol. 4 p. 54, 74, 542.

Reichart, A., N. Blanchot, P. Y. Baures, H. Bercegol, B. Wattellier, J. P. Zou, C. Sauteret, and J. Dijon 2001. CPA compression gratings with improved damage performance. *Laser-Induced Damage in Optical Materials: 2000, SPIE* 4347, pp. 521–527.

Rouyer, C., E. Mazataud, I. L. Allais, A. Pierre, S. Seznec, C. Sauteret, and G. Mourou 1993. Generation of 50-tw femtosecond pulses In A Ti-sapphire/nd-glass chain. *Optics Letters* 18 (3): 214–216.

Sakata, H. 1973. Etude ellipsométrique du mécanisme fondamental de polissage du verre, Japanese. *Journal of Applied Physics* 12: 173.

Salleo, A., F. Y. Genin, J. Yoshiyama, C. J. Stolz, and M. R. Kozlowski 1998. Laser-induced damage of fused silica at 355 nm initiated at scratches. *Laser-Induced Damage in Optical Materials: 1997, Proceedings* 3244: 341–347.

Schindler, A., G. Boehm, T. Haensel, W. Frank, A. Nickel, B. Rauschenbach, and F. Bigl 2001. Precision optical asphere fabrication by plasma jet chemical etching (PJCE) and ion beam figuring. *Optical Manufacturing and Testing* 4451: 242–248.

Shear Stress in Magnetorheological Finishing for Glasses, LLE Review Vol. 117, 2009.

Shimotsuma, Y., P. G. Kazansky, J. R. Qiu, and K. Hirao 2003. Self-organized nanogratings in glass irradiated by ultrashort light pulses. *Physical Review Letters* 91 (24): 247405.

Smith, A. V. and B. T. Do 2008. Bulk and surface laser damage of silica by picosecond and nanosecond pulses at 1064 nm. *Applied Optics* 47 (26): 4812–4832.

Stuart, B. C., M. D. Feit, S. Herman, A. M. Rubenchik, B. W. Shore, and M. D. Perry 1996. Optical ablation by high-power short-pulse lasers. *Journal of the Optical Society of America B-optical Physics* 13 (2): 459–468.

Sun, Y., Q. Zhang, and H. Gong 1994. Polarized 10-ns frequency-doubled Nd:YAG laser-induced damage to titanium-doped sapphire. *Paper Presented at the Meeting of the Boulder Damage Symposium, SPIE* 2114, pp. 166–177.

Suratwala, T., R. Steele, M. D. Feit, L. Wong, P. Miller, J. Menapace, and P. Davis 2008. Effect of rogue particles on the sub-surface damage of fused silica during grinding/polishing. *Journal of Non-crystalline Solids* 354 (18): 2023–2037.

Suratwala, T., L. Wong, P. Miller, M. D. Feit, J. Menapace, R. Steele, P. Davis, and D. Walmer 2006. Sub-surface mechanical damage distributions during grinding of fused silica. *Journal of Non-Crystalline Solids* 352 (52–54): 5601–5617.

Suratwala, T. I., P. E. Miller, J. D. Bude, W. A. Steele, N. Shen, M. V. Monticelli, M. D. Feit et al. 2011. HF-based etching processes for improving laser damage resistance of fused silica optical surfaces. *Journal of the American Ceramic Society* 94 (2): 416–428.

Temple, P. A. 1981. Total internal reflection microscopy: A surface inspection technique. *Applied Optics* 20: 2656–2664.

Temple, P. A. and M. J. Soileau 1981. Polarization charge model for laser-induced ripple patterns in dielectric materials. *IEEE Journal of Quantum Electronics* QE-17 10: 2067–2072.

Trost, M., T. Herffurth, D. Schmitz, S. Schröder, A. Duparré, and A. Tünnermann, 2013. Evaluation of subsurface damage by light scattering techniques. *Applied Optics* 52 (26): 6579–6588.

Uteza, O., B. Bussiere, F. Canova, J. P. Chambaret, P. Delaporte, T. Itina, and M. Sentis 2007. Laser-induced damage threshold of sapphire in nanosecond, picosecond and femtosecond regimes. *Applied Surface Science* 254 (4): 799–803.

Wang, B. and Y. Li 2011. Wideband femtosecond polarizing beam splitter grating with orthogonally diffractive directions. *Journal of Optoelectronics and Advanced Materials* 13 (5–6): 609–612.

Wang, J., Y. Li, J. Han, Q. Xu, and Y. Guo 2011. Evaluating subsurface damage in optical glasses. *Journal of the European Optical Society-Rapid Publications* 6, pp. 11001-1–11001-11.

Wang, Y. Wu, Y. Dai, and S. Li 2008. Subsurface damage distribution in the lapping process. *Applied Optics* 47: 1417–1426.

Wood, R. M., P. Waite, and S. K. Sharme. The effect of surface finish on the laser induced damage thresholds of gold coated copper mirrors. *Laser-Induced Damage in Optical Materials: 1985, Proceedings* 688: 157–163.

Wuttig, A., J. Steinert, A. Duparre, and H. Truckenbrodt 1999. Surface roughness and subsurface damage characterization of fused silica substrates. *Proceedings of SPIE* 3739: 369–376.

Xu, S., W. Zheng, X. Yuan, H. Lv, and X. Zu 2008, Recovery of fused silica surface damage resistance by ion beam etching. *Nuclear Instruments and Methods in Physics Research B* 266: 3370–3374.

Young, J. F., J. S. Preston, H. M. Vandriel, J. E. Sipe 1983. Laser-induced periodic surface-structure.2. Experiments on Ge, Si, Al, and Brass. *Physical Review B* 27 (2): 1155–1172.

Zhou, Y., P. D. Funkenbusch, D. J. Quesnel, D. Golini, and A. Lindquist 1994. Effect of etching and imaging mode on the measurement of subsurface damage in microground optical glasses. *Journal of American Ceramic Society* 77 (12), 3277–3280.

13. Introduction to Optical Coatings and Thin Film Production

H. Angus Macleod

13.1 Introduction

An optical component generally consists initially of a body of suitable optical material worked so that the surfaces are smooth and have a prescribed shape. We describe their properties as specular, from the Latin speculum meaning a mirror. These surfaces are designed to transmit and reflect the incident light in a particular manner according to their shapes and their optical properties. The emphasis in the design of the component is the desired manipulation of the direction of the light, and this takes precedence over other aspects of performance. As a result, the quality of the light that results from the interaction with the surface is rarely ideal. Modification of such properties of the surfaces as reflectance or transmittance, without compromising their directional qualities, is the major purpose of an optical coating. The coatings consist of one or more thin layers of material

Laser-Induced Damage in Optical Materials. Edited by Detlev Ristau © 2015 CRC Press/Taylor & Francis Group, LLC. ISBN: 978-1-4398-7216-1.

Chapter 13

and operate through a mixture of fundamental material properties and interference. The sequence of layer materials and thicknesses is designed to give, as far as possible, the desired properties, but interference depends on path differences and these vary with wavelength and with angle of incidence. Thus, while achieving a required performance over a quite limited range of wavelengths and/or angles of incidence is usually straight-forward, required performance over wide ranges is much more difficult and compromises are frequently necessary. To compound the design problem, optical coatings are often required also to improve the environmental resistance of the treated components. An unfortunate but inevitable feature of an optical coating is that its application is frequently the final task in a series of complex and expensive operations, with consequent implications.

13.2   Fundamentals

Light is a propagating electromagnetic disturbance. The disturbance has both an electric field and a magnetic field associated with it that are transverse to the direction of propagation, and both fields are necessary for the transport of energy. Indeed, the instantaneous propagating power density is the product of the two fields. The interactions we deal with in this chapter are linear so that the total response to a sum of stimuli is just the sum of the individual responses. This allows us to represent our general light wave as a set of spectral components, each of which may be treated separately, easing our problems of calculation. We are very used in optics to the ideas of spectral decomposition and of spectral response to the extent that we hardly think about it. In this chapter, our spectral component is a plane, linearly polarized, harmonic wave, the simplest type of spectral component. This component possesses some attributes that we need to quantify.

For this initial discussion, let us choose our propagation direction as the z-axis. Since this is a plane wave, it will have no variation in either the x- or y-direction and will, therefore, be a function of distance, z, and time, t, only. Because we are dealing with linear effects, we can represent this wave in complex form for its electric field as

$$E = \mathcal{E} \exp\left[i\left(\omega t - \kappa z \right) \right], \tag{13.1}$$

with a similar expression for the magnetic field, H. In a simple isotropic medium E, H and the direction of propagation are mutually perpendicular and form a right-handed set. The amplitude $\mathcal{E}$ is complex and so contains any relative phase, as does the magnetic field amplitude, $\mathcal{H}$. This expression contains an implied sign convention, because the phase term could equally well be written ($\kappa z - \omega t$). The velocity of the wave is given by ω / κ and in vacuo is a constant, c. Interaction with materials alters the velocity of the wave to v, and the greater the interaction, the slower is v. This change in velocity is expressed by the parameter refractive index, n, that is given by c/v. In linear processes, the frequency is invariant and so the wave vector or wave number, κ, will vary according to the material. κ is normally given by $2\pi/\lambda$, where λ is the actual wavelength. However, we use wavelength to characterize the wave and a variable wavelength would cause problems for us. We therefore define wavelength λ as the value it would have in vacuo and write κ as $2\pi n/\lambda$. A convenient way of handling absorption with waves in the complex form is to replace n by $(n - ik)$, where k is known as the extinction coefficient. This introduces an exponential decay into the wave as it propagates.

At optical frequencies, the interaction of the electromagnetic wave with any material is through the electric field only. Any magnetic effects are vanishingly small. Thus, parameters such as phase, or polarization, are conventionally referred to the electric field. However, the magnetic field is important both in the transport of energy and in the boundary conditions at any interface. The electric and magnetic fields of a harmonic wave are proportional to each other, the constant of proportionality being a material parameter. In the optical regime, this parameter is defined as

$$y = H/E \tag{13.2}$$

and is known as the characteristic admittance. The inverse, called the characteristic impedance, is more usual at microwave frequencies and below. At optical frequencies, the absence of magnetic effects implies a simple relationship between y and $(n - ik)$:

$$y = (n - ik)\mathscr{Y}, \tag{13.3}$$

where $\mathscr{Y}$ is the admittance of free space, $1/376.73$ siemens. In most of what we do, the characteristic admittance can be normalized by expressing it in units of $\mathscr{Y}$; thus,

$$y = (n - ik) \text{ free space units.} \tag{13.4}$$

The instantaneous power density transported by the wave is given by the product of the electric and magnetic fields and so fluctuates at twice the frequency of the wave. The mean of this, known as irradiance, is the important quantity. This is a nonlinear operation that should use real fields, but fortunately, the irradiance can be put into a very simple form with complex waves:

$$I = \frac{1}{2}\text{Re}(EH^*), \tag{13.5}$$

where we are using the nonstandard symbol I for irradiance rather than the SI symbol E that would lead to confusion with electric field.

13.3   Surfaces

When light is incident on a surface between two optical media, experience tells us that a portion is transmitted through the surface while another portion is reflected. The directions of the waves are coplanar with the surface normal, but the angles between the directions and the normal are fixed by Snell's Law:

$$n_0 \sin \vartheta_0 = n_1 \sin \vartheta_1, \tag{13.6}$$

Chapter 13

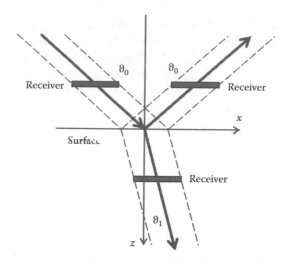

FIGURE 13.1   In the measurement of reflectance and transmittance with infinite plane waves, the receiver must always be parallel to the interface so that equal areas are subtended at it. The upper, incident, medium has refractive index n_0 and the lower, emergent, medium, n_1. (Courtesy of Thin Film Center, Tucson, AZ.)

and the law of reflection, that is the angle of reflection is equal to the angle of incidence but on the other side of the surface normal. Our plane waves are infinite in their extent. Therefore, so that in the absence of absorption the reflectance, R, and transmittance, T, must sum to unity, or 100%, they must be defined as the appropriate ratios of the components of irradiance normal to the surface, Figure 13.1. These irradiances are given by the components of electric and magnetic fields parallel to the boundary, the tangential components. These are the very same components that are involved in the boundary conditions, which are that the total tangential components are continuous across the boundary. Thin-film calculations take advantage of this by defining the amplitude reflection and transmission coefficients in terms of these tangential components. The coefficients are not the Fresnel components, which involve the complete amplitudes, but they lead to a welcome simplification. Before deriving them, however, we need briefly to discuss some questions of polarization that are important at oblique incidence.

Once again, the problem has its origin in refraction. An arbitrarily linearly polarized incident wave will have a component on the surface arbitrarily tilted with respect to the plane of incidence. The reflected wave with similar polarization can have the same component direction but, to conform, the refracted wave must have a different orientation of polarization because of its different angle of propagation. To avoid this difficulty, we invariably express the waves in terms of two special orthogonally polarized components, known as the polarization eigenmodes, because they are completely independent of each other. They are defined as s-polarization with its electric field normal to the plane of incidence, and p-polarization with electric field parallel to the plane of incidence. So that they are consistent with each other and with the situation at normal incidence where there is no plane of incidence, their positive electric field directions are defined as in Figure 13.2.

Note that the p-polarization reflection convention is the opposite of that used in ellipsometry. The electric field in s-polarization and the magnetic field in p-polarization are both already parallel to the surface, and the other fields simply have to be multiplied by

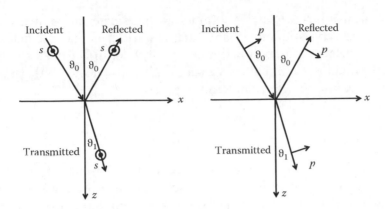

FIGURE 13.2 Sign convention for the positive electric field directions in *s*- and *p*-polarizations. (Courtesy of Thin Film Center, Tucson, AZ.)

the appropriate cosine of the angle of propagation to give the tangential components. The ratio of these tangential components is then

$$\frac{H_s}{E_s} = \eta_s = \frac{H\cos\vartheta}{E} = y\cos\vartheta,$$

$$\frac{H_p}{E_p} = \eta_p = \frac{H}{E\cos\vartheta} = \frac{y}{\cos\vartheta},$$

(13.7)

where η is known as the tilted admittance and collapses to y at normal incidence.

Once we have settled the details of our sign convention, we can then apply the boundary conditions at a simple interface to find the amplitude reflection coefficient, ρ, and amplitude transmission coefficient, τ, that are defined as the ratios of the appropriate tangential amplitudes rather than, as in the Fresnel coefficients, the total amplitudes. These two parameters are complex and contain any phase change at the surface:

$$\rho = \frac{\eta_0 - \eta_1}{\eta_0 + \eta_1} \quad \text{and} \quad \tau = \frac{2\eta_0}{\eta_0 + \eta_1}.$$

(13.8)

On the emergent side of this simple surface, there is only one wave and it is propagating away from the surface. Thus, at the surface, the total tangential and magnetic fields are those that belong to this wave. Their ratio is an admittance that we can write as Y, a property of the surface that we can call the surface admittance, although in this case, we know that it is equal to η_1. Now, let us suppose that our knowledge of the structure is limited to the incident side of the surface so that what happens beyond it is unknown but that we do have a value for Y. Knowing Y we can still write an expression for ρ:

$$\rho = \frac{\eta_0 - Y}{\eta_0 + Y} = \frac{\eta_0\mathcal{E} - \mathcal{H}}{\eta_0\mathcal{E} + \mathcal{H}},$$

(13.9)

where $\mathcal{E}$ and $\mathcal{H}$ are the total tangential components at the surface. Both expressions, Equation 13.9, are of great significance. A fruitful way of treating a thin film is as a transformer of the surface admittance or, more generally, of the total tangential fields.

Chapter 13

Y, the surface admittance, is a very important parameter in its own right. At any interface, it is the ratio of the total tangential magnetic and electric fields. It can be thought of as a surface property, and not just of any real surface but also of any plane, even one in the middle of a thin film. We shall make use of it later in this chapter.

Reflectance can now be defined in terms of ρ as

$$R = \rho\rho^* = |\rho|^2. \tag{13.10}$$

However, there is a caveat. If our incident medium suffers from absorption, a curious coupling between the irradiances of the incident and reflected waves occurs through what is known as the mixed Poynting expression. Amplitudes are not affected, and Equations 13.8 and 13.9 are completely valid. But, to apply Equation 13.10 requires a real η_0. The limitation is not a serious one because the measurement of reflectance in any reasonably absorbing medium presents impossible difficulties, while the error involved in Equation 13.10 in any medium of light absorption is vanishingly small. Since we limit our investigations inside any coating to expressions of amplitudes, our coating performance calculations are unaffected. Rigor demands, however, that the limitation be acknowledged.

Because of the form of η_p in Equation 13.7, it is possible to find an angle of incidence at any simple surface between two dielectric materials where ρ_p is actually zero and p-polarized light is, therefore, transmitted without loss. This angle is known as the *Brewster Angle* and can be shown to be given by

$$\vartheta_B = \arctan\left(n_H/n_L\right), \tag{13.11}$$

where ϑ_B is the propagation angle in the low-index material. There are many applications of this phenomenon, the commonest probably being the construction of cell windows that do not require antireflection coatings. We shall return to the Brewster angle when dealing with plate polarizers.

13.4 Thin-Film Coating Calculations

Our ideal model of a thin-film coating is of a homogeneous and isotropic slab of material with exactly parallel and featureless surfaces that support complete multiple-beam interference. Of course, real thin films may depart more or less from such a model, and additional tools handle any defects and their perturbations of performance. There are many different techniques for the calculation of performance, but a particularly useful one that we adopt here involves the transformation of the total tangential fields at the emergent surface into those at the entrance surface of the series of layers. First, we need to examine an important parameter of our thin film, the phase thickness.

In any interference calculation, we simply add the appropriate electric and magnetic fields. Although it may seem self-evident, any such addition must involve fields at the same designated point. Addition of a field at one point with that at another is meaningless. Our first task, therefore, is to identify two points, one on the front and one on the rear surfaces of the film where we will perform the addition. The points we

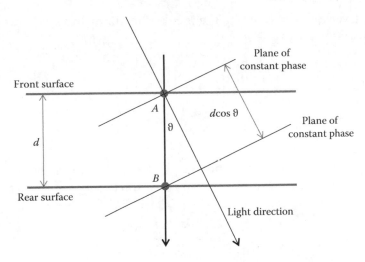

FIGURE 13.3   The planes of constant phase passing through A and B are a physical distant $d\cos\vartheta$ apart so that the optical distance is $nd\cos\vartheta$ and, expressed in terms of phase, $2\pi nd\cos\vartheta/\lambda$. (Courtesy of Thin Film Center, Tucson, AZ.)

choose, therefore, are the simplest possible, the intersections of the z-axis with the film surfaces. This ensures the same change of phase in either direction. Let the index of the film (complex if necessary) be n and the physical thickness be d. The direction of propagation of the light in the film we take as ϑ, and the same angle applies to forward and backward propagation in the film. Our chosen points are marked as A and B in Figure 13.3, and the change in phase between them, calculated as the space between the relevant planes of constant phase, is clearly

$$\delta = \frac{2\pi nd\cos\vartheta}{\lambda}. \tag{13.12}$$

This value is the same for either eigenmode of polarization. Note that δ decreases with increasing ϑ.

The films are completely characterized by their values of η and δ. We can suppose that the optical coating is bounded by an incident medium on one side and an emergent medium on the other, both being semi-infinite. Usually, but not necessarily always, the emergent medium and the supporting substrate are the same so that we can assign by convention the characteristic admittance y_{sub} to the emergent medium (even if the support is actually the incident medium). The beam that emerges into the emergent medium through the final surface then propagates away from it and is not matched by any returning wave. At the final interface, therefore, Y is equal to η_{sub} with the appropriate polarization. Since the effects we are dealing with are linear, we can choose the actual tangential fields at the emergent surface as we wish and a good choice is to make E unity when H becomes η_{sub}. For most purposes, the units of H are unimportant, and so η_{sub} can be left in free space units. SI units can readily be recovered by multiplying by the admittance of free space. The transformation from the rear surface tangential components to those at the front surface is then a simple operation that can readily be written in terms of a matrix multiplication. For a multilayer of

q layers with layer number 1 next to the incident medium and q next to the emergent medium (substrate), the expression is

$$\begin{bmatrix} B \\ C \end{bmatrix} = \left\{ \prod_{j=1}^{q} \begin{bmatrix} \cos\delta_j & \dfrac{i\sin\delta_j}{\eta_j} \\ i\eta_j\sin\delta_j & \cos\delta_j \end{bmatrix} \right\} \begin{bmatrix} 1 \\ \eta_{sub} \end{bmatrix}, \tag{13.13}$$

where B and C are the tangential components of electric and magnetic fields, respectively, at the front surface. The most important parameters to be derived from Equation 13.13 are

$$\rho = \frac{\eta_0 B - C}{\eta_0 B + C},$$

$$R = \rho\rho^*, \tag{13.14}$$

$$T = \frac{4\eta_0 \operatorname{Re}(\eta_{sub})}{(\eta_0 B + C)(\eta_0 B + C)^*}.$$

Note that the phase shift derivable from the expression for ρ is referred to the front surface of the coating. Equation 13.13 appears to be quite simple and straightforward, but the volume of tedious calculation when a modest multilayer is involved is immense, and nowadays, such calculation is invariably computer based. We shall, however, look shortly at some simple cases that will help with visualization.

The ratio C/B can be recognized as the surface admittance, Y, which was defined in Equation 13.3. Since we know that C and B are continuous across any interface, Y also must be continuous. This means that as we slide a reference plane continuously through a coating structure from rear surface to front, the admittance referred to that surface is continuous. We shall return to this record of surface admittance in Section 13.10 where we consider admittance loci.

13.5 Coherence

Interference is simply the consequence of a combination of two or more waves of identical frequency. If only two plane waves of identical polarization and propagating in the same direction are concerned, then the resultant irradiance is given by

$$I = I_1 + I_2 + 2\sqrt{I_1 I_2}\,\cos\varphi, \tag{13.15}$$

where φ is their relative phase. The cosine term is known as the interference term, and it can clearly be positive or negative. Now, let more waves be involved in the operation. These additional waves may not be of exactly the same frequency, nor may they have consistent phase differences. The resulting interference term will be a combination of all the individual interference terms. If an undiminished interference term survives this combination, then the situation is described as coherent. Complete disappearance of the interference term is described as incoherent. A diminished but not zero term is described as partially coherent.

Coherence length is a particularly useful parameter. The coherence length is compared with the interference path difference. If the coherence length is rather less than the path difference, then we have the incoherent case. If the coherence length is rather greater than the path difference, then the case is coherent. In between these two cases, we have the partially coherent case. The concept of coherence length is frequently applied to a light source but is better thought of as a system property. It depends on the spectral bandwidth, the range of angles of incidence, the variations in thickness of a component, and so on. A particularly simple expression exists if the limiting parameter is the spectral bandwidth, $\Delta\lambda$. Then, the coherence length is given by $\lambda^2/\Delta\lambda$. In most cases, it is usual to assume complete coherence in the calculation of the optical properties of the thin-film system, Equation 13.13, but complete incoherence in the case of the much thicker substrates that support the coating. We recall that complete incoherence is calculated by summing the irradiances.

13.6 Substrate

Substrate thickness is usually measured in millimeters, and so substrates are very thick compared with a wavelength. The effective light will usually have sufficient variation in wavelength and angle of incidence, and the substrate surfaces over the illuminated area will usually exhibit sufficient variation that the beam combination is completely incoherent. Given complete incoherence, there are then two limiting cases. Either all beams reflected back and forth between the perhaps coated substrate surfaces may be collected by the receiver, when we can describe this as parallel, or any multiply reflected beams may walk out of the system that we can describe as wedged. Let the transmittances of the two surfaces be T_1 and T_2 and the resultant transmittance be T. Let there also be no loss in the system. Then, the two cases become

$$\text{Parallel: } T = \frac{1}{1/T_1 + 1/T_2 - 1},$$

(13.16)

$$\text{Wedged: } T = T_1 T_2.$$

The differences become significant when T_1 and T_2 are small and, hence, reflectance is large.

When we include absorption in the substrate, the calculation becomes a little more complicated. Extinction coefficient is not a good parameter for the calculation of substrate losses, because it is usually quite small. What may seem a miniscule error in its magnitude, in terms of the losses in a thin film of nanometer proportions, can become of major importance when the material is millimeters thick in the form of a substrate. A better parameter is absorption coefficient, α, usually quoted in units of cm^{-1} (although sometimes in mm^{-1}) and related to extinction coefficient through

$$\alpha = \frac{4\pi k}{\lambda},$$

(13.17)

Chapter 13

a relationship that should be used with caution in deriving α from k. After passage of thickness d of material, an initial irradiance, I_0, becomes

$$I = I_0 \exp(-\alpha d). \tag{13.18}$$

An alternative measure that is frequently used by material suppliers is the internal transmittance of a given thickness of material. This is the transmittance of a given thickness with the effect of the surfaces at either end completely removed. If the thickness of the material is d, then the internal transmittance is given by

$$T_{int} = I/I_0 = \exp(-\alpha d). \tag{13.19}$$

T_{int} is usually expressed in percent for a given thickness.

Including absorption in the substrate, and in the coatings, the transmittance then becomes

$$\text{Parallel: } T = \frac{T_1 T_2 T_{int}}{\left(1 - R_1 R_2 T_{int}^2\right)},$$

$$\text{Wedged: } T = T_1 T_2 T_{int}, \tag{13.20}$$

where R_1 and R_2 are the internal reflectances at the two surfaces.

We did issue a caveat in Section 13.3 that the incident medium for a statement of reflectance had to be free of absorption, and this may seem to invalidate Equation 13.20. However, if T_{int} is sufficiently large to yield a measurable transmittance, then the extinction coefficient will be sufficiently small to avoid any significant error.

13.7 Optical Properties of Materials

In this chapter, we are concentrating on the optical properties of materials, but this should not obscure the fact that almost any property of a thin film material, thermal, mechanical, and chemical, can influence important aspects of the behavior of an optical coating and especially the laser damage threshold.

Electric and magnetic fields can interact with charged particles, but while an electric field exerts a force on a stationary particle, movement of the charged particle is required for any magnetic interaction, and for any appreciable interaction, the speed must be very high. At the high frequencies of optical waves, any movement of charged particles is reversed so quickly that magnetic effects are virtually completely inhibited. Thus, only interactions with the electric field are responsible for the optical properties of materials. The electrons are very light compared with the heavy positively charged parts of atoms and molecules, and so we can think of the electrons as moving while the positively charged masses are stationary. Movement of the electron against the bond that holds it to the positively charged mass forms a dipole that vibrates at the frequency of the incoming electromagnetic wave. This vibration takes energy from the wave, but the vibrating dipole also emits energy, that is, scatters it, all at the same frequency. The close packing of the units in a solid implies a close phase relationship between the various oscillating dipoles, resulting in what is called coherent scattering of the input light where the scattered light cancels in every direction but the primary one. Because of the transfer of energy first into the dipole oscillations and then out, the scattered field lags behind the primary field in phase and when the two are combined into a resultant, it is found to move more slowly through the solid. This is the origin of the refractive index that we recall is the ratio of the velocity in free space to the resulting velocity in the solid. The packing does not vary greatly from one solid to another, and so the degree of interaction is largely a matter of the strength of the electron bond: the weaker the bond, the higher the refractive index. However, the interaction between electromagnetic waves and particles is somewhat more complicated. Energy is delivered by the electromagnetic wave in discrete amounts, known as photons. In the linear regime where we are working, the electron gets one photon or none. The photon energy is proportional to the wave frequency and as it rises so does the photon energy. Eventually, depending on the bond strength, the electron can be torn loose and can then move through the material, destroying the dipole and failing to redeliver its acquired energy. From the solid-state point of view, the electron has been elevated from the valence band to the conduction band. From the macroscopic point of view, the material has suddenly started to absorb rather than transmit. Very roughly speaking, we can see that the high-index materials will tend to have weaker bonds and therefore an earlier onset of absorption as the frequency rises. This explains why as we move from the infrared to the ultraviolet, our highest index materials gradually fade away till in the ultraviolet, we have no really high-index materials at all, and why it is pointless to seek them. Although we have been assuming that the heavy positively charged parts are essentially stationary, this is no longer quite correct at low frequencies, or long wavelengths. The long-wavelength limit of transparency tends to be defined by resonances involving the heavy parts. Many of the materials with transparency to the longest wavelengths in the infrared thus consist of heavy-metal compounds with low resonant frequencies and their accompanying unfortunate toxicity.

These materials that we have described are largely classified as dielectric or semiconductor, the semiconductors having such weak bonds that thermal vibrations at room temperature are enough to assure some electrons in the conduction band. In the thin-film field, we generally lump semiconductors with dielectrics since their behavior is similar. However, there is another class of material where the packing is a little denser than the dielectrics. These materials tend to be heavier as a result, but the primary property that interests us is that the closeness of packing causes interactions between the various electrons that results in large numbers being forced into the conduction band without any stimulation by an impressed electromagnetic field. Now, the incident electromagnetic wave experiences high absorption just like the dielectrics at short wavelengths. These materials are, of course, the metals, and they are characterized by very high extinction coefficients that are usually proportional to wavelength. Electrons in the conduction band are said to be free, because they are quite mobile and can conduct electricity, but some materials have freer electrons than others. Very free electrons imply very little dipole moment and very small refractive index n. The small n in the complex refractive index and characteristic admittance, $n - ik$, implies an electric field that is almost 90° out of phase with the magnetic field and a wave in the metal that, therefore, carries little energy. Although the exponential decay of the fields is rapid, the actual absorbed energy turns out to be surprisingly small. A rough figure of merit is the product nk. The smaller this product, the lower is the loss, and the more useful the metal to us in multilayer optical coatings. Silver, gold, and aluminum are of especially high performance in this respect and so are much used in coatings.

13.8 Quarter-Wave Rule

Equation 13.13 is the basis for almost all of our optical calculations of optical coatings. A general calculation involving a number of layers, some of which may be absorbing, turns out to be of such complexity and volume that we invariably, these days, carry it out by computer. Computers are simple calculation machines with no knowledge of, or insight into, what they do. To make effective use of computers in design and analysis of coatings, we do need to have some understanding of the results. Of major help in such understanding is knowledge of simple cases.

Let us suppose, first, that we are limited to perfect dielectric materials, that is, materials without loss. We take Equation 13.13, and we look for any simplifications.

A particularly easy calculation involves a δ of zero. However, since this represents a layer of zero thickness, and the resulting characteristic matrix is the unit matrix, it does not take us forward. The first really useful and simple value of δ is $\pi/2$ or 90°. This implies an optical thickness nd of $\lambda/4$, or, if tilted, $nd\cos\vartheta$ of $\lambda/4$, that is a quarter-wave, and the result is known as the quarter-wave rule. Equation 13.13 in single-layer form becomes

$$\begin{bmatrix} B \\ C \end{bmatrix} = \begin{bmatrix} 0 & \dfrac{i}{\eta} \\ i\eta & 0 \end{bmatrix} \begin{bmatrix} 1 \\ \eta_{sub} \end{bmatrix} \tag{13.21}$$

so that

$$Y = \frac{C}{B} = \frac{\eta^2}{\eta_{sub}}. \tag{13.22}$$

This is a particularly simple result, making assemblies of quarter-waves very easy to calculate and understand.

For the moment, let us stick to normal incidence. The phase thickness δ is given by

$$\delta = \frac{2\pi nd}{\lambda}, \tag{13.23}$$

and this is a nonlinear function of wavelength, λ. We can simplify the interpretation of the expression for δ by introducing the idea of a reference wavelength, λ_0, that we will use to normalize the important quantities in δ. We write δ as

$$\delta = 2\pi\left(\frac{nd}{\lambda_0}\right)\left(\frac{\lambda_0}{\lambda}\right) = 2\pi\left(\frac{nd}{\lambda_0}\right)g, \tag{13.24}$$

where both quantities in the brackets are dimensionless, and g, ($= \lambda_0/\lambda$), is linearly related to δ.

13.9   Optical Coatings

Let us take a simple example of a material of index and characteristic admittance 2.00 on a substrate of index and characteristic admittance 1.50, typical of a crown glass, and in an incident medium of 1.00. The uncoated substrate reflectance is, therefore, 4%. The quarter-wave rule tells us that the admittance presented by a quarter-wave thick film is 4/1.5, that is, 2.6667 and a reflectance of 20.66%. The full line in Figure 13.4 shows the resulting fringes plotted against g. If we change the characteristic admittance of the film to 1.38, the surface admittance becomes $1.38^2/1.50$, that is, 1.270 and the reflectance falls to 1.41%, shown as a broken line in Figure 13.4. The reflectance of the uncoated substrate is 4.00% and at values of g of 2.0 and 4.0, both films are half-waves and the reflectance is that of the uncoated substrate. For these films, the fringes are virtually sine or cosine function.

Wavelength is inversely proportional to g. The fringes are symmetrical in terms of g and so will be distorted when plotted in terms of wavelength. The visible region runs from 400 to 700 nm. If we want to position the peaks of the fringes in the visible region so that the reflectances at the limits of the region are equal, then the reference wavelength, λ_0, should be chosen as the center of the $1/\lambda$ scale, that is, 510 nm.

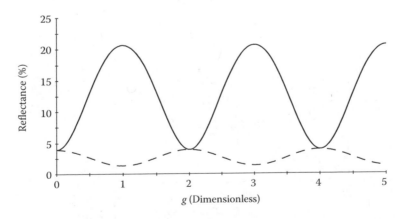

FIGURE 13.4   Reflection fringes as a function of g for a quarter-wave films of characteristic admittance 2.0 (full line) and 1.38 (broken line) over a substrate of characteristic admittance 1.50 in air of admittance 1.00. (Courtesy of Thin Film Center, Tucson, AZ.)

Chapter 13

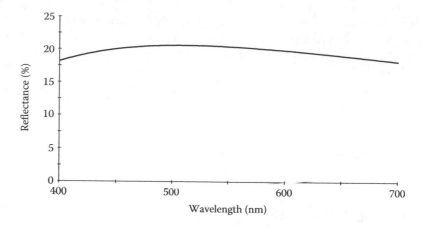

FIGURE 13.5 The quarter-wave of high index from Figure 1.4 can act as a useful beam splitter reflecting around 20% over the visible region. (Courtesy of Thin Film Center, Tucson, AZ.)

Once we choose 510 nm as the reference wavelength and plot the characteristics over the visible region, we can see that the high-index layer is a quite useful beam splitter, reflecting roughly 20% of the incident light, while the low-index layer acts as an anti-reflection coating, reducing the 4% reflection loss of the uncoated substrate surface to rather less than 2%. Magnesium fluoride that this layer represents is used to an enormous extent as a single-layer antireflection coating for the visible region. On crown glass, the performance is much as in Figures 13.5 and 13.6, but on higher-index flints, the reflectance is even lower. The minimum can readily be calculated by the quarter-wave rule.

From the quarter-wave rule, we can immediately see that the characteristic admittance required for the perfect antireflection coating is $y_f = \sqrt{(y_0 y_{sub})}$. For normal crown glass with its index and characteristic admittance of perhaps 1.52 and an incident medium of air at 1.00, this implies a characteristic admittance of 1.23. We have no materials with such a low value of index or characteristic admittance. It is, of course, possible to introduce sufficient void volume into a material of higher index so as to

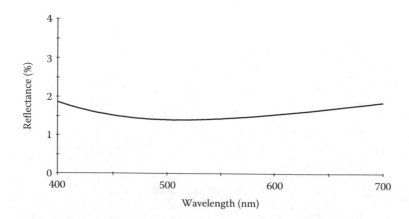

FIGURE 13.6 The quarter-wave of low index from Figure 1.4 acts as a useful single-layer antireflection coating over the visible region. (Courtesy of Thin Film Center, Tucson, AZ.)

depress the resulting index to the required value, but such a material is mechanically weak. Most antireflection coatings are exposed to the environment, and so such solutions are confined to components that are well shielded inside systems or in controlled environments. If the lowest index available is that of magnesium fluoride at 1.38, then to achieve zero reflectance, we will need more parameters, in other words, more layers.

We recall that we number the materials from the incident medium. The incident material will have subscript *0*, the layer next to it, *1*, and so on to the substrate or emergent medium that has the subscript *sub*.

The quarter-wave rule makes it clear that for two quarter-waves, the relationship between the admittances for zero reflectance at the reference wavelength must be

$$\frac{y_1}{y_2} = \sqrt{\frac{y_0}{y_{sub}}}. \tag{13.25}$$

Values of 1.00 for y_0, 1.52 for y_{sub}, and 1.38 for y_1 yield 1.70 for y_2. The performance of a coating with these values is shown in Figure 13.7. The single zero and the shape of the characteristic have led to such coatings being known by the name V-coat.

The material with index at 1.70 presents some difficulties. Aluminum oxide in thin film form usually has an index around 1.66 that is a little low and magnesium oxide at 1.74 is a little high. Cerium fluoride has sometimes been used. A mixture of praseodymium oxide and aluminum oxide that is transparent from 300 nm to around 9 μm can give the correct value (Friz et al. 1992) and also lanthanum oxide mixed with aluminum oxide (Friz 1995).

The limited range of the V-coat means that it is not a good replacement for the single-layer antireflection coating if the whole of the visible region is involved. To do better, we need more adjustable parameters, that is, more layers. However, the performance at 510 nm is very satisfactory and should not be changed and that suggests that any extra layer should have a thickness of one half-wave at 510 nm. A little trial and error arrives

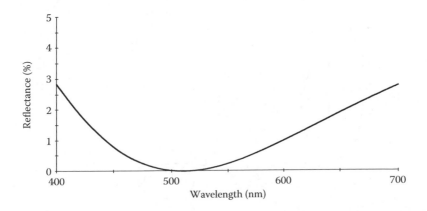

FIGURE 13.7   The performance of the two-layer antireflection coating with a reference wavelength of 510 nm. This is a special case of what is usually called a V-coat because of the shape of the characteristic. (Courtesy of Thin Film Center, Tucson, AZ.)

Chapter 13

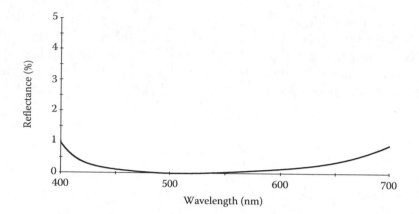

FIGURE 13.8 The three-layer coating usually known by the name quarter-half-quarter because of the construction. (Courtesy of Thin Film Center, Tucson, AZ.)

quickly at a high-index half-wave layer inserted in between the two existing quarter-waves. An index of around 2.15 is good and corresponds to a material like tantalum pentoxide. Figures 13.8 and 13.9 show some calculated results.

The V-coat based on two quarter-wave layers demands accurate indices of refraction for the layers that must satisfy Equation 13.25. This is a serious limitation because we do not have a continuous set of indices available since they are tied to our preferred materials. However, it can be shown that for a normal range of indices, provided that y_L^2/y_0 lies between y_{sub} and y_H^2/y_{sub}, a two-layer coating with y_H next to the substrate can be designed to give zero reflectance. There are some other existence conditions valid in less usual cases. Normally, the high-index layer will have index greater than that of the substrate and then the coating consists of a high-index layer, less than a quarter-wave, next to the substrate, followed by a low-index layer, of thickness in excess of a quarter-wave, next to the incident medium. Theory can give us exact thicknesses, but computer refinement is an effective and fast technique. Again, because this coating has

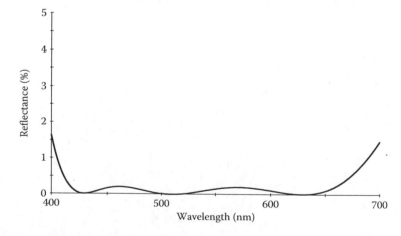

FIGURE 13.9 Increasing the thickness of the intermediate index layer next to the substrate to three quarter-waves alters the performance as shown. The performance in the center is improved at the expense of that at the periphery. (Courtesy of Thin Film Center, Tucson, AZ.)

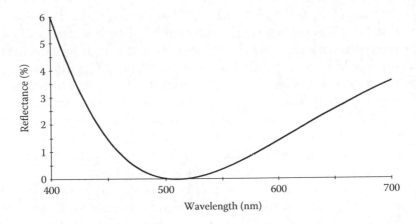

FIGURE 13.10   The V-coat designed using a high-index layer. (Courtesy of Thin Film Center, Tucson, AZ.)

one reflectance zero, it is known by the name V-coat (Figure 13.10). Its advantage is that well-tried and reliable materials can be used.

The V-coat based on a high- and low-index materials can be flattened in the same way as that based on the two quarter-waves. However, the outermost layer should remain as a quarter-wave. Some slight refinement is usually necessary to achieve the optimum performance. Since refinement is going to be used, it is not necessary to start with the exact thicknesses necessary for the V-coat. A design of the form

Air│ L HH 0.25L 0.25H │Glass

will be found adequate. Here, we are using a shorthand notation where H indicates a quarter-wave of high index and L a quarter-wave of low index. Additional factors are used to multiply the following thicknesses as in the two layers next to the glass substrate (Figure 13.10). Figure 13.11 shows the performance of such a coating with y_H of 2.15 and y_L of 1.45 on a substrate of y_{sub} 1.52 corresponding once again to tantalum pentoxide, silicon dioxide, and glass, respectively. The incident medium is air.

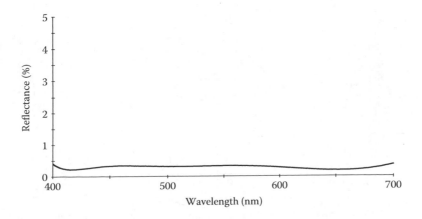

FIGURE 13.11   The performance of a four-layer coating based on the V-coat of Figure 13.10 flattened with an inserted half-wave and refined. (Courtesy of Thin Film Center, Tucson, AZ.)

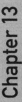
Chapter 13

What might represent the perfect antireflection coating? Clearly as long as there are abrupt interfaces between layers with different admittances, there will be nonzero amplitude reflection coefficients. We can reduce these coefficients by reducing the index contrast across the interfaces. A transition layer exhibiting a continuous variation of admittance from substrate to incident medium would suppress these admittance jumps completely. However, that does leave a variation of admittance that becomes more rapid as the thickness of the total transition layer decreases. Clearly as the thickness of the layer is reduced to zero, the reflectance will rise to the level of the uncoated substrate. It can be demonstrated that provided the transition is smooth and monotonic and the total optical thickness of the transition layer is greater than one half of a wavelength, then excellent antireflection properties will result. This is as close to perfection as we are able to achieve. The optimum profile is a matter of some discussion, but since dispersion is unavoidable, any profile will be a slowly varying function of wavelength and profiles like sine or cosine function that avoid the sharp corners at the ends of a linear profile are suitable. Figure 13.12 shows the performance of such a coating with quarter-wave thickness at $g = 1.0$. The thickness is one half of a wave at $g = 2.0$, and it can be seen that from there onward, the structure is an efficient antireflection coating.

Unfortunately, in the great majority of cases, the antireflection coating is faced with an incident medium of air with unity admittance. Constructing such an antireflecting layer, therefore, from completely solid materials is impossible, and it is usual to etch the surface of the substrate to produce needle-like cones so that the resulting combination of solid and introduced voids has roughly the correct admittance profile. It is important that the surface features, that is, the cones, should be small enough to avoid significant scattering losses over the performance region. Sometimes, a suitable coating is applied over the surface, and then, it is the coating that is subsequently etched.

It can be seen from Figure 13.12 that the reflectance of the substrate surface is only slightly affected if the coating is thin enough, say, around one-tenth of a wave. This implies that we can introduce tapers at interfaces in multilayer optical coatings that will not affect the performance too much as long as their thicknesses are small compared with a wavelength. Since the interfaces in an optical coating are usually rather disturbed

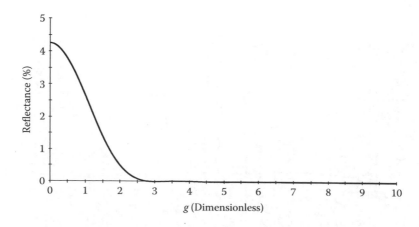

FIGURE 13.12 The reflectance as a function of g of a cosine profile tapered from 1.52 to 1.00 with total optical thickness of 0.25 waves at $g = 1.0$. (Courtesy of Thin Film Center, Tucson, AZ.)

regions, this technique is sometimes used to reduce such effects as scattering losses, or to improve slightly laser damage threshold.

The quarter-wave rule also shows us that a series of quarter-waves of alternate high and low indices will yield a high reflectance that increases with the number of layers. If there are x high-index layers and $(x-1)$ low index, then the reflectance is given by

$$R = \left[\frac{y_0 - \dfrac{y_H^{2x}}{y_L^{2(x-1)}} y_{sub}}{y_0 + \dfrac{y_H^{2x}}{y_L^{2(x-1)}} y_{sub}} \right]^2. \tag{13.26}$$

This high reflectance results from the interference of the various beams in the coating. For a quarter-wave stack with a high-index layer outermost, the waves as they emerge from the top surface are all exactly in phase, leading to very strong constructive interference. This phase coincidence, however, does not persist when the wavelength is varied from the reference, or the value of g departs from unity. The constructive interference is soon replaced by largely destructive interference, and the reflectance falls to near zero. Further wavelength variation returns to a partially constructive interference and then to a destructive condition, but the phases are now becoming jumbled and the interference is gradually weakened so that we see a series of fringes of gradually decreasing amplitude. However, if we add a thickness of exactly one half-wave to each layer, we effectively add a full wave path difference to each double traversal of the layer. This means that the value of reflectance at $g = 2.0$ must be exactly the same as that at $g = 0.0$. In fact, increasing any value of g by 2.0 will not change the interference condition. Thus, the variation in reflectance from $g = 0.0$ to $g = 2.0$ exactly repeats itself from $g = 2.0$ to $g = 4.0$, from $g = 4.0$ to $g = 6.0$, and so on, Figure 13.13.

The width of the high-reflectance zone increases with the ratio of the high to low admittances but is always narrow compared with the repeat interval of 2.0 in g. We can perhaps form a better idea of the width of the zone in Figure 13.13 by plotting

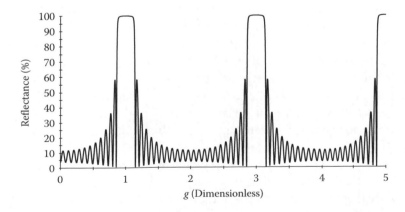

FIGURE 13.13   The reflectance as a function of g of a 23-layer quarter-wave stack based on material of admittances 2.15 and 2.45 on glass of admittance 1.52 in air as incident medium. (Courtesy of Thin Film Center, Tucson, AZ.)

Chapter 13

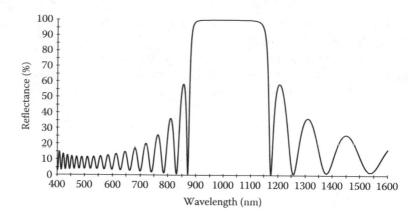

FIGURE 13.14 The performance of Figure 1.13 plotted in terms of wavelength with λ_0 of 1000 nm. (Courtesy of Thin Film Center, Tucson, AZ.)

the characteristic in terms of wavelength with a reference wavelength of 1000 nm, Figure 13.14. This structure is the basic high-reflectance coating, capable of almost 100% reflectance, provided the layers have very low losses. If k_L and k_H are the extinction coefficients of the two materials, then the loss in reflectance of the entire coating at λ_0 can be shown to be

$$A = \frac{2\pi n_0 \left(k_H + k_L\right)}{\left(n_H^2 - n_L^2\right)}, \tag{13.27}$$

where n_0 is the index of the incident medium. Basic laser mirrors use the quarter-wave stack structure.

The regions of low reflectance, albeit containing somewhat intrusive interference fringes, can be turned to advantage in the construction of edge filters. An edge filter is one with a rapid transition from transmission to reflection, or rejection. The unwanted fringes are usually termed *ripple*, and their removal turns out to be a not too difficult design task. They can be considered to arise from a mismatch between the properties of the multilayer and those of the surrounding media, and what is required for their reduction is a kind of antireflection coating. There are some analytical techniques that can assist in this, but the normal way of dealing with the problem is simply to use automatic computer methods. Automatic methods work well when there is a clearly defined solution. Figure 13.15 shows the result of converting such a quarter-wave stack into a short-wave pass filter. Note the almost invariable increase in the amplitude of the remaining fringes in the regions beyond the matching. This has sometimes been called the *toothpaste tube effect*.

The higher order reflectance zones of the quarter-wave stack that can be seen at values of g of 3.0, 5.0, and so on in Figure 13.13 are a consequence of interference conditions that are exactly that of $g = 1.0$ with the addition of some wavelengths of path difference. Suppression of these higher orders implies suppression of the reflection coefficients at every interface. The antireflection coating of Figure 13.12 is a good antireflection coating at g values of 3.0, 5.0, 7.0, and so on but not at $g = 1.0$. Since it is a quarter-wave thick

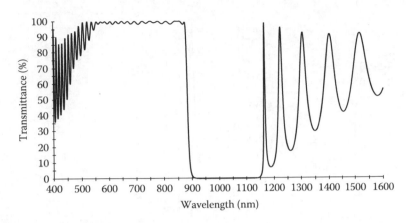

FIGURE 13.15   A quarter-wave stack similar to that of Figure 1.14 after addition of matching assemblies on either side to reduce the ripple on the shortwave side from 550 nm. (Courtesy of Thin Film Center, Tucson, AZ.)

at $g = 1.0$, it is just possible to insert it at every interface without altering the pitch of the index variation. The resulting structure with its sinusoidal or cosinusoidal variation of index is known as *rugate*. Such rugate structures exhibit the fundamental reflection peak but lack the higher orders and are much used in applications such as laser protection.

We have already alluded to a useful shorthand notation that helps in specifying designs where quarter-wave optical thicknesses at λ_0 are indicated by a capital letter. When only two materials are involved, it is useful to use H for the high-index layer and L for the low. Fractional thicknesses can be handled by a multiplying constant such as 0.5 H for an eighth wave of high index or 2L or LL for a half-wave of low index. A group of symbols raised to a power represent a repeat of group, the number of repeats being given by the power. Substrates and incident media can be added as

$$\text{Air}\left|\left(\text{HL}\right)^{11}\text{H}\right|\text{Glass},$$

representing a 23-layer quarter-wave stack with glass substrate and air incident medium. There is no rule concerning the direction of incidence. The preceding formula uses the convention that the incident medium should be on the left but just as often it can be on the right, an order often preferred by those who are mainly concerned with filter manufacture rather than design.

It is fairly easy to see that a structure given by

$$\text{Air}\left|\left(\text{HL}\right)^{5}\left(\text{LH}\right)^{5}\right|\text{Glass} = \text{Air}\left|\text{HLHLHLHLH LL HLHLHLHLH}\right|\text{Glass},$$

where we have expanded the formula to emphasize that the central layer is one half wavelength thick, is actually one large absentee structure. If we eliminate the LL from the calculations, then we see that the center now becomes HH and so on so that at λ_0 all layers simply turn into half-waves and cancel out. The two major structures making up this coating, however, are strong quarter-wave stacks. They quickly assert themselves so

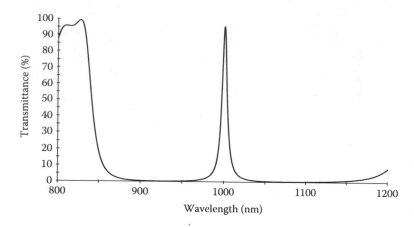

FIGURE 13.16 The performance of the single-cavity narrowband filter. (Courtesy of Thin Film Center, Tucson, AZ.)

that the region where the structure acts as an absentee is quite limited and the performance curve of Figure 13.16 is obtained. There the high and low indices used in the calculation were 2.15 and 1.45 that would correspond roughly to tantalum pentoxide and silicon dioxide, materials commonly used for high-performance filters. The structure with its central half-wave layer acts rather like a tuned cavity, and so the central layer is known as a cavity layer and the structure as a single-cavity filter, although there are some older terms, most notable being the Fabry–Perot filter because the assembly mimics the Fabry–Perot etalon in its arrangement. We can see instinctively that adding layers to the quarter-wave stacks to increase their reflectance will induce a more rapid fall in the transmittance on either side of the peak and so will reduce the passband width. Adding additional half-wave layers to the cavity will also reduce the bandwidth, because the additional material will narrow the fringes, but that will be a smaller effect. These two adjustments together give us the possibility of a fine-coarse control over passband width, although the discrete nature of the layers in the structure implies that the control cannot be continuous.

Unfortunately, the single-cavity filter has some deficiencies in its performance. The passband shape is very triangular, and the rejection outside the passband is poor and resists adjustment independent of the width. Improved performance can be achieved by coupling cavity structures together into what are known as multiple cavity filters. A structure

$$\text{Air}\left|\left(\text{HL}\right)^5\left(\text{LH}\right)^5 \text{L}\left(\text{HL}\right)^5\left(\text{LH}\right)^5\right|\text{Glass}$$

combines two of the cavities into a double-cavity (or two-cavity) filter. The central L layer is usually known as the coupling layer and is necessary in order to avoid certain spurious transmission peaks that otherwise appear. A three-cavity structure can have the design

$$\text{Air}\left|\left(\text{HL}\right)^5\left(\text{LH}\right)^5 \text{L}\left(\text{HL}\right)^5\left(\text{LH}\right)^5 \text{L}\left(\text{HL}\right)^5\left(\text{LH}\right)^5\right|\text{Glass}.$$

The performance of these designs is shown in Figure 13.17.

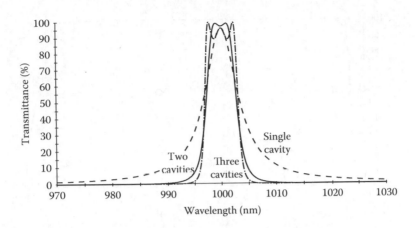

FIGURE 13.17 Multiple-cavity filters based on the same single-cavity structure. (Courtesy of Thin Film Center, Tucson, AZ.)

The quarter-wave stack is rather limited in its width. Wider regions of high reflectance are afforded by front surface metal coatings, but these are vulnerable to high-power-induced damage because of their relatively high losses. Even freshly deposited silver that has the highest reflectance of any metal in the visible region exhibits around 4% absorption loss. Low-loss broadband reflectors, therefore, must be based on interference structures. The normal approach is to arrange that different parts of the coating should reflect the different parts of the spectrum. In the simplest case, we place one quarter-wave stack under another, the inner one reflecting where the outer one fails. A typical design might be

$$\text{Air}\,|\,(H'L')^{q}\,H'L''(HL)^{q}\,H\,|\,\text{Substrate},$$

where the symbols H′, L″, and H, etc., indicate different reference wavelengths. The central L″ layer, of intermediate thickness, is necessary to avoid a central half-wave layer that would result in a deep minimum in the center of the coating. Obviously, the light reflected from the inner structure has to pass twice through the outer, and so it is prudent to arrange that the double passage should coincide with the region of the lowest absorption in the materials. For the visible region, this implies that the thinner structure should be outermost. Figure 13.18 shows a typical example. An alternative approach is to use a tapered structure. The principal problem with the extended zone reflector is the double traversal of the outer structure when the inner is reflecting and we shall return to that in the next section.

At oblique incidence, the characteristics of the coatings change from their form at normal incidence. We recall that there are two eigenmodes of polarization, *s*-polarization and *p*-polarization. The admittances *y* at normal incidence are replaced by the tilted admittances η_s and η_p (Equation 13.7), and the phase thickness δ shrinks in proportion to the cosine of the propagation angle (Equation 13.12). There are essentially three consequential effects.

The shrinking of δ induces a shift of the characteristic to shorter wavelengths.

The changes in η that are opposite for the two polarizations cause a weakening of the *p*-performance and a strengthening of the *s*-performance, leading to a sensitivity to polarization known as *polarization splitting*.

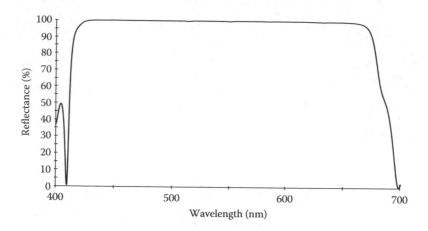

FIGURE 13.18   A simple extended zone reflector consisting of a design similar to that in the text with y_H of 2.15, y_L of 1.45, and q of 11. The outer stack has λ_0 of 470 nm and the inner of 590 nm. (Courtesy of Thin Film Center, Tucson, AZ.)

Snell's law yields a smaller propagation angle for high-index films than it does for low index, and this detunes the layers in the design, sometimes leading to distortion of the characteristics.

There are both advantages and disadvantages in this behavior. The shifting of the characteristic with angle of incidence can be harnessed in tuning the characteristic to a more suitable wavelength. This is probably most frequently employed in the case of small adjustments to the peak of narrowband filters. The polarization splitting can be employed in designing polarization-sensitive coatings such as polarizers. However, reducing the sensitivity to polarization can be a considerable problem, and the shift with angle limits the aperture of the system, especially in the case of narrowband filters.

A simple polarizer, known usually as a *plate polarizer*, is much used in laser applications and involves a quarter-wave stack tilted to a large angle of incidence. The broadening of the high-reflectance zone for s-polarization and the narrowing for p-polarization result in a narrow zone where the s-reflectance is high and the p-reflectance low. The transmitted light is, therefore, p-polarized. It is convenient to arrange that the tilt angle should correspond to the Brewster angle (see Section 1.3) at the rear of the substrate. Since it is p-polarization that is transmitted, this removes the need for an antireflection coating at the rear surface of the component. The p-transmittance can be further enhanced by removing any ripple in the appropriate region. With our material indices of 2.15 and 1.45, the width of the split region is only just wide enough, but some additional refinement that widens slightly the low-index layers and narrows the high can increase slightly the polarization splitting, yielding the performance shown in Figure 13.19.

The advantage of such polarizers is that there is essentially no limit to their size except the capacity of the manufacturing machines, and so they are frequently used in very large laser systems.

Although the performance will appear to increase without limit with the number of layers, the ultimate performance of such thin-film polarizers is governed by geometrical

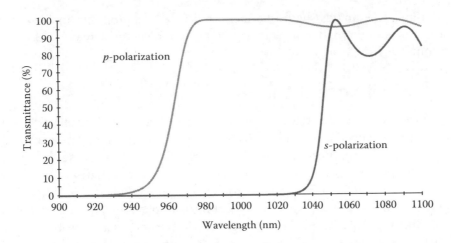

FIGURE 13.19 Thirty-nine layers of 2.15 high index and 1.45 low over a substrate of index 1.52 at 56.66° incidence in air yield the performance shown. (Courtesy of Thin Film Center, Tucson, AZ.)

considerations. The *p*- and *s*-directions are defined by the local plane of incidence and that varies throughout any illuminating cone. Thus, there is leakage from *s*- to *p*-polarization, and it is this that limits the possible performance. It can be shown that for cone apex semi-angles up to around 8° the leakage, *L*, is given by

$$L = \frac{\Omega^2}{4\tan^2\varphi}, \tag{13.28}$$

where

φ is the angle of incidence
Ω is the apex semi-angle of the cone, which should be in radians

For a 2° cone at 56.66° incidence, the leakage can be calculated as 0.013%. This is illustrated in Figure 13.20. Note that this property applies to any thin-film polarizer using

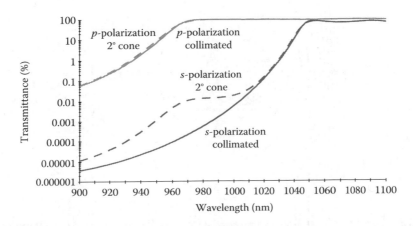

FIGURE 13.20 The performance on a logarithmic scale of the polarizer of Figure 1.19 in collimated light and illuminated by a cone of 2° semi-angle. (Courtesy of Thin Film Center, Tucson, AZ.)

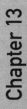
Chapter 13

isotropic films. The angles should be measured in the appropriate incident medium because this is purely a matter of geometry. Note, too, that as φ increases, the leakage for a given cone apex angle falls.

13.10 Admittance Diagram

We introduced the idea of surface admittance, Y, in 1.3 and 1.4. Y is the surface admittance and can be defined as C/B where B and C, the normalized total tangential electric and magnetic fields, are those calculated at the appropriate plane or surface. Since it is clear from the analysis in 1.4 that B and C depend only on what is beyond the appropriate plane and not what is in front, that is, toward the incident medium, then so too does Y depend only on what is behind. Y is generally complex, and if we have a complete record of Y throughout a particular coating, then, in the complex plane, this variation of Y will form a continuous locus known as the *admittance locus* (Figure 13.21).

The admittance locus of a dielectric layer is a circle centered on the real axis and cutting it in two points that obey the quarter-wave rule. The circle is described clockwise. One semicircle from real axis back to real axis represents one quarter-wave and a full turn one half-wave. To divide the locus into smaller sections, draw a circle with center the origin through the point y_f on the real axis, y_f being the characteristic admittance of the layer, and this will cut the semicircles into eighth waves. The complete set of circles for a particular material form a nested set, nested about the point y_f.

The surface admittance together with the admittance of the incident medium determines the reflectance. Because adding incident material to any surface does not alter the reflectance, admittance circles for incident medium material must coincide with isoreflectance contours. Such a set of contours are shown in Figure 13.22 for an incident medium of air.

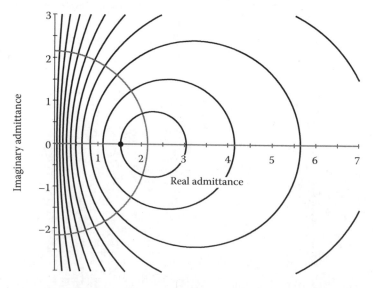

FIGURE 13.21 A set of admittance circles drawn for material with characteristic admittance 2.15. The black dot marks a starting point of 1.52. The circle centered on the origin together with the real axis cuts the loci into eighth waves. (Courtesy of Thin Film Center, Tucson, AZ.)

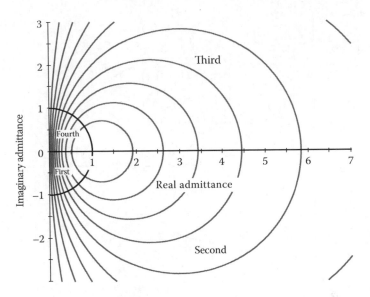

FIGURE 13.22 Isoreflectance circles and the boundaries of the quadrants of the phase shift on reflection drawn for an incident medium of unity admittance. The point 1.0 on the real axis represents zero reflectance, and the circles moving out from that point represent 10%, 20%, 30%, and so on. The quadrants of the phase shift on reflection are marked. (Courtesy of Thin Film Center, Tucson, AZ.)

It is clear from Figure 13.21 that with one material all that can happen as material is added is that the admittance will remain on the circle on which it starts. To move around the admittance diagram, a second material with its set of circles nested about a different point must be added. The admittance locus for the resulting coating will consist of a continuous path made up of joined arcs of circles from each nested set. For an antireflection coating, the locus must terminate at the point corresponding to the incident medium admittance, usually 1.00. For a high-reflectance coating, the termination point should be as far away as possible from the point corresponding to the incident medium.

The form of the circles implies that the extrema of reflectance as dielectric material is added to a substrate must coincide with the points of intersection with the real axis. Also, provided that the material has characteristic admittance greater than that of the incident medium, the loop under the real axis represents reflectance falling with increasing thickness and that above rising. This helps the diagram to give valuable insight into monitoring processes.

Although the diagram is derived from completely accurate calculations with no approximations and could, therefore, be used for calculation, its principal use is in understanding and in that role we usually draw only approximate loci. High-reflectance loci tend to walk off the diagram, and so we frequently distort the loci in a way that does not alter their essential properties. For example, we can take the quarter-wave stack. Here, the quarter-wave rule shows us that after only a few layers of normal indices, the real admittance can reach several hundreds or more in size. We, therefore, sketch the locus as shown in Figure 13.23, and this is sufficient to give us the details that we usually need.

Clearly when we compare with Figure 13.22, we see that the phase shift on reflection is 180°. Decreasing wavelength lengthens the locus and pushes the end point below the

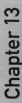

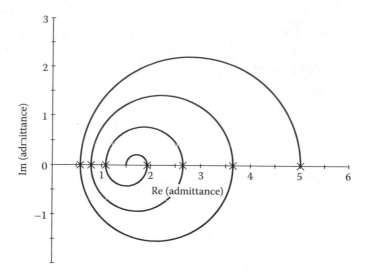

FIGURE 13.23 The admittance diagram of a quarter-wave stack. This locus is deliberately drawn for layers that have a smaller contrast of their admittances than would be usual in such a structure. However, the spiral form of the locus with high-index layers above the real axis and low-index layers below is quite typical. (Courtesy of Thin Film Center, Tucson, AZ.)

axis and into the second quadrant of phase shift on reflection. Increasing wavelength shortens it, and the phase shift enters the third quadrant. If we need more accurate information, we can turn to the computer, but this knowledge derived from the admittance locus is usually enough for us to check the integrity of computed results.

We can use the diagram to explain the construction of the V-coat that uses the materials we have rather than prescribe indices that we have difficult in achieving, Figure 13.24. Let us take 2.15 and 1.45 as the two materials. A useful way of proceeding is to assume that we have solved the design, and then the 1.45 locus will terminate at the point 1.00 on the real axis. The 2.15 locus will start at the substrate of 1.52. If the two circles intersect, then a continuous locus is possible and the coating is feasible. Rough values of thickness can then be derived from the diagram and refined to yield the final design that is shown in bold in the diagram. There is, of course, another obvious

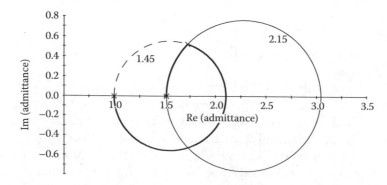

FIGURE 13.24 The 1.45 circle passing through the point 1.00 and the 2.15 circle through 1.52 intersect; therefore, a solution is possible and the thinner solution is shown in bold. A rough estimate gives a design 1.00 | 1.25L 0.25H | 1.52 that can be used as the starting design for refinement. Note that the origin shown for the real axis is 0.5. (Courtesy of Thin Film Center, Tucson, AZ.)

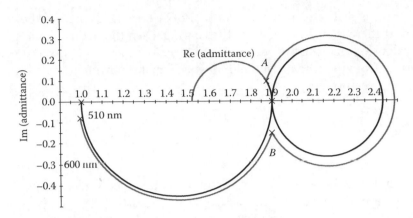

FIGURE 13.25   The admittance locus for the quarter-half-quarter coating at 510 and 600 nm. The reentrant features at *A* and *B* are responsible for retaining the end point near 1.00 and near-zero reflectance. *Note*: The origin of the real axis is at 0.9 and not at 0. (Courtesy of Thin Film Center, Tucson, AZ.)

solution involving the other point of intersection, but that solution is a thicker one and, therefore, will produce a narrower characteristic.

The diagram helps us to understand the operation of the half-wave layers in our broadband antireflection coatings. The diagram for the quarter-half-quarter coating of Figure 13.8 is shown in Figure 13.25. It is clear that the half-wave layer, which is the rightmost circle, operates because of the reentrant features at *A* and *B* that compensate for the general shortening of the locus (Figure 13.26).

Standing waves occur whenever there are counterpropagating waves. This applies to all interference structures. The standing waves exhibit variations in the associated electric field amplitude and interaction with the material, whether it be absorption or scattering goes as the square of the electric field amplitude. A useful additional feature

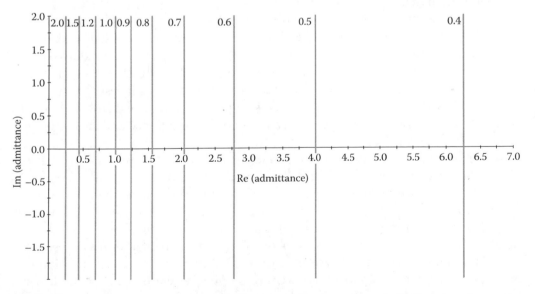

FIGURE 13.26   Typical contours of constant electric field amplitude normalized so that the contour through the point 1.0 on the real axis is 1.0. (Courtesy of Thin Film Center, Tucson, AZ.)

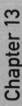

Chapter 13

of the diagram is that the relative distribution of electric field through the coating can be derived. If there is no absorption in the coating, then the net irradiance, that is, the net power density, entering the coating (what is incident less what is reflected) must be what emerges from the rear surface and must be constant through the coating. The net irradiance is given by

$$I_{\text{net}} = \frac{1}{2}\operatorname{Re}\left(\mathcal{E}\mathcal{H}^{*}\right), \tag{13.29}$$

where $\mathcal{E}$ and $\mathcal{H}$ are the total tangential components. Since we are going to scale the results in terms of the incident fields, we can write this as

$$I_{\text{net}} = \frac{1}{2}\operatorname{Re}\left(BC^{*}\right) = \frac{1}{2}\operatorname{Re}\left(BY^{*}B^{*}\right) = \text{Constant}, \tag{13.30}$$

B being the normalized electric field amplitude. This implies that

$$E \propto \frac{1}{\sqrt{\operatorname{Re}(Y)}}, \tag{13.31}$$

and so contours of constant electric field amplitude are lines normal to the real axis and with values varying as the inverse of the square root of the intercept with the real axis. This permits an assessment of the variation of electric field amplitude through any coating as soon as the admittance locus is plotted. The maximum field will occur at and be given by the point on the locus nearest the imaginary axis and the lowest field at, and by, that point furthest away from the imaginary axis.

Since most antireflection coatings have their entire loci to the right of the termination point at the incident medium, the highest electric field occurs at the front surface. The quarter-wave stack, on the other hand, provided it ends with a high-index layer, will have its termination point, Figure 13.23, as the point furthest from the imaginary axis, and so the lowest electric field will be at the front surface. The highest electric field, it can be seen, will occur at the boundary between the first and second layers. Should the outermost layer be of low index, then this condition will be inverted and the highest field will be at the front surface.

The extended zone reflector behaves much as the quarter-wave stack for that part of the high-reflectance zone that is assured by the outer structure. However, in that part of the zone where the inner part contributes the reflectance, the outer part circles around the admittance diagram and as the wavelength varies whips past and very close to the origin. At those wavelengths, the field at the front surface can be exceptionally high. Since the sensitivity of the structure to contamination is a function of the field at the front surface, we can see that there can be great variation from one type of structure to another and even from one wavelength to another. The internal field can also show great variation, causing abrupt changes in absorption and in scattering.

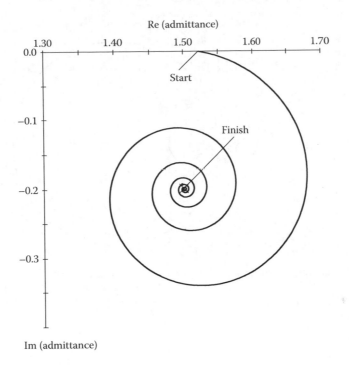

FIGURE 13.27   This is the admittance locus of a thick absorbing layer $(1.5 - i0.2)$ on a substrate of 1.52 admittance. (Courtesy of Thin Film Center, Tucson, AZ.)

Gain, too, can be included in the admittance diagram. The easiest way to include this is to change the sign of the extinction coefficient when loss then switches to gain. Figure 13.27 shows the clockwise shrinking spiral locus that characterizes an absorbing layer.

When gain is included, the spiral reverses and grows, although it is still described clockwise at this stage. However, if the layer is thick enough, it reaches and passes through the imaginary axis. At the crossing point, the reflectance becomes 100% and then greater than 100%. Beyond the imaginary axis in the second and third quadrants of the complex plane, the rules are similar to those in the first and fourth quadrants except that the directions of the loci become counterclockwise. Dielectric layer loci still form circles that obey the quarter-wave rule. The isoreflectance contours are circles centered on the real axis but nested about the point $-y_0$ and since the point $-y_0$ represents infinite reflectance, the smaller the radius of the isoreflectance circle, the higher the reflectance it represents. Now on the other side of the imaginary axis, the spiral associated with the gain layer switches to counterclockwise rotation and the spiral shrinks instead of growing. The spiral then converges on the point $-y_f$ where y_f is $(n + ik)$ and will reach it eventually if thick enough. As the thickness grows, the end point of the locus will vary in its proximity to the point $-y_0$. The reflectance will, therefore, vary, and should the locus end point actually coincide exactly with $-y_0$, then the reflectance will reach infinity. In any practical case, the gain would be affected by the output power and the reflectance would ultimately saturate at some finite, but high value (Figure 13.28).

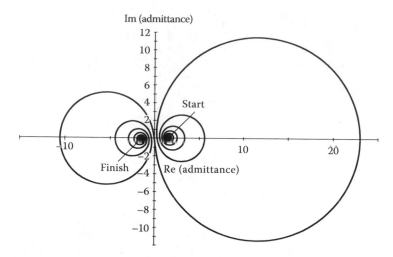

FIGURE 13.28 The admittance locus of a layer with gain. The optical constants are $(1.5 + i0.1)$, the gain coefficient being set at a very high value of 0.1, better to illustrate the form of the locus. The start point is at an admittance of 1.52, and it ends eventually at $(-1.5 - i0.1)$. (Courtesy of Thin Film Center, Tucson, AZ.)

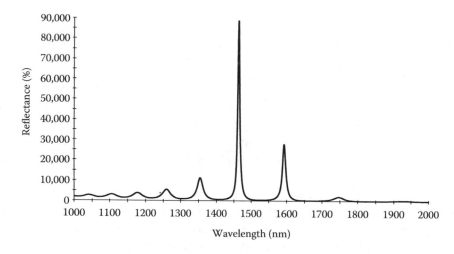

FIGURE 13.29 The reflectance as a function of wavelength of the film of Figure 1.28 with a physical thickness of 6 μm. The resulting resonance at 1464 nm can clearly be seen. (Courtesy of Thin Film Center, Tucson, AZ.)

Figure 13.29 shows the reflectance of the gain film of Figure 13.28 with its physical thickness limited to 6 μm. Now, the end point of the locus passes very close to $-y_0$ when the wavelength is 1463.87 nm and the resulting reflectance is exceedingly high. The film is acting as a resonator, and the reflectance peak is very narrow. As the thickness of the film increases, the resonances tend to disappear (actually if there is no dispersion, they move to longer and longer wavelengths) and the ultimate reflectance becomes

$$R = \frac{\left(n_0 + n_f\right)^2 + k_f^2}{\left(n_0 - n_f\right)^2 + k_f^2}.$$

(13.32)

A semi-infinite slab of material with optical constants $(n_f + ik_f)$, that is, with gain, will yield a reflectance of

$$R = \frac{\left(n_0 - n_f\right)^2 + k_f^2}{\left(n_0 + n_f\right)^2 + k_f^2}. \tag{13.33}$$

These two results, Equations 13.32 and 13.33, the inverse of each other, have led to much confusion. Equation 13.33 gives a reflectance less than 100% while Equation 13.32 yields a reflectance greater than 100%. The question of whether or not a semi-infinite slab of a gain medium can amplify an externally arriving wave has been the subject of a number of studies with differing results and opinions. The source of the confusion is essentially in the nature of the two results. If we truly have a semi-infinite slab of completely homogeneous gain material, then the problem reduces to the reflectance of a simple surface and Equation 13.33 applies with no apparent gain. However, if there is the slightest disturbance deep in the gain medium, even as small as digital noise in a model, then the spiral of Figure 13.28 opens and then closes, yielding the result in Equation 13.32. The Equation 13.33 expression is an unstable result while Equation 13.32 is stable. The situation is still more confusing at oblique incidence, which is a little beyond the scope of this chapter, but the two solutions, stable and unstable, still exist. In all cases, the solution with gain is the stable one.

There is much more to the admittance diagram, especially at oblique incidence and when metal films are included.

13.11 Deposition and Processes

The number of different processes that can or have been or are currently used for the production of optical coatings is enormous, and a full description requires a capacity well beyond a single section of a chapter. Fortunately, the commonest processes involve condensation of material from the vapor phase, and we shall concentrate on those. Most of the processes we deal with operate under vacuum. This avoids disturbances to the vapor, turbulence, reaction with atmospheric components, contamination, and other perturbing influences. The materials may also be toxic, and the vacuum process inherently isolates them from the surrounding environment. We can divide most of the important processes into two rough groups. In the first group, the vapor is essentially already in the form of the desired thin film material so that the vapor and condensing film largely share their essential composition. Such processes are known as *physical vapor deposition*. In the second group, the materials in the vapor phase are rather more remote from the desired film composition and the composition of the growing film is the result of a reaction that is induced in the components of the vapor. This second group of processes is known collectively as *chemical vapor deposition*. We shall concentrate on these two groups and particularly physical vapor deposition in what follows, but we mention briefly here the additional process of etching that removes rather than adds material. Etching treats the surface of a component, or sometimes a film that has been deposited over the surface, to remove material and leave a structure that behaves rather like the transition layer of Figure 13.12. Almost invariably, the object of etching is an

Chapter 13

antireflecting effect. Many different agents of material removal can be used, including aggressive chemicals, beams of energetic ions, gaseous plasma, and so on.

The various physical vapor deposition processes differ in the way in which the material is placed in the vapor phase. In the middle of the nineteenth century, William Grove (1852) reported experiments on what was later called sputtering, but the process of thermal evaporation, introduced by Pohl and Pringsheim (1912) rather later in 1912, became the process used for almost all optical coating until recently. This simple method heats the material until it evaporates. The crucible holding the material is frequently constructed of refractory metal and is heated by passing a current through it. So that the electrodes can readily be attached, the crucible is generally made long and narrow with flat lands on the ends, and has the appearance of a punt, a type of boat. It is, therefore, usually called a boat. Since the charge in the crucible is often molten, it must be arranged so that the vapor travels upward and the substrates to be coated are held above it. Because of the vacuum, the vapor travels very smoothly toward the substrates so that there is a consistency in the deposited thickness that follows roughly the inverse square law together with the cosine terms of the Lambertian illumination. Mostly this variation in thickness is beyond what can be tolerated in an optical coating, and so the substrates are moved during the deposition to achieve the required thickness uniformity. The movement can range from simple rotation around the center of the system over offset evaporation sources (Figure 13.30) to double rotation where the substrates spin rapidly about a local axis while rotating together around the center of the machine. The term *planetary* is often used to describe this double rotation. There are even more complex systems where a rocking motion is applied as the substrates turn. Masks are commonly used to trim the thickness further. There can often be elements of trial and error mixed with considerable skill and experience in the setting up of a machine for uniform coatings.

FIGURE 13.30 The substrate carrier (calotte) of an APS machine SyrusPro 1100 showing a number of different substrates. Not all are from the same deposition run. APS is derived from advanced plasma source and is a variant of ion-assisted deposition using a proprietary plasma source for the generation of the bombarding ions. (The APS SyrusPro is a product of Leybold Optics GmbH, Siemensstrasse 88, 63755 Alzenau, Germany. Photo courtesy of Laser Zentrum, Hanover, Germany.)

Boat sources are simple and comparatively inexpensive, but they do have some disadvantages. One is connected with the aggressive corrosion that can often be exhibited by the molten charges. A boat failure during a deposition run can cause rejection of the entire batch. Removal of coatings is a much more difficult process than applying them, and often the substrates are irrecoverable. Then, distortion of the boat can affect the uniformity of the deposited films. Such problems have largely been overcome by the use of electron beam heating of the materials. Here, the crucible holding the material to be evaporated is kept cool, usually by a supply of cooling water, and the heat is injected by bombardment by a beam of energetic electrons, usually accelerated by some kilovolts and carrying several amperes of current. Although the initial capital cost of an electron-beam source is high compared with boat sources, the stability and reliability that it brings are so attractive that it is nowadays the source of choice in thermal evaporation processes.

Heating a material till it boils is a quite aggressive process, and the stoichiometry of many materials, particularly the oxides, can be adversely affected. A serious consequence of the loss of oxygen in the oxides is the appearance of absorption in the layer. Restoration of the oxygen is required to remove the unwanted absorption. The extra oxygen can be supplied by raising the background pressure of oxygen in the chamber so that it reacts with the growing film. Such processes are termed reactive, and reactive evaporation is an important method for producing films with low loss.

It has gradually emerged that thermal evaporation produces films with a columnar structure of somewhat lower than bulk density. The pores inherent in the loose columnar structure adsorb atmospheric moisture by capillary condensation, and their refractive index is raised in consequence, causing a dependence of the optical properties on the nature of the environment. This can be particularly serious when environmental conditions are changing. The coatings exhibit an optical instability that in the past was sometimes referred to as settling. A simple empirical relationship (Kinosita and Nishibori 1969) involving packing density is often used to predict the optical behavior:

$$n = (1-p)n_v + pn_s,$$
(13.34)

where
    n is the index of the film
    p the packing density (defined as the ratio of solid volume to total volume)
    n_s is the solid index
    n_v is the void index (1.00 for vacuum or air and 1.33 for water)

It was found that the columnar structure could be disrupted by adding momentum to the particles in the film by a process of bombardment. The bombardment could be by a beam of energetic ions created for this purpose in a broad-beam ion source or could even be a consequence of increased energy of the condensing species. Ion-assisted deposition is the term normally used for the ion bombardment and ion plating for the process where the evaporant itself carries the increased energy. Such processes have had an immense influence on coating production. In many cases, they can virtually completely eliminate the unfortunate effects of atmospheric moisture.

Chapter 13

Although it is strictly a momentum-driven effect, the processes are usually known as energetic processes. They are much used in a reactive role where the bombardment may be by oxygen or nitrogen ions. The ions are positive and cause charge buildup over dielectric films. To avoid this effect, electrons are added to the ion beam so as to neutralize the charge on the growing film, the beam being termed, rather confusingly, a neutral ion beam.

A remaining problem, not solved satisfactorily as yet, is that the bombardment inherent in the energetic processes causes a depletion of fluorine in the fluorides that, because of the difficulty in restoring it, are unable, therefore, to support the levels of bombardment possible in oxides and nitrides. Various attempts have been made to introduce extra fluorine into the process but with rather mixed success.

Meanwhile in the last couple of decades, there has been a gradual increase in the use of sputtering in optical coating. Sputtering has developed greatly since the already mentioned Grove paper. There are three major variants now used in optical coating.

Sputtering involves the bombardment of a source of material, the target, with energetic particles, ions, that transfer momentum to the molecules of the target, causing them to be ejected with considerable energy. These ejected molecules then travel to the substrate to be coated and form part of the growing film, transferring momentum to it in the manner of the energetic processes already mentioned. The simplest way of arranging such a process is to generate the ions in a discharge excited by the target as cathode and the substrate as anode.

The first problem is that the ions that accomplish the sputtering are heavy positive ions that are attracted to the target by its negative potential and, if the target is dielectric, its small capacitance implies that it rapidly charges to a potential that repels further bombardment. The cure is to discharge the cathode, and since the capacitance is small and the time constant short, this discharging has to be done at radio frequency, a process known as radio frequency or rf sputtering. The process works well and is used to an extent in industry, but a slightly different way of overcoming the charging problem is gaining ground.

The broad-beam ion source already mentioned in the ion-assisted deposition context can emit a beam of positive ions that can accommodate sufficient externally added electrons to be essentially neutral in its charging effect. Thus, such a beam can sputter a dielectric material without charging and without the complications of radio frequency (although the broad-beam ion source may contain a microwave-excited plasma). Beginning in the late 1970s, ion-beam sputtering, as it is termed, has been very successfully used, first in the production of mirrors of high quality for such applications as laser gyros and then later of telecom quality narrowband filters, from which it has spread into precision optics in general. Certain materials become amorphous under bombardment (Naguib and Kelly 1975), and this is true of tantalum pentoxide, niobium pentoxide, titanium dioxide, and silicon dioxide, all very popular high-quality materials. Amorphous materials have no grain boundaries and their interfaces are featureless, so scattering losses can be exceptionally low. A variant of ion-beam sputtering includes a second ion source that bombards the growing film directly, mainly to modify and control to some extent its stress. This variant is known as dual ion-beam sputtering. The ion-beam sputtering process is rather slower than most of the alternative physical vapor deposition processes, and the area that can be coated is limited. It tends, therefore, to be

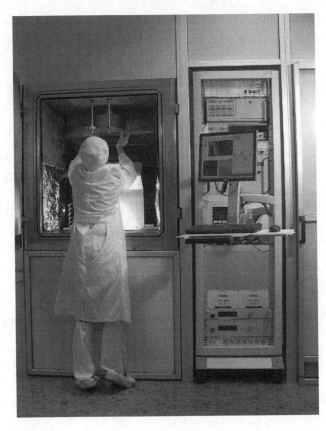

FIGURE 13.31   Loading of an ion-beam sputtering machine in a clean room. (Courtesy of Laser Zentrum, Hanover, Germany.)

used for demanding coatings, especially those where quality and performance are of great importance and often more important than cost (Figure 13.31).

A different, but very significant, innovation known as magnetron sputtering also dates from the 1970s. In the early forms of sputtering, the target and substrate form the cathode and the anode of the discharge, respectively. The plasma from which the positive ions are derived appears between cathode and anode, and the sputtered material must pass through the plasma to reach the anode. This passage causes scattering of the material and a reduction in its energy. To reduce this effect, the pressure must be reduced, but unless the ionization efficiency of the electrons is improved, the plasma cannot be maintained. The path length of the electrons can be greatly lengthened by adding a crossed magnetic field to the electric field. This forces the electrons to move in cycloids with much greater path length and much greater ionization efficiency so that the pressure can safely be reduced. To produce the crossed magnetic field, permanent magnets with alternating poles are added behind the plate of target material such that their magnetic field crosses the electric field in front of the target. The intense plasma that forms there causes the greatest erosion of the target just underneath it, creating a track around the target that looks a little like a map of a racetrack. The eroded area is, therefore, called the racetrack. Normally, these magnetrons are operated at zero frequency, and so their targets should be conducting. To produce dielectric material,

Chapter 13

a process of reactive deposition, where the reacting gas is fed into the chamber and is partially ionized in the plasma, is used. The original magnetron targets were massive affairs used to coat flat glass, but the process has migrated into the precision optics field. It can reach very high rates of deposition.

Unfortunately, there are still problems. One is usually called target poisoning. The positive ions of the reactive gas interact with the cathode to form an insulating film. In the racetrack, the constant bombardment scrubs out the insulating film although there can be some resulting hysteresis in the source characteristic, but this is not the major effect. Outside the racetrack, the insulating film tends to grow. Because the film is thin, its capacitance is high and it can carry a very large charge. The film is not a reliable capacitor and breaks down from time to time, the discharge forming an arc that can cause local melting of the target and ejection of metal globules into the film. The time constant is long, and so the problem can be cured by periodic discharge of the capacitor by reversal of the power polarity, and specialized power supplies are arranged to do this. However, an innovation from the 1990s has been widely adopted and is known as midfrequency sputtering.

Midfrequency sputtering involves two similar magnetron targets. They are connected to opposite poles of a midfrequency power supply operating at around 40 kHz so that while one target is the anode, the other is the cathode and vice versa. Thus, both targets have their capacitors discharged during every cycle. There is another problem associated with reactive sputtering known as the disappearing anode problem. The anode for the single magnetron is often the structure of the machine, although sometimes a rod is provided to act as anode. All of this is coated with the film along with the substrates, and as a result, the anode is gradually insulated and, effectively, disappears. Since the second magnetron is now acting as anode, the problem is solved. Midfrequency sputtering has become common in the precision optics field.

The key to the success of the energetic processes is the transfer of momentum in the bombardment of the growing film, compacting it and eliminating the pore-shaped voids. However, the deposition and bombardment roles frequently clash because the conditions are rarely simultaneously ideal for both. In 1975, Schiller and his colleagues published a very significant paper (Schiller et al. 1975). The effects of the bombardment reach a certain depth in the film, and so it is not strictly necessary that the bombardment involved in the energetic process should be continuous. Instead, the authors proposed an arrangement where increments would be added in thickness, each then followed by bombardment. Thus, the two activities would be completely separated. A simple way of realizing this was to mount the substrates on the periphery of a drum that rotated first past a source where an increment of material could be added, and then past a bombardment stage. It took a certain time for the utility of this approach to be recognized, but we now see it echoed in many of our processes. Usually, the bombardment will also be arranged to make up any deficiency in oxygen or nitrogen in the growing film, and the separation avoids the poisoning problems of reactive sputter deposition.

A different process that uses the same intermittent sputter deposition is radical-assisted sputtering or RAS. (The RAS process is a product of Shincron Co. Ltd., 4-3-5 Minato Mirai, Nishi-ku, Yokohama City, 220-8680 Japan.) Here, the metallic increments of film are treated with atomic oxygen. There is essentially no bombardment, but since the rapid oxidation of the metal causes a swelling of the film, the packing density is high as in the energetic processes.

Another interesting development is in what is termed closed field magnetron sputtering (Gibson et al. 2008). Here, the cylindrical substrate carrier rotates past a magnetron source that adds the increment of film material. But the cylinder is surrounded by an even number of similar sources constructed to have the polarity of their magnets arranged so that all around the cylinder, there is an alternating SNSNSN... arrangement. This traps the plasma that forms a sheath all around the cylindrical substrate carrier and, therefore, assures high efficiency of the plasma treatment.

Sputtering, unfortunately, has never worked well with the fluorides, the staple materials for the ultraviolet particularly between 100 and 200 nm. The problem is their loss of fluorine due to the bombardment. Although many techniques have been proposed to overcome this problem, none have so far been so satisfactory that they have significant influence on processes normally used in production. Thermal evaporation is still overwhelmingly used for the fluorides.

We are seeing much more use in precision optics of load lock systems where the coating chamber is completely isolated from the outside world and opened only for cleaning and maintenance. Sputtering processes are particularly well suited for such an approach, but evaporation processes can be similarly sealed if arrangements to replenish them under vacuum can be included. This tendency is part of a general trend to limit the extent of human intervention in the manufacturing process and so improve the control of the coating environment.

Chemical vapor deposition is a completely different process. Here, the substrate to be coated is situated in a reactor. A carrier gas transports components, known as precursors, into the reactor where the precursors are involved in a chemical reaction, with one of the products as the material that is to be deposited. In the traditional form of the process, the reaction was thermally induced. Nowadays, a plasma-driven reaction is more common. The reaction is often so rapid that the material is produced faster than the growing film can accommodate it. This results in a very loose, poorly adherent film unsuitable for optical purposes. The cure is to pulse the reaction, by pulsing the plasma, so that the film arrives in small increments that can be readily accommodated and that form a solid closely packed film. The plasma can be produced in various ways, but both microwave and radio frequency plasmas are common and the technique is known, for example, as pulsed plasma-enhanced chemical vapor deposition with a whole series of inevitable acronyms. The process is very attractive for the manufacture of long runs of exactly similar components. The reactor usually has to be modified for each different component in order that acceptable uniformity should be achieved.

A variant of chemical vapor deposition is atomic layer deposition where the precursors are admitted into the reactor one at a time and are adsorbed over the substrate. Conditions are arranged so that roughly, a monolayer is adsorbed, any surplus being flushed out of the reactor. When the second precursor enters, it reacts with the first on the substrate to form what would ideally be a monolayer but in practice usually departs somewhat from that. The second precursor is then flushed and the first returns so that the processes of adsorption and subsequent reaction continue. Here, the surface is much more involved in the determination of film thickness, whereas in the conventional chemical vapor deposition, it is much more the flow pattern in the reactor.

Chapter 13

Thus, the uniformity of deposition over complex surface shapes is excellent in atomic layer deposition, and the consistency of the increments of thickness in each cycle makes for accurate control of layer thickness. Another attractive feature is the absence of pinholes. The process has had considerable success with dense wavelength division multiplexing filters, but a current barrier to further progress is the lack of a sufficiently reliable low-index material. At the time of writing, the low-index material of choice is aluminum oxide. Efforts are being made to devise a similarly useful technique for silica, and success here would open up great possibilities for this process.

The sol-gel process is a rather specialized method that has considerable importance in large laser systems. The name sol-gel refers to a solution that is transformed into a gel. An organometallic compound is hydrolyzed by mixing with water in an appropriate mutual solvent, the pH of which is adjusted so that the material forms an oxide polymer with liquid-filled pores. Dipping an optical component into this gel spreads it over the surface. Heat treatment removes the liquid in the pores and densifies the material, complete densification requiring rather high temperatures, perhaps as high as 1000°C. For laser applications, the film is normally based on silica and temperatures are kept low so that little densification results. The film is extremely porous, consisting largely of a silica skeleton with sufficiently low index of refraction to act as an efficient antireflection coating even for fused silica. A typical starting material is TEOS [tetra-ethylorthosilicate, $Si(OC_2H_5)_4$] that is dissolved in ethanol and hydrolyzed by adding a little distilled water. The coating is rather weak mechanically, and so it is little used in general optical coating. However, it was discovered that such sol-gel-deposited antireflection coatings had exceptionally high laser damage threshold (Thomas 1993), and so the technique has become much used in producing antireflection coatings for components in very large lasers for fusion experiments where all components are in a very protected environment.

There are, of course, problems still to be solved. Energetic processes for the fluorides cause serious loss of fluorine under bombardment, so most cannot be used. Even ion-assisted deposition has to be kept so light that only a slight improvement in stability results. The fluorides are the only useful materials for the 100–200 nm region of the ultraviolet, and they are also needed for the far infrared. The energetic processes, too, tend to induce in those materials that become amorphous, high compressive strain energy. Apart from distorting the substrate, high strain energy encourages failures, such as delamination, where strain energy can be converted into surface energy, and the trigger for the failure can be a shock, a scratch, but, in the context of this book, illumination by an energetic laser. At the present time, there seems no way of assuring high packing density without accompanying high compressive strain energy. Then, there are defects in the layers, the commonest being the nodule, an inverted conical growth emanating from a small defect known as a seed that can be almost any slight departure from perfection. The nodules are usually in poor thermal contact with the surrounding film material, implying large thermal gradients under irradiation and an enhanced tendency toward destruction. Then, it sometimes appears that those layers that best withstand high pulse energy are those of disappointingly low packing density. There is still much to be done.

References

Friz, M. 1995. Vapor-deposition material for the production of optical coatings of medium refractive index. Merck GmbH, US Patent 5,415,946.

Friz, M., F. Koenig, and S. Feiman. 1992. New materials for production of optical coatings. In *35th Annual Technical Conference Proceedings*, Baltimore, MD, *Society of Vacuum Coaters*. pp. 143–148.

Gibson, D. R., I. T. Brinkley, E. M. Waddell, and J. M. Walls. 2008. Closed field magnetron sputter deposition of carbides and nitrides for optical applications. In *51st Annual Technical Conference Proceedings. Society of Vacuum Coaters*, Chicago, IL. pp. 487–491.

Grove, W. R. 1852. On the electro-chemical polarity of gases. *Philosophical Transactions of the Royal Society* B142:87–101.

Kinosita, K. and M. Nishibori. 1969. Porosity of MgF2 films—Evaluation based on changes in refractive index due to adsorption of vapors. *Journal of Vacuum Science and Technology* 6:730–733.

Naguib, H. M. and R. Kelly. 1975. Criteria for bombardment-induced structural changes in non-metallic solids. *Radiation Effects* 25:1–12.

Pohl, R. and P. Pringsheim. 1912. Über der Herstellung von Metallspiegeln durch Destillation im Vakuum. *Verhandlungen Deutsche Physikalische Gesellschaft* 14:506–507.

Schiller, S., U. Heisig, and G. Goedicke. 1975. Alternating ion plating—A method of high-rate ion vapor deposition. *Journal of Vacuum Science and Technology* 12:858–864.

Thomas, I. M. 1993. Sol-gel coatings for high power laser optics: Past present and future. *Proceedings of the Society of Photo-Optical Instrumentation Engineers* 2114:232–243.

14. High-Power Coatings for NIR Lasers

Christopher J. Stolz

Chapter 14

Laser-Induced Damage in Optical Materials. Edited by Detlev Ristau © 2015 CRC Press/Taylor & Francis Group, LLC. ISBN: 978-1-4398-7216-1.

14.1 Introduction

Thin films are an integral part of any laser design. Coatings are used for reducing surface losses and ghosts (antireflection), pulse trapping and beam combination (polarizers), and beam steering (mirrors). Unfortunately, near-infrared (NIR) lasers are typically fluence limited by the coatings in the optical system. These limitations can impact laser design by necessitating an increase in optic size to reduce the laser fluence. Brewster's angle windows have been incorporated into commercial NIR lasers to eliminate the need for fluence-limiting antireflection coatings. Commercial NIR lasers are also typically limited not only in fluence, but the lifetime of reflective coatings.

Optical coatings can be permanently altered by intense photon irradiation. This alteration can be either positive, such as laser cleaning or absorption reduction, or it can be detrimental including pitting, blistering, delaminating, scalding, crystalline phase transformations, and fracturing. The topic of laser damage to optical thin films is complex (Walker et al., 1981; Ristau et al., 2009). Factors such as film stoichiometry, microstructure (pores, columns, etc.), crystalline structure, composition, defects (from electronic to micron size), and bandgap can all play a role in laser damage (Bloembergen, 1974; Gallais et al., 2011). Laser properties such as pulse length, wavelength, and repetition rate can significantly impact which mechanism or combinations of mechanisms are responsible for laser damage. Coating design can impact laser resistance by modifying the standing-wave electromagnetic field (SWEF) profile, thus causing light intensification.

14.2 Electromagnetic Fields

Multilayer high-reflective mirror coatings use constructive optical interference to achieve a reflectivity typically exceeding 99.5%. In the extreme case, a reflectivity of 99.99984% has been demonstrated on mirrors for the Laser Interferometer Gravitational Wave Observatory (Lalezari et al., 1992). A result of the optical interference is the creation of a SWEF within the multilayer structure. The laser intensity within the multilayer coating is proportional to the square of the electromagnetic field. The most efficient design for a high reflector is constructed of alternating high and low refractive index materials with quarter-wave optical thicknesses. The design starts and ends with a high refractive index material to maximize the refractive index differential with the substrate and incident medium.

A drawback of this constructive interference structure for mirrors is the significant SWEF prevalent near the top of the coating, as illustrated in Figure 14.1. For typical laser-resistant coating materials, the SWEF peak of the incident beam is equivalent to the electromagnetic field peak within the multilayer. Since almost all of the light is reflected, the SWEF peak of the exiting beam is nearly twice that of the incident beam.

14.2.1 Reduction Design Techniques

Design techniques have been developed to reduce the SWEF profile within a coating. The concept is relatively simple. Typically, low refractive index materials withstand a higher SWEF strength than high refractive index materials (Bettis et al., 1979).

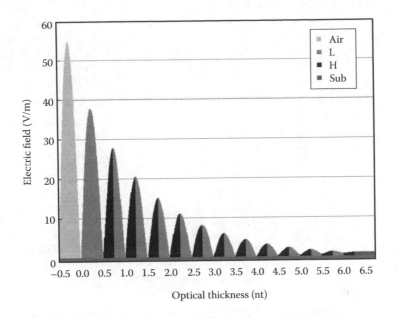

FIGURE 14.1 Standing wave electric field profile of a 24-layer quarter-wave high reflector coating.

Therefore, design techniques are employed that reduce the SWEF strength in the high refractive index material at the expense of increasing the SWEF strength within the low refractive index material.

The interfaces between the coating layers are particularly complicated. At these interfaces, coating materials are intermixed. Polycrystalline layers typically begin as amorphous materials at interfaces and transition to a polycrystalline material with increased thickness. In these regions, atomistic defects are prevalent. Therefore, the SWEF reduction technique is focused on reducing the SWEF within the high refractive index material and at the interface. The alternating interfaces where the low refractive index layer is above the high refractive index material are at a SWEF minimum.

The SWEF peak is reduced in the high refractive index material and the interface by reducing the thickness of the high refractive index material and consequently increasing the thickness of the low refractive index material. By maintaining a half-wave thickness of the combined layer pair, the electric field is reduced in the high refractive index material and the perturbation of the SWEF is minimized in the rest of the coating. This design technique leads to a slight reduction in reflection efficiency of the reflector as a result of the non-quarter-wave thicknesses. Addition of extra layer pairs easily recovers the reduced reflection. This strategy originally was graphically developed to manufacture low absorption interference coatings (DeBell, 1972) and later analytically developed (Apfel, 1977; Gill et al., 1978) to modify the thicknesses of the outer layer pairs, so that the SWEF peak was no greater than that in the last quarter-wave layer pair, as illustrated in Figure 14.2.

The reduction in SWEF strength of the high refractive index material is given by

$$\frac{C}{\sqrt{(m+1)C^2 - m}},$$ (14.1)

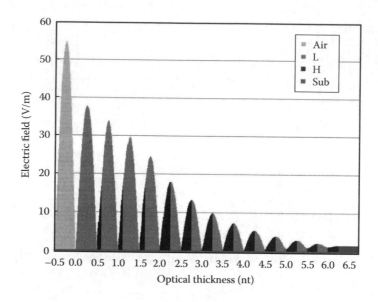

FIGURE 14.2 Standing wave electric field profile of a high reflector coating that has been designed to reduce the standing wave electric field amplitude in the exterior high refractive index layers.

and the increase in the SWEF strength of the low refractive index material is given by

$$C\sqrt{\frac{mC^2-(m-1)}{(m+1)C^2-m}},\qquad(14.2)$$

where

    m is the number of layer pairs that are modified
    C is the contrast ratio between the high, n_h, and low, n_l, refractive index materials

$$C=\frac{n_h}{n_l}.\qquad(14.3)$$

An alternate approach to modifying the layer thicknesses is to use a three-material coating design where a higher laser-resistant medium-index material is substituted for the high-index material on the top layers of the multilayer structure. This hybrid structure overcomes a significant drawback of the non-quarter-wave method, which is the SWEF peak sensitivity to spectral and angular centering. Specifically, poorly centered coatings with non-quarter-wave reduction designs can have significantly higher SWEF peaks than standard quarter-wave designs.

SWEF reduction techniques have also successfully been applied to polarizers (Zhu et al., 2011), which have the complication of not only a high reflector structure, but also a high transmission structure at the opposite polarization. Polarizers can be used in two directions, either forward propagating light going from air through the film or rear propagating light going from the substrate through the film, making the optimization significantly more difficult. To further complicate the optimization, multiple design

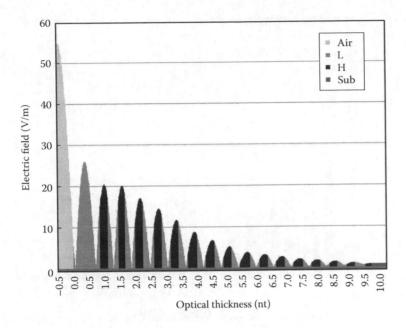

FIGURE 14.3   *S*-polarized standing wave electric field profile of a 32-layer long-wave-pass Brewster angle plate polarizer coating.

techniques can be used to create polarizing structures including a short-wave pass, long-wave pass, and the Fabry–Perot design.

Most coating design strategies consist of a series of compromises. This is certainly true when designing a polarizer. The optimum starting design for minimum SWEF strength within the high-index layers is a short-wave pass, as illustrated in Figure 14.3. Unfortunately, this design strategy has a greater film thickness and smaller polarizer spectral bandwidth than a long-wave pass design. A more optimum SWEF reduction design consists of a gradual transition from thin high-index layers and thick silica layers at the top and bottom of the coating stack transitioning to a more even thickness ratio between the two materials in the center of the stack, as illustrated in Figure 14.4.

Polarizers constructed with HfO_2 and SiO_2 tend to have the highest laser resistance. Unfortunately, the low refractive index contrast between these two materials results in a narrow polarizer region. Combining short-wave and long-wave pass designs can increase the polarization region. By coating the short-wave under the long-wave pass structure, the electric field is minimized, resulting in higher laser resistance (Zhang et al., 2013).

14.2.2   Interfacial Damage

Interfacial NIR laser damage in multilayer thin films is typically a flat-bottom pit morphology consisting of a small (tens of nanometers) divot at a SWEF peak interface (Dijon et al., 1999). The divot is typically surrounded by a circular delamination zone, usually in the tens of microns in diameter. This morphology is the result of a small interfacial defect that is laser heated to a plasma, which then blisters the coating above the plasma, as illustrated in Figure 14.5. The blistering-induced film stretching exceeds the

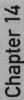

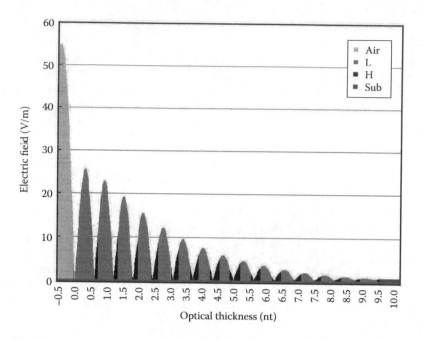

FIGURE 14.4 *S*-polarized standing wave electric field profile of a 30-layer short-wave-pass Brewster angle plate polarizer coating designed for low electric field in the exterior high refractive index coating layers.

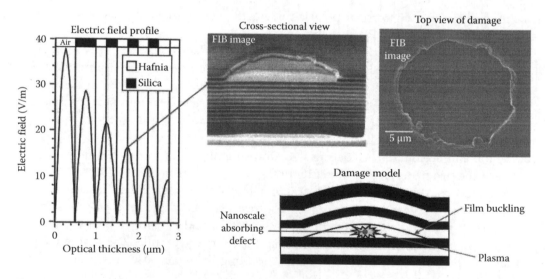

FIGURE 14.5 Flat-bottom pit laser damage morphology occurs when nanoscale absorbing defects at SWEF peak interfaces are laser heated to a plasma, resulting in film buckling, material yielding, and blistering.

coating material yield strength, leading to a complete delamination of the blistered film. Nondelaminated flat-bottom pit blistering has been observed at subejection thresholds, adding validity to the proposed damage mechanism. Flat-bottom pits are typically no deeper than three layer pairs because of the rapid decay of the SWEF strength and the increase in the mechanical durability of the film with increasing thickness.

Plasma formation during the damage event leads to flat-bottom pits that are usually surrounded by a plasma scald outside of the delamination zone. Interfacial damage due to UV laser exposure can have multiple initiation sites or small divots within the flat-bottom pit, indicating a greater density of UV-absorbing interfacial defects compared to IR-absorbing interfacial defects.

The laser damage growth threshold of flat-bottom pits is usually higher than the initiation fluence of the flat-bottom pit (Génin et al., 1997). If the NIR laser system can tolerate the additional scattering due to this damage morphology, then flat-bottom pits may not be the fluence-limiting morphology of the coating. Transitioning from oxide to metallic starting high-index materials tended to significantly reduce or eliminate this damage morphology (Weakley et al., 1999). By burying a half-wave silica layer into the coating design, the SWEF peaks can be shifted from the high over low refractive index material interface to the low over high refractive index material interface (Wang et al., 2013). Damage tests illustrated that the high over low refractive index material interface tends to form flat-bottom pit damage while the opposite interface tends to be much more laser resistant.

The Brewster angle polarizers are particularly susceptible to interfacial damage or delamination of the top layer. A standard long-wave-pass design is $(.5HL.5H)^n$ where H is the high refractive index material of quarter-wave optical thickness, L is the low refractive index material of quarter-wave optical thickness, and n is the number of layer pairs. Therefore, a simple long-wave pass consists of a standard high reflector of alternating layers of high and low refractive index materials with quarter-wave optical thickness that are bracketed by eighth-wave optically thick high refractive index materials on the top and bottom of the coating design. The thin hafnia layer on the top and bottom of the coating stack admittance matches the incident and exit medium to the polarizing stack, maximizing the P transmission. Addition of a silica capping layer (see Section 14.3.1) and optimizing the design for maximum polarization extinction and bandwidth typically result in a thin silica layer (<100 nm) that easily delaminates at low fluence. Delamination can be eliminated by increasing the silica layer thickness by an additional half-wave optical thickness (Stolz et al., 1997).

The angular dependent spectral properties of the Brewster angle polarizers make it an ideal coating type to explore the impact of a wide range of electromagnetic field profiles for a single polarization on the laser resistance of an optical coating. The use of a single coating tested at multiple angles significantly reduces the number of variables that can impact laser damage threshold results. As the incident angle increases of a Brewster angle polarizer, the SWEF peak moves further into the coating. The laser damage morphology is consistent with the angularly dependent SWEF profile, with the severity of top layer delamination reducing with increasing incident angle (Génin et al., 1997).

14.2.3   Nodular Defects

Electromagnetic field intensification can also occur within multilayer structures due to nodular defects. A nodule starts as a seed, a particulate that becomes overcoated by the multilayer structure, thus becoming an inclusion (Tench et al., 1994). For deposition

Chapter 14

processes that have a wide range of deposition angles, the diameter of the nodule increases with increasing film thickness according to the following relation:

$$D = \sqrt{Cdt},$$

(14.4)

where

D is the nodule diameter
C is a constant (typically eight for a wide range of deposition angles)
d is the seed diameter
t is the total film thickness

For a unidirectional deposition process, the nodule diameter remains constant. Nodules behave as microlenses, focusing light within a coating, which reduces the laser resistance of the coating (Dijon et al., 1999). Because nodules are domed, as illustrated in Figure 14.6, a wide range of incident angles occur that can exceed the angular reflection bandwidth of the coating, making the coating defects more efficient lenses (Stolz et al., 2004). Interference caused by alternating coating materials also has a significant impact on the SWEF intensification profile within the defect.

The SWEF profile within a multilayer nodular structure is complex. Instead of an analytic solution, finite-difference time-domain analysis software that solves Maxwell's equations over a three-dimensional gridded domain has been used. These simulations have illustrated a few general trends. The SWEF peaks within nodules can be significantly greater than the SWEF profile of a perfect multilayer structure (DeFord and Kozlowski, 1993). Previously published work shows light intensification as high as 35× in the presence of nodular defects

1 mm inclusion, 24 layers, angle matched to 45°

S polarization
12.6 × light intensification

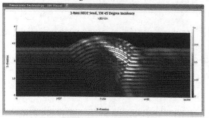

P polarization
20.4 × light intensification

FDTD simulation

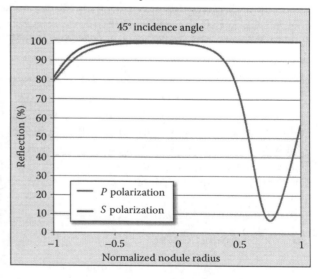

FIGURE 14.6 Nodular coating defects have a wide range of incidence angle, resulting in high SWEF enhancement when laser irradiated.

at 1064 nm and as high as 63× at 351 nm (Stolz et al., 2008). The SWEF tends to increase with inclusion diameter; therefore, the laser resistance is reduced with increasing nodule diameter (Cheng et al., 2011). The closer the nodule is to the top of the multilayer, the higher the SWEF, although the special case of inclusions located on the substrate also demonstrates a high SWEF. For equal diameter and depth inclusions, ellipsoidal nodules with steeper parabolic walls ($C < 8$) tend to have a higher SWEF and lower laser resistance (Cheng et al., 2014). Experimental investigations have demonstrated that highly absorbing inclusions tend to damage at a lower fluence than weakly absorbing inclusions for comparable defect inclusion and nodule geometries (Cheng et al., 2011; Wei et al., 2012).

For nodular defects, the laser-induced damage morphology is typically pitting of the coating due to nodular ejection. The severity of the pitting depends on the laser fluence, the inclusion depth, and the mechanical adhesion of the nodular defect. For nodules that are mechanically unstable and irradiated at a low fluence, nodule ejection tends to leave smooth pits that remain stable above the ejection fluence. This laser-conditioning process is described in more detail in Section 14.4.1.

14.3   Plasma Damage

Surface plasmas can initiate from multiple sources including laser-heated interfacial defects (flat-bottom pits), coating defects (nodular ejection pits), scratches, and surface contamination. A visible surface plasma will have a temperature exceeding 10,000 K. The plasma is composed of ionized coating material, incident medium, and, if present, contamination. The surface plasma is laser heated and lasts longer than the laser pulse (Demos et al., 2013). The morphology of plasma damage is typically scalding of the outer coating material, leading to a microroughening of the surface (Liu et al., 2011).

Surface plasmas can create interfacial defects due to the presence of UV light and x-rays, causing color center formation within the multilayer. These color centers can absorb in the NIR. Color center defects located at interfacial SWEF peaks will tend to damage with IR laser exposure; therefore, surface plasmas can reduce the laser resistance of multilayer mirrors (Dijon et al., 1999).

14.3.1   Overcoats

The plasma durability of high-index materials tends to be less than lower-index materials such as silica. High-index materials tend to become substoichiometric when laser heated and in the presence of a plasma. This increase in absorption can fuel creation of surface plasmas on subsequent laser pulses, leading to laser damage growth. SiO_2 is significantly less reactive to surface plasmas; hence, a half-wave overcoat of this material is a popular design strategy for high-fluence laser mirrors in the NIR. A half-wave is spectrally absentee, so the high reflectivity is maintained while the surface plasma durability of the coating is improved.

14.3.2   Plasma Scalds

Plasma scalds in silica layers, which are an erosion of the overcoat thickness, tend to be a stable damage morphology, and typically do not grow until exposed to fluences that are significantly higher than the initiation threshold (Liu et al., 2011). An adequate overcoat thickness must be selected to prevent complete erosion of the silica layer, which would

Chapter 14

expose the much less plasma durable high refractive index material (Walton et al., 1996). Although this morphology is laser resistant, it is not necessarily benign as has been previously published. The microroughening of the surface can lead to small-angle scatter-induced contrast in the forward propagating laser beam. This additional contrast or beam modulation can cause damage in downstream optics. Contrast values exceeding 8% have been measured for transport mirrors laser conditioned to 18 J/cm² (1064 nm, 3 ns pulse scaled), with plasma scald fractions of 25% (Schmidt et al., 2008). The plasma scald fraction is the total surface area where a plasma event was observed during laser conditioning. In this particular example, the laser fluence would have to be reduced by 8%, so that downstream optics would not be damaged.

14.4　Damage Mitigation

One approach to increasing the laser resistance or increasing optic lifetime is the use of mitigation strategies. Arresting growth of a laser damage site extends the optic lifetime if the existing damage site does not cause damage to downstream optics in the laser system. Another mitigation strategy is laser conditioning where subtle laser damage is created at low fluence, but the damage is stable at fluences higher than the initiation threshold.

14.4.1　Laser Conditioning

Laser conditioning is a standard technique used in commercial and experimental laser facilities. The lasers are initially activated at low fluence and slowly increased in fluence until achieving the desired operational limits (Wolfe et al., 1990; Sheehan et al., 1994). Within multilayer mirror coatings, it has been shown that nodular ejection at low fluence tends to create very smooth pits that can tolerate laser fluences significantly higher than the ejection fluence (Fornier et al., 1994). Nodules that are exposed to laser energies significantly higher than the ejection fluence tend to damage catastrophically, with severe microcracks radially emanating from the ejection pit. The ejection process can also remove sections of the multilayer, leading to a larger-diameter damage site. These nodular ejection sites tend to grow at significantly lower fluence, thus limiting the operational fluence or the optic lifetime.

It has been shown that laser-conditioning-induced smooth ejection pits from fractures created along the nodular boundary (Shan et al., 2010) can have significantly reduced absorption (Papandrew et al., 2001). The electromagnetic field enhancement due to the nodule is eliminated leaving a pit that although it has electromagnetic field enhancement above the pit; the absorption of the incident medium is significantly lower than within the optical coating materials (DeFord and Kozlowski, 1993; Shan et al., 2011). Papandrew also demonstrated that photon-induced annealing reduced the absorption of nonejected defects as a result of laser conditioning. Other laser-conditioning mechanisms have been proposed and studied such as laser cleaning, electronic-defect annealing, and surface smoothing (Bercegol, 1999).

14.4.2　Laser Mitigation

Significant advancements have been made in the field of laser damage mitigation in uncoated fused silica surfaces, as described in Chapter 11. UV-initiated laser damage can be micromachined using CO_2 lasers (Bass et al., 2010). The micromachined shallow conical pits are large enough to remove damaged material and the surrounding cracks. Care must be made when selecting the proper pit geometry to prevent excessive downstream modulation. When done properly, the micromachined pit can have a substantially higher UV laser resistance than the original laser damage site. This is a very interesting application of continuous-wave IR laser damage (laser ablation), creating features that are laser resistant for nanosecond pulse UV operating lasers.

This same approach has been applied to multilayer interference mirror coatings. However, mirror coatings have multiple materials (high refractive index, low refractive index, and substrate materials), which typically have very different thermal expansion and optical absorption coefficients. Therefore, a thermal machining process using either CO_2 or UV lasers is not effective and, in fact, creates additional fracturing at a coating damage site. A mechanical solution has also been shown to be ineffective due to electric field intensification from wave guiding caused by mitigation pit walls with high incident angles.

Electromagnetic field modeling of micromachined mitigation sites in optical coatings, illustrated in Figure 14.7, shows that near vertical walls in the mitigation pit tend to yield the lowest electromagnetic field enhancement for a wide range of incident

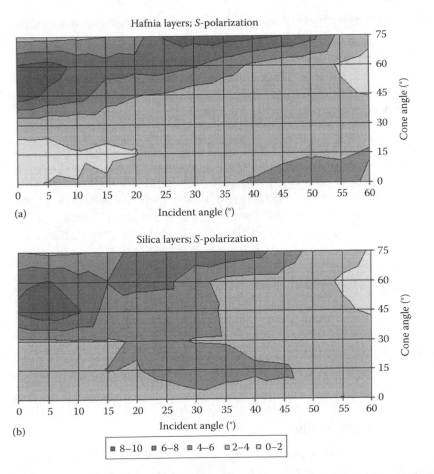

FIGURE 14.7 Modeled light intensification of mitigation pits as a function of pit incident angle illustrates that a minimum pit cone angle tends to yield the lowest light intensification over a wide range of mirror incident angles in the (a) hafnia and (b) silica layers.

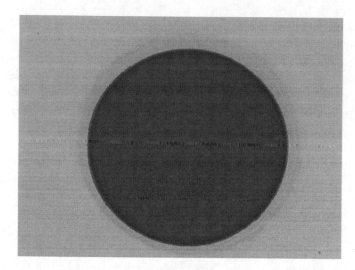

FIGURE 14.8   Micrograph of a 1 mm diameter femtosecond laser mitigation site machined into a high reflector coating.

angles (Qiu et al., 2011). This electromagnetic field modeling also successfully predicts a polarization-dependent spatial difference in the origination of laser damage to mitigation pits. *S*-polarized light tends to create a waveguide structure within the multilayer on the opposite side of the pit from the incoming beam. This waveguide structure leads to electromagnetic field enhancement and thus optical damage. At *P* polarization, the side of the pit near the incoming beam has the highest electromagnetic field intensification, hence yielding the most significant damage. Micromachining pits mechanically with vertical walls into interference coatings has proven to be a difficult task. Athermal processes that can micromachine steep vertical walls into optical interference coatings is femtosecond laser machining, as illustrated in Figure 14.8. This ablative process has been very successfully applied to creating mitigation pits that survive high laser fluences up to 40 J/cm^2 (3 ns pulse length) at 1064 nm (Wolfe et al., 2011). This is also an interesting example of laser damage at extremely short pulses, creating stable laser damage at long pulses.

14.5   Process Contamination

Multiple forms of contamination impact the laser resistance of optical coatings. Particulates can become embedded in the film, creating nodular defects. These particulates, which can fluence limit the coating, originate from multiple sources including substrate cleaning, optic transport, pump down, rotating hardware, high-voltage-induced arcing, and source ejecta.

14.5.1   Substrate Cleaning

Obviously, inadequate substrate cleaning can leave surface particulates that can become incorporated into the coating. For a 1 cm diameter laser mirror, a 1 µm diameter defect occupies 0.1 ppb of the surface area. Substrate cleaning to these levels has proven to be a daunting task. Substrate cleaning methods prior to coating have an impact on the laser resistance of high reflector coatings (Rigatti, 2005). Manual scrubbing has been shown to be critical to reduce surface particulates. The use of ultrasonic cleaning in addition to manual cleaning had a positive impact on laser resistance, whereas ultrasonic cleaning without a manual prescrub had significantly lower laser resistance. In contrast, one of the trends observed from the *Boulder Damage Symposium* thin-film damage competitions is coatings on ultrasonically cleaned substrates tend to have higher laser resistance (Stolz, 2012). It is unclear from the participants if a manual prescrub was performed on their parts. It was also observed that substrate surfaces that were pretreated with either ion or plasma etching before coating deposition tend to have higher laser resistance. Although unstudied in detail, it is likely that these etched surfaces have improved coating adhesion, which could have a very favorable impact on the laser resistance. Alternatively, the removal of an organic surface layer, which could be optically absorptive, would be beneficial from a laser damage perspective. Substrate surface particulates can originate during transport from the cleaning station to the coating chamber, vacuum pump down (Strasser et al., 1990), shedding from the rotational hardware, shield delamination (Logan and McGill, 1991), the heaters used to elevate the substrate temperature for coating, or high-voltage-induced arcing (Reicher and Jungling, 1994).

14.5.2   Coating Materials

The wavelength, pulse length, and laser resistance requirements of a particular coating application strongly influence coating material selection. For bulk materials, the laser resistance is inversely proportional to the refractive index (Bettis, 1992). This relation has also been demonstrated for single layers of thin-film materials (Bettis et al., 1979; Jensen, 2012). The composition of the starting material is equally important in terms of both absorption and particulate generation.

Electron-beam (e-beam) deposition is prone to particulate ejection when the e-beam exposes voids or air pockets in the coating plug. Thermally-induced phase transformations within the coating material can lead to volume-induced stresses that can also eject particles. Contaminants within the coating source material can also cause particulate ejection. Ion sources used in either ion beam sputtering (IBS) or ion-assisted deposition (IAD) processes can generate particles, particularly if gridded sources are used. The use of metallic starting materials, opposed to the oxide form, has been demonstrated to reduce the number of coating defects by an order of magnitude in e-beam-deposited coatings (Stolz, 1999). Metallic sources in IBS systems have also been shown to have significantly less particulate shedding (Knollenberg et al., 1995). An added benefit of this material composition is the improved plume stability achieved when evaporating from a flat molten metallic surface as opposed to an irregular surface caused by the geometry of oxide starting materials. Finally, exotic deposition processes, such as pulsed laser deposition, are notorious for high particulate generation.

14.5.3   Filtration

A wide variety of particulate filtration techniques have been explored. Gravitational filters due to long throw distances in coating chambers are only effective for very slow moving particulates. For example, a 2.5 m throw distance will effectively filter particulates traveling only 7 m/s. Gravitational filtration has been demonstrated for pulsed laser deposition over centimeter scales (Suh et al., 1995); however, peak particle velocities from this deposition technique can exceed a kilometer per second, which would require a throw distance exceeding 50 km to effectively filter all particulates.

Rotary vane filters have been used for pulsed laser deposition by multiple groups (Barr, 1969; Cherief, 1993; Yoshitake and Nagayama, 2004). A rotary vane consists of multiple fan blades that spin at high velocity to capture particulates that travel at less than an order of magnitude slower than the molecular flow. Rotary vane experiments on e-beam coatings have demonstrated particulate velocities up to 14 m/s for hafnia and silica films (Wolfe et al., 2013). Rotary vane filters can also be a source of mechanically-induced particulates and can pose a significant safety liability due to the significant inertia of the spinning blades.

14.5.4   Arc Suppression

Arc suppression circuitry was effectively used for DC magnetron-sputtered films. A combination of a total internal reflection detection device (Williams et al., 1992) and the arc suppression circuitry provided the feedback loop to determine a low particulate deposition process. Arc suppression has been demonstrated in e-beam deposition systems by shielding of high-voltage surfaces leading into the e-beam gun.

14.5.5   Laser Ablation

A particularly elegant solution to particulate filtration is laser ablation. The concept is relatively simple. By passing the ejecting particles through a laser beam, the particulates can be either laser heated and vaporized or their momentum, and hence, direction can be altered so that they do not arrive at the substrate. Laser plume heating systems have been successfully commercialized for pulsed laser deposition applications (Koren et al., 1990; György et al., 2004). Cross-beam pulsed laser deposition systems have also demonstrated particulate reduction (Strikovsky et al., 1993; Tselev et al. 2001). However, for the particulate sizes and velocities of interest for e-beam optical coatings, the laser power requirements are currently impractical at over 50 kW CO_2 laser just to reduce 1 μm diameter silica particulates to half their initial diameter (Elhadj et al., 2012). Unfortunately, as the particulate diameter reduces in size due to laser plume heating, the laser coupling efficiency significantly drops.

14.5.6   Defect Planarization

Cleaning and filtration limitations clearly suggest that total particulate elimination remains an elusive and formidable goal. Coating mitigation can repair laser initiation damage sites for NIR operational fluences up to 40 J/cm² (1064 nm, 3 ns pulse length); however, excessive defect densities can render this technique impractical for large-aperture mirrors. Within the semiconductor industry, a defect smoothing process was developed to planarize mirror surfaces for mask imprinting (Mirkarimi et al., 2005). This planarization process has been scaled and optimized for optical laser coating materials to bury defects within the coating and smoothed to minimize the electromagnetic field enhancement within the high reflector (Stolz et al., 2014).

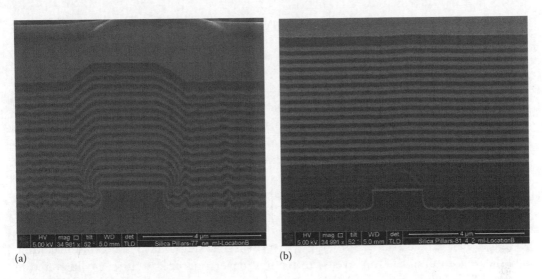

(a)   (b)

FIGURE 14.9 A planarization layer can smooth substrate defects by the use of omnidirectional deposition and angle-dependent etching of a thick silica layer. (a) No planarization over an engineered defect. (b) 2 μm thick planarization layer that effectively smooths the engineered defect.

As illustrated in Figure 14.6, a coating defect has a wide range of incident angles. Optical coating materials demonstrate an angle-dependent ion etching rate. As the incident angle increases so does the etching rate, which forces domed nodule defects to shrink in size. A process of alternating deposition and etching has been demonstrated to smooth defects. This process can be used to deposit a thick planarization layer to effectively bury defects caused prior to deposition, as illustrated in the planarized defect cross section in Figure 14.9. Mirrors deposited with this technique have survived single-shot NIR laser fluences exceeding 100 J/cm^2 (1053 nm, 10 ns pulse length). This is an improvement factor of over 20× in laser resistivity compared to large overcoated defects with no planarization. Applying this deposition and etch process during deposition of the multilayer structure can planarize defects caused during the deposition process.

The planarization process requires an ion gun with sufficient ion flux to etch the deposited coating material. Unidirectional deposition processes, such as IBS, are ideally suited for defect planarization since the diameter of the nodular defect tends to remain constant with normal incidence deposition. In contrast, deposition processes like e-beam tend to have a wide deposition angular range or nearly omnidirection deposition, resulting in a rapid increase in nodular diameter with increased coating thickness.

14.6 Coated Surface Contamination

Once the surface is coated, surface contamination can significantly lower the laser resistance of the optical coating (Guch and Hovis, 1994). The SWEF strength is highest in the air above the substrate; therefore, clean optics are crucial, particularly for high-reflective coatings. Surface contamination can be in the form of particulates or molecular contamination. In both of these cases, laser ablation of the surface contaminate can lead to plasma formation that will modify the optical surface resulting in plasma scalds, flat-bottom pits, delamination of the top coating layer, or catastrophic damage. Each of these morphologies grows at different laser fluences, so understanding of the laser operating environment and the potential contaminates is critical.

14.6.1 Particulates

To help understand laser cleaning of particulates, finite element electromagnetic field calculations of a sphere on a reflective surface show significant SWEF enhancement (Luk'yanchuk et al., 2004). Because of the high SWEF enhancement, both transmissive (see Figure 14.10) and highly absorbing particulates can be ablated at the operating laser wavelength. Ionization of these particulates during laser exposure can result in surface plasmas exceeding 10,000 K. These plasmas cause plasma surface scalding, which can induce interfacial defects that absorb at the laser wavelength. Subsequent laser exposure creates flat-bottom pit laser damage due to these absorbing interfacial defects.

Larger particulate sources in large-aperture laser systems have led to both scalding and delamination of the top layer within a multilayer mirror structure, as illustrated in Figure 14.11. Damage tests of particulates have shown a strong dependence on particulate size (Génin et al., 1997) and particulate composition (Norton et al., 2006). These subscale tests have also demonstrated catastrophic damage as a result of particulate contamination.

Typical laser systems employ a number of methods to reduce surface contamination. Many commercial laser systems are sealed to minimize contamination of the optics. Manual cleaning of optics within these commercial lasers occurs during service calls. Large laser systems, such as Omega at the Laboratory of Laser Energetics, are open architectures, which are contained in clean rooms; however, manual cleaning of the optical surfaces and regular optics inspections are part of routine operations. Other large laser systems such as Laser Interferometer Gravitational Wave Observator (LIGO), Laser MegaJoule (LMJ), and National Ignition Facility (NIF) are all enclosed systems.

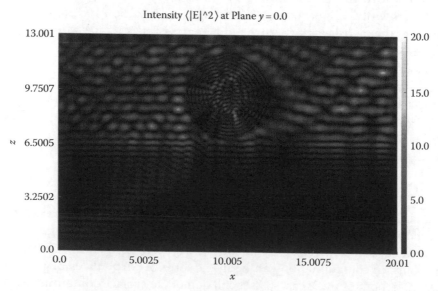

FIGURE 14.10 Electric field calculation by Roger Qiu at Lawerence Livermore National Laboratory illustrates significant SWEF enhancement within a nonabsorbing spherical contaminant on a highly reflective surface.

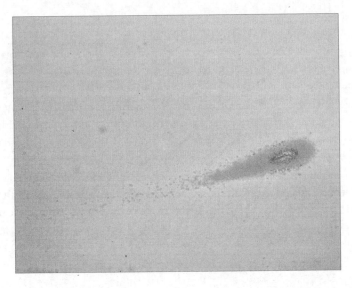

FIGURE 14.11   Contamination-induced plasma scalding and delamination of an operational high reflector mirror.

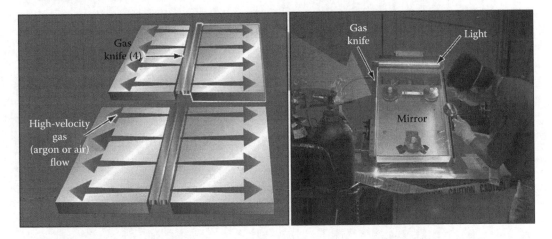

FIGURE 14.12   Air knife used to blow off particulate contamination on upward-facing transport mirrors in the National Ignition Facility.

Even though these systems are sealed, particulates do accumulate predominately on upward-facing optical surfaces. One solution implemented at NIF is the use of air knives to blow debris off the final upward-facing transport mirrors (Gourdin et al., 2005). An air knife is a tube that runs the length of the optic with a slit for high-pressure gas to flow over the optical surface at grazing angles as illustrated in Figure 14.12.

14.6.2   Molecular Contamination

Contamination outgassing is particularly problematic for high surface area thin films such as sol gel coatings (Burnham et al., 2000) and diffraction gratings (Murakami et al., 2012). Molecular contamination can fill pores between sol gel spheres, thus

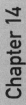

degrading the antireflection characteristics of the coating (Pryatel et al., 2007). The laser resistance of these porous films also tends to be reduced, compared to coatings deposited from densified processes such as IBS (Jensen et al., 2007). Molecular contamination levels as low as a part per million have been shown to degrade the laser resistance of optical coatings (Hovis et al., 1994). Various surface treatments have been developed to make sol gel coatings less susceptible to contamination including ammonia hardening (Thomas et al., 1999) and by modifying the sol chemistry (Suratwala et al., 2003). Fluorinated coatings have also been shown to repel organic contaminants on fused silica surfaces (Weiller et al., 2012).

Even in clean vacuum systems, the laser resistance of optical coatings can be reduced by the formation of substoichiometric surfaces (Burnham et al., 2000; Ling, 2011). Care must be taken in evaluating the differences between air and vacuum for porous high reflector coatings to make sure that humidity-induced spectral shifting and the accompanying change in SWEF profiles are properly angularly tuned. One advantage of vacuum is that the ionizing components are minimized (Giuliano, 1973).

14.7   Antireflection Coatings

Antireflection coatings are designed to destructively interfere with the surface reflection of an optic. For a single layer of quarter-wave thickness and appropriate refractive index, a 180° out-of-phase wave with equal amplitude to the surface reflection can be achieved as per the following equation:

$$n_f = \sqrt{n_s n_i}$$

(14.5)

where
    n_f is the refractive index of the thin film
    n_s is the refractive index of the substrate
    n_i is the refractive index of the incident medium

14.7.1   Hard Dielectrics

NIR lasers typically use fused silica as window and lens substrate materials due to their high laser resistance. Real coating materials do not have a low enough refractive index for a single layer; therefore, multilayer AR coating designs are commonly used. Although AR coatings tend not to have the high electric fields experienced within high reflector structures, the electric field is significantly higher at the substrate-film interface. Therefore, substrate polishing and cleaning play a significantly greater role in the laser resistance of this coating type compared to high reflectors. AR coatings also tend to be very thin (hundreds of nanometers), so they are prone to blistering in the presence of a laser-heated surface plasma that is initiated at an interfacial defect (Carniglia, 1981). Very small defects of only a few tens of nanometers in size can create delaminated blisters that are tens of microns in diameter. Fortunately, these delaminated blisters or flat-bottom pits tend to be stable when exposed to laser fluences higher than their initiation threshold (Génin et al., 1997).

14.7.2   Sol Gel

The materials available for physical deposition coatings have too high of a refractive index to be effective single-layer antireflection coatings. A more porous coating technology that makes excellent high laser-resistant single-layer antireflection coatings is sol gel. These coatings consist of a porous structure of submicron silica spheres (Stöber et al., 1968). By tuning the porosity and physical thickness, an effective refractive index can be achieved, which will yield a perfect antireflection coating.

Sol gel coatings are fabricated by either dipping in a solution with a controlled insertion and pull rate, spin coating, or meniscus coating to achieve the desired thickness. Ammonia hardening (Thomas et al., 1999) is a chemical process that cross-links the spheres, leading to a more robust coating that can tolerate moderate cleaning without degrading its laser resistance.

14.7.3   AR Surface Microstructures

Antireflection microstructures are not optical coatings; however, from a spectral perspective, they behave like an antireflection coating by maximizing the laser transmission through the optic. Surface microstructures can be designed to have very regular forms, which have fairly narrow spectral bandwidths or random structures that have significantly greater spectral bandwidth. These microstructures can be etched into the substrate, thus eliminating the need for high-index coating materials. Additionally, the laser-induced blistering issues observed with hard dielectric thin films do not occur with AR surface microstructures. Impressive laser damage thresholds have been demonstrated with these structures (Hobbs and MacLeod, 2007).

14.8   Applications

14.8.1   Gratings

High laser fluence gratings are typically used to expand and compress pulses for short pulse petawatt class lasers (Strickland and Mourou, 1985). Metallic grating structures, such as gold, have been replaced by dielectric high reflector multilayer mirrors with thick overcoats that are etched with grating structures for higher laser resistance (Nguyen et al., 2005). SWEF peaks within these grating structures can be reduced by changing the grating height and duty cycle. A direct correlation between the laser resistance and the SWEF peaks has been clearly demonstrated (Neauport et al., 2007). The laser damage morphology aligns with the SWEF peak locations within the grating structure. Typically, the SWEF strength within the coating materials along a line of constant diffraction efficiency decreases with increasing grating height and reducing duty cycle, which results in thinner grating lines (Britten et al., 2004). Because the SWEF peaks tend to be in the air space between grating lines, these structures are particularly sensitive to surface contamination, so proper cleaning is critical for high laser resistance (Ashe et al., 2007). Fine-tuning of the grating efficiency can also occur through chemical etching processes that impact grating height and duty cycle, which can also lead to improved laser resistance (Nguyen et al., 2010).

Chapter 14

14.8.2   Rugates

Rugate filters are coatings that gradually transition from one coating material to the other, leading to a sinusoidal refractive index profile instead of discrete interfaces. The spectral advantage of this coating design technique is suppression of higher-order spectral harmonics (Southwell, 1988). Reduced absorption has also been observed (Mende et al., 2011). Coating stress is reduced in IBS coatings when cosputtering high and low refractive index coating materials (Pond et al., 1989). This leads to reduced wavefront distortion for a coating process that tends to have very high compressive stresses.

These structures can be manufactured by IBS through the use of zone targets that consist of two different coating materials with a clear boundary between the two materials. The target is translated back and forth to go from 100% of material A and gradually to 100% of material B (Demiryont, 1985). The combination of zone targets and online monitoring systems creates the necessary conditions to readily fabricate IBS rugate structures (Lappschies et al., 2005). For 1064 nm high reflector coatings, substantial improvements in the laser resistance of TiO_2/SiO_2 mirrors were observed between rugate and traditional quarter-wave stack coatings (Jupé et al., 2007). Later work has shown comparable laser resistance between rugate and traditional quarter-wave stack Ta_2O_5/SiO_2 coatings, although the damage morphology was less severe at the higher fluences for the rugate style coating (Qiao et al., 2013). Similarly, no significant differences in the laser resistance between HfO_2/SiO_2 rugate and traditional quarter-wave stack coatings has been reported, even for nanolaminate coatings that contain a large number of very thin layers to create pseudo-rugate structures. However, high refractive index mixed materials based on the IBS zone target rugate deposition technology for materials such as HfO_2 and Al_2O_3, do show improved multilayer mirror laser resistance (Mende et al., 2011). In addition to lower absorption and lower coating stresses, these material mixtures tend to have a lower coefficient of thermal expansion (Harry et al., 2007), which will reduce the laser heating stresses, thus improving the laser resistance of these coating structures.

14.8.3   Solarization

Solarization is another positive exploitation of radiation-induced damage in optical materials for mirror coatings. Solarization is the formation of color centers in glass due to solar radiation (Dotsenko et al., 1997). This leads to darkening of the glass (Faraday, 1825). Transition lenses in optical eyeglasses are a perfect example of reversible glass solarization. Multiple radiation sources from UV lasers, ion beams, x-rays, and gamma rays can form color centers in glass. Some high-power fusion lasers darken their transport mirrors as a method to protect or armor mounting structures from backscatter radiation. In the National Ignition Facility, the transport mirrors are designed to efficiently transmit backscattered light from the target to prevent damage to upstream optics (Stolz et al., 2003). Gamma radiation from a cobalt-60 source is used to darken the mirror substrates, as illustrated in Figure 14.13. Thermal bleaching of the color centers necessitates substrate darkening after coating, at least for deposition processes with elevated deposition temperatures. As a result of the radiation, the glass experiences a radiation-induced compaction gradient, which can alter the reflected wavefront of the

FIGURE 14.13 Gamma ray darkening of transport mirrors in the National Ignition Facility protects mounting hardware from target-backscattered light.

mirror (Primak and Kampwirth, 1968). Uniform irradiation of both surfaces is necessary to maintain the reflected wavefront of a mirror. Alternatively, nonuniform irradiation can be used to compensate for excessive coating stresses by balancing the stresses from nonuniform compaction to reduce after coating reflected wavefront performance. Gamma irradiation for substrate darkening has been found to have minimal impact on the 1064 nm laser resistance of nanosecond-length laser pulses; however, a significant reduction in the 1064 nm laser resistance at picosecond-length pulses has been observed. A minor degradation in the 351 nm laser resistance for nanosecond-length laser pulses has also been observed.

14.9 Conclusions

Although significant progress has been made in fabricating high-fluence NIR coatings, the field remains fertile with opportunities to advance both theoretical understanding of laser damage mechanisms as well as multiple engineering opportunities to produce improved coatings. In NIR coatings, defects and contamination play a large role in fluence-limiting optics. Laser conditioning, mitigation, coating and material process optimization, coating design, cleaning, and planarization can all play a significant role in optimizing NIR laser resistance.

References

Apfel, J.H., 1977, Optical coating design with reduced electric field intensity. *Appl. Opt.* 16: 1880–1885.
Ashe, B., Marshall, K.L., Giacofei, C., Rigatti, A.L., Kessler, T.J., Schmid, A.W., Oliver, J.B., Keck, J., and Kozlov, A., 2007, Evaluation of cleaning methods for multilayer diffraction gratings. *Proc. SPIE* 6403: 640030-1-9.
Barr, W.P., 1969, The production of low scattering dielectric mirrors using rotating vane particle filtration. *J. Phys.* [E] 2: 1112–1114.

Bass, I.L., Guss, G.M., Nostrand, M.J., and Wegner, P.J., 2010, An improved method of mitigating laser induced surface damage growth in fused silica using a rastered pulsed CO_2 laser. *Proc. SPIE* 7842: 784220-1-11.

Bercegol, H., 1999, What is laser conditioning? A review focused on dielectric multilayers. *Proc. SPIE* 3578: 421–425.

Bettis, J.R., 1992, Correlation among the laser-induced breakdown thresholds in solids, liquids, and gases. *Appl. Opt.* 31: 3448–3452.

Bettis, J.R., Guenther, A.H., and House, R.A., 1979, Refractive-index dependence of pulsed-laser-induced damage. *Opt. Letts.* 4: 256–258.

Bloembergen, N., 1974, Laser-induced electric breakdown in solids. *IEEE J. Quant. Electron.* 10: 375–386.

Britten, J.A., Molander, W.A., Komashko, A.M., and Barty, C.P.J., 2004, Multilayer dielectric gratings for petawatt-class laser systems. *Proc. SPIE* V5273: 1–7.

Burnham, A.K., Runkel, M., Demos, S.G., Kozlowski, M.R., and Wegner, P.J., 2000, Effect of vacuum on the occurrence of UV-induced surface photoluminescence, transmission loss, and catastrophic surface damage. *Proc. SPIE* 4134: 243–252.

Carniglia, C.K., 1981, Oxide coatings for one micrometer laser fusion. *Thin Solid Films* 77: 225–138.

Cheng, X., Ding, T., He, W., Zhang, J., Jiao, H., Ma, B., Shen, Z., and Wang, Z., 2011a, Using engineered nodules to study laser-induced damage in optical thing films with nanosecond pulses. *Proc. SPIE* 8190: 819002-1-9.

Cheng, X., Ding, T., Wenyan, H., Zhang, J., Jiao, H., Ma, B., Shen, Z., and Wang, Z., 2011b, Using monodisperse SiO_2 microspheres to study laser-induced damage of nodules in HfO_2/SiO_2 high reflectors. *Proc. SPE* 8168: 816816-1-6.Cheng, X., Tuniyazi, A., Zhang, J., Ding, T. Jiao. H., Ma, B., Wei, Z., Ki, H., and Wang, X., 2014, Nanosecond laser-induced damage of nodular defects in dielectric multilayer mirrors. *Appl. Opt.* 53: A62–A69.

Cherief, N., Givord, D., Liénard, A., Mackay, K., McGrath, O.F.K., Rebouillat, J.P., Robaut, F., and Souche, Y., 1993, Laser ablation deposition and magnetic characterization of metallic thin films based on rare earth and transition metals. *J. Magn. Magn. Mater.* 121: 94–101.

DeBell, G., 1972, The design and measurement of low absorptance optical interference coatings. University of Rochester, Institute of Optics, PhD thesis, New York, (unpublished).

DeFord, J.F. and Kozlowski M.R., 1993, Modeling of electric-field enhancement at nodular defects in dielectric mirror coatings. *Proc. SPIE* 1848: 455–470.

Demiryont, H., 1985, Optical-properties of SiO_2-TiO_2 composite films. *Appl. Opt.* 24: 2647–2650.

Demos, S.G., Negres, R.A., Raman, R.N., Rubenchik, A.M., and Feit, M.D., 2013, Material response during nanosecond laser induced breakdown inside of the exit surface of fused silica. *Laser Photonics Rev.* 7: 444–452.

Dijon, J., Poulingue, M., and Hue, J., 1999a, Thermomechanical model of mirror laser damage at 1.06 μm. Part 1: Nodule ejection. *Proc. SPIE* 3578: 387–396.

Dijon, J., Ravel, G., and André, B., 1999b, Thermomechanical model of mirror laser damage at 1.06 μm. Part 2: Flat-bottom-pits formation. *Proc. SPIE* 3578: 398–407.

Dotsenko, A.V., Glebov, L.B., and Tsekhomsky, V.A., 1997, *Physics and Chemistry of Photochromic Glasses.* CRC Press, Boca Raton, New York.

Elhadj, S., Qiu, S.R., Monterrosa, A.M., and Stolz, C.J., 2012, Heating dynamics of CO_2-laser irradiated silica particles with evaporative shrinking: Measurements and modeling. *J. Appl. Phys.* 111L 093113-1-5.

Faraday, M., 1825, Sur la coloration produite par lumière, dans une esèce particulière de carreaux de vita. *Ann. Chem. Phys.* 25: 99–100.

Fornier, A., Cordillot, C., Ausserre, D., and Paris, F., 1994, Laser conditioning of optical coatings: Some issues in the characterization by atomic force microscopy. *SPIE* 2114: 355–365.

Gallais, L., Mangote, B., Zerrad, M., Commandré, M., Meininkaitis, A., Mirauskas, J., Jeskevic, M., and Sirutkaitis, V., 2011, Laser-induced damage of hafnia coatings as a function of pulse duration in the femtosecond to nanosecond range. *Appl. Opt.* 50: C178–C187.

Génin, F.Y., Michlitsch, K., Furr, J., Kozlowski, M.R., and Krulevitch, P., 1997, Laser-induced damage of fused silica at 355 and 1064 nm initiated at aluminum contamination particles on the surface. *Proc. SPIE* 2966: 126–138.

Génin, F.Y., Stolz, C.J., and Kozlowski, M.R., 1997a, Growth of laser-induced damage during repetitive illumination of HfO_2-SiO_2 multilayer mirror and polarizer coatings. *Proc. SPIE* 2966: 273–282.

Génin, F.Y., Stolz, C.J., Reitter, T.A., Kozlowski, M.R., Bevis, R.P., and von Gunten, M.K., 1997b, Effect of electric-field distribution on the morphologies of laser-induced damage in hafnia-silica multilayer polarizers. *Proc. SPIE* 2966: 342–352.

Gill, D.H., Newnam, B.E., and McLeod, J., 1978, Use of non-quarter-wave designs to increase the damage resistance of reflectors at 532 and 1064 nanometers. In: Glass, A.J. and Guenther, A.H. (eds) *Laser-Induced Damage in Optical Materials: 1977.* NBS SP 509: pp. 260–270.

Giuliano, C.R., 1973, The relation between surface damage and surface plasma formation. NBS SP 372, 46–54.

Gourdin, W.H., Dzenitis, E.G., Martin, D.A., Listiyo, K., Sherman, G.A., Kent, W.H., Butlin, R.K., Stolz, C.J., and Pryatel, J.A., 2005, In-situ surface debris inspection and removal system for upward-facing transport mirrors of the National Ignition Facility. *Proc. SPIE* 5647: 107–117.

Guch Jr., S. and Hovis, F., 1994, Beyond perfection: The need for understanding contamination effects on real-world optics. *Proc. SPIE* 2114: 505–511.

György, E., Mihailescu, I.N., Kompitsas, M., and Giannoudakos, A., 2004, Deposition of particulate-free thing films by two synchronized laser sources: Effects of ambient gas pressure and laser fluence. *Thin Solid Films* 446: 178–183.

Harry, G.M. et al., 2007, Titania-doped tantala/silica coatings for gravitational-wave detection. *Class. Quant. Grav.* 24: 405–415.

Hobbs, D.S. and MacLeod, B.D., 2007, High laser damage threshold surface relief micro-structures for anti-reflection applications. *Proc. SPIE* 6720: 67200L-1-10.

Hovis, F.E., Shepherd, B.A., Radcliffe, C.T., Bailey, A.L., and Boswell, W.T., 1994, Optical damage at the part per million level: The role of trace contamination in laser-induced optical damage. *Proc. SPIE* 2114: 145–152.

Jensen, L., Jupé, M., Mädebach, H., Ehlers, H., Starke, K., Ristau, D., Riede, W., Allenspacher, P., and Schroeder, H., 2007, Damage threshold investigations of high power laser optics under atmospheric and vacuum conditions, *SPIE Proc.* 6403: 64030U-1-10.

Jensen L.O. and Ristau, D., 2012, Coatings of oxide composites. *Proc. SPIE* 8530: 853013-1–853013-14.

Jupé, M., Lappschies, M., Jensen, L., Starke, K., and Ristau, D., 2007, Laser induced damage in gradual index layers and rugate filters. *Proc. SPIE* 6403: 640311-1-12.

Knollenberg, R.G., Long, D., and Lopez S., 1995, An in-situ fiber optics sensor for monitoring particle microcontamination during an IBS optical coating process, in *Optical Interference Coatings*, Vol. 17, 1995. OSA Technical Digest Series (Optical Society of America, Washington, DC), pp. 124–126.

Koren, G., Baseman, R.J., Gupta, A., Lutsyche, M.I., and Laibowitz, R.B., 1990, Particulates reduction in laser-ablated $YBa_2Cu_3O_{7-\delta}$ thin films by laser-induced plume heating. *Appl. Phys. Lett.* 56: 2144–2146.

Lalezari, R., Rempe, G., Thompson, R.J., and Kimble, H.J., 1992, Measurement of ultralow losses in dielectric mirrors. *Optical Interference Coatings Technical Digest* (Optical Society of America, Washington, DC,) Vol. 15, pp. 331–333.

Lappschies, M., Görtz, B., and Ristau, D., 2005, Optical monitoring of rugate filters. *Proc. SPIE* 5963: 59631Z-1-9.

Ling, X., 2011, Nanosecond multi-pulse damage investigation of optical coatings in atmosphere and vacuum environments. *Appl. Surf. Sci.* 257: 5601–5604.

Liu, X., Zhao, Y., Li, D., Hu, G., Gao, Y., Fan, Z., and Shao, J., 2011, Characteristics of plasma scalds in multilayer dielectric films. *Appl. Opt.* 50: 4226–4231.

Logan, J.S. and McGill, J.J., 1991, Study of particle emission in vacuum from film deposits. *J. Vac. Sci. Technol. A* 10: 1875–1878.

Luk'Yanchuk, B.S., Wang, Z.B., Song, W.D., and Hong, M.H., 2004, Particle on surface: 3D-effects in dry laser cleaning. *Appl. Phys. A* 79: 747–751.

Mende, M., Jensen, L., Ehlers, H., Riggers, W., Blaschke, H., and Ristau, D., 2011, Laser induced damage of pure and mixture material high reflectors for 355 nm and 1064 nm wavelength. *Proc. SPIE* 8168: 816821-1-11.

Mirkarimi, P.B., Spiller, E., Baker, S.L., Robinson, J.C., Sterarns, D.G. Liddle, J.A., Salmassi, F., Liang, T., and Stivers, A.R., 2005, Advancing the ion beam thin film planarization process for the smoothing of substrate particles. *Microelectron. Eng.* 77: 369–381.

Murakami, H. et al., 2012, Influences of oil-contamination on LIDT and optical properties in dielectric coatings. *Proc. SPIE* 8530: 853024-1-6.

Neauport, J., Lavastre, E., Razé, G., Dupuy, G., Bonod, N., Balas, M., de Villele, G., Flamand, J., Kaladgew, S., and Desserouer, F., 2007, Effect of electric field on laser induced damage threshold of multilayer dielectric gratings. *Opt. Expr.* 15: 12508–12522.

Nguyen, H.T., Britten, J.A., Carlson, T.C., Nissen, J.D., Summers, L.J., Hoaglan, C.R., Aasen, M.D., Peterson, J.E., and Jovanovic, I., 2005, Gratings for high-energy petawatt lasers. *Proc. SPIE* 5991: 59911M-1-7.

Chapter 14

Nguyen, H.T., Larson, C.C., and Britten, J.A., 2010, Improvement of laser damage resistance and diffraction efficiency of multilayer dielectric diffraction gratings by HF-etchback linewidth tailoring. *Proc. SPIE* 7842: 78421H-1-8.

Norton, M.A., Stolz, C.J., Donohue, E.E., Hollingsworth, W.G., Listiyo, K., Pryatel, J.A., and Hackel, R.P., 2006, Impact of contaminates on the laser damage threshold of 1ω HR coatings, *Proc. SPIE* 5991: 599100-1-9.

Papandrew, A.B., Stolz, C.J., Wu, Z.L., Loomis, G.E., and Falabella, S., 2001, Laser conditioning characterization and damage threshold prediction of hafnia/silica multilayer mirrors by photothermal microscopy, *Proc. SPIE* 4347: 53–61.

Pinot, B., Leplan, H., Houbre, F., Lavastre, E., Poncetta, J., and Chabassier, G. 2001, Laser Megajoule 1.06 μm mirrors production, with very high laser damage threshold. *Proc. SPIE* 4679: 234–241.

Pond, B.J., DeBar, J.I., Carniglia, C.K., and Raj T., 1989, Stress reduction in ion beam sputtered mixed oxide films. *Appl. Opt.* 28: 2800–2805.

Primak, W. and Kampwirth, R., 1968, The radiation compaction of vitreous silica. *J. Appl. Phys.* 39: 5651–5658.

Pryatel, J.A., Gourdin, W.H., Hampton, G.J., Behne, D.M., and Meissner, R., 2007, Qualification of materials for applications in high fluence lasers, *Proc. SPIE* 6403: 640329-1-11.

Qiao, Z., Ma, P., Liu, H., Pu, Y., and Liu, Z., 2013, Laser-induced damage of rugate and quarter-wave stacks high reflectors deposited by ion-beam sputtering. *Opt. Eng.* 52: 086103-1-5.

Qiu, S.R., Wolfe, J.E., Monterrosa, A.M., Feit, M.D., Pistor, T.V., and Stolz, C.J., 2011, Searching for optimal mitigation geometries for laser-resistant multilayer high-reflector coatings. *Appl. Opt.* 50: C373–C381.

Reicher, D.W. and Jungling K.C., 1994, Contamination of surfaces prior to optical coating by in-situ total internal reflection microscopy. *Proc. SPIE* 2114: 154–165.

Rigatti, A.L., 2005, Cleaning processes versus laser-damage threshold of coated optical components, *Proc. SPIE* 5647: 136–140.

Ristau, D., Jupé, M., and Starke, K., 2009, Laser damage thresholds of optical coatings. *Thin Solid Films* 518: 1607–1613.

Schmidt, J.R., Runkel, M.J., Martin, K.E., and Stolz, C.J., 2008, Scattering-induced downstream beam modulation by plasma scalded mirrors. *Proc. SPIE* 6720: 67201H-1–67201H-8.

Shan, Y., He, H., Wang, Y., Li, X., Li, D., and Zhao, Y., 2010, Electrical field enhancement and laser damage growth in high-reflective coatings at 1064 nm. *Opt. Commun.* 284: 625–629.

Shan, Y., He, H., Wei, C., Wang, Y., and Zhao, Y., 2011, Thermomechanical analysis of nodular damage in HfO₂/SiO₂ multilayer coatings. *Chin. Opt. Letts.* 9: 103101-1–103101-4.

Sheehan, L., Kozlowski, M., Rainer, F., and Staggs, M., 1994, Large-area conditioning of optics for high-power laser systems. *Proc. SPIE* 2114: 559–568.

Southwell, W.H., 1988, Spectral response calculations of rugate filters using coupled-wave theory. *J. Opt Soc. Am. A* 5: 1558–1564.

Stöber, W., Fink, A., and Bohn, E., 1968, Controlled growth of monodisperse silica spheres in the micron size range, *J. Colloid. Int. Sci.*, 26: 62–69.

Stolz, C.J., 2012, Boulder damage symposium annual thin-film damage competition. *Opt. Eng.* 51: 121818-1-7.

Stolz, C.J., Génin, F.Y., and Pistor, T.V., 2004, Electric-field enhancement by nodular defects in multilayer coatings irradiated at normal and 45 degree incidence. *Proc. SPIE* 5273: 41–49.

Stolz, C.J., Génin, F.Y., Reitter, T.A., Molau, N.E., Bevis, R.P., von Gunten, M.K., Smith, D.J., and Anzellotti, J.F., 1997, Effect of SiO₂ overcoat thickness on laser damage morphology of HfO₂/SiO₂ Brewster's angle polarizers at 1064 nm. *Proc. SPIE* 2966: 265–272.

Stolz, C.J., Hafeman, S., and Pistor, T.V., 2008, Light intensification modeling of coating inclusions irradiated at 351 and 1053 nm. *Appl. Opt.* 47: C162–C165.

Stolz, C.J., Menapace, J.A., Génin, F.Y., Ehrmann, P.R., Miller, P.E., and Rogowski, G.T., 2003, Influence of BK7 substrate solarization on the performance of hafnia and silica multilayer mirrors. *Proc. SPIE* 4932: 38–47.

Stolz, C.J., Sheehan, L.M., Von Gunten M.K., Bevis, R.P., and Smith, D., 1999, The advantages of evaporation of Hafnium in a reactive environment to manufacture high damage threshold multilayer coatings by electron-beam deposition. *Proc. SPIE* 3738: 318–324.

Stolz, C.J., Wolfe, J.E., Adams, J.J., Menor, M.G., Teslich, N.E., Mirkarimi, P.B., Folta, J.A., Soufli, R., Menoni, C., and Patel, C., 2014, High laser resistant multilayer mirrors by nodular defect planarization. *App. Opt.* 53: A291–A296.

Strasser, G., Bader, H.P., and Bader, M.E., 1990, Reduction of particle contamination by controlled venting and pumping of vacuum loadlocks. *J. Vac. Sci. Technol. A* 8: 4092–4097.

Strickland, D. and Mourou, G., 1985, Compression of amplified chirped optical pulses. *Opt. Expr.* 15: 12508–12522.

Strikovsky, M.D., Klyuenkov, E.B., Gaponov, S.V., Schubert, J., and Copetti, C.A., 1993, Crossed fluxes technique for pulsed laser deposition of smooth $YBa_2Cu_3O_{7-x}$ films and multilayers. *Appl. Phys. Lett.* 63: 1146–1148.

Suh, J.D., Sung, G.Y., and Kang, K.Y., 1995, Effects of oxygen pressure and target-substrate distance on the density of particulates on pulsed laser deposited $YBa_2Cu_3O_{7-x}$ thin film surfaces. *J. Matter. Sci. Letts.* 14: 832–836.

Suratwala, T.I., Hanna, M.L., Miller, E.L., Whitman, P.K., Thomas, I.M., Ehrmann, P.R., Maxwell, R.S., and Burnham, A.K., 2003, Surface chemistry and trimethylsilyl functionalization of Stöber silica sols. *J. Non-Cryst. Solids* 316: 349–363.

Tench, R.J., Chow, R., and Kozlowski, M.R., 1994, Characterization of defect geometries in multilayer optical coatings, *J. Vac. Sci. Technol. A* 12: 2808–2813.

Thomas, I.M., Burnham, A.K., Ertel, J.R., and Frieders, S.C., 1999, Method for reducing the effect of environmental contamination of sol-gel optical coatings. *Proc. SPIE* 3492: 220.

Tselev, A., Gorgunov, A., and Pompe, W., 2001, Cross-beam pulsed laser deposition: General characteristic. *Rev. Sci. Instrum.* 72: 2665–2672.

Walker, T.W., Guenther, A.H., Nielsen, P., 1981, Pulsed laser-induced damage to thin-film optical coatings-Part I: Experimental & Part II: Theory, *IEEE J. Quant. Electron.* QE-17: 2041–2065.

Walton, C.C., Génin, F.Y., Chow, R., Kozlowski, M.R., Loomis, G.E., and Pierce, E., 1996, Effect of silica overlayers on laser damage of HfO_2-SiO_2 56° incidence high reflectors. *Proc. SPIE* 2714: 550–558.

Wang, Z., Bao, G., Jiao, H., Ma, B., Zhang, J., Ding, T., and Cheng, X., 2013, Interfacial damage in a Ta_2O_5/SiO_2 double cavity filter irradiated by 1064 nm nanosecond laser pulses. *Opt. Expr.* 21: 30623–30632.

Weakley, S.C., Stolz, C.J., Wu, Z.L., Bevis, R.P., and von Gunten, M.K., 1999, Role of starting material composition in interfacial damage morphology of hafnia silica multilayer coatings. *Proc. SPIE* 3578: 137–143.

Wei, C., Yi, K., and Shao, J. 2012, Influence of composition and seed dimension on the structure and laser damage of nodular defects in HfO_2/SiO_2 high reflectors. *Appl. Opt.* 51: 6781–6788.

Weiller, B.H., Fowler, J.D., and Villahermosa, R.M., 2012, Contamination resistant coatings for enhanced laser damage thresholds. *Proc. SPIE* 8530: 85302A-1–85302A-6.

Williams, F.L., Petersen, G.A., Carniglia, C.K., and Pond, B.J., 1992, In situ characterization of thin-film defect generation using total internal reflection microscopy. *J. Vac. Sci. Technol. A* 10: 1472–1478.

Wolfe, C.R., Kozlowski, M.R., Campbell, J.H., Rainer, F., Morgan, A.J., and Gonzales, R.P., 1990, Laser conditioning of optical thin films. *Proc. SPIE* 1438: 360–375.

Wolfe, J.E., Stolz, C.J., and Falabella, S., 2013, Velocity determination of particles ejected during electron beam deposition. *Optical Interference Coatings Technical Digest* (Optical Society of America, Washington, DC.) FA.2.

Wolfe, J.E., Qiu, S.R., and Stolz, C.J., 2011, Fabrication of mitigation pits for improving laser damage resistance in dielectric mirrors by femtosecond laser machining. *Appl. Opt.* 50: C457–C462.

Yoshitake, T. and Nagayama, K., 2004, The velocity distribution of droplets ejected from Fe and Si targets by pulsed laser ablation in a vacuum and their elimination using a vane-type velocity filter. *Vacuum* 74: 515–520.

Zhang, J., Xie, Y., Cheng, X., Ding, T., and Wang, Z., 2013, Broadband thin-film polarizers for high-power laser systems. *Appl. Opt.* 52: 1512–1516.

Zhu, M., Yi, K., Fan, Z., and Shao, J., 2011, Theoretical and experimental research on spectral performance and laser induced damage of Brewster's thin film polarizers. *Appl. Surf. Sci.* 257: 6884–6888.

15. Coatings for fs Lasers

Marco Jupé and Detlev Ristau

15.1 Introduction

In various aspects, the application of ultrashort pulse (USP) laser optics differs from the short pulse (nanosecond pulses) as well as from the continuous wave laser application. This fact can be traced back to the properties of USP matter interaction. In many cases, the extreme peak intensity of the ultrashort laser pulses is used to create nonlinear effects in the sample. Because of the short pulse durations, the peak power is achieved at smaller pulse energies in comparison to nanosecond pulses. Consequently, during the interaction of laser irradiation with matter, a small quantity of heat is implemented in the material. Additionally, the energy transfer from the irradiation field to the solid occurs on the femtosecond scale, which is much faster than the interaction time of the phonons. Because of the assumption that the interaction time for an optical breakdown is small in comparison to the timescale of thermal processes, there is no thermal interaction, and the ablation processes are assumed to be a nonthermal process.

From the special properties of the irradiation, two demands result for the optics, which place the highest requirements on the coating process. On the one hand, the optics have to

Chapter 15

Laser-Induced Damage in Optical Materials. Edited by Detlev Ristau © 2015 CRC Press/Taylor & Francis Group, LLC. ISBN: 978-1-4398-7216-1.

handle extreme peak power up to the petawatts range. This peak power can cause non-linear effects in components. Inside the optics, these nonlinear effects change the optical and electronic properties of the materials and the dielectric coatings, consequently. Often these nonlinearities are used directly in the application, but often, these effects are unwanted and lead to the damage of the optic.

On the other hand, the components often demand special optical properties, which are typically not necessary in the ranges of longer pulses as well as in the continuous wave operation range. Important examples are pulse stretching or pulse compressing elements like chirp mirrors, grids, or prisms. Additionally, the broad spectral width of femtosecond laser pulses causes higher difficulties for the design of non-phase-sensitive optical components like windows and mirrors. Consequently, the manufacturing of USP optics demands for high precision especially with respect to the process control (Starke et al. 2000, Groß et al. 2001, Ristau et al. 2003, Lappschies et al. 2004, 2004a, Janicki et al. 2005, Ristau et al. 2008).

Usually, the complex setups of USP lasers and USP application require a plurality of most different components. Mostly the dielectrics optics and dielectric coatings constitute the main part in the systems. From the point of view of laser-induced damage (LID), a separation between thin film optics and bulk optics is reasonable. This classification respects the different nonlinear effects in the dielectric material. As mentioned in the chapter *LID in Femtosecond Regime*, the damage in the USP regime can be traced back to the generation of free carriers in the conduction band. Obviously, the damage can be understood as a pure electronic process. These characteristic properties of the damage process allow to simulate the damage behavior, precisely. These modulations can only be applied in the case of thin layers, because additional nonlinearity has to be respected, and the quantitative modeling of the effects requires a very detailed knowledge of the linear and nonlinear material parameters. In some cases, the processes start from statistical noise.

In other words, the unity of optical components can be divided in the part, which can be described easily, and the second part, which has to be analyzed, empirically.

Consequently, the first part of the explanation is addressed to dielectric layers, and the second part is focused on complex optics and a short discussion of the nonlinear effects taking part in bulk material damage.

15.2 Dynamics of Damage Processes in Dielectric Layers

Damage in the USP regime is driven by electronic interactions. As a consequence of this electronic damage process, the properties of the process vary significantly from the thermal processes. For the classification and for the optimization of femtosecond optics, the deterministic damage behavior is of the major interest. Such a deterministic damage behavior is defined by a negligible transition fluence ranging from clearly nondamaged to certainly damaged, and the deterministic damage behavior leads to a well-defined damage threshold. This behavior is caused by the electronic nature of the

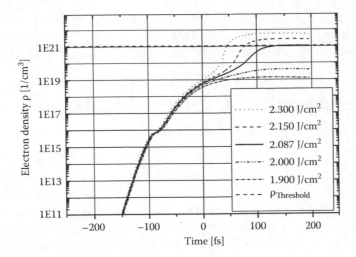

FIGURE 15.1   Calculated growth of electron density for different fluencies for $\tau_p = 100$ fs and $\lambda = 800$ nm in quartz.

damage process. The nonlinear ionization causes a drastic growth of the electron density for fluences close to the threshold.

Figure 15.1 displays the calculated increase in electron density in fused silica ($E_{gap} = 7.5$ eV) under titanium:sapphire laser irradiation for various pulse energies close to the damage threshold. For a fluence of 1.9 J/cm², a density of 10^{19} electrons/cm³ is inducted. This electron density is few orders of magnitude lower than the critical value of 10^{21} electrons/cm³ in the conditions, respectively, but an increase in the fluence of less than 10% suffices to achieve the critical level. This behavior is typical for the USP damage. Figure 15.2 shows a typical survival curve of a thin film in the respective measurements regime.

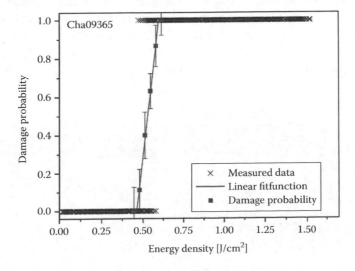

FIGURE 15.2   Typical experimental survival curve of an USP LIDT measurement.

This fundamental property of the USP damage causes a high predictability of the thresholds of single materials, as well as the damage properties of thin film stack systems. Following, the feature is applied in various applications in thin film designs.

15.3 Approximation of Damage Thresholds in Dielectric Layer Stacks

The damage of dielectric, thin film optics by USP laser irradiation is well understood compared to the damage in the short pulse or in the continuous wave range. In the USP regime, the damage is an electronic process. The characteristic of electronic damage processes is of deterministic nature, and the damage properties can be traced back to the electronic properties of the material. This deterministic character of the electronic damage is caused by the rapid increase in the electron density in the conduction band within an impact of energy of the irradiation close to the threshold of the material. The dominating material parameter is the gap of the dielectric, representing the potential barrier between valence band and the conduction band. For ionization, the electrons have to transit this barrier, and because of the photon energy, a nonlinear excitation is necessary. These facts are discussed in detail in chapter *LID in Femtosecond Regime*.

The material classification is important for the understanding of the damage in a complex coating, whereas the material gap has to be applied for a classification. Typically, the index of refraction and effective mass of the electrons in the excited states also affect the threshold of the material marginally, but for the design of a coating, the index of refraction is of the main interest. However, the gap and the index of reflection are influencing each other, and also the material gap and the laser induced damage threshold (LIDT) are correlated in a linear function (Table 15.1).

Generally, the material with a higher band gap has a higher damage threshold, but also a lower refractive index. According to the current state of the art, USP optics are manufactured from oxide material. Consequently, the refractive index contrast between the high- and the low-index materials (usually SiO_2) decreases for components with a high damage threshold. For the design of complex functional optics, the highest possible

Table 15.1 Properties of Single Layers

Material	Coating Process	Material Gap	Index of Refraction	Damage Threshold λ at 800 nm τ_{pulse} at App.100 fs
TiO_2	IBS	3.3	2.39	0.58
Ta_2O_5	IBS	3.8	2.17	0.82
HfO_2	IBS	5.1	2.09	1.29
Al_2O_3	IBS	6.5	1.65	2.06
SiO_2	IBS	8.3	1.50	2.82

Sources: Mero, M. et al., *Proc. SPIE*, 4932, 202, 2003; Mero, M. et al., *Opt. Eng.*, 44(5), 051107-1-7, 2005.

contrast of high and low refractive index materials is necessary. Obviously, these two requirements conflict with each other. Therefore, an optimization problem results for each high-power component in the USP regime, whereas typically no limitations in the spectral characteristic are expectable. Because of the deterministic damage behavior, an approximation of the LIDT of a component is possible and also optimizations were demonstrated (Britten et al. 1995, Jovanovic et al. 2004, Abromavicius et al. 2007, Neauport and Bonod 2008b, Jupe et al. 2008).

Before the calculation can be described, a general law of the laser damage has to be proposed.

The most sensitive part of a component defines the damage threshold of the whole component.

This proposition seems too canonic, but the consequent application of this theorem allows understanding the damage of a component in detail. Furthermore, the calculation of the LIDT of a component requires this law.

In a second version of the damage law, the following can be predicted:

The mechanism or the combination of damage mechanisms, which requires the lowest energy, will lead to the damage.

This postulation is considered in the majority of cases, but in the USP regime, the process time is an important parameter, which can also influence the damage of components.

However, the calculation of the damage threshold of a complex multiple layer stack is based on three characteristic physical values, which are dependent on each other. The first parameter represents the damage threshold of the material itself. This value is called *material threshold* and has been determined from the threshold of the single layer H and normalized by the electrical field strength inside the single layer $|F|$, respectively: $H * |F| = const.$

Additionally, the sequence of layers and the field strength have to be applied, whereas one value influences the other, respectively. The coaction of the influencing values is displayed in Figure 15.3.

The calculation of the internal characteristic inside the layer system has to respect the influence of the refractive index in the stack. Therefore, the energy flux E has to be applied for the calculation of the threshold characteristic in the stack.

Figure 15.4 displays an example for a quarter-wave stack (Quarter Wave optical Thickness [QWoT]) made from TiO_2/SiO_2 coated with ion beam sputtering (IBS). The LIDT of the component is given by the lowest internal LIDT, which is observed exactly at the interface between the first and second layers on the TiO_2 side.

Although the simplest possible design of a high-reflecting mirror was used, Figure 15.5 demonstrates the complex behavior inside the stack. In the next section, the presented algorithm is applied to estimate the LIDT of more complex components. Previously, the applied material system is expanded from the binary to the ternary oxides. Thereby, a much higher flexibility can be achieved. This flexibility is used for the design of components with a higher power-handling capability.

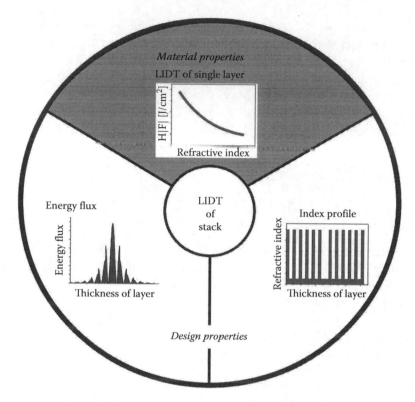

FIGURE 15.3 Concept for the calculation of laser power resistance of USP optics.

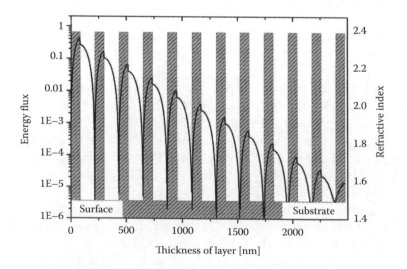

FIGURE 15.4 Energy flux in a standard QWoT stack.

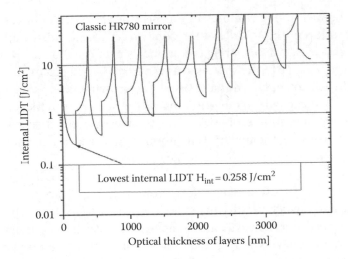

FIGURE 15.5   Calculated internal threshold level of a HL stack form TiO_2/SiO_2. (From Jupe, M. et al., *Proc. SPIE*, 6720, 67200U, 2008.)

15.4   Application of Ternary Oxides in USP Optical Components

Typically, the optic manufacturing of USP optical components places the highest technical demands on the precision and homogeneity of the coating process. The manufacturer apply established sputter processes (ion beam sputtering and magnetron sputtering), and also well-established materials are used. In the current state of the art, the dielectric optics are produced on the basis of binary oxides, but the continuous growth of the requirements in the spectral probabilities and in the power handling capability leads to an intense search for alternative materials or material classes. The close link between material and damage properties leads to the situation that the choice of materials is often the only way for an improvement of the LIDT of USP optics. Especially for these applications, ternary compounds establish a new dimension for the design of USP optics. The main advantage is reasoned to be the possibility of creating an unlimited number of materials with optical and electronic properties in between both, the pure and binary materials. This advantage leads to a rapid growth of investigations in properties of ternary oxides (Nguyen et al. 2008, Jensen et al. 2010, Melninkaitis et al. 2011, Mangote et al. 2012).

These ternary oxides can be manufactured in the ion beam cosputter procedure, which is a modified ion beam sputter (IBS) process. The noble gas ions sputter the atoms of a metal or a metal oxide target toward the samples in a rotating calotte where the oxide coating is formed by a condensation process in a reactive gas atmosphere. The main modification of the modified IBS process in comparison to the standard IBS coating plant is the construction of the target mount (Lappschies et al. 2004a, 2005, Jupe et al. 2007). In the presented setup, the two spatially separated targets were replaced by a zone target mount. This zone target is divided into two areas, silicon for the low-refracting component on one side and a dielectric metal part (titanium, niobium, tantalum, hafnium, zirconium, scandium, or aluminum or their oxides, respectively) for the formation of the high-index component on the other side (Mangote et al. 2012). Both targets join seamless in the middle of the zone target. Using this setup, the content of the respective materials can be varied continuously between 0% and 100% by vertically moving the target relative to the sputtered area of the focused plasma beam. The respective content of the two sputtered materials in the growing layer is well defined by the target position. A homogeneous distribution of the sample is achieved by placing the rotating calotte at an optimized position in the superpositioned sputter distributions. Because of the mixture of the materials in atomic scale, each single alloy is a new material with clear-defined optical and electronic properties.

Typically, the IBS coating process allows the manufacturing of designs of the highest complexity over a large spectral range. This property is important for the realization of the complex designs, which are discussed later in this section. In the presented experiments, a broad band monitor system is employed to measure complete transmission spectra of the growing layers during each revolution of the substrate holder. From the spectra, the actual thickness values and the deposition rate of the growing layer are calculated, and in the end of each layer, the spectrum is saved. These measured spectra can be used for a recalculation of the design and an error handling.

However, the technique is able to create the material in the required quality. Figure 15.6 shows the shift of the absorption gap of $Ti_xSi_{1-x}O_2$ layers. Obviously, the material gap shifts with an increasing content of silica to higher energies. Consequently, the material threshold also increases with an increasing content of silica.

It is more practical for the calculation to apply the refractive index instead of the material gap as an independent parameter. This choice is necessary with respect to state-of-the-art thin film calculation and optimization procedures. In this way, the damage properties of mixture materials simply link to the design by the refractive index. Figure 15.7 shows the normalized damage behavior of $Ti_xSi_{1-x}O_2$ single layers versus index of refraction, respectively.

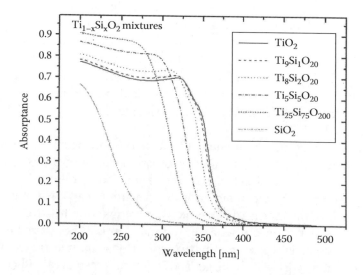

FIGURE 15.6   Spectral loss curves of $Ti_xSi_{(10-x)}O_{20}$ ($0 \leq x \leq 10$). The absorption gap is shifting as a function of the content of silica.

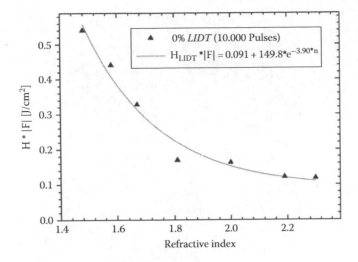

FIGURE 15.7 LIDT of single layers versus refractive index. (From Jupe, M. et al., *Proc. SPIE*, 6720, 67200U, 2008.)

On the basis of this measurement, the theoretic discussion is applied for selected optical components and the calculated thresholds are compared with experimental values. Starting from a simple design of a two-layer antireflecting (AR) coating, the complexity of the design will be increased to a gradual high reflector.

15.5 USP Damage Properties of Antireflecting Optics

Antireflecting optics are suitable examples for the demonstration of the optimization of the power-handling capability of thin film optics by oxide mixtures. In this example, V-coatings are of the main interest. These optical layers are specified to reduce the reflectance in a small wavelength range around a central wavelength. Therefore, these kinds of coatings are often called single-wavelength AR coatings, which is not fully correct, because the components typically can also be applied for USP application with a significant spectral range of the laser irradiation.

The simplest design can be achieved by a single layer of a material with a lower refractive index than the substrate material. In the past, this technique was often used for the AR coating of magnesium fluoride on YAG crystals or on nonlinear crystals for frequency conversion processes. The application of single layers on low refractive index materials is limited by the contrast of the refractive index between the materials. For instance, for a quartz substrate, follows from the amplitude condition (for an AR coating is $n_1 = \sqrt{n_0 n_s}$ with the refractive indices n_0 of the ambient medium, n_1 of the single layer, and n_s of the bulk) a maximum index of refraction n_{layer} of app. 1.2 at 800 nm of the required single-layer, which is not available. A common applied material is MgF_2 with $n_{layer} = 1.37$, and it leads to a useful AR coating for a material with a refractive index higher than $n_s = 1.88$. Obviously, the single-layer AR coating is a useful alternative only for laser components of a high refractive index of substrates or crystals. With respect to the rules of the USP damage, the material of the single-layer AR coating is typically not the most sensitive part of the component, because the refractive index of the substrate

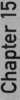

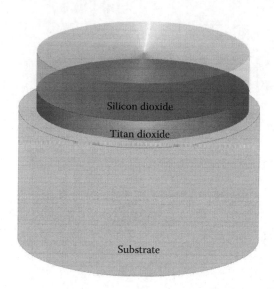

FIGURE 15.8 Schematic of the AR coating with two layers.

or bulk material is higher than the index of the layer, and consequently, the material gap of the layer material is significantly higher than the material of the substrate.

The flexibility of AR coatings can be expanded with an increasing number of layers. Mostly applied in single wavelength, AR layer system is a coating with two thin films, whereas a high refractive index layer (here titania) is covered by the substrate and a low refractive index layer (typically SiO_2). A schematic draw of the design is displayed in Figure 15.8. The reflectance of a two-layer design can be calculated by

$$R = \frac{n_0}{4n_s}\left(\left(S_1\varphi_1 + S_2\varphi_2\right)^2 + \left(S_3\varphi_3 + S_4\varphi_4\right)^2\right) \tag{15.1}$$

$$\partial_n = 2\pi n_n \frac{d_n}{\lambda}$$

$$S_1 = \frac{n_s}{n_0} - 1 \quad S_2 = \frac{n_1}{n_2} - \frac{n_s n_2}{n_0 n_1} \quad S_3 = \frac{n_s}{n_1} - \frac{n_1}{n_0} \quad S_4 = \frac{n_s}{n_2} - \frac{n_2}{n_0}$$

$$\varphi_1 = \cos(\partial_1)\cos(\partial_2) \quad \varphi_2 = \sin(\partial_1)\sin(\partial_2) \quad \varphi_3 = \sin(\partial_1)\cos(\partial_2) \quad \varphi_4 = \cos(\partial_1)\sin(\partial_2)$$

In these equations are n_0 the index of the surrounding medium, n_s the refractive index of the substrate, and $n_{1,2}$ the index of refraction of the layers, respectively. Equation 15.1 is a typical eigenvalue problem, and consequently, an infinitesimal number of solutions are available for each combination of a fixed substrate, fixed materials, and a fixed ambient medium. Nevertheless for the applications, only those two solutions with the minimum layer thicknesses are of practical interest, and usually, the design with the lower thickness of the high refractive index layer is applied. The minimum of reflectance can be determined, numerically.

As mentioned, the application of ternary oxide mixture allows varying the refractive index of the high refractive index material, continuously. Consequently, in contrast to binary oxides, a variation of the refractive index as well as the material gap is possible. For the optimization of the double-layer AR coating, the material composition of the high refractive index layer can be changed. The optimized threshold is expected for the material with the lowest refractive index, but the coating has to achieve a negligible reflectance. The solutions of the task are published by Macleod (1986). Figure 15.9 displays the results of the numerical calculation for double-layer AR coatings on different substrate materials. In this plot, the optical thickness of the low refractive index layer (silica) is displayed in dashed lines and optical thickness of the high refractive index layer (variable index) is drawn as a solid line. The following optimization is reduced to the material properties and can be focused on the optimization of the refractive index n_1. For a minimum of the refractive index contrast between n_1 and n_2, the solution is a reduced amplitude condition: The phase condition can be neglected, and the design becomes a QWoT structure. In this case, Equation 15.1 is strongly simplified, and the reflectance is given by

$$R = \frac{n_0}{4n_s}\left(\frac{n_1}{n_2} - \frac{n_s n_2}{n_0 n_1}\right)^2 \tag{15.2}$$

and the optimal refractive index of n_1 follows for $R \rightarrow 0$ is

$$n_1^{opt} = \frac{\sqrt{n_0 n_s}}{n_0}n_2 \tag{15.3}$$

An additional reduction of the refractive index of n_1 leads to an increase in the reflectance of the AR coating. Figure 15.10 displays the reflectance increase with a decreasing

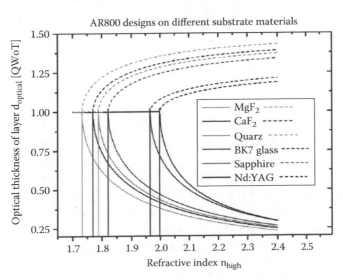

FIGURE 15.9 Numerical solution of double-layer AR coatings for different refractive indices for the high refractive index layer on different substrate materials (solid lines: thickness of the high material layer and dashed line: thickness of the high material layer). The minimal refractive index of the high layer for the different substrate materials is marked by the perpendicular solid lines. (From Jupe, M. et al., *Proc. SPIE*, 6403, 64031A, 2006.)

Chapter 15

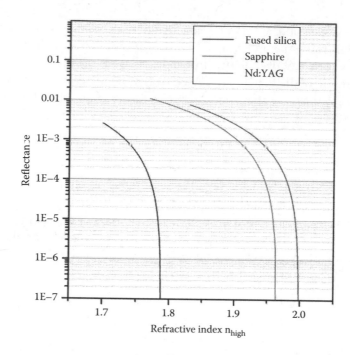

FIGURE 15.10 Theoretical reflectance at 800 nm of the double-layer AR coating on quartz, sapphire, and Nd:YAG substrate in the range of the refractive index contrast below the critical contrast.

contrast of n_1 and n_2. In the QWoT range, the reflectance can be calculated in accordance to Equation 15.2. In Figure 15.11, both the theoretical and experimental transmission and the theoretical reflectance of AR coatings are shown. Obviously, the reduction of the n_2 does not limit the optical performance of the coating (Figure 15.11).

Figure 15.12 shows the calculation of the internal damage threshold of the classic AR design and the AR-coating including the optimized ternary composite in the high refractive index layer. The characteristic damage curve of the corresponding LIDT measurement in Figure 15.13 is in perfect agreement with the calculated data. There is an increase in LIDT from 0.145 J/cm² to 0.495 J/cm², which corresponds to an increase of 340%.

In the end of the discussion of AR optics, it has to be clarified if an increase in the design complexity by additional layers will result in a further growth in LIDT. To achieve these objectives, a further decrease in n_1 is necessary. Therefore, the design has to compensate the loss of refractive index contrast. Table 15.2 shows the resulting Fresnel losses of a design study applying the refractive indices $n_{high} = 1.6$ for the high refractive index layers, $n_{low} = 1.48$ for the low refractive index layers on fused silica substrate $n_{sub} = 1.45$. Within these requirements, six layers are needed to achieve a negligible reflectance. Figure 15.14 shows the theoretical, internal LIDT of a double-layer AR coating and a six-layer AR coating. The double layer achieves the higher LIDT of about 1.4 J/cm², but the coating exhibits the Fresnel losses of about 1%, and is not suitable as an AR coating. In contrast to the double-layer AR coating, the six-layer AR coating meets the specifications and the theoretical threshold with 1.16 J/cm² is significantly higher than the optimized double-layer AR coating. For the application, it is arguable if it is worthwhile to further optimize the AR coating when the LIDT of the substrate is smaller than the LIDT of the

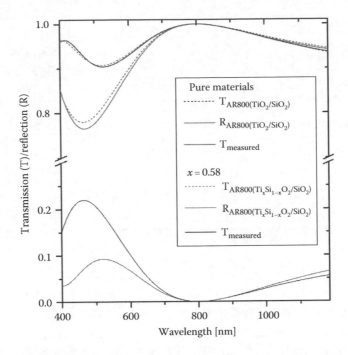

FIGURE 15.11 Theoretical reflectance and theoretical and experimental transmissions of the double-layer AR coating on quartz substrate.

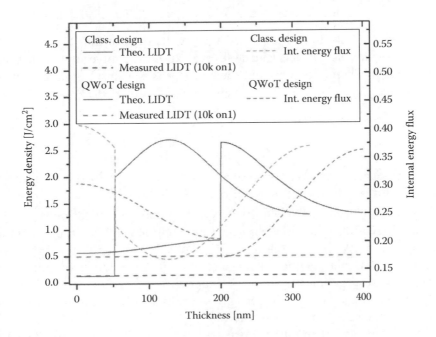

FIGURE 15.12 Internal energy flux, the resulting internal LIDT, and the experimental LIDT level of the optics of a double-layer AR coating.

Chapter 15

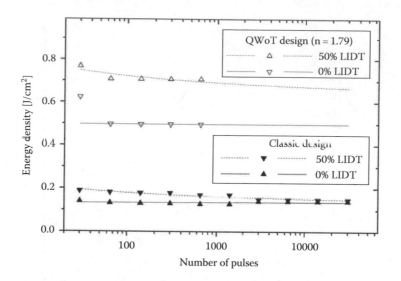

FIGURE 15.13   Characteristic damage plots of double-layer AR coatings in the QWoT design (red) and in the classic design (blue). (From Jupe, M. et al., *Proc. SPIE*, 6720, 67200U, 2008.)

Table 15.2   Fresnel Losses Dependent on the Number of Layers

Number of Layers	Fresnel Losses
2	1,23 E-02
4	1,11 E-03
6	0

coating. Therefore, it has to be mentioned that the damage of the substrate is strongly influenced by self-focusing effects (Turowski et al. 2009). A simple approximation shows that the critical irradiation power can be achieved, easily, using USP laser sources. The critical power P_{crit} for a Gaussian beam is given by Marburger (1971):

$$P_{crit} = \frac{3.77\lambda^2}{8\pi n_0 \hat{n}_2} \tag{15.4}$$

The most high-quality optics apply fused silica as substrates, which have a linear index of refraction of $n_0 = 1.45$ and a nonlinear refractive index of $\hat{n}_2 = 3 \times 10^{-16} \left[cm^2/W \right]$ for a wavelength $\lambda = 800$ nm (Brodeur 1999). Consequently, a critical power of 2.2 MW results from these parameters. The peak power of commercial USP lasers is in the range of some giga Watt (GW), and therefore, self-focusing plays a key role for high-power USP laser applications. Generally, the influence of the nonlinear effects on optics increases with increasing propagation length in the material. Self-focusing can lead to a collapsing beam. As a consequence, the fluency grows dramatically and transparent optics are mainly damaged on the beam exit surface.

Usually, in the case of moderate focused beams, additional nonlinear effects are observed before the optical breakdown occurs. These effects lead to a loss of pulse energy. In the case of the supercontinuum generation, the laser irradiation is converted into broad spectrum. This effect can be used for the generation of short pulses, but in

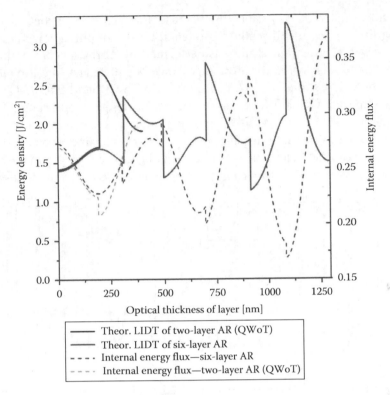

FIGURE 15.14 Comparison of internal energy flux, theoretical internal LIDT of a two and a six layer AR design.

most optical components, the effect is unwanted. Also the nonlinear absorption leads to a loss in pulse energy, but the main critical effect is the self-focusing itself, because the focused beam can also damage other components in the beam line after the focusing component. Generally, it is recommended to minimize the thickness of transparent optics in the USP range.

15.6   USP Damage Properties of High-Reflecting Optics

The second types of discussed components are high-reflecting optics. The main advantage in investigation of reflectors is the negligibility of substrate influences during the LIDT measurement in the subpicosecond range. Generally, there are degrees of freedom for the optimization of high reflectors. On the one hand, the design can be optimized, and on the other hand, the material combination can be improved. Both possibilities are discussed separately, but the optimization strategies can also be combined. In the first case, a minimized energy flux in the high refractive index material has to be achieved. It should be mentioned that the energy flux in the complete high refractive index layer has to be minimized, and it is not suitable to minimize the energy flux at the interfaces, which is often discussed for short pulse optics. In the USP regime, the material properties dominate the damage behavior and not the interface defects. However, the design optimization is a well-known and applied technique. The main advantage is the possibility to optimize the properties using the pure materials only, and every combination of high and low refractive index materials is possible. In response, the flexibility of the design increases by applying this strategy. This method is often applied, but only a few publications are known. However, a systematic study was performed by Abromavicius et al. (2007) and Melnikaitis (2009), displaying the advantages and the challenges that are related to the method. In the study, standard QWoT designs from Ta_2O_5/SiO_2, HfO_2/SiO_2, and ZrO_2/SiO_2 were compared with a field strength optimized design, respectively. The samples were manufactured in an ion-assisted deposition process, and the thickness control was a quartz crystal deposition monitor. Figure 15.15 shows the same results of the study for Ta_2O_5/SiO_2 and HfO_2/SiO_2 designs. The reduction of the field strength in the high refractive index layers was achieved by a change of the layer thickness in the first seven layers.

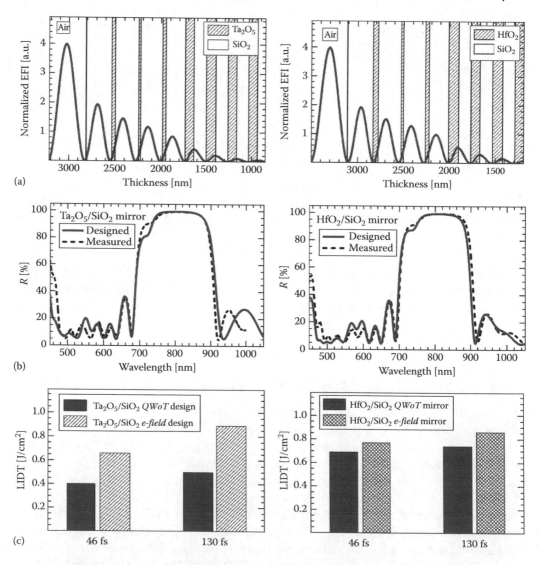

FIGURE 15.15 Design, spectral behavior, and resulting LIDT of Ta_2O_5 on the column b /SiO_2 on the column a stacks. (a) Field distribution of the optimized design, (b) reflection of the design, respectively, and (c) resulting LIDTs. (From Melninkaitis, A., Optical resistance of dielectric coatings to multi-pulse femtosecond laser radiation, PhD thesis, Vilnius University, Vilnius, Lithuania, 2009.)

Figure 15.15a and b shows that the field strength in the red hatched high refractive index layers is significantly lower than in the low refractive index layer in between. In a classic QWoT stack, the maximum field strength has been observed at the interface between the high and low refractive index layers, which is demonstrated in Figure 15.4.

The most significant effort in this study was demonstrated by the optimization of Ta_2O_5/SiO_2 coatings. In the study, the variation of the design leads to an increase in LIDT of 65%–80%. For the material combination of HfO_2/SiO_2, the optimization leads to a gain of 12%–16%. In ZrO_2/SiO_2 coatings, a design failure in layer thickness led to a change in the field strength distribution and no improvement could be observed in LIDT. Small differences between the target design and realized coatings can lead to a rapid decrease in LIDT. Usually, the discrepancies can be observed in the spectral behavior. Also the designs of the Ta_2O_5/SiO_2 and HfO_2/SiO_2 coatings display small variations between target and realized spectra in Figure 15.15. In these cases, the differences do not change the stabilization effects significantly.

From the application of mixed materials, a higher flexibility results. The strategy of the material variations applying titania/silica mixtures and QWoT designs are discussed. Initially, Figure 15.16 shows the characteristic damage curve (30.000on1) of HR780 mirrors applying pure TiO_2 or a negligible SiO_2 concentration in the high refractive index layers of the stack. Coatings of TiO_2/SiO_2 exhibit a small damage threshold, which is in conjunction to the experience of the fundamental research chapter. The calculated value (see Figure 15.5) confirms the experimental data. Therefore, titania/silica components are usually used when a very high contrast of the low refractive index and the high refractive index is needed, but this is the typical case in USP applications. However, for most of the USP applications, a significantly higher threshold is desired. The stabilization can be achieved by targeted variation of the material content of the high refractive index material. The simplest approach is the increase in the silicon content in the titania layers, but a loss of refractive index contrast and a loss of reflectivity per high/low (HL) pair have to be compensated. Consequently, a higher number of layers are necessary. Additionally, the spectral width of the reflectance band decreases and the range of applications of the component is limited. A significant improvement follows, if the internal

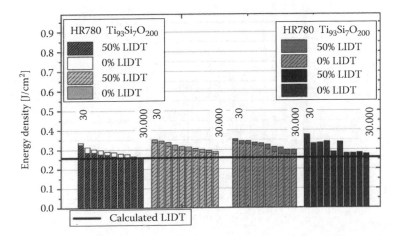

FIGURE 15.16 Measured damage threshold of a classical HL stack. (From Jupe, M. et al., *Proc. SPIE*, 6720, 67200U, 2008.)

Chapter 15

field distribution also is taken into account. In the stabilization concept, the material of the layers, which may be affected by the highest field strength, has been stabilized by applying a high content of silica. In such a high reflector, the maximum of field strength is located at the interface between the first and second layers and decreases step by step from layer pair to layer pair, which is displayed in Figure 15.4.

Compensating the field strength in a classical high-reflecting stack from the outer surface, the system begins with the high refractive index material followed by the low refractive material. Applying the optimized concept, the content of silica, which is considered to be the material with the higher power-handling capability, has to be high in the layers exposed to high field strength. Consequently, from the point of view of the coating manufacturer, the **Refractive Index StEps Down** (RISED concept) from the substrate to the surface. Therefore, the design of the RISED mirror starts with a high content of silica in the high refractive index layer, and the silica concentration steps down with respect to the field strength in the depth of the layer structure.

Initially, this design was developed for high-reflecting mirrors, and therefore, the general idea is described for a high-reflecting mirror QWoT system in the following. The RISED design of the $Ti_{1-x}Si_xO_2/SiO_2$ mirror is depicted in black bars in Figure 15.17. In addition, the designs of a pure TiO_2/SiO_2 high reflector and a $Ti_3Si_7O_{20}/SiO_2$ are compared. In Figure 15.6, all HR780 designs are QWoT stacks, and, with respect to the lower refractive index, the number of layers and thus the physical thickness are changed to achieve a comparable reflectance of the mirror. The maximum content of titanium oxide in the high refractive index layer of the presented RISED mirror is 85%, and the lowest content is 5% at the surface of the layer structure. The spectral reflectances of a standard titanium oxide HR780, the mixture HR780 $Ti_3Si_7O_{20}/SiO_2$, and the RISED

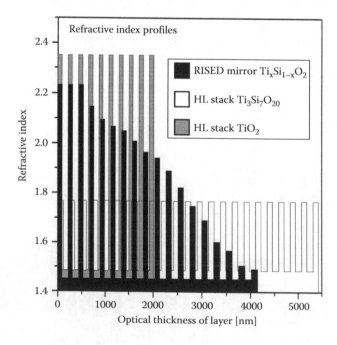

FIGURE 15.17 Refractive index profiles of standard HR780 TiO_2/SiO_2, a HR780 $Ti_3Si_7O_{20}/SiO_2$, and HR780 RISED concept. (From Jupe, M. et al., *Proc. SPIE*, 6720, 67200U, 2008.)

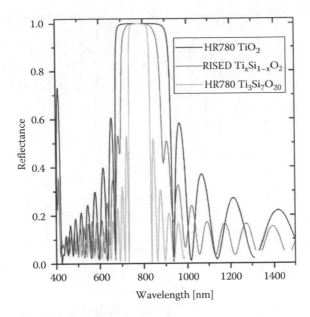

FIGURE 15.18   Spectral curves of standard HR800 as TiO_2/SiO_2, a HR780 as $Ti_3Si_7O_{20}/SiO_2$, and as RISED, respectively. (From Jupe, M. et al., *Proc. SPIE*, 6720, 67200U, 2008.)

mirror are compared in Figure 15.18. The reflectivity of all mirrors is above 99.9%, and the spectral band width is sufficient for a large number of USP lasers.

The first experimental results of an empirically designed RISED mirror are shown in Figure 15.19 (Jupe et al. 2006, 2008). The thresholds of stabilized components are higher than those of a classic component by more than a factor of two higher.

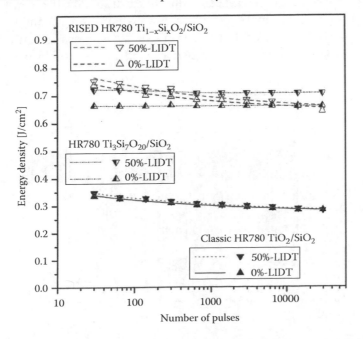

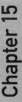

FIGURE 15.19   Characteristic damage curves of a classical TiO_2/SiO_2 mirror, a mixed HR with $Ti_3Si_7O_{20}/SiO_2$, and a RISED mirror $Ti_xSi_{1-x}O_2/SiO_2$.

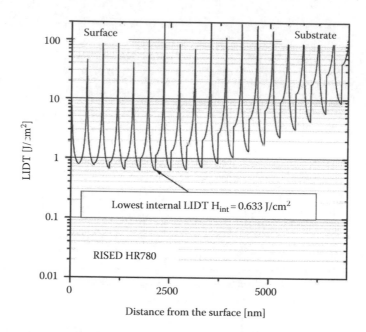

FIGURE 15.20 Internal LIDT of an empirically designed RISED mirror. (From Jupe, M. et al., *Proc. SPIE*, 6720, 67200U, 2008.)

The discrepancy between components applying a constant silica content of 70% in the high refractive index layers and the RISED mirror can be neglected. An advantage of the RISED components is the significant lower thickness of the stack, and consequently, the cost of production has been reduced. Additionally, the RISED design offers the possibility of further improvement in power-handling capability. Figure 15.20 displays the internal LIDT inside the RISED stack, whereas the minimum threshold is located in layer no. 11 from the surface and the threshold is calculated to be

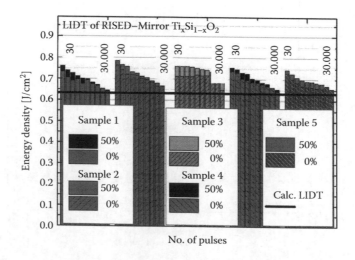

FIGURE 15.21 Measured damage threshold of a RISED mirror and theoretic threshold of 10,000 on 1 LIDT. (From Jupe, M. et al., *Proc. SPIE*, 6720, 67200U, 2008.)

0.633 J/cm². Obviously, the design shifts the position of the highest sensitivity from the layer no. 1 into the stack by stabilizing the layers close to the surface and, as expected, the threshold of the component increases. In Figure 15.21, the comparison of the measured and calculated values confirms a perfect match, and therefore, an optimization of the design can be performed, following.

15.7 Optimization of Power-Handling Capability of RISED Stacks

The optimization of the stack uses the calculation of the damage threshold. The algorithm is equivalent to the design optimization of thin film stacks with respect to the optical or phase properties. The main problem of the optimization is that the parameters depend on each other, respectively. Thus, changing one parameter affects the other parameters. For demonstration, a small example is described. The start design is a classic high low stack. For the stabilization, the silica content in the first layer is increased and consequently, the index of refraction decreases. Therefore, the energy flux in the layer increases, and a greater strain is also placed on the following layers, resulting in a decrease in the internal LIDT in the respective layer. This one has to be stabilized by a change of the content of silica in high refractive index layers, additionally, and consequently, the design has to be changed. A schematic drawing of the procedure is displayed in Figure 15.22. In other words, a design is defined, the most sensitive layer is determined, the respective layer is stabilized, and the internal LIDT is recalculated. As a consequence of this procedure, the energy flux penetrates more and more to the layer stack and the threshold converges more and more to a stable level, but finally, the design

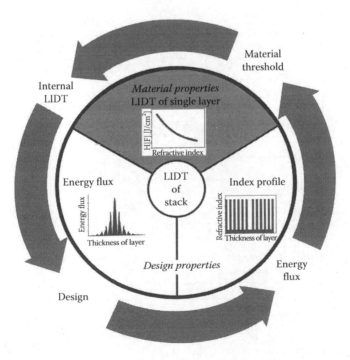

FIGURE 15.22 Concept for the optimization of laser power resistance of USP optics.

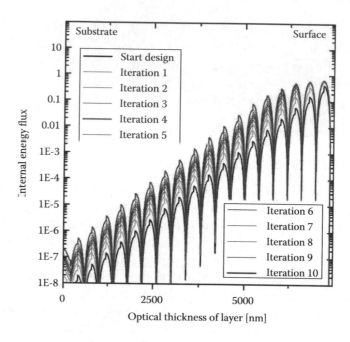

FIGURE 15.23   Penetration of energy flux into the stack after different RISED optimization iterations.

converts to a stable state. The penetration of the energy flux into the stack from iteration to iteration step is displayed in Figure 15.23. In this case, the start design is a classical HR780 applying pure TiO_2 and SiO_2. The resulting threshold of the component is 0.74 J/cm^2 according to the minimum principle. Figure 15.24 shows the calculated internal LIDT in the stack.

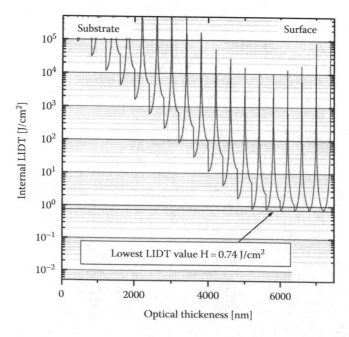

FIGURE 15.24   Theoretical LIDT of an optimized RISED design (start design: classic HR780).

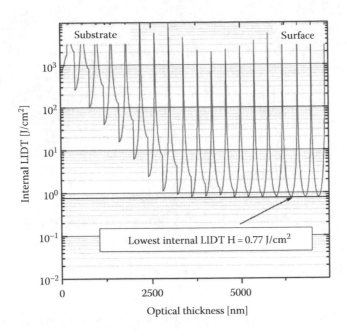

FIGURE 15.25 Theoretical LIDT of an optimized RISED design (start design: empirical RISED).

The LIDT of the manufactured mirror was measured with $H_{0\%}^{10.000on1} = 0.74\,J/cm^2$ and $H_{50\%}^{10.000on1} = 0.76\,J/cm^2$ for the onset and for 50% damage probability of the 12.000on1 LIDT measurement, respectively. The spectral characteristic of the mirror corresponds perfectly to the classic HR780 from pure materials. In detail, there is no reduction in the spectral width of the mirror, but an increase in LIDT of about 0.5 J/cm².

However, it is a well-known fact that the result of a thin film optimization can be strongly dependent on the start design. Therefore, the empirically designed RISED mirror was applied as start design, which results in an additional small increase in the LIDT (see Figure 15.25). More interesting than the LIDT of the component is the way of the waveform development of the internal LIDT. As a result of the optimization, the single minimum in the internal LIDT of the mirror does not exist anymore; rather, a localized minimum in internal LIDT after each layer has developed. Further on in a perfect design, all minima are meeting each other on the same level. Consequently, the optimum LIDT of the QWoT RISED component is limited by the power-handling capability of the first high refractive index layer, and this threshold is always lower than the threshold of the pure silica. A further improvement should be achieved, if the curve progression inside the layer is respected. This consideration leads to a gradual design finally, which is discussed in Section 15.8. Nevertheless, in Figure 15.26, the characteristic damage curves of the QWoT stacks are plotted in direct comparison.

15.8 Optimization of Power-Handling Capability Applying Rugate Designs

The canonical further development of the design can be achieved, if the design differs from the QWoT design. In this section, a study applying binary oxides in the non-QWoT configuration is discussed. The use of ternary mixtures in the coating also offers a new

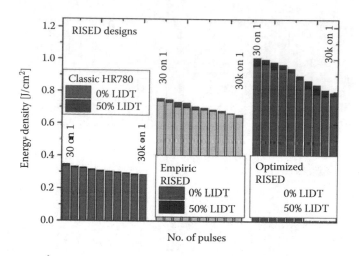

FIGURE 15.26 Improvements in LIDT applying the RISED concept with and without optimization.

dimension of design strategy in the case of non-QWoT stacks. The perfect implementation can be achieved by a 100% consideration of the stack design with respect to the internal LIDT.

Figure 15.4 displays a continuous change of the energy flux inside the stack. Consequently, a gradual system, or a rugate filter is the best application for a mirror with the highest power-handling capability. The manufacturing of rugate components was demonstrated by applying different techniques (Elder et al. 1991, Berger et al. 1997, Lappschies et al. 2005, Lorenzo et al. 2005, Jun-Chao et al. 2012). The IBS-Co-Sputter procedure has proven high thresholds for mirrors of nanosecond applications (Lappschies et al. 2005). In the case of USP applications, improvement could be achieved, reasoned by the influence of the material and by avoiding abrupt changes of the material. In fact, the whole layer structure was manufactured in one step without any interruption of the film growth. The index profile has been achieved by a gradual change of the material concentration during the material deposition. This special design is optimized for the requirements of optical components for the nanosecond regime, and obviously, the design is not optimized for the needs in USP optics. Consequently, the investigation can show the difference of the design requirements in nanosecond versus USP optics.

However, Figures 15.27 and 15.28 display the profile of the refractive index and the spectral behavior of a rugate notch filter manufactured by $Ti_xSi_{1-x}O_2$ mixture. Primary, the design was developed by generating a perfect band pass filter without sidebands and higher-order reflectance bands. The spectral characteristic in Figure 15.28 shows the successful result for the calculated design as well as for the manufactured filter. Hereby, the index profile of the filter displays that the requirements for a mirror with a high power-handling capability are realized in a natural way. The index of refraction decreases from the layers in the center of the coating to the surface, and this will stabilize the mirror. Additionally, the complete coating is manufactured in a single step without any interruption during the coating process and consequently, no interfaces exist inside the layer. It means the rugate filter is a single-layer component applying a

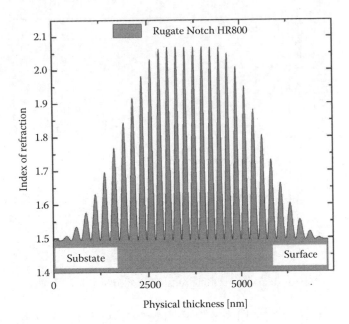

FIGURE 15.27   Index profile of a rugate notch filter.

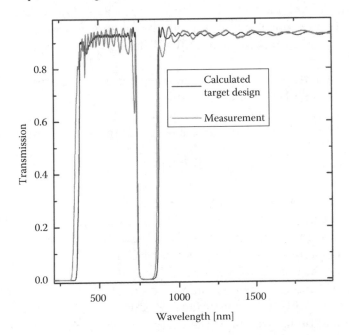

FIGURE 15.28   Spectral transmission of a rugate notch filter HR800.

changing index of refracting. These gradual layer structures exhibit no interfaces, which reduce the probability of the inclusion of absorbing defects in the alternating refractive index layer. Consequently, a high damage threshold was proven in the thermal induced damage regime (Jupe et al. 2007). The interaction time of pulse and material was calculated to be longer than 10 ps (Stuart et al. 1996). Consequently, this fact is not dominating in the discussed applications.

Chapter 15

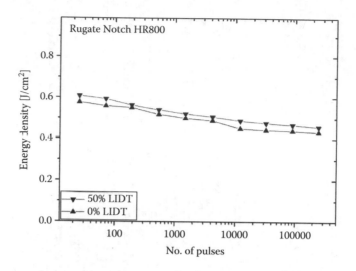

FIGURE 15.29   Characteristic damage curve of a rugate notch filter.

However, the rugate notch filter corresponds to the empirical RISED applying a gradual layer. The damage threshold of the component is displayed in Figure 15.29. The LIDT is significantly higher than the conventional TiO_2/SiO_2 stack, but the threshold of the RISED design is not achieved. It has to be mentioned that the periodicity of the design structure is still approximately a QWoT, but it is no longer expedient to call the design QWoT design.

The possibility of the stack optimization using a gradual design has to be discussed. In contrast to the RISED optimization, with respect to the distribution of the energy flux in the respective layer, the single layer is divided into thin layers, and the content of silica in the high refractive index layers is tuned. In the first studies, the discretization of the single layer is not offering a fine resolution. Nevertheless, the characteristic of the optimization can be discussed on the basis of the calculations and experiments (see Figure 15.30).

Obviously, the resulting design of the gradual optimization varies significantly in the rugate notch design, and the gradual RISED is more similar to the field-strength-optimized designs using pure materials. Consequently, the requirements of the layer thickness control monitor of the coating procedure are similar to, or higher than in, the manufacturing of the non-QWoT designs.

However, in a theoretical approximation, Figure 15.31 shows the potential of a further improvement in LIDT for the gradual RISED design, and a finer discretization should lead to additional growth, if the design can be realized in a coating process. The recommended design was manufactured experimentally, but a thickness failure has affected a lower threshold. Figure 15.32 displays the resulting characteristic damage curve. Nevertheless, the online monitor saved the spectral behavior of all layers and allowed the recalculation of the coated design. This approximation reproduced the measured LIDT perfectly. Figure 15.33 shows the internal threshold of the retracted filter.

According to the current state of the art, the gradual RISED designs promise to be the best coating designs of the highest LIDT. The manufacture places extremely high demands on precision of the coating process, but actual coating plants with high

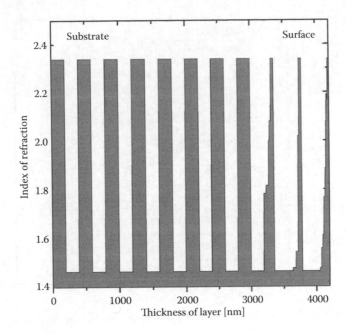

FIGURE 15.30   Index profile of gradual RISED.

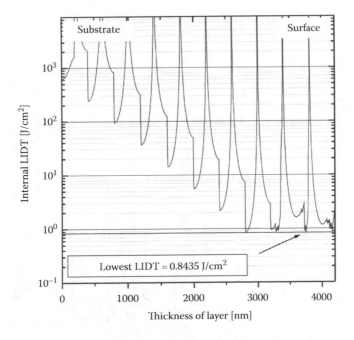

FIGURE 15.31   Theoretical LIDT of an optimized gradual RISED design.

developed monitoring systems can achieve the needed accuracy. The main question for the coating manufacturer will be the acceptance of the coating threshold compared to other demands on optical coatings, phase properties, or losses. Therefore, it is to be assumed that the main requirements on the damage threshold will be focused on the phase-sensitive optics in future, which is briefly discussed in Section 15.9.

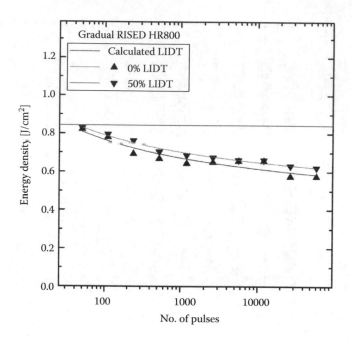

FIGURE 15.32   Characteristic damage curve of a gradual RISED filter.

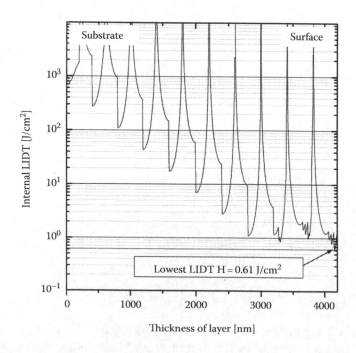

FIGURE 15.33   Threshold of the recalculated design.

15.9 Pulse Shaping Optics

Generating ultrashort laser pulses is a complex matter, and therefore, a broad spectrum of optics has been developed. Following, only two important examples will be discussed.

The highest peak power intensities of the USP application are realized by the chirped pulse amplification techniques, which are based on compression and stretching of laser pulses with well-defined phase properties. To realize stretches and compressors, different setups are available. Primarily, the concepts differ in the choice of the phase-dispersive elements. One the one hand, devices are applied by using the geometric shape of the dispersive elements. Accordingly, there is a distinction between grating and prism compressors and stretchers, whereas combinations are also used. In the last decade, dielectric gratings were developed with a focus on the power-handling capability.

One the other hand, pulse forming mirrors are used in an ever-broadening field of application, and these components are discussed in the beginning. Especially for pulses much shorter than 100 fs, these components are essential for the application. Chirped mirrors allow compensating the positive chirp, whose pulses sustain during the transition through a positive dispersive medium. Additionally, the chirped mirrors are able to stretch or to compress laser pulses.

Figure 15.34 displays schematical drawings of diffractive and refractive pulse shaping components. In the case of the refractive chirped mirror, the chirp is generated by different penetration depths for different frequency components of the beam. Figure 15.34a demonstrates the process for three wavelengths. The total chirp is achieved by multiple reflections, for instance, in a double pass as shown in Figure 15.34b. The number of reflections defines the total chirp and consequently the pulse duration. In the setup, different well-defined chirred mirrors are often used, optimizing the total group delay dispersion (GDD) of the chirped mirrors by compensating unwanted GDD ranges in a single mirror (Pervak et al. 2009).

Realizing the chirp, the mirrors have to exhibit well-defined group delay dispersions and a 50% larger band width in comparison to standard reflectors, meaning that the requirements of design and production are much higher for chirped mirrors. Therefore, the key issue of investigation is the realization of chirped mirrors. The state of investigations of the phase dispersive components is at the beginning. Nevertheless, the demand of laser manufacturers for high laser irradiation resistance components leads to growing numbers of systematic investigations in LIDT.

The following example describes some major facts and challenges. Figure 15.35 displays the refractive index profile of a chirped mirror with negative GDD of approximately—500 fs^2 in a wavelength range from 735 to 845 nm. For the design, 89 layers with a total thickness of the stack of more than 10 μm were necessary, and the individual layer thicknesses of the included single layers were between 35 and 210 nm. Obviously, the design differs significantly from the QWoT—stacks, and demands the highest precision in the coating process. In contrast to the previous stacks, the spectral parts of the laser pulse are reflected in different penetration depths. Because of the negative chirp, the short wavelengths are reflected in the layers close to the surface and the longer wavelengths are reflected in the depth of the stack. Consequently, the pulse is separated in the stack and the field penetration is dependent on the wavelength (Figure 15.35).

Chapter 15

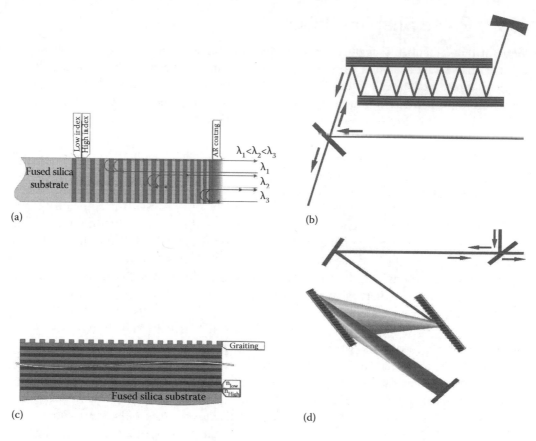

FIGURE 15.34 Schematic drawings of setups to pulse shaping components (a) internal structure of chirped mirror, (b) setup of a double-pass compressor applying a chirped mirror pair, (c) internal structure of a dielectric high-reflecting grating, and (d) setup of a double-pass compressor applying reflecting gratings.

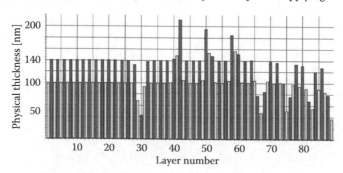

FIGURE 15.35 Refractive index profile of a chirped mirror. (Courtesy of Pervak Volodymyr.)

Figure 15.36 shows the distribution of field strength with respect to the wavelength and penetration depth. These internal separations lead to a complex behavior, which prohibits applying the calculation and optimization of the last sections. Additionally, the pulse forming leads to the situation that the pulse duration of the incoming pulse differs from the duration of the outgoing pulse. Concluding, according to the actual state of the art, the damage threshold of chirped mirrors has to be measured, and

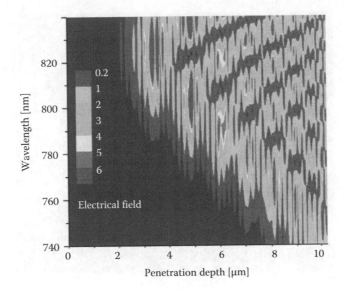

FIGURE 15.36 Distribution of field strength with respect to the wavelength and penetration depth. (Courtesy of Pervak Volodymyr.)

optimizations can be performed empirically. The presented mirror has been designed for a titanium–sapphire laser, and a pulse duration of 19.1 fs was achieved. The LIDT of the mirror was also measured in an Son1- procedure at a central wavelength of 790 nm and pulse duration of 30 fs (Angelov et al. 2011) with 0.3 J/cm². The currently published results suggest that the influence of the material can be transferred for the chirped mirror fabrication. Chirped mirrors with high gap materials in the high refractive index layers show a higher LIDT than the comparable materials with a lower gap in the respective layer. This was shown for mixed materials (Angelov et al. 2011).

The second possibility for pulse forming is using grating and/or prism compressors and stretchers. Hereby, the pulse is spectral separated, and the chirp is caused by the geometric setup of gratings and prisms (Strickland 1985, Diels and Rudolph 2006). The damage of prisms is extremely linked to self-focusing, filamentation, and supercontinuum generation. The propagation length through the prism changes for different wavelengths. Consequently, a discussion of prism damage is linked to the damage properties of the bulk material of the prism. Therefore, the following discussion is restricted to gratings.

However, generally, transmission and reflection gratings are established. Generally, the diffractive dispersion of the grating is achieved by a well-defined periodic structure with respect to the wavelength of the laser irradiation. Figure 15.37 shows a TEM picture of a grating structure on fused silica. This structure is manufactured by chemical or ion etching processes. Because of resulting effective index of refraction from this structure, an AR effect results on the substrate. For a transparent grating, the backside surface has to be AR coated for the respective incident angle or the grating structure can be optimized for a Brewster angle of substrate. Transparent gratings do not appear perfect for high-power applications. Because of the longer propagation length of the laser beam in the material, the mentioned nonlinear effects play a more important role in transparent gratings.

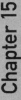

Chapter 15

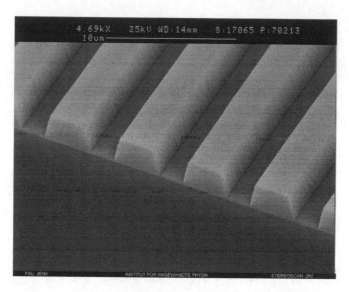

FIGURE 15.37 Electron microscopic image of a diffractive structure of a grating. (Courtesy of Ernst-Bernhard Kley and Marcel Schulze.)

However, the development of optical components with the highest power-handling capability is forced by large research organizations. Consequently, it is not surprising that numerous publications in LIDT optimization are published by the multiple petawatt laser projects. Over the last decade, a trend from megajoule lasers using nanosecond pulses to femtosecond kilojoule laser sources could be observed (Jovanovic et al. 2004, Martinez et al. 2005, Ditmire 2010). Though the pulse power of the actual system is in the same order of magnitude, the specifications of these lasers are impressive. In such facilities, pulse energies of a few kilojoules are reached and compressed by grating compressors. With respect to the megajoule laser facilities, the amplification of the pulses are stretched to a few nanoseconds, which does not put high demand on the LIDT of the components, but the pulse compressing plants place extremely high requirements on the power-handling capability of the compression gratings, where the pulse durations are reduced to few 100 fs. Figure 15.34d displays one possible setup of a compressor with two reflection gratings (Strickland and Mourou 1985). Dielectric reflection grating is the state of the art in the highest power application. The dimension of the grating has to be in a fabricable size with significantly more than 80×40 cm^2 (Britten et al. 2004, Jovanovic et al. 2004). Consequently, an LIDT of the grating has been required in the range of 3–5 J/cm^2. Caused by the requirement in power-handling capability, the traditional gold gratings are replaced by dielectric or hybrid components (Neauport et al. 2010). The damage threshold of gold gratings has been measured to be approximately 0.5–1 J/cm^2 at 1053 nm (Raze et al. 2007) and 0.5 ps, whereas dielectric grating concepts have achieved up to 5 J/cm^2 under the same conditions (Neauport and Bonod 2008b). Equivalent to the optimization of dielectric mirrors, the designs have to be optimized for achieving this goal.

In the case of the reflection grating, the structure is processed on top of a high-reflecting mirror. In reflecting gratings, only the field strength has to be respected. Possibly, this is one reason for the mainly use of multiple-layer dielectric (MLD) gratings in extreme power laser applications. Usually, the component is based on an HfO_2/SiO_2 stack (Neauport et al. 2007, 2010), but designs applying a TiO_2/SiO_2 in combination with HfO_2/SiO_2 (Dai et al. 2006), Ta_2O_5 (Britten et al. 2003), and first experiments applying ternary mixtures (NbTaO) (Martz et al. 2009) are also known. Usually, this stack is covered by an SiO_2 layer, which is structured. In the case of a chemical etching procedure, a significant influence on damage threshold of the chemical residues (Ashe et al. 2007) has been proven. The damage mechanism of the contaminants is not solved.

Generally, the influence of the field strength on the grating structure, as well as on the covered multiple layer stacks, has been demonstrated. The optimization of the field strength |F| of the component is usually combined with a loss in efficiency. Gratings can achieve efficiencies η of more than 99.5%, but in the setups, an efficiency of more than 95% is specified. In this context, the merit function can be defined:

$$MF = \frac{\eta}{|F|^2} \qquad (15.5)$$

For the optimization of the field in the design, the geometry of the groove structure has been varied. In detail, the duty cycle (ratio of width of ridge and the periodicity of the grating) and the depth of the grooves are changed. Exemplary, a microscopic cross-section image of a grating and a field strength distribution are displayed in Figures 15.37 and 15.38. However, in a study of CEA–PETAL (Neauport and Bonod 2008a), the field strength influence of four gratings with different groove structures on LIDT was measured under angles of incidence of 77.2° and 64.05°, whereby the field strength has been tuned. The plot of the measured corresponds to the scaling law of thin films (Jupe et al. 2007a). This fact indicates that the MLD gratings belong to the general context of

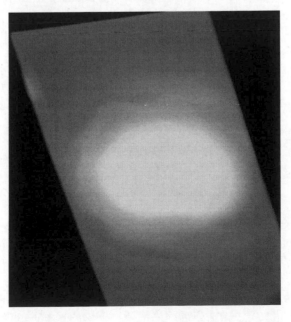

FIGURE 15.38 Supercontinuum after a transmission grating closed to the LIDT.

dielectric coatings with respect to the damage behavior. Of course, the damage behavior of transparent dielectric gratings and metal gratings will be different from the behavior of MLD gratings.

15.10 Summary and Outlook

USP optics are exposed to extreme burdens, and the laser damage fluences of the materials are one to two orders of magnitude smaller than the ones in the short pulse regime. In addition, the demands in spectral and pulse forming of the components are more complex than in the optic application, which are specified for longer pulses. Indeed, the damage process is driven by electronic processes in the solid, and therefore, it is strict deterministic. Because of this deterministic damage behavior, the LIDT of the components can be calculated and optimized. The calculation is based on the single-layer threshold, the design, and the energy flux inside the stack. Applying this algorithm, optimizations of very complex layer structures are possible, and the calculated and measured results correspond. For the optimization of layer stacks, binary and ternary materials can be applied, but ternary mixtures allow to choose the perfect material for a selected layer, and therefore to optimize the design. For simple layer AR coatings applying one or two layer designs, the perfect mixture can be calculated analytically. For complex designs, a numerical problem has to be solved. This optimization method is concluded in the RISED concept.

According to the current state of the art, optimizations of the LIDT were demonstrated mainly for silica–titania ternary oxides. Nevertheless, empirical studies show that the concept is also working for tantala and niobia. That is why the application of the RISED concept is still in development. Especially, gradual coating designs are expected to improve the LIDT of future components. However, the limitation of optimization will be defined by the low refractive index material, because this one is characterized by the highest material gap in the coating.

Equally to the layer stacks, the gratings can be optimized by minimizing the electric field strength in the stack, too. In case of the gratings, the optimizations have been performed by changing the groove structure. The investigation indicates that the experience from the dielectric layer can also be applied for MDL gratings. The usage of mixture materials in grating stacks is only known with respect to the reduction of stress-induced cracks.

The largest challenge is revealed in the case of chirped mirrors, since here the material dependency has been proven. Because of the complexity of the designs, the biggest potential for optimization is assumed in this field.

References

Abromavicius, G.; Buzelis, R.; Drazdys, R.; Melninkaitis, A., and Sirutkaitis, V. 2007. Influence of electric field distribution on laser induced damage threshold and morphology of high reflectance optical coatings. *Proceedings of the SPIE* 6720: 67200Y-1-8.

Angelov, I.; Trushin, S. et al. 2011. Investigation of the laser-induced damage of dispersive coatings. *Proceedings of SPIE* 8190: 1–6.

Ashe, B.; Giacofei, C.; Myhre, G., and Schmid, A. 2007. Optimizing a cleaning process for multilayer-dielectric- (MLD) diffraction gratings. *Proceedings of SPIE* 6720: 1–8.

Berger, M.; Arens-Fischer, R.; Thönissen, M. et al. 1997. Dielectric filters made of PS: Advanced performance by oxidation and new layer structures. *Thin Solid Films* 297: 237–240.

Britten, J.; Perry, M.; Shore, B.; Boyd, R.; Loomis, G., and Chow, R. 1995. High-efficiency dielectric multi-layer gratings optimized for manufacturability and laser damage threshold. *Proceedings of SPIE* 2714: 511–520.

Britten, J. A.; Molander, W. A.; Komashko, A. M., and Barty, C. P. 2003. Multilayer dielectric gratings for petawatt-class laser systems. *Proceedings of SPIE* 5273: 1–7.

Britten, J. A.; Nguyen, H. T.; JonesII, L. M. et al. 2004. First demonstration of a meter-scale multilayer dielec-tric reflection grating for high-energy petawatt-class lasers. *Optics Letters*: 1–16.

Brodeur, A.; Chin, S. L. 1999. Ultrafast white light continuum generation and self-focusing in transparent condensed media. *Optical Society of America* 16: 637–650.

Cheriaux, G.; Chambaret, J., and Canova, F. 2005. 100 TW ultra-intense femtosecond laser systems. *Proceedings of SPIE* 599: 1–13.

Dai, Y.; Liu, S.; He, H.; Shao, J.; Yi, K., and Fan, Z. 2006. Multilayer dielectric gratings for ultrashort pulse compressor. *Proceedings of SPIE* 6403: 1–9.

Diels, J. and Rudolph, W. 2006. *Ultrashort Laser Pulse Phenomena*. Amsterdam, the Netherlands: Academic Press.

Ditmire, T. 2010. High-power lasers. *American-Scientist* 98: 394–401.

Elder, M.; Jancaitis, K. S.; Milam, D., and Campbell, J. 1991. Optical characterization of damage resistant "kilolayer" rugate filters*. *Proceedings of the SPIE* 1441: 1–10.

Groß, T.; Lappschies, M.; Starke, K., and Ristau, D. 2001. Systematic errors in broadband optical monitoring. *Optical Interference Coatings, OSA Technical Digest*: ME4.

Janicki, V.; Lappschies, M.; Görtz, B. et al. 2005. Comparison of gradient index and classical designs of a nar-row band notch filter. *Proceedings of SPIE* 5963: 59631O.

Jensen, L.; Mende, M.; Blaschke, H. et al. 2010. Investigations on SiO_2/HfO_2 mixtures for nanosecond and femtosecond pulses. *Proceedings of SPIE* 7842:78207.

Jovanovic, I.; Brown, C.; Stuart, B. et al. 2004. Precision damage tests of multilayer dielectric gratings for high-energy petawatt lasers. *Proceedings of SPIE* 5647: 34–42.

Jun-Chao, Z.; Ming, F.; Yu-Chuan, S.; Yun-Xia, J., and Hong-Bo, H. 2012. Design and fabrication of broad-band rugate filter. *Chinese Physics B* 21: 054219-1-5.

Jupé, M.; Lappschies, M.; Jensen, L.; Starke, K., and Ristau, D. 2006.Improvement in laser irradiation resis-tance of fs- dielectric optics using silica mixtures. *Proceedings of SPIE* 6403: 64031A.

Jupé, M.; Lappschies, M.; Jensen, L.; Starke, K., and Ristau, D. 2007. Laser induced damage in gradual index layers and rugate filters. *Proceedings of SPIE* 6403: 640311.

Jupé, M.; Lappschies, M.; Jensen, L., and Ristau, D. 2007a. Applications of mixture oxide materials for fs optics. *Optical Interference Coatings, OSA Technical Digest (CD) (Optical Society of America)*: TuA6.

Jupé, M.; Lappschies, M.;Jensen, L., and Ristau, D. 2008. Mixed oxide coatings for advanced fs-laser applica-tions. *Proceedings of SPIE* 6720: 67200U-67200U-13.

Lappschies, M.; Görtz, B., and Ristau, D. 2004a. Application of optical broad band monitoring to quasi-rugate filters by ion beam sputtering. *Optical Interference Coatings, OSA Technical Digest*: TuE4.

Lappschies, M.; Görtz, B., and Ristau, D. 2005. Optical monitoring of rugate filters. *Proceedings of SPIE* 5963: 1Z1–1Z9.

Lappschies, M.; Groß, T.; Ehlers, H., and Ristau, D. 2004. Broadband optical monitoring for the deposition of complex coatings. *Proceedings of SPIE* 5250: 637–645.

Lorenzo, E.; Oton, C.J.; Capuj, N. E. et al. 2005. Fabrication and optimization of rugate filters based on porous silicon. *Physica Status Solidi* 9: 3227–3231.

Macleod, H. A. 1986.*Thin-Film Optical Filters*. Boca Raton, FL: Taylor & Francis.

Mangote, B.; Gallais, L.; Commandré, M. et al. 2012. Femtosecond laser damage resistance of oxide and mixture oxide optical coatings. *Optics Letters* 37: 1478–1480.

Marburger, J. 1971. Theory of self-focusing for fast non-linear response. *National Bureau of Standards Special Publication* 356: 1–10.

Marburger, J. H. 1975. Self focusing: Theory. *Progress in Quantum Electronics* 4: 35–110.

Martinez, M.; Gaul, E.; Ditmire, T. et al. 2005. The Texas petawatt laser. *Proceedings of SPIE* 5991: 1–9.

Martz, D. H.; Nguyen, H. T., and Patel, D. 2009. Large area high efficiency broad bandwith 800 nm dielectric gratings for high energy laser pulse compression. *Optics Express* 17: 23809–23816.

Melninkaitis, A. 2009. Optical resistance of dielectric coatings to multi-pulse femtosecond laser radiation. PhD-thesis, Vilnius University, Vilnius, Lithuania.

Chapter 15

Melninkaitis, A.; Tolenis, T., and Mažulė, L. 2011. Characterization of zirconia—And niobia–silica mixture coatings produced by ion-beam sputtering. *Applied Optics* 50: 188–196.

Mero, M.; Liu, J.; Sabbah, A. et al. 2003. Femtosecond pulse damage and pre-damage behavior of dielectric thin films. *Proceedings of SPIE* 4932: 202–212.

Mero, M.; Clapp, B.; Jasapara, J. et al. 2005. On the damage behavior of dielectric films when illuminated with multiple femtosecond laser pulses. *Optical Engineering* 44(5): 051107-1-7.

Neauport, J. and Bonod, N. 2008a. Pulse compression gratings for the Petal project—A review of technologies. *Proceedings of SPIE* 5991: 1–9.

Néauport, J. and Bonod, N. 2008b. Pulse compression gratings for the PETAL project: A review of various technologies. *Proceedigns of SPIE, Laser-Induced Damage in Optical Materials* 7132: 1–9.

Neauport, J.; Bonod, N.; Hocquet, S.; Palmier, S., and Dupuy, G. 2010. Mixed metal dielectric gratings for pulse compression. *Optics Express* 18: 23776–23783.

Neauport, J.; Lavastre, E.; Razé, G. et al. 2007. Effect of electric field on laser induced damage threshold multilayer dielectric gratings. *Optics Express* 15: 12508–12522.

Nguyen, D.; Emmert, L. A.; Cracetchi, I. V. et al. 2008. $Ti_xSi_{1-x}O_2$ optical coatings with tunable index and their response to intense subpicosecond laser pulse irradiation. *American Institute of Physics* 93: 261903.

Pervak, V.; Ahmad, I.; Trushin, S. et al. 2009. Chirped-pulse amplification of laser pulses with dispersive mirrors. *Optics Express* 17: 19204–19212.

Raze, G.; Néauport, J.; Dupuy, G.; Mennerat, G., and Lavastre, E. 2007. Short pulse laser damage measurements of pulse compression gratings for petawatt laser. *Proceedings of SPIE* 6720: 1–9.

Ristau, D.; Ehlers, H.; Schlichting, S., and Lappschies, M. 2008. State of the art in deterministic production of optial thin films. *Proceedings of SPIE* 7101: 1–14.

Ristau, D.; Groß, T.; Lappschies, M., and Starke, K. 2003. Towards rapid prototyping in thin technology. In *Proceedings of the Workshop on Optical Coatings: Theory, Production, and Characterisation*, eds. E. Masetti; D. Ristau; and A. Krasilnikova, p. 29–35. Erice, Sizilien: International School of Quantum Electronics.

Strickland, M. 1985. Compression of amplified chirped optical pulses. *Optics Communications* 56: 219.

Strickland, D and Mourou, G. 1985. Compression of amplified chirped optical pulses. *Optics Communications* 56: 219–221.

Starke, K.; Groß, T.; Lappschies, M., and Ristau, D. 2000. Rapid prototyping of optical thin film filters. *Proceedings of SPIE* 4094: 83–92.

Stuart, B. C.; Feit, M. D.; Herman, S.; Rubenchik, A. M.; Shore, B. W., and Perry, M. 1996. Nanosecond to femtosecond laser induced breakdown in dielectrics. *Physical Review B* 53 Nr. 4: 1749–1761.

Turowski, M.; Jupé, M.; Jensen, L., and Ristau, D. 2009. Laser-induced damage and nonlinear absorption of ultra-short laser pulses in the bulk of fused silica. *Proceedings of SPIE* 7504: 75040H.

16. Spaceborne Applications

Wolfgang Riede and Denny Wernham

16.1 Introduction

The space environment presents unique challenges for the operation of optics and optical coatings as part of laser systems. Components must be designed to survive long-term operation under variation of temperature, under exposure to ultraviolet (UV) and other ionizing radiations, under the presence of atomic oxygen, and vibration loads during launch. Furthermore, vacuum operation of lasers can be accompanied with contamination of the involved optics, especially in the ultraviolet spectral range. Hence, a reliable, long-term

Chapter 16

Laser-Induced Damage in Optical Materials. Edited by Detlev Ristau © 2015 CRC Press/Taylor & Francis Group, LLC. ISBN: 978-1-4398-7216-1.

operation necessitates comprehensive testing to analyze space-induced effects and to develop risk mitigation procedures. As typical satellite mission durations range from 3 to 10 years, there is also a fundamental problem in testing for sufficient durations in order to make proper predictions on the end-of-life (EOL) performance. This has necessitated forms of accelerated testing whose validity must be carefully analyzed.

The encountered impact of space environment on optics is multifaceted, and their combined, synergistic effects might be detrimental. An important phenomenon observed for porous coatings is the shift of spectral performance to lower wavelengths in vacuum and a reduction of the laser-induced damage threshold. This is especially deleterious for pulsed lasers where the margins for safe operation are smaller than for continuous wave (cw) lasers. Atomic oxygen present in low Earth orbits (LEO) causes erosion of coatings. Space also has a harsh radiation environment, which can cause absorptive losses in optics due to color center activation. Coatings exposed to solar radiation are subject to UV fixation, that is, increased adhesion to the surface, of outgassing contaminants. An additional factor is the thermal excursion that coatings can experience in space. It is typically ranging from −50°C to +80°C, notwithstanding deep space missions that involve cryogenic temperatures where coatings have to withstand temperatures down to −270°C. Furthermore, in missions to the inner planets, coatings may have to survive temperatures in excess of 300°C.

This chapter gives a general overview of the effects of the space environment on optical coatings, optics, and subsystems, and describes test procedures for screening and qualifying optical components and for process improvements. As a rule, equipment that operates during specific space exposure conditions shall be operated and monitored during the test itself.

Besides testing components and subsystems, the extended testing of complete laser systems (e.g., flight modules) is an effective way to determine the reliability and long-term stability of laser systems, and to mitigate the mission risk. An example of such a test is detailed by Seas et al. (2010), where the ICESat (ice, cloud, and land elevation satellite) GLAS (geoscience laser altimeter system) laser A has been operating for over two and a half billion shots in air while the second system has accumulated half a billion shots while operating in a vacuum environment.

16.2 Past and Future Space Laser Missions

Due to its directionality, laser radiation is well suited to bridge the large distances encountered in space to gather information of planetary surfaces and atmospheres. LIDAR (light detection and ranging) technology based on high-power compact and pulsed laser sources is a cornerstone of several actual and upcoming space missions measuring earth and atmospheric features like aerosol height profile, wind velocity fields, and trace gas concentrations. In addition, LIDARs are widely used for planetary surveying, like altimetry. Space LIDAR systems typically operate high-energy pulsed lasers in the infrared (IR), green, or UV spectral range. In the past, several of these missions have suffered from anomalous performance losses or even failure after short operation times. Table 16.1 summarizes important space laser missions launched during the last two decades by NASA and other space agencies (Yu 2009). The laser instruments of LITE (LIDAR In-Space Technology Experiment), Mars Global Surveyor, and ICESat suffered from contamination-related phenomena, preventing fulfillment of their

Table 16.1   Past and Present Important NASA, JAXA (Japan Aerospace Exploration Agency), and CNES (Centre National d'Etudes Spatiales) Space Laser Missions

Instrument (Mission)	Operating Wavelength/ Repetition Rate/Pulse Energy	Pulses [Mio]	Launch	Agency
LITE (STS-64)	355 nm/10 Hz/400 mJ at 1064 nm	4–5	1994	NASA
MOLA 2 (Mars Global Surveyor)	1064 nm/10 Hz/48 mJ at 1064 nm	671	1996	NASA
GLAS (ICESat)	532 nm/40 Hz/110 mJ at 1064 nm	2000	2003	NASA
CALIOP (CALIPSO)	532 nm/20 Hz/100 mJ	Laser 1: 1600	2006	NASA/CNES
LALT (SELENE)	1064 nm/1 Hz/100 mJ		2007	JAXA

Source: Seas, A. et al., *Proc. SPIE*, 7578, 757806-1, 2010.
Note: The follow-on mission ICESat-2 is planned for launch in 2016.

full mission goals. LITE has been the first demonstration of a LIDAR technology for atmospheric studies from space, and was a project to measure atmospheric parameters from a space platform. MOLA (Mars Orbiter Laser Altimeter), as part of the Mars Global Surveyor mission, conducted range measurements for attaining the precise topography of the Mars surface. The GLAS (Geoscience Laser Altimeter System) was part of the ICESat mission and conducted measurements of land and sea ice elevations. Successfully operating is the CALIOP (Cloud-Aerosol LIDAR with Orthogonal Polarization) laser aboard the CALIPSO mission satellite, providing data on high-resolution vertical profiles of aerosols and clouds. CALIOP is pressurized, and after a switch to an identical laser after 3 years is now still operating. The Laser Altimeter (LALT) aboard JAXA's lunar explorer (SELENE) captured data on the entire lunar surface height with a range accuracy of ±5 m. Further important space laser missions like NASA Mercury Laser Altimeter (MLA), NASA Lunar Orbiter Laser Altimeter (LOLA), and ChemCam on NASA's Curiosity rover can be mentioned here. Continuous wave lasers are used in laser communication terminals like in the Tesat laser communication terminal (LCT) for installing intersatellite data links or data links to ground stations.

In the near future, ESA (European Space Agency) will realize two ambitious missions within the Earth Explorer Core Mission program (Durand et al. 2008, Table 16.2): ADM-AEOLUS, which is aiming at a global measurement of three-dimensional wind speed fields,

Table 16.2   Upcoming ESA Space Laser Missions

Instrument (Mission)	Operating Wavelength/ Repetition Rate/Pulse Energy	Pulses [Mio]	Predicted Launch	Agency
ALADIN (ADM-Aeolus)	355 nm/100 Hz/100 mJ	3,000	2015	ESA
ATLID (EarthCARE)	355 nm/74 Hz/30 mJ	10,000	2016	ESA
BELA (BepiColombo)	1064 nm/10 Hz/50 mJ	300	2015	ESA/JAXA

Chapter 16

and EarthCARE, which is a joint European–Japanese mission addressing the need for a better understanding of the interactions between cloud and radiative and aerosol processes. Both missions will deliver inputs to climate modeling. The expected mission duration is 36 months, and up to 10 billion of laser pulses will be emitted in orbit. The laser instruments in both missions (ALADIN [Atmospheric Laser Doppler Instrument] and ATLID [Atmospheric LIDAR]) will operate in the UV wavelength range and emit high-power UV laser pulses. Hence, rigorous qualification is a prerequisite for long-term failure-free operation. Another important upcoming ESA mission is the BepiColombo Laser Altimeter (BELA), which will be the first European planetary laser altimeter system (Thomas et al. 2007). It is scheduled for launch in August 2015 and is planned to arrive at Mercury in January 2022.

16.3 Space Environment

With the notable exception of high-energy radiations and space vacuum in the case of vented systems, the vast majority of laser optics is generally protected from the natural space environment. Of course, there is necessarily one optical surface, which by necessity requires to be fully exposed to space (laser output window) as well as telescope optics, which are generally contained within open structures. The natural space environment is defined by the particulate and nonparticulate radiations, residual atmosphere, plasmas, and microparticles (small-sized debris and micrometeoroids). The accompanying effects will be radiation damage, charging, erosion, and impact damage of exposed optics. A laser system will generally be operated inside of a protective housing, which will shield its interior from direct exposure to space. The housing—on the other hand—will change the energy spectrum of particle radiation to lower energies, it will induce bremsstrahlung due to deceleration of charged particles, and change the relevant contamination environment, to mention just a few consequences.

Figure 16.1 provides an overview of the main effects of the induced space environment that can affect optical coatings and their performance.

Each of the space environment effects is briefly described in the following sections along with some examples of the impacts that they have on space optics. An overview of the space environment is given by Fortescue et al. (2004) and in the relevant European Cooperation for Space Standardization (ECSS) documents (ECSS-E-ST-10-04C 2008).

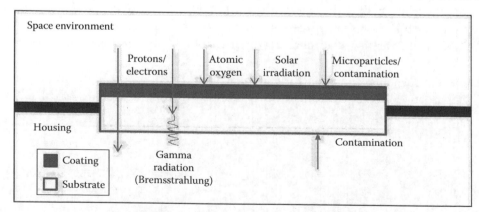

FIGURE 16.1 Schematic showing the main space environment effects that can impact the performance of space optics.

16.4 Impact of Space Vacuum

16.4.1 Vacuum Shift

Space vacuum involves pressures between 10^{-4} and 10^{-7} Pa for a typical satellite LEO environment and pressures of 10^{-14} Pa for interplanetary space (Gargaud 2011).

For a highly densified coating, vacuum has little impact, but for porous coatings, having a columnar microstructure, we see the shift of the spectral response of the coating to lower wavelengths (Stolz et al. 1993, Leplan et al. 1995, 1996, Jensen et al. 2007). The vacuum shift is attributed to effects of water vapor from ambient being desorbed in the coating, vice versa, water removal under vacuum from the voids between columns will decrease the optical thickness because of the difference of the real part of the refractive index between water (n = 1.33) and air (n = 1.00). Figure 16.2 shows the vacuum shift for

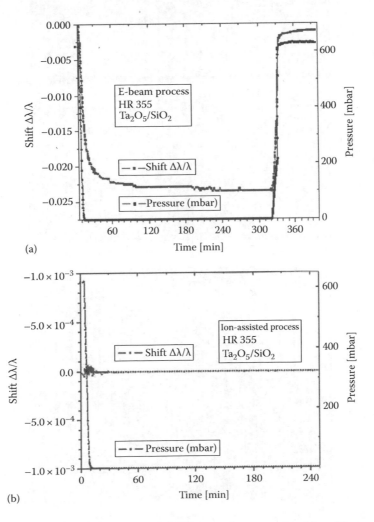

(a)

(b)

FIGURE 16.2 Charts displaying the vacuum shift for an e-beam coating (a) and ion-assisted coating (b) as a function of time during evacuation and refill with ambient atmosphere. The vacuum shift curve is resulting from the monitoring of a pronounced edge in the transmission spectrum during pressure change in the vacuum chamber.

Chapter 16

both, an e-beam coating and ion-assisted coating as a function of time during evacuation and refill with ambient atmosphere.

A shift to shorter wavelengths is typically found with e-beam coatings (with a relative shift $\Delta\lambda/\lambda$ on the order of a few percent), and an almost immediate response to the falling and rising pressure in the vacuum chamber is observed. Principally, the process is reversible, but much longer timescales of several hours have to be taken into account to reach the initial value after complete refill. On the other hand, ion-assisted coatings show a very small or not measurable vacuum shift, with $\Delta\lambda/\lambda$ being typically less than 10^{-3}, depending on the ion dose (electric charge per mass units).

The microstructure of the coatings considered earlier is illustrated in Figure 16.3. The porous coating on the right of Figure 16.3 has been deposited by classical nonassisted e-beam evaporation, whereas for the highly densified coating on the left of Figure 16.3,

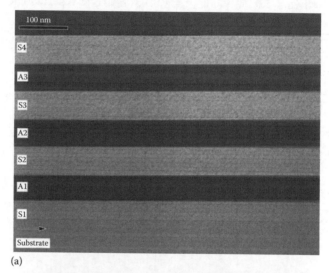

(a)

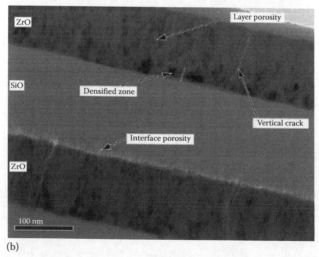

(b)

FIGURE 16.3   Micrographs of TEM cross-sectional views showing the structure of an high-reflection (HR) ion-assisted (a) and electron-beam-coated layer stack (b). The TEM was operating in bright-field mode. The porosity of the electron beam coating (HR45 at 355 nm) is clearly visible. The length of the black scaling bar is 100 nm.

assisted sputtering was used. In Figure 16.3a, one can see from the cross-sectional transmission electron microscopy (TEM) micrographs that this coating has a much less porous structure, resulting in very little observable shift. The TEM pictures displayed were obtained by preparing an electron transparent lamella, using focused ion beam milling. The thickness of the lamella is approximately 100 nm. A larger porosity of the e-beam coating is clearly visible in the ZrO_2 layers. Conversely, the classical unassisted e-beam evaporated coating clearly shows a large degree of porosity, particularly at the interfaces and within the high-index material, and the air–vacuum shift is clearly evident.

The impacts on instrument performance caused by the air–vacuum shift can be significant. In the case of the ALADIN laser (cf. Table 16.2), the air–vacuum shift of a dielectric polarizer within the master oscillator of the laser caused a complete failure of the Q-switching mechanism caused by an increase in the passive losses of the laser cavity. It is clear that coatings in space should be fully densified to avoid this effect, in addition to allowing proper performance testing on ground that can be related to in-orbit performance. The lesson learned in this case is to test the air–vacuum shift on the coatings that will be exposed to space vacuum, and not to simply accept that coatings are vacuum compatible from the manufacturer.

16.4.2 Bending Tests

Adsorption of water molecules in a porous coating is accompanied by changes in the stress level, that is, it will generally lead to an increase in tensile stress (Stolz et al. 1993). A very straightforward real-time test of stress-induced bending in e-beam-coated optics is the monitoring of the bending of a thin, single-side highly reflective coated lamina when vacuum is applied. The setup is depicted in Figure 16.4a. The samples under test are HR e-beam-coated slices of borosilicate cover glass (laminas) of 70 μm thickness. The laminas are tightly clamped on a holder inside the vacuum chamber, and the deflection of the free end is measured using a monitoring laser beam. The tests have shown that the evacuation produces a considerable tensile stress, which bends the coated surface to a concave shape on timescales of minutes (Figure 16.4a). The vacuum is, due to the small chamber volume, settled within a few seconds to the mbar range. The effect is completely reversible when repressurizing the chamber and no bending is observed with uncoated samples.

The bending radius on the laminas (21 layer stack, thickness 3.2 μm) that was measured with an HR0 355 coating was R = 0.6 m in air and R = 0.2 m under vacuum. The stress that was induced by evacuation was found to be tensile (due to the concave shape change of the sample).

By making use of Stoney's equation

$$\sigma = \frac{Y_s t_s^2}{6(1-v)t_f}\frac{1}{R},$$

where
    R is the radius of curvature
    Y_s is the Young modulus
    t_s/t_f is the substrate/film thickness

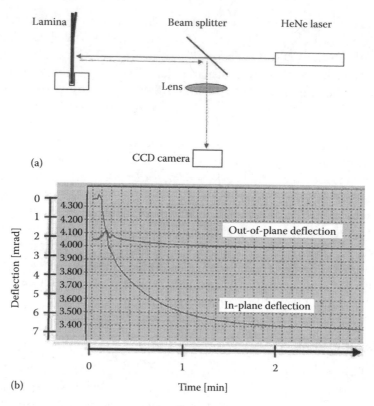

FIGURE 16.4   Vacuum bending test setup (a) and lamina deflection (b) as a function of time. (Green line: in-plane deflection; red line: out-of-plane deflection.)

We can calculate an average tensile film stress of 43 MPa under ambient air and ~130 MPa under vacuum, which is responsible for a change in radius of curvature of the sample.

This change of stress in the coatings caused by the evacuation of water vapor from the pores can have a deleterious impact for instruments containing alignment sensitive focal planes, due to an unacceptable change in the position of the focal plane due to the induced bending of the optical surface. In some cases, the coating stress can overcome the adhesion to the substrate, resulting in stress failures and, in the worst cases, complete delamination of the coating.

16.4.3   Vacuum Laser Damage Testing

16.4.3.1   Introduction

This section addresses the issues of performing laser-induced damage threshold (LIDT) tests in vacuum environment to cover the majority of cases under which currently spaceborne lasers are operated. It might also be adequate to apply small partial pressures of oxygen or gas mixtures at higher pressure levels. Furthermore, newer developments of spaceborne lasers tend to have a hermetically sealed gas filled package necessitating a low leakage rate operation. Laser damage testing under vacuum is an extension of the S-on-1 tests in ambient conditions according to ISO 21254-2 where the sample is mounted inside a vacuum chamber. The whole chamber is scanned along a test grid

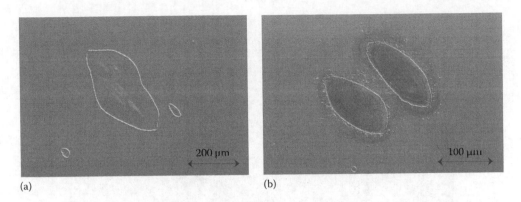

FIGURE 16.5 Comparison of the damage morphologies of a V-AR (antireflective) coating for a wavelength of 1064 nm on BK7 substrate irradiated under high-vacuum conditions (b) and under air (a). The air above the sample has a decelerating effect during the ablation, and is responsible for scattering and enhanced redeposition of debris near the damaged site.

pattern. VITON (registered trademark of DuPont performance elastomers) o-ring sealing allows for lightweight, high-vacuum chamber that can be moved with standard DC or stepper motors. The chamber windows must be sufficiently separated from the sample itself to prevent any damage on them.

When qualifying laser components being used as part of space lasers, it is important to reproduce identical conditions in the test to that of the final operational conditions. This includes the spectral and temporal features of the laser source as well as sample orientation and coating type.

Vacuum LIDT testing has the disadvantage of direct accessibility of the sample during the test, but on the other hand offers a second channel for damage monitoring based on detection of a transient pressure signal due to material ablation (ISO 21254-4). Furthermore, the redeposition of ablative debris in the neighborhood of the damaged site is marginal in the case of vacuum, whereas in the case of air, tiny melt droplets emitted from the damage spot are redirected back to the surface (Wu 2008). This effect is illustrated in Figure 16.5.

The occurrence of laser-induced contamination needs to be checked for regularly (especially when operating in the UV spectral range and testing mounted optics) by using ultraviolet light-emitting diode (UV LEDs) to induce fluorescence light emission, for example, on the chamber windows. Laser-induced fluorescence is a sensitive means to monitor nanometric deposits of hydrocarbons. Preferably, vacuum outgassing under a slight nitrogen flow can be used as a remedy of chamber contamination.

16.4.3.1.1 Vacuum Laser Damage Test Setup and Procedure The basic setup and the test procedure can be directly adapted from Chapter 2.1 with the extension of providing a vacuum environment for the sample (Allenspacher and Riede 2005). The core component of the vacuum system is a stainless steel vacuum chamber like the one depicted in Figure 16.6. To reduce the volumes of hidden voids in the inner walls, the surface should be pretreated by glass bead blasting. The chamber is connected via flexible tubing with a two-stage oil-free, dry vacuum pump system consisting of a fore-pump (e.g., membrane type) and a main turbo molecular pump. The chamber is preferably equipped with various ports to allow for pressure sensor adaptation, purging

Chapter 16

FIGURE 16.6 Lightweight high-vacuum chamber equipped with pressure sensor and optical ports for in situ inspection.

and evacuation, and optical ports for in situ inspection. Heating ribbons can be used for vacuum bake-out. A multifunctional construction enabling tests of different sample types (under various angles of incidence) and in reflection and transmission configuration is recommended. The chamber windows should be selected in diameter such that the sample inside the chamber can be irradiated without clipping. The entrance and exit windows must be separated from the sample itself to reduce their fluence level within the main beam caustic.

The evacuation procedure is started by purging the vacuum chamber with ultrapure, dry nitrogen after mounting the sample and closure of the chamber, followed by the evacuation. A vacuum dwell time of at least 12 h preceding the test is recommended to guarantee representative conditions during the actual irradiation test. The margin for the dwell time is based on vacuum shift measurements. During the tests, the vacuum is maintained permanently by dynamic pumping.

As online damage assessment techniques, collinear scatter probing with a cw HeNe laser beam and transient pressure monitoring can be used (cf. Figure 16.7, Allenspacher

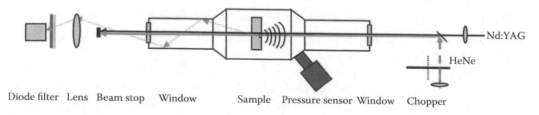

Diode filter   Lens   Beam stop   Window   Sample   Pressure sensor   Window   Chopper

FIGURE 16.7 Collinear scatter probe setup including pressure sensing. The pressure sensor is placed in the vicinity of the sample under test exposed to vacuum. Ablated material is emitted and preferably detected in upstream direction. The solid line indicates an unscattered light path, whereas the dotted lines represent a possible light path for light scattered at the damage site.

and Riede 2005). The signal to noise ratio of the collinear scatter probing can be greatly improved with lock-in detection, necessitating chopping of the HeNe laser beam. For optimum sensitivity, the HeNe laser is focused down to an diameter on the sample, which will cover at least the assumed damage area. In practice, an HeNe beam diameter of half of the diameter of the Nd:YAG beam is recommended. An increase in amplitude of scattered HeNe light measured with a photodiode after passing a filter stack indicates damage having occurred either on the sample front or back surface, or in the bulk. As soon as the scatter amplitude exceeds a predefined threshold, the irradiation sequence should interrupt the beam to prevent massive damage. The detection limit in terms of damaged area dimension on the sample can be as low as a few micrometers in diameter.

High-vacuum conditions offer the possibility for transient pressure sensing by making use of a pressure sensor (e.g., cold or hot cathode type). It has to be mounted in the neighborhood of the sample surface under investigation (Figure 16.7). Ionization gages sense the pressure indirectly by measuring the number of electrical ions produced when the gas is bombarded with electrons. Hence, neutral ablation products will be ionized, or emitted ions will be detected directly as a current from the cathode. When the pressure in the chamber is small enough ($<10^{-4}$ mbar), the mean free path length is larger than typical chamber dimension and particles ablated from or molecules emitted off the sample surface can propagate to the pressure sensor directly, without being scattered along their trajectory. The response time of the pressure rise signal can be reduced down to 10 ms or less, when the pressure sensor is mounted in emission direction of the plume. As the volume is pumped down continuously, the pressure inside the chamber will be recovered fast, that is, in less than 1 s.

The pressure-sensing technique has proved to have several advantages. Firstly, it is insensitive on interference with optical signals and vibration, which may deteriorate the scatter probe signal. Furthermore, damage testing of inherently strong scattering surfaces like light trap absorbers is possible. Pressure sensing is suitable also for testing of curved surfaces. Finally, it is especially useful for detecting the damage onset when working with top hat beam profiles on samples, as a large area that has to be monitored by a scatter probe beam reduces its sensitivity because the location of damage initiation is unknown in advance.

16.4.3.1.2 Vacuum Coating Performance

As already mentioned, space optics qualification tests have to be performed in a representative environment. This includes the application of high-vacuum conditions to the samples under test. The baseline is usually the performance of optics under atmospheric conditions. For this reason, split area tests were chosen to identify differences in air–vacuum performance. Here, half of the sample area was tested under air and the residual area subsequently under vacuum to force identical test conditions. Such split area tests were performed for different coating technologies, wavelengths, and types of optics. The results are being summarized in Figures 16.8 and 16.9.

As a general rule, degradation in the characteristic damage curve was found for tests of porous, that is, e-beam coatings for the transition of air to vacuum. An example is depicted in Figure 16.8. Here, a bending mirror for 355 nm based on e-beam coating technology was the sample under test. All fluence values given are on axis fluence values

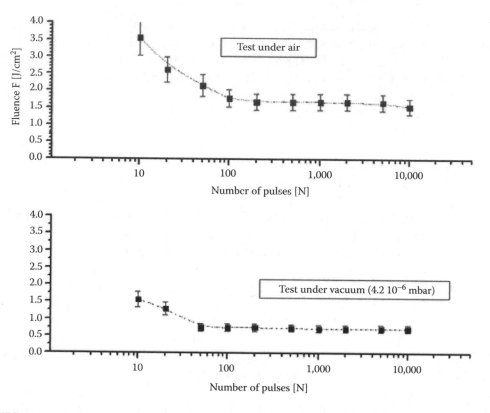

FIGURE 16.8   Characteristic damage curves for LIDT tests in air (upper curve) and vacuum (lower curve). The test sample was an e-beam evaporated coating, HR45 for 355 nm on Lithosil substrate, with test parameters: $1/e^2$ beam diameter 320 μm; 122 sites irradiated. The dashed line is connecting the data points. The length of the error bars is due to the fluence measurement error of 15%.

for a Gaussian beam profile. The fluence error is assessed from error propagation due to the variation of laser beam. In Figure 16.8, a strong degradation for large pulse numbers was observed. It must be mentioned that this degradation is a function of the dwell time of the sample in vacuum prior to the test (Allenspacher and Riede 2005). Dry nitrogen backfill and repeating the test did not show recovery.

In Riede (2008), it was reported that this effect was due to the release of water and the resulting increase in the tensile stress of the coating.

On the other hand, no measurable degradation of dense coatings (ion-assisted deposition [IAD], ion beam sputtering [IBS]) under vacuum was observed. In Figure 16.9, a test on dense coating showed no difference within the error bar length. This was confirmed in various tests where we found this effect to be independent of parameters like coating design (AR or HR), test wavelength (1064 or 355 nm), coating constituents, substrate, angle of incidence, and coating supplier. It is concluded that water dehydration under vacuum is responsible for tensile stress increase, leading to a lowering of the LIDT for porous coatings and not affecting the dense coating types.

16.4.3.2   Raster Scan Test Procedure

The characteristic damage curve is the main outcome of an S-on-1 test according to ISO 21254-2, allowing for the prediction of the LIDT of a sample at large pulse numbers

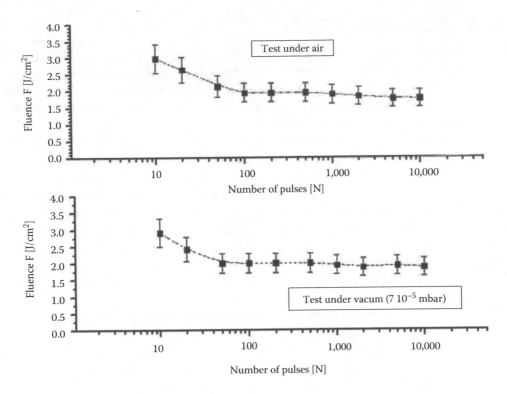

FIGURE 16.9 Characteristic damage curves for LIDT tests in air (upper curve) and vacuum (lower curve). The test sample was an ion-assisted deposition (IAD) coating, AR 355/532/1064 on Lithosil substrate, with test parameters: $1/e^2$ beam diameter 300 μm; 101 sites irradiated. The dashed line is connecting the data points. The length of the error bars is due to the fluence measurement error of 15%.

(i.e., higher than applied in the test itself) to estimate the real lifetime of the optical component. The drawback of this test is the small area coverage. This is due to the minimum separation between the test sites, which is recommended in ISO 21254-1. There, the distance between the test sites shall be larger than three times the laser spot diameter, to prevent debris-related cross-talk or other effects. Hence, a low-density, low-threshold distribution of damage precursor might not be recognized during the test.

For further operational risk mitigation, which is very stringent for space laser operation, the sample should be subjected to a test, which is representative in terms of the tested area relative to the intended application (Schwartz et al. 1999, Lamaignere et al. 2007). As the test is considered to be nondestructive by itself, a minimum test site separation, as mentioned earlier, is not adequate.

The raster scanning is done under a constant fluence level, which can be selected either application oriented, that is, above a certain fluence margin deduced from the actual fluence load of the optic, or based on the result of the S-on-1 test.

Energy drift compensation must be implemented in order to ensure a constant level of fluence for all the time of the test. The drift compensation shall be appropriate to stabilize the on sample fluence to a level which is less than the 1 sigma error bar length.

The test is generally intended to be nondestructive; nevertheless, a scatter probe monitoring of damage occurring during time is needed to avoid propagation. The test can either validate a batch, that is, a witness sample test, or validate a flight optics sample itself.

16.4.4   Thermal Vacuum Tests

Coatings in space are often cyclically subjected to large thermal excursions. Deleterious effects to be anticipated during thermal vacuum tests include outgassing and redeposition, cracking or fracture of materials or assemblies due to sudden dimensional changes by thermal expansion, contraction or pressure, and overheating of materials or assemblies due to change in convection and conductive heat transfer characteristics.

Optics are usually—besides operational thermal cycles subjected in orbit to non-operational thermal cycles during orbit acquisition, and when the satellite goes into safe mode as a result of an operational failure. This will typically invoke six or more thermal cycles in the range of −40°C to 150°C for the qualification of optics operating in a LEO. Optics that are used on inner planet or solar missions can experience much higher thermal excursions as the satellite is exposed to up to 20 solar constants, leading to temperature peaks of several hundred °C and extremely low temperatures during eclipses (−150°C). Optics that are designed for deep-space applications or are used as part of cryogenic detectors will need coatings that will have to operate at even lower temperatures.

One major test objective is to detect latent material, design, and workmanship defects, which become obvious due to thermal stress. Failures of coatings due to vacuum thermal cycles occur frequently, due to the change in stress caused by water release as well as the stresses caused by the different coefficients of thermal expansion of the materials, which make up the layers of the coating and the substrate. A typical failure due to vacuum thermal cycle of a space optic is shown in Figure 16.10. The key for these coatings is to ensure that they are qualified with sufficient margins (generally ±5°C of the nominal values in extension of the acceptance temperature range), paying attention to the potential rate of change of the thermal cycles, in addition to the actual levels. The equipment must be tested at a pressure of 10^{-5} hPa or less.

It is important to differentiate between thermal cycling in vacuum and thermal cycling carried out under high-pressure conditions as coatings that survive in the latter might possibly fail under vacuum testing.

(a)

(b)

FIGURE 16.10   Example of a coating failing under vacuum thermal cycle. (a) The photograph shows severe delamination of the coating. (b) The micrograph on the right displays a close-up picture of cracking due to stress of the coating.

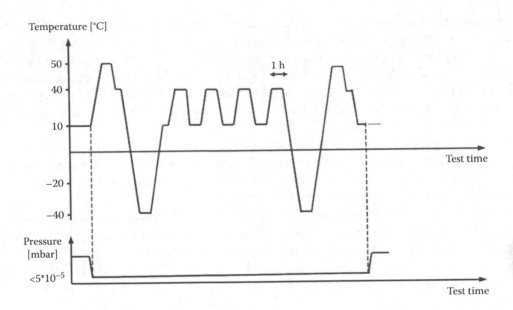

FIGURE 16.11 Chart showing a typical thermal vacuum cycling profile. The sample temperature is cycled between the extremes −40°C and +50°C. The temperature gradient is selected in the range of 0.5–2 K/min.

In a typical thermal vacuum test (ECSS-Q-70-04A), one discriminates between operational and nonoperational cycles. Generally, a test is considered to be successful, if the no visible damage or degradation has occurred and if the thermal loads under vacuum condition have been applied within the defined tolerances. In addition, no degradation in transmission, efficiency, and laser damage threshold of the sample under test should be noticed after applying nonoperational temperature extremes and within the operational temperature limits.

A thermal vacuum cycling profile for the operating and nonoperating temperatures is shown in Figure 16.11. In total, six operating and two nonoperating temperature cycles will be performed. Each of the two nonoperating cycles includes an operating cycle with the corresponding hold points.

16.4.5   Radiation Exposure

16.4.5.1   Introduction

Prior to usage in space, optical components and especially nonlinear crystals have to be tested for radiation hardness. The typical operation time in orbit ranges up to 5–10 years, during which period, the effects of the exposure to highly energetic cosmic radiation and charged particles like electrons, protons, and electromagnetic radiation have to be considered. The particle type and particle flux strongly depend on the orbit selected for the mission and also on the shielding of the satellite. These charged particles spiral around Earth's magnetic fields, reflecting back and forth between the magnetic poles to define the Van Allen belt. For orbital heights of less than 6000 km, the energetic particles are dominated by protons in the energetic range of 1–100 MeV, whereas at heights of 15,000–25,000 km, primarily electrons can be found. Gamma radiation is due to deceleration of incident particle radiation and can be simulated experimentally by gamma rays by making use of a cobalt-60 source. Depending on the orbit, the doses can reach a value of up to 100 krad of gamma irradiation and up to 10^{12} protons/cm^2 cumulating over the system lifetime. It should be noted that 1 rad corresponds to 10 mJ of energy absorbed by 1 kg of material. For planetary missions, the accumulated proton exposure might be higher and reach up to 10^{13} protons/cm^2 (Thomes et al. 2011).

Unlike electrical components that are usually most sensitive to radiation that can cause large amounts of displacement damage from neutrons or heavy ions, optical components are often most sensitive to energetic particles (protons or gamma photons) that excite electrons, leading to color center formation. Color centers can be due to recharging regular lattice ions or recharging of impurity ions (Grachev et al. 2007). Displacement damage comes from collisions between an incoming particle and a lattice atom, which subsequently displaces the atom from its original lattice position.

There are means to improve the radiation hardness of optical components, for example, by doping of the cover glass with cerium. The optimum level of cerium doping appears to be between 1% and 2% by weight (Haynes 1970).

In the following, the effects of energetic radiation mainly on nonlinear converters and electro-optic crystals are detailed.

16.4.5.2 Gamma Irradiation

Ionization effects in laser optics can conveniently be induced by making use of a cobalt-60 gamma radiation source, which emits gamma quanta with energies of 1.17 and 1.33 MeV, and which allows for a control of the dose rate via an adjustment of the sample distance to the source. The general test philosophy of a high-energy radiation tests is to apply an equivalent orbital dose of gamma radiation within several hours to the optical components. According to Tylka et al. (1997), an overall dose of ~100 krad can be reached during a 5-year period in orbit. Up to now, there are only few investigations about the influence of gamma irradiation on the optical properties of nonlinear crystals (Alampiev 2000, Roth et al. 2002, Ciapponi et al. 2011).

In the following, typical gamma radiation tests performed at the ESTEC (European Space Research and Technology Centre) radiation facility in Noordwijk on a selected number of nonlinear crystals are detailed. These components are candidate crystals for upcoming ESA LIDAR missions. All irradiations were conducted in air and at room temperature. The tests intend to demonstrate that the coated nonlinear optical crystals are able to cope with the total ionizing dose radiation conditions expected in orbit without serious degradation. The aim of the tests is to measure and classify any deterioration effects of the coated crystals caused by gamma radiation through visual inspection, transmission, and conversion efficiency measurements. The penetration range for the 1.17 and 1.33 MeV photons is much longer than the length of the crystals. Therefore, an isotropic bulk effect is expected for the transmission degradation.

Three irradiation runs with 10, 30, and 60 krad were applied successively to all crystals. A fourth irradiation of 100 krad was additionally applied to LBO, BBO, and BiBO crystals.

The transmission of borate crystals (BiBO, LBO, and BBO) was totally unaffected by the applied gamma radiations. Even after a total dose of 200 krad, the crystals

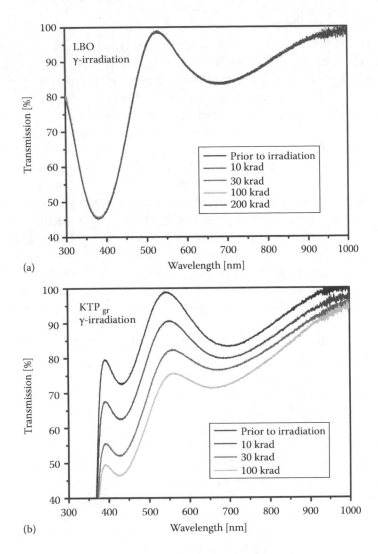

FIGURE 16.12 Transmission spectra of (coated) LBO (a) and KTP (b) before and after gamma irradiation. The crystals had a length of 20 and 6.2 mm, respectively.

showed no transmission degradation in the whole spectral range from 300 to 1000 nm. In Figure 16.12a is displayed, as an example, the transmission chart representative for an LBO crystal prior and after application of the dose. No radiation-induced drop in the spectrum was monitored.

On the other hand, phosphate and arsenate crystals (RTP, KTP, and KTA) show a strong reduction of transmission after the gamma irradiations. The radiation-induced absorption has a broad maximum between 350 and 550 nm. For KTP, the transmission was reduced from 98% to 73% at 532 nm (cf. Figure 16.12b). Above a wavelength of 700 nm, the transmission loss was less than 10% and above a wavelength of 1000 nm, less than 5%. The study of the influence of the dose rate revealed that the degradation effect is determined by the total dose and independent of the dose rate.

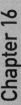
Chapter 16

16.4.5.3 Proton Irradiation

The proton irradiation tests reported here were performed at Paul Scherrer Institute in Switzerland. The energy levels selected for testing were 11, 100, and 230 MeV. The irradiations were executed in two runs: two low-energy irradiations at 11 MeV with a total dose of 6.5×10^{11} p+/cm^2 and two high-energy irradiations at 100 and 230 MeV with a total dose of 7.2×10^{11} p+/cm^2. The protons were accelerated in a cyclotron and reduced in energy by a degrader consisting of aluminum plates of different thickness.

For the borate crystals (BiBO, LBO, and BBO), there was no degradation detectable within the measurement uncertainty of 0.5% in the whole spectral range between 300 and 1000 nm after low- and high-energy proton irradiations (cf. Figures 16.13b, 16.14a, and 16.15a). In contrast to the borates, the titanyle crystals (KTP$_{gr}$, KTA, and RTP) showed clear transmission degradation both for the low- and high-energy irradiations (cf. Figures 16.14 and 16.15b for KTP). Similar to the degradation effects observed after gamma irradiation, the proton radiation induced an increasing absorption with decreasing wavelength. A partial self-annealing at room temperature of the transmission degradation was observed by measuring the transmission several days after the irradiation tests. The recovery was complete after a 2 h annealing at 150°C in vacuum.

The mean penetration depth of the protons in the different crystal materials was estimated with stopping and range of ions in matter (SRIM) routine (Ziegler et al. 2008). The mean penetration depth is the average value of the depth to which the protons will penetrate in the course of slowing down to rest. It is measured along the initial direction of the protons. Crystals used in the irradiation tests are listed in Table 16.3 and the mean penetration depths are summarized in Table 16.4. There is a strong increase in deposited dose at the end of the slowing-down process before the ions are completely stopped. For the low-energy irradiation, the penetration depth is clearly shorter than the thickness of the crystal. The protons are totally stopped in the sample. Due to the Bragg peak effect, most of the dose is concentrated in a small volume at the end of the path. For 100 and 230 MeV irradiations, the proton range is larger than the crystal length. The Bragg peak effect is not relevant, and potential defects are generated nearly homogenous along the proton path during the crystal.

(a)

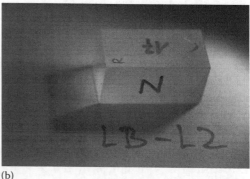

(b)

FIGURE 16.13 KTA (a) and LBO (b) after low-energy proton irradiations (11 MeV). The direction of the proton beam is indicated by arrows. A darkening of the proton exposed facet of KTA can be inferred. The crystals had a length of 15.3 and 20 mm, respectively. No darkening was seen with LBO.

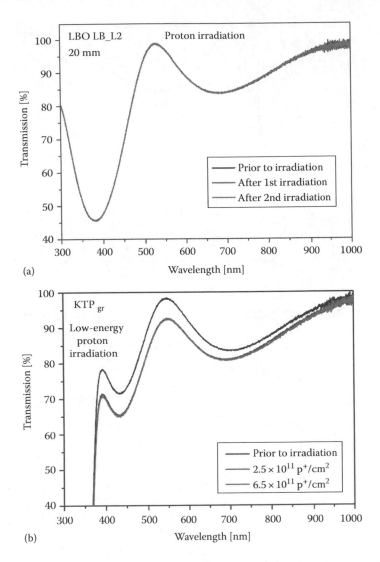

(a)

(b)

FIGURE 16.14 Transmission spectra for LBO (a) and KTP$_{gr}$ (b) crystals before and after low-energy (11 MeV) proton irradiations. The crystals had a length of 20 and 6.1 mm, respectively. With KTP$_{gr}$, saturation in the transmission loss with increasing proton dose was observed (the subscript gr denotes gray-tracking resistance of the crystal).

The borate crystals (LBO, BBO, and BiBO) showed no degradation after low- and high-energy proton irradiations in the range of measurement accuracy of ±1%.

Radiation-induced absorption found for the titanyle crystals after gamma or proton irradiation is decreasing with increasing wavelength. At 1000 nm, the radiation-induced transmission losses are below 7%. The operation of these crystals in the IR wavelength region would be only slightly affected by radiation effects.

16.4.5.4 Solar Irradiance

For components exterior to the satellite housing, the solar spectrum, especially, the UV spectral range, has to be considered. It can cause darkening of coatings due to color center formation or polymerization of organic contaminants. In this case, there is a

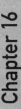

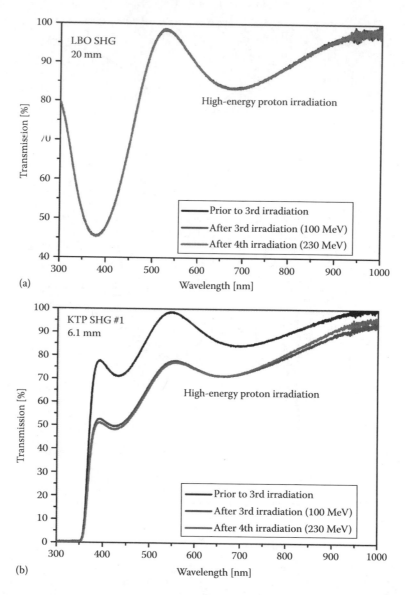

FIGURE 16.15 Transmission spectra of LBO (a) and KTP (b) crystals before and after high-energy proton irradiations (100 and 230 MeV). The crystals had a length of 20 and 6.1 mm, respectively.

Table 16.3 Crystals Used for Irradiation Tests

Crystal		Applications	Transmission Range [nm]
KTP	KTiOPO$_4$	SHG, OPO, OPA	350–4500
RTP	RbTiOPO$_4$	Q-switch	350–4500
KTA	KTiOAsO$_4$	OPO, OPA	350–5300
LBO	LiB$_3$O$_5$	SHG, THG, OPO, OPA	160–2600
BBO	BaB$_2$O$_4$	SHG, THG, 4HG, 5HG, OPO, OPA	190–3500
BiBO	BiB$_3$O$_6$	SHG, THG, OPO, OPA	290–2500

Table 16.4 Mean Penetration Depth in Millimeter for 11, 100, and 230 MeV Protons in (Uncoated) Nonlinear Crystals, Calculated with SRIM Routine

	11 MeV [mm]	100 MeV [mm]	230 MeV [mm]
BBO	0.6	30	130
LBO	0.7	36	160
BiBO	0.5	25	100
KTP	0.6	32	140
KTA	0.6	30	130
RTP	0.6	29	120

typical increase in the absorption of the coatings, which is greatest at the UV end of the spectrum and becomes less prevalent as we move to higher wavelengths.

This degradation in performance is particularly serious for solar cell cover glasses (Haynes 1970, Russell and Jones 2003, Gusarov et al. 2007) and optical solar reflectors used in radiators. In the former case, there can be a significant reduction in the conversion efficiencies of the cells and a resulting loss in power generation of the array. In the latter, there is a loss in the efficiency of the radiator resulting in excessive heating. Degradation of a solar cell cover glass tested with UV irradiation in vacuum is shown in Figure 16.16. There is an annealing effect that can reverse the degradation caused by UV exposure in vacuum when the optics are reexposed to ambient conditions. This, of course, does not happen in space, meaning that performance degradation effects can be underestimated if an in situ measurement is not used during testing. It should be noted that great care is required in performing vacuum testing for space optics where chamber contamination due to outgassing has to be minimized as far as possible.

For optical solar reflectors, degradation leads to an increase in the solar absorbance, α, of the coating, meaning that the radiative efficiency (directly proportional to α/ε, where ε is the emissivity) is degraded, leading to an overall increase in the temperature of the satellite.

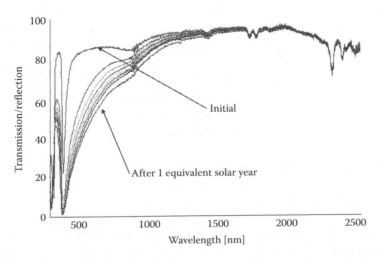

FIGURE 16.16 Degradation of solar cell cover glasses, measured in situ, during a UV test in vacuum.

FIGURE 16.17 Phase matching mount for nonlinear frequency conversion mounted on a shaker.

16.4.6 Vibration Tests

Spaceflight hardware will be exposed to intense vibration levels during launch and release of upper stages when approaching the final orbit. To ensure that the functionality is not impaired by these environments, mimicking of vibration levels during preparative tests is essential, especially for mounted optical components. These tests might reveal latent material defects that become obvious due to vibration loads, comparing low-level pre- and postvibration runs. Furthermore, the angular displacement of the sample under test mounted in a subassembly must not exceed certain system margins. These can be rather stringent, that is, on the order of tens of microradians or less (e.g., for mounted frequency converters).

The qualification test level is usually selected for a sine vibration test with a factor of 1.25 up to 1.5 above the limit of the load spectrum. ECSS-E-10-03B, a typical test sequence will be initiated by a resonance search tests at a lower load (0–150 Hz/0.5 g), followed by sinus vibration test at acceptance level and a final sinus vibration test at qualification level. The tests are performed in a vibration test facility (shaker) with the corresponding data acquisition system (cf. Figure 16.17). The shaker shall be capable to apply the required loads. The vibration tests must cover all the three mutually orthogonal directions. Two acceleration sensors have to be used, where the first sensor detects the excitation acceleration and the second the response of the unit.

The test parameters for vibration tests are summarized in Table 16.5.

16.4.7 Product Assurance for Space Coatings

16.4.7.1 Introduction

The general principles of space product assurance relevant to space optics are to properly define and control the materials and processes used in the manufacture, assembly, integration, and test of the components; to verify that the component meets its operational requirements with sufficient margins to account for any uncertainties ensuring

Table 16.5 Test Parameters for Vibration Tests

Frequency Range [Hz]	Resonance Search	Acceptance Level Sweep Rate: 4 oct./min Duration: 1 Sweep Up	Qualification Level Sweep Rate: 2 oct./min Duration: 1 Sweep Up
11–25	0.5 g	8 mm	10 mm
25–100	0.5 g	20 g	25 g
100–150	0.5 g	4 g	5 g

performance over lifetime; and to formally track any nonconformances and ensure that these do not have an unacceptable impact on the final performance. These are usually defined beforehand in a dedicated product assurance plan for the deliverable item.

16.4.7.2 End-to-End Traceability

It is of particular importance for space that the quality chain from source materials to finished product remains unbroken and includes all of the materials and processes in between. The main principle behind this is that if a problem does occur during subsequent testing phases, or in orbit, that this problem is traceable and the root cause for any anomaly can be determined and corrected efficiently (or at the very least not repeated). Thus, all of the materials used for the manufacture of optics and coatings should have a valid certification of composition, which is traceable to the batch used. This is usually tracked with a declared materials list (DML) that is supplied to the customer along with the finished optics. This should include materials that are used in the manufacture and then removed such as waxes, rosins, pitch, and polishing compounds used in the forming of optical substrates, along with the solvents used to remove them. It is also important to review the packaging materials used for the transportation and containment of space optics to ensure that these are compatible, particularly given the long durations that optics often reside in the packaging container prior to being integrated into instrument or satellite assemblies.

Residual contaminants from organic compounds, if not properly removed, can result in subsequent failure of the optics in humidity, thermal vacuum, and adhesion tests. They can also darken in the space environment via the interaction with the radiations described previously. Residual polishing compounds, such as cerium dioxide, can have a detrimental effect on the damage threshold of high-power laser optics especially at shorter wavelengths where they can have significant absorption. It is often the case that materials that are used in the manufacturing process, which are subsequently removed, are omitted from the DML, only to be found later on during an anomaly investigation.

Certain materials are not permitted for use in space due to their incompatibility with the vacuum environment (for reference: ASTM-E595-07, ASTM-E1559-09, ISO 15388:2012, ISO 14644-8:2013, ECSS-Q-ST-70-02C). These include mercury, cadmium, and zinc due to their propensity to evaporate in vacuum, PVC that contains a large proportion of plasticizers that are released in vacuum, and pure tin which can grow tin whiskers in space. Silicones should be generally avoided for space optics due to the fact that they are very difficult to remove and will darken if exposed to UV or low-energy proton radiation.

Chapter 16

It is usually the responsibility of the goods-receiving party, for example, coating manufacturer receiving a substrate, to have the same level of product assurance in their supply chain, and so on. For example, if there is a delamination failure of a coating, it is all too often the case that the failure cannot be attributed to problems at the coating manufacturer or the substrate supplier, because the quality chain has been broken and both the substrate and the coating are potentially implicated in the failure.

16.4.7.3   Process Control

The DML is usually cross-referenced with a declared process list (DPL) that uniquely identifies all of the processes used for the manufacture of the optics. This includes cleaning processes and measurement and verification processes. Log books and plant maintenance files are complementary documents to the DML and the DPL. One of the most important aspects to track is engineering change that is often implemented by experienced coating engineers, in order to stabilize or rectify drifts in coating plant performance. It is often the case that the materials are defined by the coating manufacturer in a process identification document (PID), and the various processes and measurements performed are recorded in a *traveler* that follows the product through its entire manufacturing life cycle. These documents can often take the place of the DPL.

16.4.7.4   Nonconformance Control

Nonconformances, defined as any breach of a requirement, either from the customer's specification of the delivered product or from the internal specifications defined by the manufacturer for the manufacturing processes, should be traced and subject to formal processing. Generally speaking, nonconformances are divided into major and minor depending on their degree of severity. Major nonconformances should be reported to the supplier as they are usually in a better position to make an assessment of the potential impacts at system level and judge whether they are acceptable or not. The nonconformance process usually comes under the responsibility of the quality assurance or product assurance department, who raise a nonconformance report (NCR) and then chair the subsequent nonconformance review boards (NRBs) until the issue is closed. For space products, any nonconformances associated with a product are considered to be an inherent part of its configuration. Thus, all NCRs are usually reviewed at the point of the delivery of the product, as part of the overall acceptance process.

The nonconformance control process should be regarded by the supplier as beneficial, inasmuch as it often provides valuable inputs for improvements to manufacturing processes, and eventual yield of good-quality components, along with efficient identification and rectification of recurring manufacturing problems.

16.4.7.5   Space Qualification

Qualification is the verification that the equipment meets its requirements with sufficient margins. Qualification can be by design; by analysis; by test; via similarity to components made using the same design, materials, and processes, as one that has a demonstrated space heritage; or a combination of them. The majority of space optics is qualified by test and similarity. It should be noted that in the latter case, it is often difficult to obtain direct performance data of optics from operational satellites due to the fact that the performance is often not directly measured, that the performance degradation cannot easily be traced to a single cause, and that satellite operators do not often provide detailed performance degradation data. In addition, depending on the particular satellite mission and orbit, any qualification by similarity argument has to ensure that the operation environment, for example, orbit and mission lifetime, for the component is completely enveloped by the heritage satellite. The fact that optics are flying on a particular satellite is not a demonstration that they are performing to requirements applicable to another mission, and therefore cannot be used as an argument for qualification in isolation. Thus, a certain amount of verification testing should be performed in order to augment the heritage argument and ensure that there has been no process or material changes in the interim period, which could affect the performance.

For optical coatings, witness samples are often used for qualification testing. Care has to be taken that these witness samples are not *hostile* and are truly representative of the actual flight optic that will be produced. There have been several cases where witness samples have passed tests, only for the flight optic to subsequently fail. Common differences are geometry, for example, when the witness sample is planar and the flight optic is curved; that the thermal environments in the coating plant are different between the witness sample and the flight optic due to geometry or tooling differences; in materials and processing, where different processes or materials are used for the witness sample and the flight optic; or that the witness sample is in a different coating run to that of the flight optic. For space optics, qualification is preferably at coating lot level unless there is evidence of process stability over a statistically significant number of coating runs. For critical components that can cause a single-point failure for the satellite instrument, such as those for high-power lasers, or where a small number of failures randomly distributed with a large number of components can have a large overall impact on system performance, for example, solar cell assemblies for solar arrays, qualification at coating lot level is even more important. Clearly, the qualification methodology should be based upon an assessment of the potential severity of the eventual performance impacts at system level and be agreed between the customer and the supplier beforehand.

References

Alampiev, M.V. 2000. Optical absorption of gamma-irradiated KTP crystals in the 0.9–2.5 μm range. *Quantum Electronics* 30:255.

Allenspacher, P. and Riede, W. 2005. Vacuum laser damage test bench. *Proceedings of SPIE* 5991:599128-1.

ASTM-E595-07, Standard test method for total mass loss and collected volatile condensable materials from outgassing in a vacuum environment.

ASTM-E1559-09, Standard test method for contamination outgassing characteristics of spacecraft materials.

Ciapponi, A. et al. 2011. Non-linear optical frequency conversion crystals for space applications. *SPIE* 7912:791205-1.

Durand, Y., Bézy, J.-L., and Meynart, R. 2008. Laser technology developments in support of ESA's earth observation missions. *Proceedings of the SPIE* 6871:68710G.

ECSS-E-10-03B Draft 2, Space engineering, Testing, January 2008.

ECSS-E-ST-10-04C, Space engineering, Space environment, November 2008.

ECSS-Q-ST-70-02C, Space product assurance—Thermal vacuum outgassing test for the screening of space materials.

Fortescue, P., Stark, J., and Swinerd, G. 2004. *Spacecraft Systems Engineering*, 3rd edn. Chichester, U.K.: Wiley.

Gargaud, M. (ed.). 2011. *Encyclopedia of Astrobiology*. Heidelberg, Germany: Springer.

Grachev, V. et al. 2007. Flight to Mars and radiation defect in $Li_2B_4O_7$ and $KTiOPO_4$ crystals. *Physica Status Solidi* 4:1288.

Gusarov, A. et al. 2007. Comparison of radiation-induced transmission degradation of borosilicate crown optical glass from four different manufacturers. *Optical Engineering* 46:043004.

Haynes, G. 1970. Effect of radiation on cerium-doped solar cell coverglass. NASA TN D- 6062, December 1970. Washington, DC: National Aeronautics and Space Administration, 20546.

ISO/FDIS 21254-1. Lasers and laser-related equipment—Test methods for laser-induced damage threshold—Part 1: Definitions and general principles.

ISO/FDIS 21254-2. Lasers and laser-related equipment—Test methods for laser radiation- induced damage threshold—Part 2: Threshold determination.

ISO/DTR 21254-4 Lasers and laser-related equipment Test methods for laser radiation induced damage threshold—Part 4: Inspection, detection and measurement.

ISO 15388:2012, Space systems—Contamination and cleanliness control.

ISO 14644-8:2013, Cleanrooms and associated controlled environments—Part 8: Classification of air cleanliness by chemical concentration.

Jensen, L. et al. 2007. Damage threshold investigations of high power laser optics under atmospheric and vacuum conditions. *Proceedings of SPIE* 6403:64030U-1.

Lamaignere, L. et al. 2007. An accurate, repeatable, and well characterized measurement of laser damage density of optical materials. *The Review of Scientific Instruments* 78:103105.

Leplan, H. et al. 1995. Residual stresses in evaporated silicon dioxide thin films: Correlation with deposition parameters and aging behavior. *Journal of Applied Physics* 78:962.

Leplan, H., Robic, J.Y., and Pauleau, Y. 1996. Kinetics of residual stress evolution in evaporated silicon dioxide films exposed to room air. *Journal of Applied Physics* 79:6926.

Riede, W., Allenspacher, P., Jensen, L., and Jupe, M. 2008. Analysis of the air—Vacuum effect in dielectric coatings. *Proceedings SPIE the International Society for Optical Engineering* 7132:71320F.

Roth, U. et al. 2002. Proton and gamma radiation tests on nonlinear crystals. *Applied Optics* 41:464.

Russell, J. and Jones, G. 2003. Radiation testing of coverglasses and second surface mirrors. *Proceedings of Ninth International Symposium on Materials in a Space Environment*. Noordwijk, the Netherlands.

Schwartz, S. et al. 1999. Current 3-ω large optic test procedures and data analysis for the quality assurance of National Ignition Facility optics. *Proceedings of SPIE* 3578:314.

Seas, A. et al. 2010. Extended testing of laser systems for space applications. *Proceedings of SPIE* 7578:757806-1.

Stolz, C. et al. 1993. Effects of vacuum exposure on stress and spectral shift of high reflective coatings. *Applied Optics* 32(28):5666–5672.

Thomas, N. et al. 2007. The BepiColombo laser altimeter (BELA): Concept and baseline design. *Planetary and Space Science* 55:1398–1413.

Thomes, W.J., Cavanaugh, J.F., and Ott, M. 2011. Proton radiation testing of laser optical components for NASA Jupiter Europa Orbiter Mission. *Proceedings of SPIE* 8164:816403-10.

Tylka, A.J., Adams, J.H., and Boberg, P.R. 1997. CREME96: A revision of the cosmic ray effects on micro-electronics code. *IEEE Transactions on Nuclear Science* 44: 2150–2160.

Wu, B. 2008. High-intensity nanosecond-pulsed laser-induced plasma in air, water, and vacuum: A comparative study of the early-stage evolution using a physics-based predictive model. *Applied Physics Letters* 93:101104.

Yu, A.W. 2009. Overview of space qualified solid-state lasers development at NASA Goddard Space Flight Center. *Proceedings of SPIE* 7193:719305-1.

Ziegler, J.F., Biersack, J.P., and Ziegler, M.D. 2008. *SRIM: The Stopping and Range of Ions in Matter*. Chester, U.K.: SRIM Co.

IV Applications

17. Lithography in the Deep Ultraviolet and Extreme Ultraviolet

Klaus R. Mann

17.1 Introduction: Microlithography

In the semiconductor industry, optical projection lithography is employed for the production of microchips, using ultraviolet radiation for structured exposure of photoresists on silicon wafers. This key process represents the basis of computer and information technologies and is therefore also of fundamental importance for our modern society. The development of the structure sizes in computer chips is described since 1965 by Moore's law, stating that the number of transistors on integrated circuits doubles approximately every 2 years. Due to the diffraction limit, this tremendous speed in shrinkage of chip structures is of course only possible by using shorter and shorter wavelengths for the lithographic process. Starting with the g-line ($\lambda = 436$ nm) and i-line ($\lambda = 365$ nm) of continuously emitting mercury lamps, mass production of microchips is performed since the 1990s with pulsed excimer laser radiation in the deep ultraviolet (DUV) spectral range, first at $\lambda = 248$ nm (KrF laser), and since about 10 years at $\lambda = 193$ nm (ArF laser). State-of-the-art ArF laser–based wafer stepper systems enable the generation of structure sizes smaller than 30 nm, employing immersion objectives for index matching to increase the numerical aperture and thus the achievable resolution (*193i technology*).

Laser-Induced Damage in Optical Materials. Edited by Detlev Ristau © 2015 CRC Press/Taylor & Francis Group, LLC. ISBN: 978-1-4398-7216-1.

Chapter 17

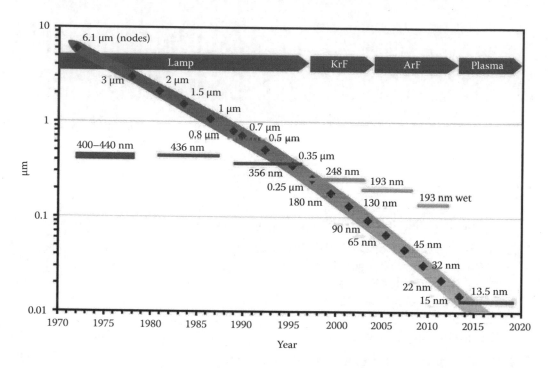

FIGURE 17.1 Lithography scaling indicating Moore's law: Shrinkage of structure sizes in high-volume microchip manufacturing according to International Technology Roadmap for Semiconductors (ITRS), as well as utilized exposure wavelengths. (From Armbrust, D., Technology & manufacturing innovations: New SEMATECH initiatives, 2011; http://www.sematech.org/meetings/archives/symposia/10187/Keynote-Plenary/01_Armbrust.pdf (accessed February 1, 2014). Reprinted with permission of SEMATECH, Albany, New York.)

An overview of the *road map* of the semiconductor industry is displayed in Figure 17.1, indicating also the possible next-generation technology, that is, extreme ultraviolet (EUV) lithography at a wavelength of 13.5 nm for generation of structure sizes below 20 nm (SEMATECH). Currently, enormous efforts are undertaken to develop EUV wafer stepper systems for high-volume manufacturing of high-end microchips.

Regarding the ultrahigh precision of the lithographic process on the one hand, and the high photon energies in the deep UV or even extreme UV spectral range on the other, it is obvious that the requirements for the optical components to be employed for microlithography are extremely challenging. The situation is tightened even more by the fact that wafer steppers are operated at pulse repetition rates of several kHz, accumulating high pulse numbers (up to 100 billion) over an expected lifetime of 10 years. Thus, dose-dependent degradation of the optical properties from various radiation damage mechanisms can severely affect the stability of the process, and ever-increasing demands on imaging quality and throughput call for an ongoing optimization of the relevant optical parameters.

In this contribution, a review of the requirements, problems, and challenges associated with the performance and especially the stability of optical materials employed in industrial wafer steppers operating in the deep UV and in the extreme UV is given.

17.2   Stability of Optics for Deep UV Lithography

Due to the high photon energies of KrF and ArF excimer lasers of E_{ph} = 5.0 and 6.4 eV, respectively, the number of available transmissive optical materials, requiring a band gap $E_g > E_{ph}$, is strongly limited. In fact, only fused silica and, to a much smaller extent, CaF_2 are used as transmissive optics in modern DUV wafer steppers. Despite considerable quality improvements achieved for these materials in recent years, the combination of high photon energies and high radiation doses can still result in intolerable damaging processes, especially due to ever-increasing fluences and average powers for enhanced production efficiency.

Other than for high-power laser optics used in the infrared or visible spectrum, radiation-induced damage of DUV lithographic optics is in general not associated with easily detectable morphological changes or catastrophic failure, since the employed fluences (<10–100 mJ/cm^2) are well below the single-pulse damage threshold (several J/cm^2) of the optical material. Nevertheless, the high photon energies can rather lead to a slow but irreversible deterioration of the optical quality of a component. Such degradation effects of DUV optical materials can be registered in most cases only after a prolonged irradiation, applying pulse numbers in the range of several 100 millions up to billions (see, e.g., Krajnovich et al. 1992a; Libermann et al. 1999a,b).

The most severe degradation phenomena of DUV lens materials are attributed to *color center formation* (both for fused silica and CaF_2) and *compaction/rarefaction* (only fused silica). In addition to these irreversible *damage* effects, there are also reversible changes of the optical properties, which can strongly influence the process stability, in particular, *two-photon absorption, stress birefringence,* and *thermal lensing.*

In the following, the different loss mechanisms and degradation channels of DUV lithography optics shall be addressed in more detail.

17.2.1   Two-Photon Absorption

In transparent media, the band gap is larger than the photon energy of the considered wavelength, and single-photon absorption is therefore only possible by absorbing impurities or by surface and defect states within the forbidden zone. However, the combined energy of several photons may suffice to cause multiphoton absorption, a process first described theoretically in 1931 (Göppert-Mayer 1931). In compliance with the Keldysh theory, two-photon absorption in a medium is possible if its band gap is smaller than twice the photon energy (Liu et al. 1978) (cf. Figure 17.2). This is exactly the situation in the deep UV spectral range for the lithographic lens materials fused silica and CaF_2.

In order to characterize two-photon absorption, photothermal or calorimetric methods are suitable, providing a much higher sensitivity than transmission measurements: Rather than observing small changes in a large quantity, absorption-induced heating is detected directly. Laser calorimetry, so far widely used for absorbance measurements of IR optics, was first applied in 1995 by Eva in the DUV spectral range (Eva and Mann 1996). Figure 17.3 shows a sketch of the laser calorimeter setup developed for absorption measurements on DUV optics with KrF, ArF, and F_2 excimer lasers (wavelengths 248, 193, 157 nm, respectively). The sample is mounted in a thermally insulating enclosure,

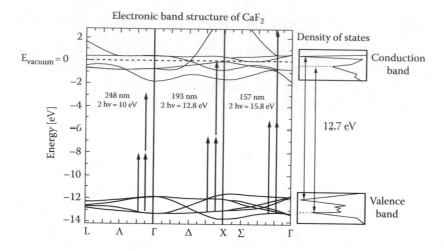

FIGURE 17.2 Electronic band structure and densities of states in CaF_2. The arrows denote the photon energy for single- and two-photon absorption. Different positions of the arrows are for clarity only. (From Görling, C. et al., *Appl. Phys. B*, 74, 259, 2002.)

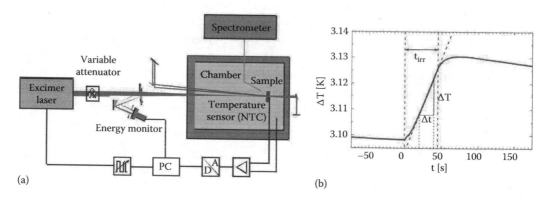

FIGURE 17.3 (a) Laser calorimeter for absorption measurement of DUV optics (cf. text) and (b) temperature signal monitored on a CaF_2 sample during 193 nm irradiation (0–45 s). The absorbance is evaluated after drift correction within the time interval of 20–40 s.

and its temperature rise during an irradiation with known laser power P_{in} for a time interval Δt (several seconds) is registered with a high-sensitivity negative temperature coefficient (NTC) resistor. According to

$$A = \frac{P_{abs}}{P_{in}} = C_{eff} \frac{\Delta T}{(P_{in} \cdot \Delta t)},$$

(17.1)

the temperature increase ΔT is directly proportional to the absorbance A, which can be measured with a sensitivity in the ppm range (P_{abs} = absorbed laser power). Absolute absorbance data can be achieved from an electrical calibration by determining the effective heat capacitance C_{eff} of sample and mount (ISO 11551:2003).

The described DUV calorimeter was extensively used to investigate nonlinear loss effects in DUV bulk media at nanosecond pulse lengths (Eva and Mann 1996; Görling et al. 2002). Depending upon the band gap of the investigated medium and the wavelength and pulse duration used, three fundamental classes of behavior can be distinguished, that is,

1. Constant absorption practically independent of any parameter
2. Two-photon absorption without degradation
3. Two-photon absorption accompanied by degradation

Examples for these different interaction mechanisms will be given in the following.

Figure 17.4 shows the calorimetric DUV absorbance data obtained for high-purity CaF_2 crystals of different thickness from the same batch. The samples were irradiated at successively increased fluence at 193 nm, each data point representing a burst of 4500 pulses at 300 Hz. Regarding the band gap ($E_g \sim 12$ eV) and the density of states of CaF_2 shown in Figure 17.2, two-photon absorption promoting an electron from the valence band into the conduction band should be possible at 193 nm ($h\nu = 6.4$ eV). Indeed, a strongly linear increase in absorbance with energy density as well as a thickness scaling of the data is evident from Figure 17.4. The increase in absorbance is transient, as low-intensity data are identical before and after high-intensity irradiation. In contrast, at 248 nm, due to the double photon energy of only 10 eV, two-photon absorption processes are not observed in high-purity CaF_2 crystals (Eva and Mann 1996; Görling et al. 2002).

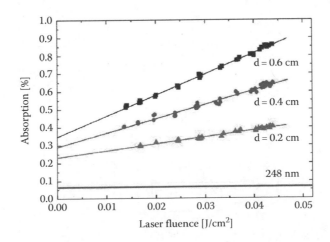

FIGURE 17.4 Reversible increase in absorbance of CaF_2 crystals of different thickness d from the same batch with laser fluence due to single- and two-photon absorption of 193 nm radiation (pulse length 14 ns, irradiation time per data point 17s). From the linear fit of the data, linear and nonlinear absorbance coefficients are determined by the intercept with the ordinate (A_0) and the slope (β), respectively. β values are in the range of a few times 10^{-9} cm/W. Corresponding absorbance data at 248 nm irradiation wavelength indicate no fluence dependence, that is, single-photon absorbance only.

Chapter 17

The results can be described by Lambert–Beer's law for optically thin samples. Taking into account surface, single- and two-photon absorption, the total absorption A reads as

$$A = \left| \frac{dI}{I} \right| = A_0 + \beta \cdot I \cdot d = A_{Surface} + \left(\alpha + \beta \frac{H}{\tau} \right) \cdot d \qquad (17.2)$$

where

A_0 describes the linear absorbance, which is given by the sum of bulk absorbance $\alpha \cdot d$ and (linear) surface absorbance $A_{Surface}$

β is the second-order nonlinear absorption coefficient (I, incident intensity; d, sample thickness; α, linear absorption coefficient; H, fluence; τ, pulse length)

Thus, from the linear slope of absorbance over intensity in Figure 17.4, the nonlinear absorption coefficient β is determined, whereas the single-photon absorption A_0 is derived from the intercept with the ordinate.

The investigations have shown, however, that the coefficient β describes not exclusively the pure two-photon absorption, as registered, for example, in femtosecond experiments. It rather represents the sum of pure two-photon excitation processes in single pulses (β_{2-Ph}) and two-step absorption mechanisms (β_{2-step}) (Görling et al. 2005). The latter require transient defect states that are populated in a first pulse and excited to the conduction band in subsequent pulses. If the lifetime of such states is of the order of milliseconds, a strong repetition rate dependence of the absorbance can be observed, as demonstrated in Figure 17.5. Obviously, transient defects of such kind would set an upper limit for the practical pulse repetition rate and are not tolerable in modern wafer stepper systems operating at 6 kHz.

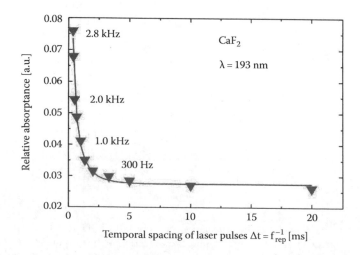

FIGURE 17.5 Influence of pulse repetition rate on absorbance of a CaF$_2$ sample obtained by laser calorimetry at 193 nm (15 ns, 33 mJ/cm^2). The relative absorbance increases with decreasing time between two consecutive laser pulses $\Delta t = f_{rep}^{-1}$, being a strong indication for a two-step absorption process. (From Görling, Ch., U. et al., *Opt. Commun.*, 249, 319, 2005.)

The data indicate that defect states or color centers, whether transient or permanent, play an extremely important role for the performance of DUV lithography optics. Permanent color center formation will be addressed in the following section.

17.2.2 Color Center Formation

In fused silica and CaF_2 materials, DUV radiation can produce defect states whose absorption maxima fall within the band gap. Regarding the underlying generation mechanism, it is widely assumed that electrons excited in a two-photon process form localized excitons, which may eventually create color centers by altering the crystalline or amorphous structure of the solid upon their decay (Liu et al. 1978). Thus, although generated in a multiphoton process, these color centers lead to a steadily increasing linear absorption, resulting in a degradation of stepper optics (Libermann et al. 1999a,b).

Whereas for crystalline CaF_2 color center formation can be attributed to purity and preexisting defects exclusively, in the case of synthetic fused silica, being by far the most common DUV optical material, also annealing history during the manufacturing process, as well OH and H_2 contents are influencing the material stability (Liu et al. 1978; Schenker et al. 1994a; Libermann et al. 1999a; Smith et al. 2000). In fused silica, the color centers are believed to arise from bonding defects in the as-manufactured amorphous glass structure. Absorption is due to radiation-induced defects that are fragments of broken Si–O–Si bonds. In particular, E′ color centers with an absorption peak at 215 nm, and nonbridging-oxygen hole centers (NBOH) absorbing around 260 nm are generated by both ArF and KrF excimer laser radiation. According to the peak positions and the band widths of these absorption bands, induced absorption at ArF wavelength (193 nm) is dominated by the E′ center, while KrF laser optics (248 nm) suffers from induced absorption due to NBOH centers (Moll and Schermerhorn 1999).

In order to enable a stepper lifetime of 10 years or more despite of this adverse effect of defect generation, large efforts have been undertaken since the 1990s to characterize and improve the material quality. Since the amount of color center formation is strongly dose dependent, long-term irradiations applying up to several billion excimer laser pulses are performed to stress the defect generation process, and to allow for predictions regarding system lifetimes (Krajnovich et al. 1992a; Libermann et al. 1999a; Moll and Schermerhorn 1999; Krajnovich et al. 1992a). These marathon tests are typically combined with ratiometric transmission measurements, that is, determining the losses by monitoring the laser power both with and without the samples under test.

Figures 17.6 through 17.8 display results of such induced absorption measurements on DUV optics at 248 and 193 nm. It turns out that, in addition to the earlier-mentioned specific material properties, the rate of color center formation depends strongly on the wavelength, the laser fluence, the pulse duration, the pulse repetition rate, and also on the sample temperature during irradiation. Based on these experimental observations, empirical models for induced absorption in fused silica at 248 and 193 nm have been proposed (Thomas and Kuehn 1997; Araujo et al. 1998) in order to be able to extrapolate the induced absorbance to pulse counts of 10^{11} and higher. However, regarding the

Chapter 17

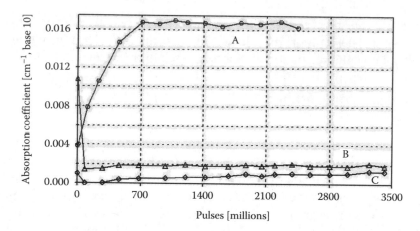

FIGURE 17.6 Representative trends of the absorption coefficients of SiO_2 samples during 193 nm irradiation as a function of laser pulses (up to 3.5 billion, fluence 1 mJ/cm²). The three curves represent different behaviors of fused silica as a function of pulse count. Whereas sample A shows initially a strong increase due to color center formation, absorption often appears to decrease during the first 10^8 pulses (traces B and C), which is attributed to partial bleaching of the bulk material. (From Liberman, V. et al., *Opt. Lett.*, 24, 58, 1999a; Reprinted with permission of MIT Lincoln Laboratory, Lexington, Massachusetts.)

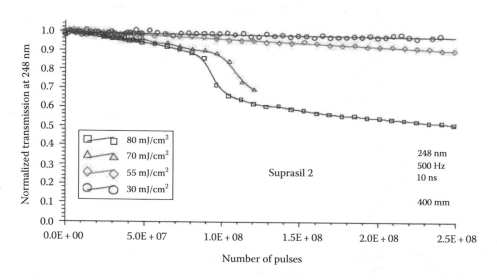

FIGURE 17.7 Normalized transmission of fused silica (Heraeus, Suprasil 2, sample length 400 mm) vs. number of 248 nm pulses. (Courtesy of St. Thomas/Heraeus Quarzglas, Thomas et al. 1997.) In the damage curves measured with 70 and 80 mJ/cm², a steeper absorption increase occurs after reaching a transmission change of ~10%, which can be attributed to a *strong absorption transition* (SAT). (From Krajnovich, D.J. et al., *Proc. SPIE*, 1848, 544, 1992b.)

described complexity of the color center generation (and bleaching) phenomena, such models and scaling laws can of course only partially describe the experimental findings. Thus, marathon stability tests, using laser irradiation at fluences close to the one expected in stepper systems, still need to be performed for further improvements of the material quality, as required for future wafer steppers operating at higher laser powers.

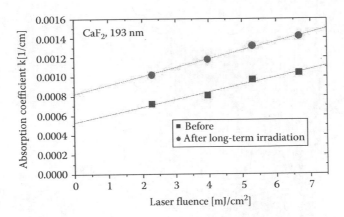

FIGURE 17.8   Absorption coefficient of CaF_2 sample at 193 nm as a function of fluence, both before (lower points) and after (upper points) a long-term irradiation of 10^9 laser pulses, indicating reversible nonlinear two-photon absorbance (cf. Section 17.2.1). Obviously, the long-term irradiation causes the generation of defects, which, although created in two-photon processes, absorb linearly. This is indicated by an increased ordinate section, but similar slope of the straight lines. (From Leinhos, U. and Mann, K. unpublished results.)

17.2.3   Compaction/Rarefaction

Since the amorphous structure of synthetic fused silica is less dense than that of a respective crystalline lattice, it can undergo modifications when exposed to x-rays, high energy particle beams, or also DUV radiation (Primak and Kampwirth 1968). This structural rearrangement of the fused silica network is assumed to originate from a local weakening of Si–O–Si bridging bonds, leading to a compaction or densification of the material (Piao et al. 2000). It was shown already that the compaction behavior from other radiation sources and DUV excimer laser pulses is similar, although in the latter case, a two-photon ionization process is believed to be responsible for the mechanism, similar to DUV-induced color center formation (Schenker and Oldham 1997).

The fused silica densification is necessarily accompanied by a permanent surface deformation, and, even more severe, an increase in the refractive index in the irradiation zone, producing image aberrations in an optical system. In the compacted volume, the increased density obviously shortens the physical path length through the material; however, the index of refraction is increased to a greater extent, so the net change is an increase in the optical path (Van Peski 2000).

As an example for this adverse effect, Figure 17.9 shows the surface topography as registered by white-light interferometry of a (low-quality) fused silica sample after irradiation with ~10 million 193 nm pulses at a fluence of 40 mJ/cm². A deformation of ~20 nm in the center of the irradiated area is obvious; shape and amount of this permanent surface distortion are found to be almost identical for both the front and back surfaces of the sample.

Due to its importance for the lifetime perspectives of *diffraction-limited* wafer stepper optics, damage of fused silica due to DUV radiation-induced compaction has been extensively studied in the last two decades (Schenker and Oldham 1997; Rothschild et al. 1989; Krajnovich et al. 1993; Schenker et al. 1994b). In long-term exposures to excimer laser radiation, the influence of laser parameters like wavelength, radiation

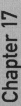

FIGURE 17.9 White-light interferogram showing 3D surface contour of low-quality fused silica after ~10^7 ArF excimer laser pulses of fluence 40 mJ/cm², indicating a surface deformation of ~20 nm (sample thickness 11 mm, field size 3.5 × 3.5 mm). A similar densification is observed on the back surface of the sample.

dose, fluence, and pulse duration has been investigated. In such experiments, one has to distinguish clearly between *unconstrained* and *constrained compaction:* When only a small area of a sample is exposed, the rest of the sample resists the shrinkage of the damaged portion (*constrained compaction*), with the result that the net densification is less than if the specimen had been uniformly exposed (*unconstrained compaction*) (Schenker and Oldham 1997).

Figure 17.10 shows the unconstrained density change $(\Delta\rho/\rho)_u$ of a fused silica sample irradiated at 193 nm as a function of the irradiation dose, applying different pulse numbers N and laser fluences H (Schenker and Oldham 1997). The data are plotted versus

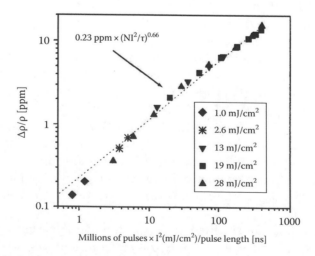

FIGURE 17.10 Unconstrained densification in fused silica (Corning 7940) for five different fluences vs. the number of pulses times the fluence squared, divided by the pulse length. (Data from Schenker, R.E. and Oldham, W.G., *J. Appl. Phys.*, 82, 1065, 1997, reproduced with permission from AIP Publishing LLC.)

a parameter defined by the pulse count times the fluence squared and divided by the pulse length τ, being proportional to the number of two-photon occurrences within the sample. Obviously, all the data points lie close to a single straight line, justifying a power law dependence of the kind

$$\left(\frac{\Delta\rho}{\rho}\right)_u = k\left(\frac{N \cdot H^2}{\tau}\right)^c. \tag{17.3}$$

For the exponent c, a value between 0.65 and 0.7 is determined, indicating a sublinear pulse count dependence, which was described by relaxation of bridging bonds upon two-photon absorption (Piao et al. 2000). The value of c ~ 2/3 is in agreement with compaction data from a number of different investigations (Primak and Kampwirth 1968; Borelli et al. 1997). A qualitatively similar behavior is observed at 248 nm for a higher fluence. Except for the two-photon absorption mechanism, Equation 17.3 holds also for fused silica irradiations with γ-rays and electron beams. On the other hand, collision-induced atomic displacements from particle beams (neutrons, ions) lead to a compaction exponent c ~ 1 (Piao et al. 2000).

Thus, other than in the case of damage due to color center formation (cf. Section 17.2.2), the situation is a bit more deterministic for fused silica compaction. Long-term predictions of dose-dependent aging are possible in certain limits, since the quality of the specific material is mainly determined by the proportionality factor k of Equation 17.3. Increasing the pulse length and maintaining the same pulse energy reduces the rate of compaction in a sample, which is the reason for the usage of pulse-extension units in modern wafer stepper systems.

Surprisingly, for extremely long DUV exposures at low fluence (≤ 0.1 mJ/cm^2), also an effect opposite to the described densification is observed, as known already from γ irradiation of hydrogen-rich fused silica (Shelby 1979). Van Peski et al. (2000) performed marathon tests with two alternately firing ArF excimer lasers at an effective pulse rate of 8 kHz, irradiating fused silica samples for pulse counts as high as 50 billions. A typical interferometer plot of a sample is shown in Figure 17.11. Whereas a high-fluence region irradiated at 3 mJ/cm^2 for 100 million pulses exhibits an increase in the optical path length of 30 nm as expected from compaction, an adjacent low-fluence region exposed to 12 billion pulses of 0.1 mJ/cm^2 produced a decrease in the optical path length (–12 nm). This results from a material expansion in combination with a diminished refractive index in the irradiated zone.

The so-called rarefaction phenomenon is similar to compaction but has the opposite sign with respect to the resulting wavefront distortions. A model based on the experimental results assumes the formation of SiOH in single-photon processes, indicating a linear dependence of the cumulative fluence (N·H) only, and no dependence on the pulse duration at all (Algots et al. 2003).

17.2.4 Stress Birefringence

Strongly related to laser-induced compaction or rarefaction is another effect that deteriorates the optical quality of DUV lenses made of fused silica, that is, stress-induced

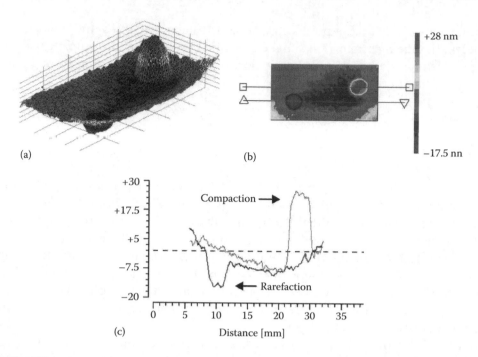

FIGURE 17.11   (a and b) Wavefront distortion of fused silica sample irradiated with 12 billion pulses of 0.095 mJ/cm² and 0.1 billion pulses of 3 mJ/cm², respectively; (c) cross sections of the distortions in nanometer, indicating an increase (compaction) and a decrease in optical path length (rarefaction), respectively. (Courtesy of Z. Bor., From Van Peski, C.K. et al., *J. Non-Cryst. Solids*, 265, 285, 2000.)

birefringence: In constrained compaction, the nonirradiated (outer) region of an optical element resists the shrinkage of the irradiated center part (cf. Section 17.2.3), inevitably leading to mechanical stress. As a consequence, axial and radial stress differences result in directional changes of the refractive index. This stress birefringence is especially large in the transition region from a compacted to a noncompacted area, that is, at the edge of the irradiating laser beam (Schenker et al. 1996).

Investigations of stress-induced birefringence in DUV optics have been performed by several groups (Borelli et al. 1997; Kühn et al. 2003). In these measurements, the samples are typically placed between crossed polarizers, and the null signal of an initially linearly polarized HeNe laser beam is monitored. Spatial distributions are recorded by raster-scanning the sample through the probe beam. An example is shown in Figure 17.12 (Kühn et al. 2003). A detailed description of compaction and its effect on stress-induced birefringence using finite element analysis can be found in Schenker et al. (1996) and Borelli et al. (1997).

Spatially resolved monitoring of birefringence induced by compaction can also be performed without raster-scanning the sample, using a camera behind the two crossed polarizers for detection of the test laser. Figure 17.13 shows the clover leaf–like intensity distribution of this test beam after passing a quartz sample, which had been irradiated with 1.2×10^6 pulses of an ArF excimer laser at a fluence of 0.1 J/cm² (Schäfer et al. 2009).

This pattern can also be described theoretically: Assuming a constant sample temperature and unsaturated compaction, the following proportionalities regarding the

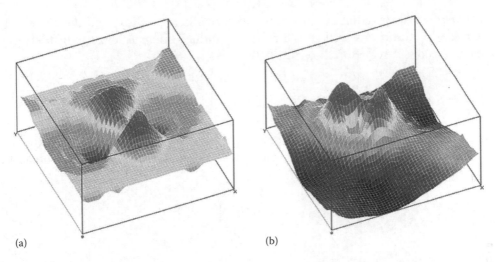

FIGURE 17.12 Stress-induced birefringence distribution in fused silica sample irradiated with 3 billion pulses at 40 μJ/cm²: (a) raw data after irradiation and (b) after subtraction of the stress-induced birefringence distribution measured prior to irradiation. The resulting stress birefringence (peak to valley) is 0.06 nm/cm. (From Kühn, B. et al., *J. Non-Cryst. Solids*, 330, 23, 2003; Reprinted with permission of Heraeus Quarzglas, Hanau, Germany.)

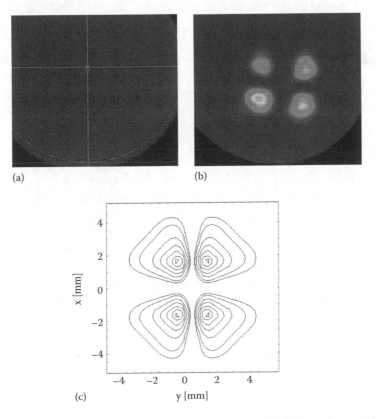

FIGURE 17.13 Test beam pattern of a quartz sample between crossed polarizers monitored with a camera before (a) and after irradiation (b) with $1.2 \cdot 10^6$ pulses ($\lambda = 193$ nm) at 0.1 J/cm². (c) Simulated test beam pattern for a flat-top beam of 3 mm × 1.5 mm cross section and a compaction of $4 \cdot 10^{-6}$. The isolines represent transmitted intensities of ~0.01%–0.1% relative to the test beam without polarizers.

refractive indexes for radial and azimuthal polarization n_r and n_φ and the corresponding stresses σ_r and σ_φ, as well as the radiation-induced density change of the sample $\Delta\rho/\rho$ can be considered (Borelli et al. 1997):

$$n_r - n_\varphi \sim \sigma_r - \sigma_\varphi \sim \frac{\Delta\rho}{\rho}. \tag{17.4}$$

Hence, the intensity distribution for crossed polarizers $I_\perp$, normalized to the test beam intensity $I_\parallel$ without sample and at parallel orientation of the polarizers, can be written as (Schäfer et al. 2009)

$$\frac{I_\perp}{I_\parallel} = \sin^2\varphi\cos^2\varphi \cdot \left(1 - \cos\left(\frac{2\pi\left(n_r - n_\varphi\right)l_c}{\lambda}\right)\right), \tag{17.5}$$

where
  φ is the azimuth
  l_c is the sample length

Figure 17.13b displays the correspondingly computed intensity distribution of the stress birefringence pattern.

17.2.5   Thermal Lensing

In addition to the radiation-induced permanent changes of the optical properties described in the last sections, there are also transient effects affecting the optical performance of DUV optics: When a transmissive optical element is irradiated by a laser beam, the absorption both in the bulk and at the surface of the material gives rise to heating, which leads, on the one hand, to a local change of the refractive index (dn/dT) and, on the other, to a thermal expansion of the sample. Both effects together are responsible for thermal lensing, that is, the distortion of a wavefront transmitted through this optical element. The associated formation of a transient and power-dependent lens (in general positive) can severely deteriorate the image formation in lithographic projection optics and has to be taken into account thoroughly in the design of wafer steppers (Beak et al. 2013; Fujishima et al. 2013).

The situation is illustrated in Figure 17.14 for a circular laser beam traveling in z-direction through a cylindrical optical element of length l. In addition to the laser power P and the bulk and surface absorption, the transient temperature rise $\delta T(r, z, t)$ encountered during irradiation depends on the sample geometry and well-known thermomechanical properties of the optical material. Given sample dimensions and material parameters, the temperature change can be computed by a numerical solution of the heat flow equation (Schäfer et al. 2009). As a consequence of the inhomogeneous temperature distribution, the refractive index varies locally according to

$$n\left(\delta T(r, z, t)\right) = n_0 + \frac{\partial n}{\partial T}\delta T(r, z, t) \tag{17.6a}$$

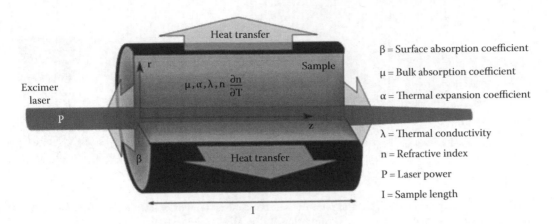

FIGURE 17.14 Geometry and notations characterizing a cylindrical optical element during laser irradiation.

and the elongation of the sample $\delta l(r, t)$ writes

$$\delta l(r,t) = \alpha \cdot l \cdot \delta T(r,t) \tag{17.6b}$$

where α represents the thermal expansion coefficient. According to Equations 17.6a and 17.6b, the wavefront of a well-collimated probe beam, traveling parallel to the specimen z-axis, picks up a spatially and temporally varying wavefront deformation $\delta w(r, t)$

$$\delta w(r,t) = \left(\frac{\partial n}{\partial T}(T) + \alpha(n(T)-1) \right) \cdot l \cdot \delta T(r) \propto P \cdot \left(A_{bulk} + A_{surf} \right), \tag{17.7}$$

being proportional both to the applied average laser power P and the overall absorbance of the sample, that is, the sum of bulk (A_{bulk}) and surface (A_{surf}) contributions.

In order to estimate the influence of thermal lensing on optical quality, and, possibly, taking it into account in the design of lithography objectives, a setup for registration and quantitative evaluation of this transient wavefront deformation on DUV optical materials was developed (Schäfer et al. 2009): As shown in Figure 17.15, a collimated ArF excimer laser irradiates a sample collinearly; the probe beam, a 639 nm fiber coupled diode laser, intersects the specimen under a small angle and is registered by a high-sensitivity Hartmann–Shack wavefront sensor, consisting of a digital charge-coupled device (CCD) camera behind an orthogonal quartz microlens array. The oxygen pressure in the chamber is kept below 100 ppm during the measurements by purging with nitrogen in order to prevent wavefront distortions due to O_2/O_3 absorption of 193 nm radiation.

A typical photothermal wavefront distortion of a quartz sample recorded by the Hartmann–Shack sensor during ArF excimer laser irradiation with 50 mW/cm² is shown in Figure 17.16a. As compared to the unirradiated sample for which the transmitted test laser wavefront is absolutely flat, this wavefront exhibits a peak-valley deformation of

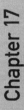

Chapter 17

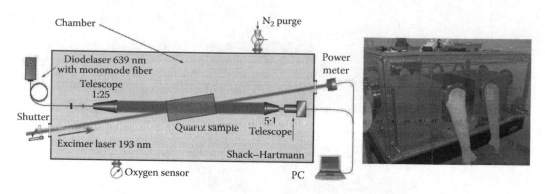

FIGURE 17.15 Schematic drawing and photo of setup for measurement of laser-induced photothermal wavefront deformation of DUV optics. (From Schäfer, B. et al., *Opt. Expr.* 17, 23025, 2009.)

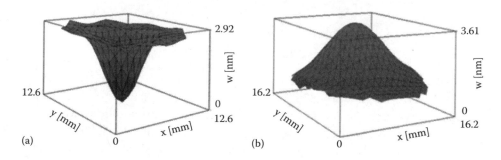

FIGURE 17.16 Wavefront deformation of fused silica (a, l = 40 mm) and CaF$_2$ (b, l = 20 mm) during excimer laser irradiation (power density 50 mW/cm^2). The observed sign reversal of the thermal lens stems from the positive or negative dn/dT value, respectively.

~2 nm after ~10 s of irradiation, which corresponds to a defocus term of approximately 5 km for a spherical surface.

For comparison, Figure 17.16b shows the wavefront deformation of a CaF$_2$ sample recorded under identical conditions. In contrast to fused silica, which always exhibits a positive lens effect, the wavefront deformation in CaF$_2$ shows a reversed sign representing a negative lens. This can be explained by the negative dn/dT value of CaF$_2$. It is interesting to note that, despite an almost 40 times larger expansion coefficient as compared to quartz, the negative change of refractive index with temperature for CaF$_2$ still overcompensates the specimen expansion. However, for longer CaF$_2$ samples, the expansion term will dominate, leading to a positive lens effect as for quartz.

The results of photothermal wavefront measurements for two different fused silica samples irradiated at 193 nm at different laser fluences are displayed in Figure 17.17. From the wavefront distributions w(x, y) and measured laser powers, the w_{rms}/P data are evaluated (w_{rms} = root mean square value of w(x, y)), which are proportional to the absorbance according to Equation 17.7. Apparently, a reversible linear increase with fluence H is observed, which can be attributed to efficient two-photon absorption of ArF laser-irradiated fused silica (cf. Section 17.2.1). As for absorbance data obtained from

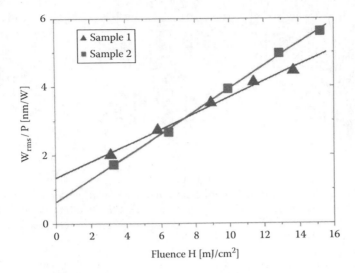

FIGURE 17.17   Transient wavefront deformations normalized to the incident laser power P for two fused silica samples vs. applied ArF laser fluence from 3 to 15 mJ/cm² (sample length l = 50 mm, λ = 193 nm, 1 kHz).

calorimetry (Eva and Mann 1996; Görling et al. 2002), the behavior can be described by the intensity-dependent Lambert–Beer law, leading to

$$\frac{W_{rms}}{P} \propto A = A_{surf} + k_0 \cdot dz + k_{2-Ph} \cdot H \cdot dz, \tag{17.8}$$

where

A_{surf} denotes the (linear) surface absorbance
k_0 is the linear absorption coefficient (base 10)
k_{2-Ph} is the two-photon absorption coefficient

In order to gain absolute values of absorption coefficients, the setup can be calibrated, for example, by electrical heating similar to calorimetry (ISO 11551:2003). The resulting linear bulk absorption coefficients for the samples of Figure 17.17 are in the range of several 10^{-4} cm⁻¹.

Wavefront measurements can also be employed to visualize thermally induced deformations of mirror surfaces. Such deformations of a reflected collimated test laser beam were recorded for different types of mirrors during 193 nm irradiation with a Hartmann–Shack sensor, as demonstrated in Figure 17.18. A highly absorbing aluminum mirror on fused silica exhibits a large *thermal bump* of $w_{pv} = 650$ nm, while a low absorbing dielectric mirror on fused silica yields $w_{pv} = 1.7$ nm only. Interestingly, for a similarly low absorbing dielectric mirror on a CaF_2 substrate, a much higher wavefront deformation of $w_{pv} = 71$ nm is registered (Leinhos et al.). Obviously, although the substrate is almost totally shielded against DUV radiation by the dielectric mirror layers, it is indirectly heated due to coating absorption, and the much higher expansion coefficient of CaF_2 leads to a larger thermal bump as compared to fused silica.

The main advantages of the photothermal approach can be summarized as follows: On the one hand, it provides quantitative absorbance data for flexible sample geometries

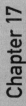

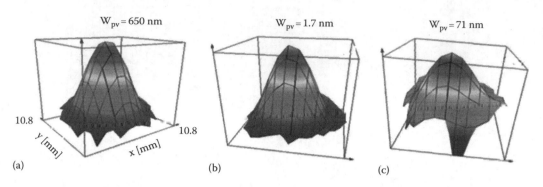

FIGURE 17.18 Wavefront deformations of different mirrors recorded with the Hartmann–Shack sensor during 193 nm irradiation: an aluminum mirror on fused silica exhibits a peak-valley deformation w_{pv} = 650 nm (a), a dielectric mirror on fused silica w_{pv} = 1.7 nm (b), while for a dielectric mirror on CaF_2, w_{pv} = 71 nm is registered (c).

in much shorter measurement times as compared to calorimetry, on the other, it accomplishes direct monitoring of the imaging performance, that is, the wavefront distribution (optical path difference [OPD]) transmitted through an optical element under radiation load, and in particular its stability over time.

17.3 Stability of Optics for Extreme UV Lithography

17.3.1 Introduction

EUV sources emitting radiation at a wavelength of 13.5 nm are constantly under development since the late 1990s. They shall be employed for the next-generation wafer stepper systems in the semiconductor industry, addressing high-volume manufacturing at feature sizes below 20 nm. In order to meet the requirements on wafer throughput in the microlithographic process, the average power of these laser-produced (LPP) or discharge-produced (DPP) plasma sources is steadily increasing, aiming for in-band EUV powers much larger than 100 W in the intermediate focus. The most promising EUV source concepts for high-volume wafer production tools are based on a tin droplet plasma generated by a multi-kW CO_2 laser (Fomenkov et al. 2012). This process of EUV plasma generation is inevitably associated with an extremely harsh environment for the EUV optical elements operated in close vicinity to the source, in particular the condenser mirror (see Figure 17.19 and Feigl et al. 2012). Thus, despite sophisticated debris mitigation techniques employed to inhibit damage of the mirror surfaces, beam guiding optics of increased resistance to thermal degradation and radiation-induced damage are strongly required.

Starting with the collector, the reflective optical system of LPP-based EUV wafer steppers will be composed exclusively of near-normal incidence multilayer mirrors for 13.5 nm radiation. They consist of very thin alternating layers of molybdenum and silicon, attaining reflectivities of nearly 70% at the wavelength of 13.5 nm (Feigl et al. 2008). These mirrors are currently being optimized in terms of thermal resistance and reflectivity close to the theoretical limit. However, only very few damage threshold investigations have been presented up to now for such optical elements (Barkusky et al. 2010; Khorsand et al. 2010; Müller et al. 2012). The topic of radiation-induced damage

FIGURE 17.19   Collector substrate for EUV wafer stepper. (From Feigl, T. et al., *Proc. SPIE*, 8322, 832217, 2012; Copyright: Fraunhofer IOF, Jena, Germany.)

to Mo/Si multilayer mirrors is of special interest also for actinic mask inspection, where the peak fluences are even higher than those applied in wafer stepper optics.

Recently, a laser-based EUV source in combination with a Schwarzschild objective for generation of a microfocus of high fluences at $\lambda = 13.5$ nm was developed and employed for investigations on radiation-induced damage thresholds of relevant EUV optical materials (Barkusky et al. 2009). The experimental setup consists of a table-top laser-based EUV source and a separate optics chamber adapted to this source (cf. Figure 17.20). EUV radiation is generated by focusing an Nd:YAG laser (wavelength 1064 nm, pulse energy 700 mJ, pulse duration 8.8 ns) onto a solid Au target, yielding a plasma diameter of ~50 μm (FWHM).

In order to achieve high EUV fluences on a test sample, a Schwarzschild objective with a demagnification of 10× at a maximum numerical aperture of 0.4 is employed, consisting of two spherical, annular mirror substrates coated with an Mo/Si multi-layer system (reflectivity R ~ 0.65 per mirror at 13.5 nm). A plane Au-coated mirror

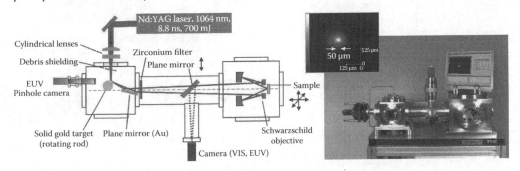

FIGURE 17.20   Schematic drawing and photo of the EUV source and optics system used for laser-induced damage threshold (LIDT) tests at a wavelength of 13.5 nm (Barkusky et al. 2009). The inset shows a pinhole camera image of the EUV plasma.

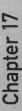

Chapter 17

is positioned between source and objective in order to protect the Mo/Si multilayers against contamination from debris of the laser plasma, and a thin zirconium foil can be inserted in the beam path, blocking effectively all radiation above 25 nm. Thus, spectrally pure EUV radiation at a maximum fluence of 1.16 J/cm² (6.6 J/cm² without Zr filter) can be achieved. The size of the EUV spot in the focus of the Schwarzschild objective is ~5 μm × 2 μm (FWHM).

17.3.2 Damage of Mo/Si Multilayer Optics

A 1-on-1 damage test according to ISO 11254 Part I (ISO11254-1:2000) was performed with the described setup for an Mo/Si multilayer mirror, irradiating 10 positions with single 13.5 nm pulses of constant fluence. The damage probability as a function of EUV fluence is displayed in Figure 17.21 together with the typical differential interference contrast (DIC) micrographs of irradiated sites. At fluences larger than 800 mJ/cm², spot-like damage occurs (DIC image 1), merging into crater damage at 1.7 J/cm². In all cases, small craters with depths of 40–70 nm are generated, mostly with a small bump in the center. In case of spot-like damage, several micrometer-sized pits are detectable, independent from the plasma intensity. This effect may be attributed to inhomogeneities or defects in the multilayer structure, leading to locally higher absorption and therefore reduced damage threshold. At higher fluences, the irradiated multilayer surface is damaged completely, resulting in a single crater. From the surface morphology, the damage mechanism is apparently driven by thermal heating of the multilayer structure (Barkusky et al. 2010).

In this context, it is interesting to compare the aforementioned results with measurements performed on Mo/Si mirrors with femtosecond pulses at the same EUV wavelength of 13.5 nm, using the free electron laser (FEL) FLASH in Hamburg (Khorsand et al. 2010). As displayed in Figure 17.22, the strongly reduced pulse duration leads to a much smaller damage threshold (10 fs pulses: H_D = 45 mJ/cm², 8 ns pulses: H_D = 800 mJ/cm²).

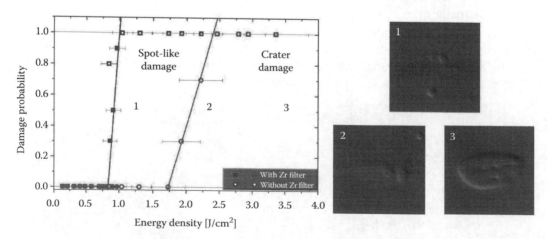

FIGURE 17.21 1-on-1 damage probability plot of Mo/Si mirror (at λ = 13.5 nm, 60 double layers, incidence angle 20°, silicon substrate) and corresponding Nomarski (DIC) images for selected fluences (image sizes 15 μm × 15 μm).

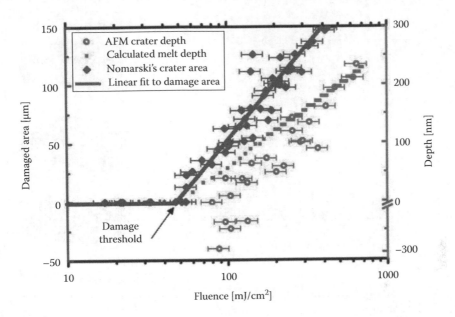

FIGURE 17.22 Fluence dependence of the damaged area/depth on an Mo/Si multilayer mirror irradiated at FLASH (DESY/Hamburg, wavelength of 13.5 nm, pulse duration 10 fs). The damaged area was measured by DIC microscopy while the depth of the craters was determined by atomic force microscopy (AFM). The negative values of the AFM correspond to the height of the hills. (Courtesy of R. Sobierajski; From Khorsand, A.R. et al., *Opt. Expr.*, 18, 700, 2010.)

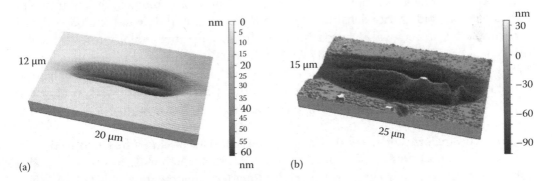

FIGURE 17.23 AFM micrographs of EUV damage sites on Mo/Si multilayer mirror: (a) tested at FLASH (pulse duration 10 fs, courtesy of R. Sobierajski [Khorsand et al. 2010]) and (b) tested with laboratory based LPP source (cf. Figure 17.20, pulse duration 8 ns). The wavelength used in the tests was 13.5 nm in both cases.

However, the damage morphology is rather similar to the nanosecond case as depicted in Figure 17.23, indicating surface structures with depths of ~60 nm. According to Khorsand et al. (2010), compaction of the multilayer mirror due to intermixing of molybdenum and silicon is involved in the damaging process.

17.3.3 Damage Testing of Other EUV Optical Materials

The LPP source with the EUV Schwarzschild objective could also be employed to perform first nanosecond damage experiments on other optical materials relevant for future

Chapter 17

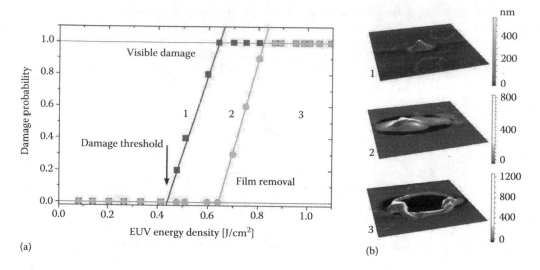

FIGURE 17.24 EUV damage test of gold layer (69 nm thickness) on glass substrate: 1-on-1 damage probability (a) and AFM micrographs for selected energy densities, showing increasing damage effects up to complete film removal (b).

EUV lithography systems. In particular, grazing incidence metal mirrors were investigated, which are used, for instance, as nested collector mirrors for DPP-based devices.

Single-pulse EUV damage data obtained for a gold coating (thickness 69 nm) on a BK7 glass substrate are compiled in Figure 17.24. The 1-on-1 damage probability plot indicates threshold energy densities of 420 mJ/cm² for initial damage and 650 mJ/cm² for complete film removal. Obviously, from the damage morphology (molten structures in AFM images, cf. Figure 17.24b), a thermally induced damage mechanism can be deduced. This is supported by the observed linear relation between ablation threshold and film thickness (Barkusky et al. 2010), which is well known already from ablation experiments on metal coatings in the deep UV wavelength range (Matthias et al. 1994).

Since the majority of EUV optical elements consists of thin films or multilayers deposited on substrates like fused silica or silicon, it is worthwhile to investigate also substrate damage by EUV pulses. Especially, long-term degradation effects like oxidation or compaction should be considered here due to the usually quite low fraction of photons penetrating through the coating and reaching the substrate. However, also single-pulse damage thresholds have to be known to make sure that the substrate does not limit the usable EUV fluence range.

As an example, EUV damage experiments were conducted on fused silica, silicon, and calcium fluoride crystals. The results of all single-pulse EUV damage experiments are compiled in Figure 17.25, indicating that fused silica shows a higher damage threshold (3.2 J/cm²) than the reflective coatings, leading to very smooth ablation craters. For silicon, the highest damage threshold fluences (up to 5 J/cm²) were determined, mainly caused by the low absorption of 13.5 nm radiation. The observed surface damage with a depth of 1–2 nm may be attributed to a thin natural oxide layer.

The experiments demonstrate that single-pulse EUV damage thresholds of substrate materials are higher than those of thin reflective coatings, as known already for higher UV

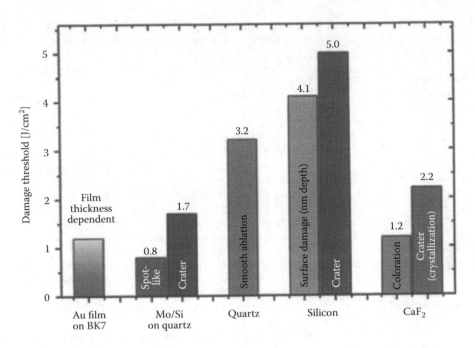

FIGURE 17.25 Comparison of single-pulse damage thresholds for EUV optics and substrates.

and visible wavelengths. Thus, metal or multilayer films immediately exposed to EUV radiation will clearly be the limiting factor for high-fluence applications of 13.5 nm radiation.

It can be concluded that the interaction of EUV radiation with thin gold films is primarily governed by thermal effects, whereas for multilayer coatings, there are indications for a defect-induced damage mechanism. This is supported by recent data demonstrating that multiple-pulse damage thresholds on EUV mirrors can be up to 60% lower than the single-pulse thresholds (Müller et al. 2012).

Further studies on EUV material interactions comprising also multiple-pulse damage tests and monitoring of the reflectivity change upon prolonged EUV irradiation are definitely necessary in order to improve the damage resistance of optical elements for future high-volume manufacturing EUV wafer steppers and mask inspection tools.

References

Algots, J. M., R. Sandstrom, W. N. Partlo, P. Maroevic, E. Eva, M. Gerhard, R. Lindner, and F. Stietz. 2003. Compaction and rarefaction of fused silica with 193-nm excimer laser exposure. *Proc. SPIE* 5040:1639–1650.

Araujo, R. J., N. F. Borelli, and C. Smith. 1998. Induced absorption in silica: A preliminary model. *Proc. SPIE* 3424:2–9.

Armbrust, D. 2011. Technology & manufacturing innovations: New SEMATECH initiatives. SEMATECH. org. http://www.sematech.org/meetings/archives/symposia/10187/Keynote-Plenary/01_Armbrust.pdf (accessed February 1, 2014).

Barkusky, F., A. Bayer, S. Döring, P. Grossmann, and K. Mann. 2010. Damage threshold measurements on EUV optics using focused radiation from a table-top laser produced plasma source. *Opt. Expr.* 18:4346–4355.

Barkusky, F., A. Bayer, C. Peth, and K. Mann. 2009. Direct photoetching of polymers using radiation of high energy density from a table-top extreme ultraviolet plasma source. *J. Appl. Phys.* 105:014906.

Chapter 17

Beak, D. H., J. P. Choi, T. Park, Y. S. Nam, Y. S. Kang, C. H. Park, K. Y. Park, C. H. Ryu, W. Huang, and K. H. Baik. 2013. Lens heating impact analysis and controls for critical device layers by computational method. *Proc. SPIE* 8683:86831Q.

Borelli, N. F., C. Smith, D. C. Allan, and T. P. Seward. 1997. Densification of fused silica under 193-nm excitation. *J. Opt. Soc. Am. B* 14:1606–1625.

Eva, E. and K. Mann. 1996. Calorimetric measurement of two-photon absorption and color-center formation in ultraviolet-window materials. *Appl. Phys. A* 62:143 149.

Feigl, T., M. Perske, H Pauer, T. Fiedler, S. Yulin, M. Trost, S. Schröder et al. 2012. Optical performance of LPP multilayer collector mirrors. *Proc. SPIE* 8322:832217.

Feigl, T., S. Yulin, M. Perske, H. Pauer, M. Schürmann, N. Kaiser, N. R. Böwering, O. V. Khodykin, I. V. Fomenkov, and D. C. Brandt. 2008. Enhanced reflectivity and stability of high-temperature LPP collector mirrors. *Proc. SPIE* 7077:70771W.

Fomenkov, I. V., N. R. Böwering, D. C. Brandt, D. J. Brown, and A. N. Bykanov. 2012. Light sources for EUV lithography at the 22-nm node and beyond. *Proc. SPIE* 8322:8322N.

Fujishima, Y., S. Ishijama, S. Isago, A. Fukui, H. Yamamoto, T. Hirayama, T. Matsuyama, and Y. Ohmura. 2013. Comprehensive thermal aberration and distortion control of lithographic lenses for accurate overlay. *Proc. SPIE* 8683:86831H.

Göppert-Mayer, M. 1931. Über Elementarakte mit zwei Quantensprüngen. *Ann. Phys.* 401:273–294.

Görling, C., U. Leinhos, and K. Mann. 2002. Comparative studies of absorptance behaviour of alkaline-earth fluorides at 193 nm and 157 nm. *Appl. Phys. B* 74:259–265.

Görling, Ch., U. Leinhos, and K. Mann. 2005. Surface and bulk absorption in CaF_2 at 193 nm and 157 nm. *Opt. Commun.* 249:319–328.

ISO11254-1:2000. *Lasers and Laser-Related Equipment—Determination of Laser-Induced Damage Threshold of Optical Surfaces—Part 1: 1-on-1 Test.* Geneva, Switzerland: International Organization for Standardization.

ISO 11551:2003. *Optics and Optical Instruments—Lasers and Laser-Related Equipment—Test Method for Absorptance of Optical Laser Components.* Geneva, Switzerland: International Organization for Standardization.

Khorsand, A. R., R. Sobierajski, E. Louis et al. 2010. Single shot damage mechanism of Mo/Si multilayer optics under intense pulsed XUV-exposure. *Opt. Expr.* 18:700–712.

Krajnovich, D. J., M. Kulkarni, W. Leung, A. C. Tam, A. Spool, and B. York. 1992a. Testing of the durability of single-crystal calcium fluoride with and without antireflection coatings for use with high-power KrF excimer lasers. *Appl. Opt.* 31:6062–6075.

Krajnovich, D. J., I. K. Pour, A. C Tam, W. P. Leung, and M. V. Kulkarni. 1992b. 248-nm lens materials: Performance and durability issues in an industrial environment. *Proc. SPIE* 1848:544–560.

Krajnovich, D. J., I. K. Pour, A. C. Tam, W. P. Leung, and M. V. Kulkarni. 1993. Sudden onset of strong absorption followed by forced recovery in KrF laser-irradiated fused silica. *Opt. Lett.* 18:453–455.

Kühn, B., B. Uebbing, M. Stamminger, I. Radosevic, and S. Kaiser. 2003. Compaction versus expansion behaviour related to the OH-content of synthetic fused silica under prolonged UV-laser irradiation. *J. Non-Cryst. Solids* 330:23–32.

Liberman, V., M. Rothschild, J. H. C. Sedlacek, R. S. Uttaro, and A. Grenville. 1999b. Excimer-laser-induced densification of fused silica: Laser-fluence and material-grade effects on the scaling law. *J. Non-Cryst. Solids* 244:159–171.

Liberman, V., M. Rothschild, J. H. Sedlacek, R. S. Uttaro, A. Grenville, A. K. Bates, and C. Van Peski. 1999a. Excimer-laser-induced degradation of fused silica and calcium fluoride for 193-nm lithographic applications. *Opt. Lett.* 24:58–60.

Liu, P., W. L. Smith, H. Lotem, J. Bechtel, and N. Bloembergen. 1978. Absolute two-photon absorption coefficients at 355 and 266 nm. *Phys. Rev. B* 17:4620–4632.

Matthias, E., M. Reichling, J. Siegel, O. W. Käding, S. Petzoldt, H. Skurk, P. Bizenberger, and E. Neske. 1994. The influence of thermal diffusion on laser ablation of metal films. *Appl. Phys. A* 58:129–136.

Moll, J. and P. M. Schermerhorn. 1999. Excimer-laser-induced absorption in fused silica. *Proc. SPIE* 3679:1129–1136.

Müller, M., F. Barkusky, T. Feigl, and K. Mann. 2012. EUV damage threshold measurements of Mo/Si multilayer mirrors. *Appl. Phys. A* 108:263–267.

Piao, F., W. G. Oldham, and E. E. Haller. 2000. The mechanism of radiation-induced compaction in vitreous silica. *J. Non-Cryst. Solids* 276:61–71.

Primak, W. and R. Kampwirth. 1968. The radiation compaction of vitreous silica. *J. Appl. Phys.* 39:5651–5657.

Rothschild, M., D. J. Ehrlich, and D. C. Shaver. 1989. Effects of excimer laser irradiation on the transmission, index of refraction, and density of ultraviolet grade fused silica. *Appl. Phys. Lett.* 55:1276–1278.

Schäfer, B., J. Gloger, U. Leinhos, and K. Mann. 2009. Photo-thermal measurement of absorptance losses, temperature induced wavefront deformation and compaction in DUV-optics. *Opt. Expr.* 17:23025–23036.

Schenker, R., P. Schermerhorn, and W. G. Oldham. 1994b. Deep-ultraviolet damgae to fused silica. *J. Vac. Sci. Technol. B* 12:3275–3279.

Schenker, R. E., L. Eichner, H. Vaidya, S. Vaidya, P. M. Schermerhorn, D. R. Fladd, and W. G. Oldham. 1994a. Ultraviolet damage properties of various fused silica materials. *Proc. SPIE* 2428:458–468.

Schenker, R. E. and W. G. Oldham. 1997. Ultraviolet-induced densification in fused silica. *J. Appl. Phys.* 82:1065–1071.

Schenker, R. E., F. Piao, and W. G. Oldham. 1996. Material limitations to 193-nm lithographic system lifetimes. *Proc. SPIE* 2726:698–706.

SEMATECH. SEMATECH Lithography Program. SEMATECH.org. http://www.sematech.org/research/litho/index.htm (accessed February 1, 2014).

Shelby, J. E. 1979. Radiation effects in hydrogen-impregnated vitreous silica. *J. Appl. Phys.* 50:3702–3706.

Smith, C. M., N. F. Borrelli, and R. J. Araujo. 2000. Transient absorption in excimer-exposed silica. *Appl. Opt.* 39:5778–5784.

Thomas, S. and B. Kuehn. 1997. KrF laser-induced absorption in synthetic fused silica. *Proc. SPIE* 2966:56–64.

Van Peski, C. 2000. Behavior of Fused Silica under 193 nm Irradiation. SEMATECH.org. http://www.sematech.org/docubase/document/3974atr.pdf (accessed February 1, 2014).

Van Peski, C. K., R. Morton, and Z. Bor. 2000. Behavior of fused silica irradiated by low level 193 nm excimer laser for tens of billions of pulses. *J. Non-Cryst. Solids* 265:285–289.

Chapter 17

18. Free-Electron Lasers

Michelle Shinn

18.1 Introduction to Free-Electron Lasers

The lasers discussed so far in this book are based on bound electron systems. It is also possible to have laser action using an ensemble of free electrons. At first blush, elementary quantum mechanics would lead one to think that since a free electron has a continuous set of energy levels to choose from, there is no upper or lower energy level to transition between. However, as shown in [1], it is possible to create such a pair of states. What is notionally correct, based on elementary quantum theory, is that one can set the transition energy over an enormous range; laser action has been demonstrated at long (mm) wavelengths to as short as 0.12 nm. Since these lasers use free electrons, they are called free-electron lasers (FELs).

The first FEL was operated at 3.4 μm in 1976 [1]. A review on FELs is found in [2]. A schematic view of a particular type of FEL using superconducting linear accelerator technology to achieve high average power is shown in Figure 18.1.

As shown in the figure, bunches of free electrons (the charge per bunch is in the range of tens of pC to about an nC) are accelerated to relativistic speeds and enter a structure known as a wiggler, a periodically spaced arrangement of magnets. As the electron's trajectory oscillates, they radiate through the well-known phenomenon of synchrotron radiation [3]. This radiation interacts with the magnetic field of the wiggler to form a ponderomotive wave, which appears at rest with respect to the electrons. This wave causes the electrons to bunch in such a way as to form a distribution that longitudinally is shorter than the electron's emission wavelength, causing them to emit in phase, that is, coherently.

Chapter 18

Laser-Induced Damage in Optical Materials. Edited by Detlev Ristau © 2015 CRC Press/Taylor & Francis Group, LLC. ISBN: 978-1-4398-7216-1.

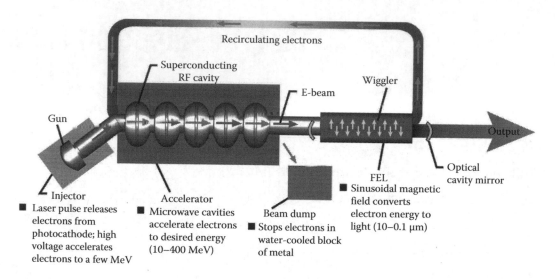

FIGURE 18.1   Schematic diagram of an FEL. (Courtesy Jefferson Lab, Newport News, VA.)

The wavelength of the laser radiation is determined by the following expression:

$$\lambda_L = \lambda_u/2\gamma^2(1 + K^2), \tag{18.1}$$

where
   λ_L is the lasing wavelength
   λ_u is the wiggler magnet's period
   γ is the relativistic parameter E/mc^2, where E is the electron beam energy
   K is the rms wiggler parameter $K = eB\lambda_u/(2\pi mc^2)$ [2], where B is the wiggler's rms
     magnetic field

As described, this arrangement has gain. In most cases, the gain is quite low for each pass through the wiggler, so mirrors are used to form a resonator, as shown in Figure 18.1. This raises the stimulating field, or thinking in terms of photons, the flux, which increases the gain. However, the gain enhancement only occurs if the photons produced by a previous electron bunch arrive in time to stimulate a fresh electron bunch in the wiggler. Thus, the optical cavity length must be precisely set, to within a few microns, so the following equation is satisfied:

$$L = \frac{nc}{2f}, \tag{18.2}$$

where
   f is the electron bunch frequency
   n = 1, 2, 3, ... to allow for longer cavity lengths that are also in synchronism

With the enhancement provided by the resonant cavity, small signal gains of the order of tens to well over 100% per pass can be achieved with short wigglers of a couple of meters length, or less.

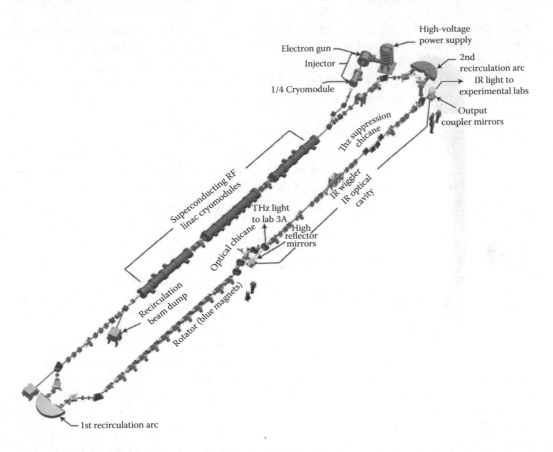

FIGURE 18.2 Schematic layout of the FEL at the Thomas Jefferson National Accelerator Facility (USA). The length of the accelerator is about 56 m.

With a superconducting radiofrequency (SRF) linac, it is possible to continuously produce fresh bunches of electrons, so the output power can be quite high. The IR Upgrade FEL at the Thomas Jefferson National Accelerator Facility has produced over 14 kW of average power at 1.61 μm, and kilowatt levels of power through the near-IR to mid-IR spectral range [4]. A schematic depiction of this machine is shown in Figure 18.2. These machines are always big, typically many tens of meters in length, so they tend to be installed as part of multiuser facilities. At present, there are over a dozen of such facilities around the world.

This is only one type of accelerator design to produce electrons for the FEL. FELs built using storage rings are in use [5], as are accelerators built with normal-conducting RF accelerator structures [1,6]. Such an accelerator can only operate in bursts with durations less than ~10 ms, as heating from absorbed RF power causes expansion of the structures and degradation of the accelerating gradient as the cavity detunes. The output of these FELs is a burst lasting up to a few ms in duration, composed of shorter pulses (ps to 100 s of fs) usually corresponding to some subharmonic of the accelerator frequency (cf. Equation IV18.1). Each burst is called a macropulse, and their repetition rate is of order 30 Hz. For example, the free electron laser Hamburg (FLASH) in Germany produces a macropulse of up to 0.8 ms in duration with the micropulse period within the macropulse set to 1, 2, 10, or 100 µs. The repetition rate is 5 Hz. Given the wavelength flexibility of FELs, there has been a growing trend to build them to produce x-rays. The proper electron bunch parameters and the use of longer wigglers allow the gain through the wiggler to reach many orders of magnitude, typically 10^3–10^7. In these cases, lasing of the self-amplified spontaneous emission (SASE) type occurs. This is how the FELs operating in the soft x-ray and higher energies are produced [6].

With high repetition rates (MHz), ultrashort pulse lengths, and tunable wavelength output, FELs are being used in science and technology to answer questions in the fields of medicine, materials research, and as a tool for materials processing. It is not clear whether stand-alone facilities for materials processing will be built solely for a specific processing activity, but clearly, it has the ability to map out a parameter space in order to optimize a process [7,8].

18.2 Laser Damage Issues

When discussing laser damage, especially, when it concerns FEL optics, be they part of the cavity (when the FEL is operated as a resonator) or optical transport, it is important to separate nondestructive changes due to contamination or color centering [9], from the more usual definition, irreversible changes to the functionality of the coating. While there is a rich literature about the former, the depth of literature citations on the latter is rather shallow, particularly in the visible and infrared. In part, this is due to the fact that FELs are traditionally associated with user facilities, and there is a strong desire to not damage optics. Given the extremely large wavelength range over which FELs operate, as the wavelength shortens, the mirror coatings run the gamut, from bare metal substrates to multilayer dielectrics, and then to layered metals and semiconductors. Hence, it is valid to ask, what causes the damage; the high peak electric fields from the individual ps (or shorter) pulses, or failure from heating due to the absorbed average power, be it as part of a macropulse, or from the quasi-continuous wave (cw) operation enabled by a superconducting RF linac? Almost two decades ago, I posed that question to David Milam at Lawrence Livermore National Lab. His reply was, "I don't know, it seems you have the worst of both worlds." I answer this question at the end of this section.

So, highlighting results from the literature that does exist, one can separate laser damage into two broad categories based on spectral range: (1) IR–visible and (2) soft x-ray and x-ray.

18.3 IR FELs

As mentioned at the beginning of this chapter, FELs initially operated in the IR region of the spectrum, so the coatings were nonoxide dielectrics like ZnSe and ThF_4 on substrates such as ZnSe, CaF_2, and Cu. The first report of laser damage to FEL optics was surface damage to the ZnSe plates used intracavity at near Brewster's angle to outcouple the laser output of the Mark III FEL [10]. The chemical vapor deposition (CVD) ZnSe laser-induced damage thresholds (LIDTs) shown in Table 18.1 and damage morphology, shown in Figure 18.3, are close to what is commonly measured with pulsed CO_2 lasers [11], which is ascribed as due to ablation or heating and ejection

Table 18.1   CVD Zinc Selenide Laser Damage Thresholds, Adjusted for Normal Incidence

Sample	Wavelength (µm)	Beam Radium (mm)	Angle of Incidence (°)	Damage Threshold 2 µs: (J/cm²)	Run Time before Onset of Damage	Type of Damage
1	3 0–4 0	1.4	61.6	11.3 ± 1.6	<10 min	(b)
2	3.0–4.0	1.4	61.6	11.3 ± 1.6	<1½ h	(c)
3	3.6	1.39	60.4	10.8 ± 0.9	<20 min	(a)
4	3.2	0.31	59.6	8.7 ± 0.6	<2 h	(a)
5	3.33	1.34	61.6	8.2 ± 0.6	<6 h	(b)

Source: Benson, S.V. et al., Laser damage on zinc selenide and cadmium telluride using the stanford mark III infrared free electron laser ((a) catastrophic coalescence of pits; (b) catastrophic fracture; (c) non-catastrophic pitting), in *Damage in Optical Materials: 1987*, NIST Special Publication 756, p. 41, 1988.

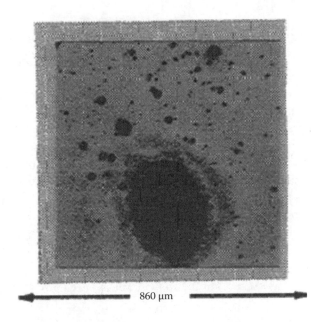

860 µm

FIGURE 18.3   Photomicrograph of damage in CVD ZnSe due to the coalescence of many small pits. (From Benson, S.V. et al., Laser damage on zinc selenide and cadmium telluride using the stanford mark III infrared free electron laser, in *Damage in Optical Materials: 1987*, NIST Special Publication 756, p. 41, 1988.)

of defects in the material. Hence, the ~1 ps pulse length of the micropulses was not a factor in the damage, and it was the power absorbed during the ~2 µs macropulse.

As discussed earlier, superconducting accelerators produce a continuous train of micropulses, usually of sub-ps duration, at repetition rates of megahertz to many tens of megahertz. Powers range from a few hundred watts (at Rossendorf) to over 10 kW (at JLab), leading to output pulse energies of a couple of 100 µJ or less. When cavity optics comprised of ZnSe and Si substrates coated with ZnSe and ThF_4 are subjected to this output format at 6 µm from the JLab FEL, the effect on the coatings was much the same as reported earlier [12], the pinhole defects in the coatings grew in diameter under

Chapter 18

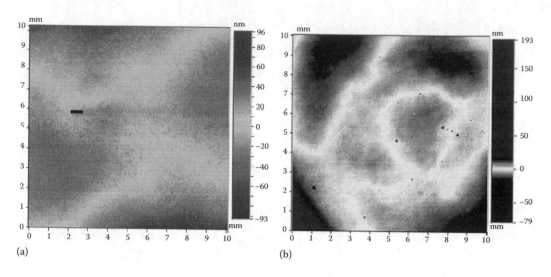

FIGURE 18.4 Profilometry of JLab FEL ZnSe outcoupler before (a) and after (b) operation at 6 µm. (From Shinn, M.D. and Gould, C., Analysis of coating defects in cavity mirrors for the IR Upgrade FEL, *Laser-Induced Damage in Optical Materials: 2004, Proceedings of SPIE*, Vol. 5647, p. 23, 2005.)

an irradiance peaking at 30 kW/cm² with the fluence of ~0.6 mJ/cm² from each micropulse [12]. This is depicted clearly in noncontact profilometry scans of the outcoupler before and after it was used in the FEL, as shown in Figure 18.4.

In addition to performing profilometry on cavity optics ex situ, the mirrors were observed in situ with visible and FLIR cameras during operation of the JLab FEL. What was noted [12] was that occasionally damage would initiate at a relatively low irradiance, ca. 10 kW/cm² as evidenced by a sudden increase in scattered coherent harmonic light as viewed by the visible camera, but with no indication of heating as viewed by the FLIR camera, nor growth in the damaged area as the irradiance was increased. This set of behaviors is consistent with cw laser damage [13]. If the damage was due to the peak E field from each micropulse, damage would occur at a well-defined threshold, and continue to grow in extent as one attempted to raise the average power.

18.4 Soft X-Ray FELs

As mentioned in the introductory paragraphs, the first soft x-ray free-electron laser (XFEL) was the FLASH facility at the synchrotron light source Deutsches Elektronen-Synchrotron (DESY) in Germany. Now operated routinely as a user facility [14], it is important to deliver as much of the output as possible to the users, yet not damage the beamline optics. Thus, understanding the damage mechanisms and thresholds is very important, especially since other facilities are delivering beam to users [15] or will be soon [16] with high pulse energies and shorter wavelengths. The coatings for such short wavelengths are quite different than those used for visible or infrared FELs; here, multilayers of molybdenum and silicon, or carbon, etc., are used. In addition, as the wavelength grows shorter, the interaction of the light with the material changes. Soft x-rays ionize the valence electrons, while x-rays ionize inner-shell electrons, with the subsequent emission of the Auger electrons.

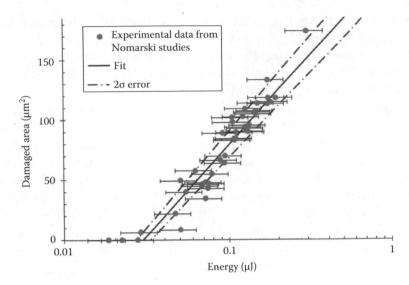

FIGURE 18.5 Single shot damage area vs. input energy on a multilayer Mo/Si mirror coated for maximum reflectivity at 13.5 nm and 28.2° from normal incidence. (From Khorsand, A.R. et al., *Opt. Ex.*, 18, 700, 2010.)

With micropulse energies of order 100 μJ possible in a 10 fs pulse propagating with a spot size of only 4 mm in diameter, the resulting fluence at the peak of the Gaussian mode profile is ~1.5 mJ/cm². Given the high beam quality and short wavelength, focusing to small spot sizes raises the fluence of each pulse by several orders of magnitude [17]. In this study, the FEL was focused to a 60 μm spot size to determine the single shot damage threshold of Mo/Si multilayer coating on a superpolished Si substrate. The damage threshold was determined to be 45 mJ ± 7/cm² as shown in Figure 18.5. While this is low compared to the damage threshold in oxide dielectrics irradiated at 800 nm of ~0.5 to 1.5 J/cm² [18], a closer analog would be a semiconductor irradiated above its bandgap. For example, the damage threshold of Si irradiated at 800 nm with 180 fs pulses is 0.17 J/cm² [19]. To determine the physical mechanisms behind the observed damage pattern, two approaches were used: one was done in situ using a two color pump-probe arrangement, and the other was a postirradiation study where the mirror was profiled using atomic force microscopy (AFM) and the Nomarski microscopy. In the later evaluation, the mirror was sectioned transversely to the elliptical-shaped damage area using a focused ion beam, polished with sputtered argon ions, then examined with scanning transmission electron microscopy (STEM) as well as energy dispersive x-ray spectroscopy (EDX). As the damaged region was scanned from the edge to the center, the well-ordered layers of Mo and Si were modified. First the Mo layers became thinner, and then the two materials became intermixed. In the pump-probe experiments, the FEL was the pump and an 800 nm Ti:S laser probed the surface reflectivity 10 ps after the pump pulse. This delay was chosen as it was anticipated that the material under the illuminated area would have totally thermalized. These experiments determined that the silicon had melted and diffused into the molybdenum layers, forming molybdenum silicide, MoS_2.

The authors conclude that the damage mechanism followed the well-established theory for damage from ultrashort pulses—electrons are initially placed in a nonequilibrium distribution through impact ionization until ~1 ps has elapsed, at which time

Chapter 18

the electrons are in thermal equilibrium with the ions. These in turn transfer their energy to phonons and melting occurs. The authors conclude that so long as the light source operates below the damage threshold and at pulse repetition rates low enough to permit diffusion of heat into the bulk, ~100 kHz, then damage will not occur.

A different damage morphology manifests itself when the wavelength is shortened into the x-ray region. In experiments performed at the Linac Coherent Light Source (LCLS), two materials commonly used in x-ray beamlines, bulk B_4C and SiC thin films, were exposed to focused FEL light at 0.83 keV, equal to 1.49 nm [20]. These materials were chosen because B_4C is used as a beam stop and SiC as a mirror coating. These experiments were a follow-up to an investigation done at FLASH the year before, where the damage threshold as a function of wavelength (35–13.5 nm) was determined for the same two materials [21]. The experimental setup was similar to that used in [17], the FEL output was focused to rather small spot sizes with areas of ~200 μm^2. As the fluence was increased, both materials exhibited similar damage morphology, as shown in Figures 18.6 and 18.7;

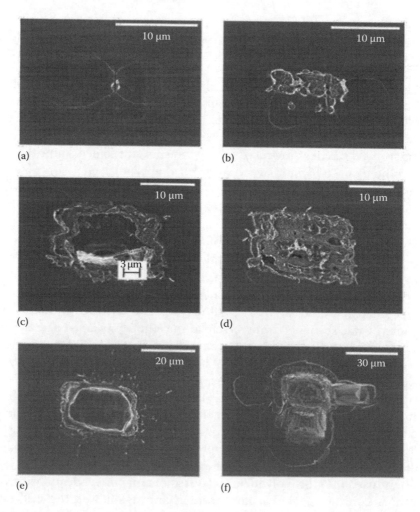

FIGURE 18.6 SEM pictures of bulk B_4C exposed to single XFEL pulses. Fluences were (a) 2.6, (b) 5.4, (c) 7.0, (d) 11.4, (e) 19.6, and (f) 90.5 J/cm². (From Hau-Riege, S.P. et al., *Opt. Ex.*, 18, 23933, 2010.)

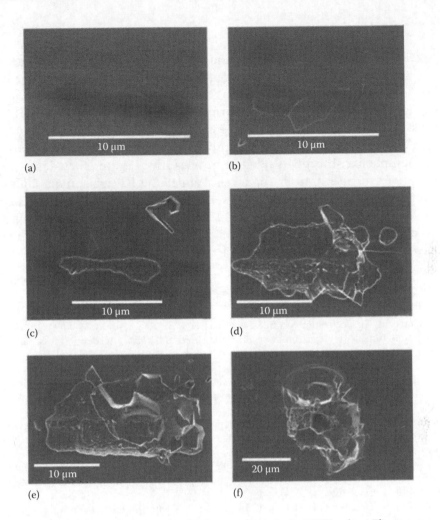

FIGURE 18.7 SEM pictures of a 1 mm SiC film exposed to single XFEL pulses. Fluences were (a) 1.0, (b) 1.6, (c) 2.9, (d) 5.8, (e) 14.8, and (f) 57.5 J/cm². (From Hau-Riege, S.P. et al., *Opt. Ex.*, 18, 23933, 2010.)

the material undergoes surface roughening at threshold and then begins to crack. The damage threshold of 2.7 J/cm² for B_4C corresponds quite well with its predicted melt threshold, the damage threshold of 0.62 J/cm² for SiC is somewhat lower than the predicted value of 1.2–1.7 J/cm². This same observation about damage threshold fluence relative to that required for melting was noted as well in the earlier study [21].

At longer wavelengths, the damage morphology was different; material was ablated and occasionally extruded from the surface, but there was no surface cracking. As an explanation for their observations, the authors propose that due to the larger (~5×) material volume traversed by the x-ray photons more stress is created that leads to fracture. Nevertheless, ultimately, the process leading to damage essentially follows the ultrafast excitation route—a nonequilibrium, critical density of electrons eventually transfer their energy to the ions that nonradiatively transfer it to phonons. However, no experiments dedicated to damage thresholds for materials exposed to a macropulse with a beam size characteristic for the path from the undulator hall to the end station have been published so far.

Chapter 18

18.5   Conclusions

In this section, we have introduced the unique output temporal format FELs are capable of, that of a macropulse composed of a comb of micropulses. We have reviewed laser damage results for operating FELs spanning three orders of magnitude in lasing energy What emerges from the experiments performed so far is a picture where the FEL micropulse peak irradiance governs the damage mechanism. If the peak intensity is higher than $\sim 10^{12}$ W/cm^2, the described ultrafast processes are at work and predominantly initiated by multiphoton absorption at longer wavelengths. If lower, then damage initiates at defects, either intrinsic or from impurities. While FELs have the unique ability to create coherent, ultrafast, tunable radiation over a wavelength range that spans some eight orders of magnitude, the challenge remains for the optics manufacturer and the laser designer to push the performance to the limits in order to achieve the end user's goal, be it a process leading to a product, or new science.

References

1. D. Deacon et al., First operation of a free-electron laser. *Phys. Rev. Lett.*, **38**, 892–894 (1977).
2. S.V. Benson, Tunable free-electron lasers, *Tunable Lasers Handbook*, Academic Press, San Diego, CA (1995).
3. H. Wiedermann, *Synchrotron Radiation*, Springer, Berlin, Germany (2002).
4. G.R. Neil et al., The JLab high power ERL light source, *Nucl. Instrum. Meth.*, **A557**, 9 (2006).
5. V.N. Litvinenko, Storage ring FELs and the prospects, *Nucl. Instrum. Meth. Phys. Res.*, **A304**, 40–46 (1991).
6. S. Schreiber, B. Faatz, K. Honkavaara, Operation of FLASH at 6.5 nm wavelength, *Proceedings of the EPAC08 Conference*, Genoa, Italy. Retrievable from http://accelconf.web.cern.ch/AccelConf/e08/papers/mopc030.pdf (accessed July 31, 2014).
7. M.D. Shinn, Experience and plans of the JLab FEL facility as a user facility, *Proceedings of the FEL07 Conference*, Novosibirsk, Russia. Retrievable from: http://accelconf.web.cern.ch/accelconf/f07/papers/thaau01.pdf (accessed July 31, 2014).
8. S.V. Benson, D. Douglas, G.R. Neil, and M.D. Shinn, The Jefferson laboratory FEL program, in New insights into the structure of matter: The first decade of science at Jefferson lab, *J. Phys. G Conf. Series*, **299**, 012014 (2011).
9. A. Gatto, N. Kaiser, S. Gunster, D. Ristau, F. Sarto, M. Trovo, and M. Danailov, Synchrotron radiation dinduced damages in optical materials, in *Laser-Induced Damage in Optical Materials: 2002 Proceedings of SPIE*, Boulder, CO, Vol. 4932, p. 366 (2003).
10. S.V. Benson, E.B. Szarmes, B.A. Hooper, E.L. Dottery, and J.M.J. Madey, Laser damage on zinc selenide and cadmium telluride using the stanford mark III infrared free electron, in *Laser-Induced Damage in Optical Materials: 1987*, Boulder, CO, NIST Special Publication 756, p. 41 (1988).
11. W. Plass, R. Krupka, A. Giesen, H.E. Reedy, M. Kennedy, and D. Ristau, Laser damage studies of metal mirrors and ZnSe-optics by long pulse and TEA-CO$_2$-lasers at 10.6 μm, in *Laser-Induced Damage in Optical Materials: 1993*, Proceedings of SPIE, Boulder, CO, Vol. 2114, p. 187 (1994).
12. M.D. Shinn and C. Gould, Analysis of coating defects in cavity mirrors for the IR Upgrade FEL, *Laser-Induced Damage in Optical Materials: 2004 Proceedings of SPIE*, Boulder, CO, Vol. 5647, p. 23 (2005).
13. Alan F. Stewart and Rashmi S. Shah, Diagnostic methods for CW laser testing, in *Laser-Induced Damage in Optical Materials: 2003 Proceedings of SPIE*, Boulder, CO, Vol. 5273, p. 50 (2004).
14. K. Tiedtke et al., The soft x-ray free-electron laser FLASH at DESY: Beamlines, diagnostics and end-stations, *N. J. Phys.*, **11**, 023029 (2009).
15. P. Emma et al., First lasing and operation of an ångstrom-wavelength free-electron laser, *Nat. Photon.*, **4**, 641 (2010).
16. T. Ishikawa et al., A compact x-ray free-electron laser emitting in the sub-ångstrom region, *Nat. Photon.*, **6**, 540 (2012).

17. A.R. Khorsand et al., Single shot damage mechanism of Mo/Si multilayer optics under intense pulsed XUV-exposure, *Opt. Ex.*, **18**, 700 (2010).

18. M. Mero et al., Femtosecond pulse damage and pre-damage behavior of dielectric thin films", in *Laser-Induced Damage in Optical Materials: 2002 Proceedings of the SPIE*, Boulder, CO, Vol. 4932, p. 202 (2003).

19. P. Allenspacher, B. Hüttner, and W. Riede, Ultrashort pulse damage of Si and Ge semiconductors, *Laser-Induced Damage in Optical Materials: 2002 Proceedings of SPIE*, Boulder, CO, Vol. 4932, p. 358 (2003).

20. S.P. Hau-Riege et al., Interaction of short x-ray pulses with low-Z x-ray optics materials at the LCLS free-electron laser, *Opt. Ex.*, **18**, 23933 (2010).

21. S.P. Hau-Riege et al., Wavelength dependence of the damage threshold of inorganic materials under extreme-ultraviolet free-electron-laser irradiation, *Appl. Phys. Lett.*, **95**, 111104 (2009).

Chapter 18

19. The High–Energy Petawatt Laser PHELIX and Its High-Power Optics

Stefan Borneis and Collaborators*

---

* B. Becker-de Mos (GSI, Helmholtzzentrum fuer Schwerionenforschung mbH, Darmstadt, Germany; European XFEL GmbH, Hamburg, Germany), A. Blazevic (GSI, Helmholtzzentrum fuer Schwerionenforschung mbH, Darmstadt, Germany), C. Bruske (GSI, Helmholtzzentrum fuer Schwerionenforschung mbH, Darmstadt, Germany), S. Calderon (Pegasus Design, Livermore, California), T. Eberl (GSI, Helmholtzzentrum fuer Schwerionenforschung mbH, Darmstadt, Germany), D. Eimerl (EIMEX, Fairfield, California), J. Fils (GSI, Helmholtzzentrum fuer Schwerionenforschung mbH, Darmstadt, Germany), E. Gaul (University of Texas at Austin, Austin, Texas), T. Hahn (GSI, Helmholtzzentrum fuer Schwerionenforschung mbH, Darmstadt, Germany), H.-M. Heuck (Leica Microsystems CMS GmbH, Wetzlar, Germany), D.H.H. Hoffmann (Technische Universitaet Darmstadt, Darmstadt, Germany), D. Kaliská (GSI, Helmholtzzentrum fuer Schwerionenforschung mbH, Darmstadt, Germany), T. Kuehl (GSI, Helmholtzzentrum fuer Schwerionenforschung mbH, Darmstadt, Germany; Johannes Gutenberg Universitaet Mainz, Mainz, Germany), S. Kunzer (GSI, Helmholtzzentrum fuer Schwerionenforschung mbH, Darmstadt, Germany), M. Kreutz (GSI, Helmholtzzentrum fuer Schwerionenforschung mbH, Darmstadt, Germany), T. Merz-Mantwill (GSI, Helmholtzzentrum fuer Schwerionenforschung mbH, Darmstadt, Germany), P. Neumayer (GSI, Helmholtzzentrum fuer Schwerionenforschung mbH, Darmstadt, Germany; ExtreMe Matter Institute EMMI, Darmstadt, Germany), E. Onkels (GSI, Helmholtzzentrum fuer Schwerionenforschung mbH, Darmstadt, Germany), M.D. Perry (General Atomics, San Diego, California), D. Reemts (GSI, Helmholtzzentrum fuer Schwerionenforschung mbH, Darmstadt, Germany), M. Roth (Technische Universitaet Darmstadt, Darmstadt, Germany), A. Tauschwitz (GSI, Helmholtzzentrum fuer Schwerionenforschung mbH, Darmstadt, Germany), R. Thiel (GSI, Helmholtzzentrum fuer Schwerionenforschung mbH, Darmstadt, Germany), P. Wiewior (Nevada Terawatt Facility, University of Nevada Reno, Reno, Nevada), K. Witte (Ludwig-Maximilians-Universitaet Muenchen, Garching, Germany).

Laser-Induced Damage in Optical Materials. Edited by Detlev Ristau © 2015 CRC Press/Taylor & Francis Group, LLC. ISBN: 978-1-4398-7216-1.

Chapter 19

19.1 Overview

PHELIX, a petawatt high-energy laser for heavy-ion experiments at GSI Helmholtzzentrum für Schwerionenforschung GmbH in Darmstadt, is Germany's largest laser installation. It was designed and built in close cooperation with the Lawrence Livermore National Laboratory (LLNL) in Livermore, USA, and the Commissariat à l'énergie atomique (CEA) in France. The project was driven by the technical and scientific achievements of Perry et al. (1999), who broke the petawatt barrier with a high-intensity beamline of the Nova laser at LLNL in 1996. The exciting discoveries enabled by the Nova petawatt laser have led to a revolution in the development of ultra-intense lasers and their applications. Scientific achievements span from laser-driven compact electron, proton, and ion acceleration (Jung et al. 2011) and advanced x-ray source developments to strongly relativistic laser–matter interactions relevant for astrophysics. One major driving force for the world's leading national laser laboratories to develop high-energy petawatt lasers and related technologies is the concept of fast ignition (FI) (Tabak 1994), where a short pulse laser beam ignites controlled thermonuclear fusion at a few hundred kilojoules of pulse energy versus a couple of megajoules in the classic *central hot-spot* scenario (Zuegel et al. 2006, Dunne et al. 2007). Within the European *High Power laser Energy Research facility (HiPER) project* (Dunne et al. 2007), cone-assisted and shock ignitions are currently investigated as alternate methods to assemble and ignite thermonuclear fuel (Baton et al. 2012).

At GSI, the plasma physics department and their international user community exploit with PHELIX the worldwide unique combination of an intense heavy-ion and a high-energy laser beam. Interaction experiments of heavy-ion beams with laser-heated dense plasma provide new insights into energy loss and charge exchange processes in laser-generated plasmas. At focused beam intensities on the order of 10^{21} W/cm^2, PHELIX may generate particle beams for proton radiography and short x-ray bursts for backlighting as well as the Thomson scattering in ion beam–generated plasmas. Ion beams may heat large volumes homogeneously. The future Facility for Antiproton and Ion Research, FAIR (Gutbrod et al. 2006), which is currently being built at GSI, will provide heavy-ion beams that are a hundred times more intense than those currently available at GSI and will deposit hundreds of kilojoule per gram specific energy into solids creating high-energy density (HED) states of matter including strongly coupled plasmas. One backbone to the success of these experiments will be advanced diagnostics by x-ray and particle sources driven by a high-energy short pulse laser.

19.2 Laser Architecture and Optics of PHELIX

One of the main technological challenges in the generation of petawatt laser pulses is the reduction of the peak intensity in today's highest damage threshold laser materials, which is on the order of 10 J/cm^2 for few nanosecond pulses. The key enabling technological advancement reducing the intensity in the laser amplifiers of PHELIX by a factor of 23,000 and thus avoiding catastrophic damage as well as deleterious nonlinear effects is chirped-pulse amplification (CPA) (Strickland and Mourou 1985). The substantial reduction of the intensity is achieved by stretching the 110-femtosecond (fs) pulses of the mode-locked Ti:sapphire oscillator of PHELIX to 2.3 nanosecond (ns) with a grating

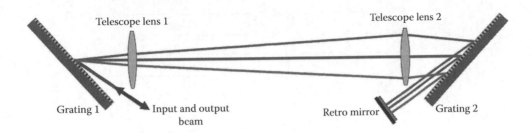

FIGURE 19.1 The concept of a grating pulse stretcher.

stretcher. It consists of an antiparallel pair of bulk diffraction gratings (1740 lines/mm) and an internal 1:1 imaging telescope, which images the first grating behind the second grating, as shown in Figure 19.1. This arrangement acts as a positive-dispersion delay line, transforming the broad-bandwidth input pulses into spectral–temporally chirped pulses (long wavelengths are leading). PHELIX employs a compact folded all-reflective stretcher design, which was developed and patented in 1999 by Banks et al. (2000).

Subsequent amplification of the stretched pulses can be accomplished at irradiance levels well below the self-focusing threshold of optical materials. After the amplification, the high-energy chirped pulse is recompressed in time by a pair of parallel diffraction gratings, producing a chirp of the same magnitude but opposite sign and canceling the temporal chirp of the stretcher. The beam is then transported and focused in vacuum onto the target to eliminate self-focusing in air. The postcompression final optics experiences both high energy and short pulse duration. The state-of-the-art method for circumventing optical damage is beam expansion prior to the compressor and the usage of meter-size optical apertures and the world's most advanced high-power laser coatings. The beam diameter of PHELIX is magnified from millimeter size at the millijoule level in the front-end to a few centimeters at a few joules in the preamplifier to finally 28 cm diameter at the kilojoule-level in the main amplifier.

PHELIX comprises two independent front-ends, which are injected into the same preamplifier and main amplifier resulting in either 250 J at 500 fs or up to 1 kJ for arbitrarily shaped ns pulses (Bagnoud et al. 2010). Figure 19.2 depicts a schematic overview of the laser architecture of PHELIX.

19.2.1 Front-Ends and Preamplifier of PHELIX

The fs front-end uses an optimized design of the LLNL petawatt laser (Perry et al. 1999) and generates up to 30 mJ pulses at a repetition rate of 10 Hz. The ns front-end is a modified prototype of the National Ignition Facility (NIF) front-end. The ns oscillator is entirely fiber based and thus hands-off and compact. A Nd:glass flashlamp-pumped ring amplifier amplifies the ns pulses to the energy level of 20 mJ at 0.5 Hz. Both front-ends seed the same preamplifier module with output energies between 5 and 10 J. The preamplifier is comprised of two 19 mm and one 45 mm diameter flashlamp-pumped Nd:glass rod amplifiers used in single-pass geometry. A lithographic serrated aperture shapes the Gaussian beam to a top-hat beam profile for efficient energy extraction in the amplifier chain. A Pockels cell between the two 19 mm amplifiers and behind the 45 mm amplifier provides protection against potential back reflections. To avoid beam distortions during the propagation of the top-hat beam profile, the beam is relay imaged

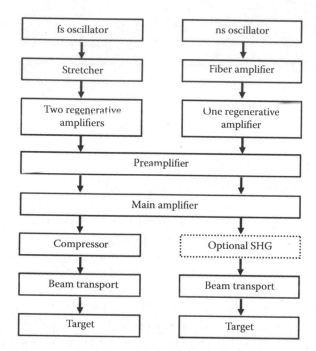

FIGURE 19.2   Schematic overview of the laser architecture of PHELIX.

FIGURE 19.3   Preamplifier module of PHELIX. Front center: 45 mm amplifier head.

through the system using four vacuum telescopes. At the exit of the preamplifier, the beam reaches energies of up to 10 J at 10 ns pulse width and a 70 mm diameter super-Gaussian beam profile. Figure 19.3 shows in the center the 45 mm amplifier head.

19.2.2   Main Amplifier of PHELIX

The two-pass main amplifier consists of a 15 m long confocal injection and exit relay telescope and five 315 mm aperture Nova amplifiers. Each amplifier houses two 43 mm thick

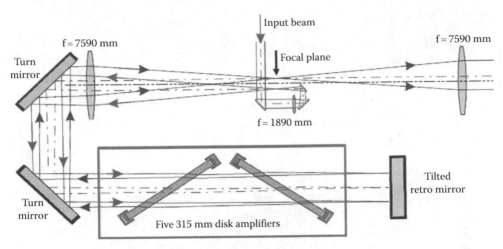

FIGURE 19.4 Schematics of the injection of the two-pass main amplifier.

Brewster-angle Nd:glass disks (Schott LG-750, laser damage threshold (LDT) ~25 J/cm² at 3 ns) pumped by 20 flashlamps with an electrical input energy of 300 kJ. The gain of the two-pass amplifier chain amounts 100. The symmetric far field angle separation of the input and the exit beam is schematically shown in Figure 19.4. The input beam is magnified by the f = 1890 mm injection lens and the f = 7590 mm injection relay telescope lens to 280 mm diameter. Two 520 mm diameter, 85 mm thick (45 kg) dielectric N-BK7 high-power turning mirrors steer the beam through the five disk amplifiers to the tilted 0° retro-reflecting mirror. This mirror steers the beam back through the amplifier chain and the exit pinhole of the injection telescope. The distance between the input and exit pinholes amounts to 30 mm. The distance of the pinholes defines the angle separation of the injected and amplified beams, while their diameters (input: 2.85 mm, exit: 3.85 mm) define the amount of unwanted, high-angle diffracted and scattered light that is scraped off (low-pass spatial filter). At the exit of the injection telescope, a turning mirror guides the beam through a 315 mm aperture pulsed Faraday isolator (Figure 19.5), employing a 20 mm thick Hoya FR-5 glass disk. FR-5 has an LDT of only ~3 J/cm² for 3 ns pulses and is thus the lowest LDT material of the beam line. The Faraday isolator has a measured extinction of about 1:1000 for retro-reflected light to eliminate damage of upstream optics by target back reflection, which experience almost the full system gain as they propagate through the unsaturated laser amplifiers. The coating of the 700 × 400 × 20 mm octagon polarizer plates of the Faraday isolator is one of the most challenging ones of PHELIX. High transmission and extinction over at least ±4 nm bandwidth require a large number of coating layers, which is detrimental for achieving a high LDT. We measured a LDT of 31.5 J/cm² for p-polarized light and 105 J/cm² for s-polarization for 11 ns pulses and an extinction of 1:2300 at an angle of incidence of 61°. The coating was fabricated by Newport Spectra-Physics in Santa Clara, CA.

When the pulsed magnetic field generating coil of the Faraday isolator is fired during a laser shot, the incoming p-polarization from the main amplifier is turned 45° clockwise, which is not compatible with the s-polarization required for the petawatt compression gratings. Gooch and Housego (Cleveland Crystals) manufactured for PHELIX a 320 mm diameter, 7.9 mm thick potassium dihydrogen phosphate (KDP) crystal, cut at θ = 0° which acts as a half-wave plate. The transmitted distortion of the sol-gel-coated KDP is 0.167

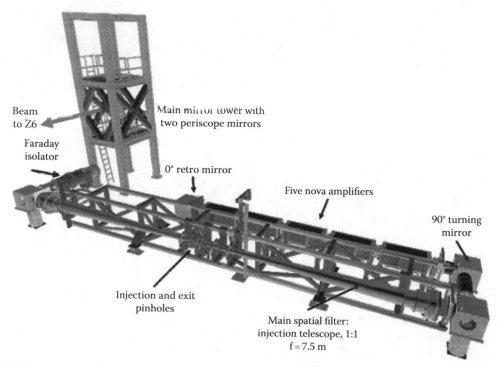

Beam
to Z6

Faraday
isolator

Main mirror tower with
two periscope mirrors

0° retro mirror

Five nova amplifiers

90° turning
mirror

Injection and exit
pinholes

Main spatial filter:
injection telescope, 1:1
f = 7.5 m

FIGURE 19.5 CAD drawing of the PHELIX two-pass main amplifier together with the Faraday isolator and the main mirror tower.

waves at 633 nm. The hygroscopic KDP is mounted in a rotation stage and installed in a sealed, clean room compatible enclosure, which is continuously flushed with dry nitrogen.

19.2.3 Beam Transport of the 280 mm Diameter PHELIX Beam

Behind the Faraday isolator, the beam may be directed either into the pulse compressor in the laser bay or over ~60 m to the Z6 target chamber at the Universal Linear Accelerator (UNILAC) for combined laser and heavy-ion beam experiments. The beam transport to Z6 requires two mirrors forming a periscope in the main mirror tower and a 28 m long transport relay telescope, as shown in Figure 19.6.

The coating layers of the 520 mm diameter turning mirrors are HfO_2/SiO_2 deposited by e-beam evaporation at Newport Spectra-Physics in Santa Clara, California, and the Laboratory for Laser Energetics (LLE) in Rochester, NY. The homogeneity of the reflectivity of the coating across the aperture is typically 1.5×10^{-3} (1σ) with a measured LDT of 35 J/cm² at 3 ns at 1064 nm. The reflected distortion at 45° of the Schott N-BK7 and Ohara S-BSL7 substrates degrades during the coating process from $\lambda/13$ peak-to-valley (p-t-v) at 1054 nm over 95% of the full aperture to typically $\sim\lambda/6$ p-t-v at 1054 nm over the central 315 mm diameter subaperture.

One of the major engineering challenges of the PHELIX project was the design of gimbal mirror mounts and supports capable of introducing less than $\lambda/15$ p-t-v wavefront distortions at 1054 nm and limiting the beam pointing instability in the focal plane of the f = 4 m final focusing lens at the target chamber of Z6 to < 30 μrad (2σ). From the injection into the main amplifier to the target chamber of Z6, the beam passes 10

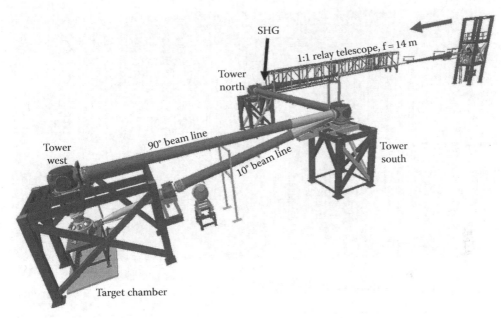

FIGURE 19.6   Beam transport to the target chamber at the experimental area Z6.

half-meter size mirrors and travels more than 100 m. This poses serious constraints onto the design of all mounts and structures even though the 1 meter thick concrete foundations of the mirror supports keep the measured power spectral density (PSD) of random vibrations below 10^{-11} g²/Hz between 1 and 100 Hz. Design of these mounts requires high stiffness to achieve natural frequencies >50 Hz in order to stay above the fundamental frequencies of the supports. To minimize self-weight deflection of the vertically oriented and the periscope mirror mounts, Pegasus Design in Livermore, California, and ROM Engineering Inc. in Tucson, Arizona, developed together with us a center of gravity edge mount for vertical orientation and a six-point whiffle-tree (Hindle-mount) back support, as shown in Figure 19.7. Flexures, which are bonded onto the mirror's rear surface provide, equal loads at 0.67 times the mirror radius from the center of the mirror to compensate gravity sag. The finite element (FE) analysis of ROM engineering predicts a reflected wavefront deformation introduced by the vertical mount of only 0.048 waves p-t-v at 633 nm. The whiffle-tree back support reduces the reflected wavefront deformation of a periscope mirror mounted in an optimal three-point support from 0.3 to 0.05 waves p-t-v at 633 nm, as depicted in Figure 19.8. These results are in excellent agreement with wavefront measurements conducted at the 18″ diameter Zygo interferometer of the Helmholtz-Zentrum Berlin fuer Materialien und Energie GmbH in Berlin, Germany. It is important to note that mounting of the whiffle-tree support does not require an interferometer and that the support is fully passive.

19.2.4   Pulse Compressor of PHELIX

To compress the spectral–temporally chirped pulses down to <500 fs, PHELIX uses a pair of multilayer dielectric (MLD) gratings in single pass configuration. The gratings are installed in a 5.8 m long vacuum chamber, as shown in Figure 19.9. The compressor may house 480 × 350 mm² gratings fabricated by Horiba Jobin Yvon as well as 800 × 400 mm²

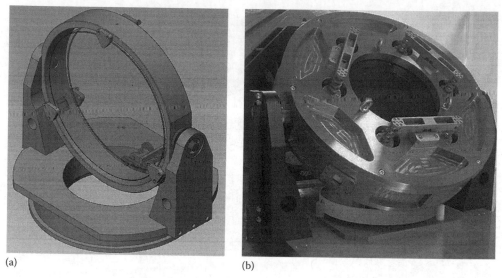

(a)   (b)

FIGURE 19.7   (a) CAD drawing of the vertical mirror mount and (b) photo of the six-point whiffle-tree mount.

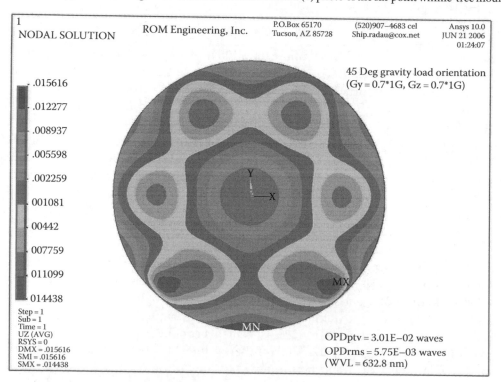

FIGURE 19.8   FE analysis of reflected wavefront of six-point whiffle-tree support, mirror tilted 45°. (Courtesy of ROM Engineering, Tucson, AZ.)

gratings manufactured by LLNL. The line density is 1740 l/mm with a diffraction efficiency of ~97% at an incident angle of 72°. Measured damage thresholds of these gratings in vacuum range from 1.2–2 J/cm² (beam fluence) at 350 fs to a *ramped* (R) damage threshold (R:1) of up to 5 J/cm² at a 10 ps pulse duration. This is more than five times the damage threshold of a similar holographic gold-overcoated photoresist grating (Britten 2008).

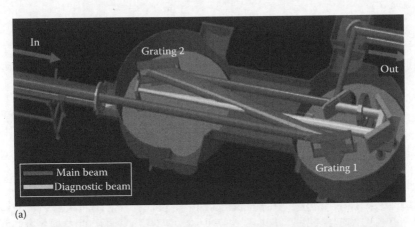

FIGURE 19.9 Three-dimensional CAD drawing (a) and photo (b) of the dual grating PHELIX pulse compressor.

19.2.5   Frequency Doubling Module of PHELIX

Interaction experiments of heavy-ion beams with laser-heated dense plasma at the experimental area Z6 have unveiled exciting new physics. Experiments have demonstrated that the stopping power of ionized matter is substantially higher than the stopping power of cold nonionized matter (Frank et al. 2010). For the heating of planar foil and hohlraum targets, wavelengths shorter than the fundamental laser wavelength of 1054 nm offer significant advantage in terms of higher penetration capability and higher absorption. To further optimize the specific experimental conditions for these types of experiments, a frequency doubling module (SHG, second harmonic generation) that converts the 1054 nm fundamental wavelength of PHELIX to 527 nm was designed and commissioned.

19.3   SHG Baseline Design

One of the main challenges of the design of the SHG module was to optimize the efficiency for 1 ns pulses at a maximum input energy of 450 J required for hohlraum experiments while maintaining the highest possible efficiency for up to 10 ns pulses

Chapter 19

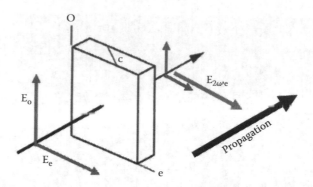

FIGURE 19.10 Type II SHG in DKDP: the induced nonlinear polarization generates from one ordinary (o) and one extraordinary (e) polarized photon at 1054 nm (1ω) one photon at 527 nm (2 ω) with e-polarization.

at a maximum energy of 1000 J for other laser interaction experiments. This factor of four in intensity requires careful design considerations because frequency conversion is a nonlinear process, scaling with the square of the input intensity to first order. Comprehensive model calculations and numerous engineering considerations such as alignment sensitivity, beam pointing stability, and temperature stability determined that a 310 mm diameter and 25 mm thick type II 70% deuterated KDP crystal (DKDP) would fulfill our requirements best. KDP is a strongly birefringent uniaxial nonlinear crystal. In type II SHG, an ordinary photon (o) polarized perpendicular to the KDP c-axis and an orthogonally polarized (extraordinary) photon (e) induce a nonlinear polarization, generating a photon with twice the frequency and extraordinary (e) polarization (Figure 19.10). Deuteration reduces the o-ray absorption from 6% to 2% per cm at 1054 nm by eliminating O-H vibration excitation. In addition to higher conversion efficiency as a consequence of the higher nonlinear coefficient d_{eff} of type II (0.338 pm/V) versus type I SHG (0.222 pm/V), type II SHG beneficially exhibits larger crystal angle tolerance (2.59 vs. 1.39 mrad cm), allowing larger alignment and beam pointing

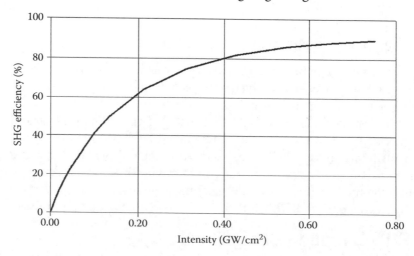

FIGURE 19.11 Calculated efficiency vs. input intensity for a 25 mm thick type II DKDP crystal, assuming a spatial beam modulation of 70% and a phase modulation corresponding to ±100 μrad local beam tilt external to the crystal.

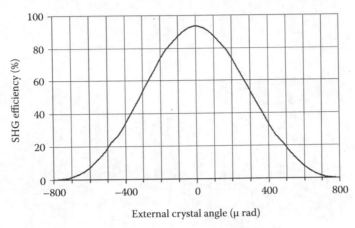

FIGURE 19.12   Calculated efficiency vs. external angle detuning for a 25 mm thick type II DKDP crystal at 1 GW/cm².

errors. Figure 19.11 depicts the model predictions of a 3D full beam calculation for a 25 mm thick DKDP crystal, assuming a spatial beam modulation of 70% and a phase modulation corresponding to ±100 μrad local beam tilt external to the crystal. Figure 19.12 shows the alignment tolerances of the crystal. The DKDP crystal of PHELIX was manufactured by Gooch and Housego, Cleveland Crystals, Cleveland, OH, USA. The LDT exceeds 7 J/cm² at 1054 nm and 3 J/cm² at 527 nm for 1.5 ns pulses. The transmitted wavefront distortion of the mounted DKDP for the central 285 mm aperture is 0.17 waves p-t-v at 633 nm.

19.3.1   Implementation of the SHG at Z6

The DKDP crystal is located behind the exit lens of the Z6 transport telescope, as shown in Figure 19.6. It is mounted in a three-axis step motor actuated precision rotation stage with tip and tilt enclosed by a sealed chamber with dry N_2 purge gas supply to prevent moisture damage of the hygroscopic crystal (Figure 19.13). The N_2 purge gas and the clean room class 100 air-flow tent enclosing the SHG module are temperature stabilized to 20°C ± 0.1°C. Aside avoiding risks associated with moisture, the DKDP crystal also requires special attention not only in terms of dangerous contamination of airborne dust particles but likewise in terms of residual gaseous contaminants inside the beam line, an issue arising from the sensitive sol-gel antireflection coating of the crystal surfaces. Inherent microscopic voids and general roughness act sponge-like on abundant gaseous contaminants, thereby deteriorating optical parameters such as residual reflectivity and, most importantly, optical damage threshold.

Extensive calculations were carried out to determine those locations behind the exit lens of the transport telescope, which prevent catastrophic optical damage of the lens, the DKDP and the chamber window. The damage, related to the so-called ghost foci, arises from the convergence of beams undergoing multiple residual reflections between the optical surfaces of all three optical components. Placing the upstream surface of the DKDP crystal 350 mm and the upstream surface of the window 550 mm from the lens keeps the intensity level below the optical damage threshold while accommodating comfortable mechanical tolerances for mounts and enclosures.

Chapter 19

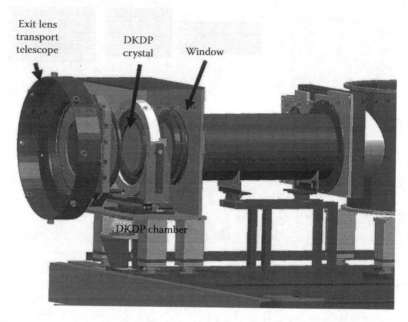

FIGURE 19.13 CAD model of SHG module.

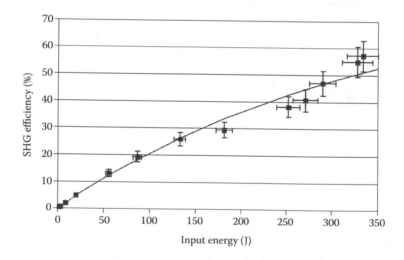

FIGURE 19.14 Measured SHG efficiency vs. input energy at 1.5-ns pulse duration.

We typically achieve a conversion efficiency to 527 nm of 60% for 1.5 ns pulses, as shown in Figure 19.14.

19.4 Conclusion

The high-energy petawatt laser PHELIX is Germany's largest laser system. It is a flash-lamp-pumped Nd:glass laser employing two independent front-ends, a preamplifier and a main amplifier, comprising five 315 mm aperture disk amplifiers from the decommissioned Nova and Phebus lasers in two-pass geometry. Using the fs front-end PHELIX may generate <500 fs pulses with powers >500 TW, while in long pulse mode, ns pulses with arbitrary pulse forms between 1.4 and 20 ns and with pulse energies reaching up to 1 kJ are provided. PHELIX was built in close cooperation with LLNL in United States and the Commissariat à l'énergie atomique (CEA) in France. The program benefited very much from the immense technology advancement that was driven by the NIF at LLNL. Rapidly grown KDP and large-aperture DKDP crystals are now commercially available. Optical materials and finishing and coating processes have dramatically improved performance, and meter-sized precision laser optics may be manufactured in large quantities and at reduced cost. In the future, facilities like LIFE (Bayramian et al. 2010), HiPER (Dunne et al. 2007), and ELI (Gerstner 2007) will push for the development of large-size optics for repetition rates of up to 13 Hz and for bandwidths up to 200 nm centered at 800 nm for Ti:sapphire-based systems.

Acknowledgments

Countless dedicated individuals have contributed to the success of PHELIX. The basis of the project was the endowment of the expensive large-diameter laser components from the former Nova and Phebus systems by the U.S. Department of Energy and the Commissariat à l'énergie atomique in France. In this matter, our sincere thanks go to J. Caird, M. E. Campbell, R. McKnight, G. Logan, A. Bettinger, M. Decroisette, and F. Kovacs. We also thank D. Habs (LMU Munich), F. Krausz (MPQ Munich), E. Moses (LLNL), W. Sandner (ELI Delivery Consortium International Association, Berlin), R. Sauerbrey (HZDR Rossendorf), and U. Wittrock (FH Muenster) for their important contributions and support. Important laser developments and access of European researchers were supported by the European International Infrastructure Initiative Laserlab, Europe.

References

Bagnoud, V., Aurand, B., Blazevic, A. et al., 2010, Commissioning and early experiments of the PHELIX facility, *Appl. Phys. B*, 100, 137.

Banks, P. S., Perry, M. D., Yanovsky, V. et al., 2000, Novel all-reflective stretcher for chirped-pulse amplification of ultrashort pulses, *IEEE J. Quant. Electron.*, 36, 268.

Baton, S. D., Koenig, M., Brambrink, E. et al., 2012, Experiment in planar geometry for shock ignition studies, *Phys. Rev. Lett.*, 108, 195002.

Bayramian, A. J., Campbell, R. W., Ebbers, C. A. et al., 2010, A laser technology test facility for laser inertial fusion energy (LIFE), *The Sixth International Conference on Inertial Fusion Sciences and Applications*, *J. Phys. Conf. Series*, 244, 032016.

Britten, J. A., 2008, Diffraction gratings for high-intensity laser applications, LLNL-BOOK-401125, https://e-reports-ext.llnl.gov/pdf/357070.pdf.

Dunne, M., Alexander, N., Amiranoff, F., 2007, HiPER technical background and conceptual design report, https://e-reports-ext.llnl.gov/pdf/357070.pdf.

Frank, A., Blažević, A., Grande, P. L. et al., 2010, Energy loss of argon in a laser-generated carbon plasma, *Phys. Rev. E*, 81, 026401.

Gerstner, E., 2007, Extreme light, *Nature*, 446, 16–18.

Gutbrod, H. H., Augustin, I, Eickhoff, H. et al., 2006, FAIR baseline technical report, http://www.fair-center.eu/fileadmin/fair/publications_FAIR/FAIR_BTR_6.pdf (accessed July 2014).

Chapter 19

Jung, D., Yin, L., Albright, B. et al., 2011, Monoenergetic ion beam generation by driving ion solitary waves with circularly polarized laser light, *Phys. Rev. Lett.*, 107, 115002.

Perry, M. D., Pennington, D., Stuart, B. C. et al., 1999, Petawatt laser pulses, *Opt. Lett.*, 24, 160.

Strickland, D., Mourou, G., 1985, Compression of amplified chirped optical pulses, *Opt. Comm.*, 56, 219.

Tabak, M. et al., 1994, Ignition and high gain with ultrapowerful lasers, *Phys. Plasmas*, 1, 1626.

Zuegel, J. D., Borneis, S., Barty, C. et al., 2006, Laser challenges for fast ignition, *Fusion Sci. Technol.*, 49(3), 453–482.

Index